Ergebnisse der Mathematik und ihrer Grenzgebiete

3. Folge · Band 2

A Series of Modern Surveys in Mathematics

William Fulton

Intersection Theory

Springer-Verlag
Berlin Heidelberg New York Tokyo 1984

William Fulton
Department of Mathematics
Brown University
Providence, RI 02912, USA

AMS-MOS (1980) Classification numbers: 14C17, 14-02, 14C10, 14C15, 14C25, 14C40, 14E10, 14M12, 14M15, 14N10, 55N45

ISBN 3-540-12176-5 Springer-Verlag Berlin Heidelberg New York Tokyo
ISBN 0-387-12176-5 Springer-Verlag New York Heidelberg Berlin Tokyo

Library of Congress Cataloging in Publication Data
Fulton, William, 1939– Intersection theory.
(Ergebnisse der Mathematik und ihrer Grenzgebiete ; 3. Folge, Bd. 2)
Bibliography: p. Includes index.
1. Intersection theory. I. Title. II. Series.
QA564.F84 1984 512′.33 83-16762
ISBN 0-387-12176-5 (U.S.)

Printed in Germany
Typesetting, printing and binding: Konrad Triltsch, D-8700 Würzburg
2141/3140-543210

Preface

From the ancient origins of algebraic geometry in the solution of polynomial equations, through the triumphs of algebraic geometry during the last two centuries, intersection theory has played a central role. Since its role in foundational crises has been no less prominent, the lack of a complete modern treatise on intersection theory has been something of an embarrassment. The aim of this book is to develop the foundations of intersection theory, and to indicate the range of classical and modern applications. Although a comprehensive history of this vast subject is not attempted, we have tried to point out some of the striking early appearances of the ideas of intersection theory.

Recent improvements in our understanding not only yield a stronger and more useful theory than previously available, but also make it possible to develop the subject from the beginning with fewer prerequisites from algebra and algebraic geometry. It is hoped that the basic text can be read by one equipped with a first course in algebraic geometry, with occasional use of the two appendices. Some of the examples, and a few of the later sections, require more specialized knowledge. The text is designed so that one who understands the constructions and grants the main theorems of the first six chapters can read other chapters separately. Frequent parenthetical references to previous sections are included for such readers. The summaries which begin each chapter should facilitate use as a reference.

Several theorems are new or stronger than those which have appeared before, and some proofs are significantly simpler. Among the former are a new blow-up formula, a stronger residual intersection formula, and the removal of a projective hypotheses from intersection theory and Riemann-Roch theorems; the latter includes the proof of the Grothendieck-Riemann-Roch theorem. Some formulas from classical enumerative geometry receive a first modern or rigorous proof here.

Acknowledgements. The intersection theory described here was developed together with R. MacPherson. The author whose name appears on the cover is responsible for the presentation of details, and many of the applications and examples, but the extent to which it forms a coherent theory derives from collaboration with MacPherson. Previously unpublished results of R. Lazarsfeld, and joint work with Lazarsfeld, and with H. Gillet, is also included. During the course of the work, many helpful suggestions were made by A. Collino, P. Deligne, S. Diaz, J. Harris, B. Iversen, S. L. Kleiman, A. Landman, Lazarsfeld, and J-P. Serre. Although other contributions and historical precedents are acknowledged in the text, many others, such as those of students and others who have responded to talks on these subjects, must be silently, but gratefully, cited.

This undertaking was made possible by the support of several foundations and institutions. The Guggenheim Foundation provided a fellowship in 1980–81, the Sloan Foundation provided support in 1981–82, and grants have been received from the National Science Foundation during six years of research and writing on this subject. The support and hospitality of several institutions and their staffs has been equally vital: Mathematisk Institut of the University of Århus, Denmark (1976–77); Institute des Hautes Études Scientifiques, Bures-sur-Yvette, France (1981); Institute for Advanced Study, Princeton (1981–82); and Brown University. A summer course in Cortona, Italy in 1980 provided a chance to test a preliminary version of the first portion of the book. Thanks are due to the staffs at the IAS and Brown, especially to K. Jacques, for expert typing, and to the publishers for their cooperation.

Contents

Introduction

A useful intersection theory requires more than the construction of rings of cycle classes on non-singular varieties. For example, if A and B are subvarieties of a non-singular variety X, the intersection product $A \cdot B$ should be an equivalence class of algebraic cycles closely related to the geometry of how $A \cap B$, A and B are situated in X. Two extreme cases have been most familiar. If the intersection is *proper*, i.e., $\dim(A \cap B) = \dim A + \dim B - \dim X$, then $A \cdot B$ is a linear combination of the irreducible components of $A \cap B$, with coefficients the intersection multiplicities. At the other extreme, if $A = B$ is a non-singular subvariety, the self-intersection formula says that $A \cdot B$ is represented by the top Chern class of the normal bundle of A in X. In each case $A \cdot B$ is represented by a cycle on $A \cap B$, well-defined up to rational equivalence on $A \cap B$. One consequence of the theory developed here is a construction of, and formulas for, the intersection product $A \cdot B$ as a rational equivalence class of cycles on $A \cap B$, regardless of the dimensions of the components of $A \cap B$. We call such classes *refined* intersection products. Similarly other intersection formulas such as the Giambelli-Thom-Porteous formulas for the degeneracy loci of a vector bundle homomorphism, are constructed on and related to the geometry of these loci, including the cases where the loci have excess dimensions.

To give an idea of the main thrust of the text, we sketch what we call the *basic construction*, from which such refined classes are derived. To a closed regular imbedding $i: X \to Y$ of codimension d, and a morphism $f: V \to Y$, with V a k-dimensional variety (or any purely k-dimensional scheme), this construction produces a rational equivalence class of $(k-d)$-cycles on $W = f^{-1}(X)$. This *intersection class*, denoted $X \cdot_Y V$, can be formed as follows. Since i is a regular imbedding, the normal cone to X in Y is a vector bundle; let N denote the pull-back of this bundle to W. The normal cone C to W in V is a k-dimensional closed subscheme of N. Using the lengths of local rings of C along its irreducible components as coefficients, C determines an algebraic k-cycle, denoted $[C]$, on N. One may construct $X \cdot_Y V$ by intersecting $[C]$ with the zero section of N. Thus a $(k-d)$-cycle $\sum m_i[Z_i]$ on W represents $X \cdot_Y V$ if $\sum m_i[N_{Z_i}]$ is rationally equivalent to $[C]$ on N, where N_{Z_i} is the restriction of N to Z_i.

Three situations show the utility of this construction: (1) If X is a d-dimensional non-singular variety, the diagonal imbedding of X in $X \times X$ is regular. With $Y = X \times X$, $V = A \times B$ the Cartesian product of subvarieties A, B of X, the construction determines the intersection class $A \cdot B$ on $A \cap B$. In particular, this determines the ring structure on the rational equivalence classes on X. (2) If $H_1, \ldots, H_d$ are effective Cartier divisors on a variety X, and V is a subvariety of X, the product imbedding of $H_1 \times \ldots \times H_d$ in $X \times \ldots \times X$ is regular. If f is the diagonal imbedding of V in $X \times \ldots \times X$, the construction determines a class on $H_1 \bigcap \ldots \bigcap H_d \bigcap V$. This is useful in enumerative geometry, where X parametrizes geometric figures, and the hypersurfaces H_i represent "simple conditions" on the figures. (3) If E is a vector bundle of rank d on a variety X, any section s of E is a regular embedding of X in E. Applying the construction to $Y = E$, $V = X$, f the zero section, this produces a class on the zero-scheme of s which represents the top Chern class $c_d(E) \cap [X]$ on X. This is used on Grassmann and flag bundles to represent determinantal formulas by cycles on degeneracy loci.

To use the basic construction one needs to verify several properties which are not obvious from this description. For example, if $f: V \to Y$ is also a regular imbedding, then there is a commutativity property: $V \cdot_Y X = X \cdot_Y V$. This assures that intersection products which could be formed in different ways, e.g. by (1) or (2) above, lead to the same classes. One also needs to know that the construction passes to rational equivalence. Precisely, with $i: X \to Y$ a regular imbedding of codimension d, $f: Y' \to Y$ any morphism, and $a = \sum n_i [V_i]$ any algebraic k-cycle on Y', define a class $i^! \alpha$ of $(k-d)$-cycles on $X' = f^{-1}(X)$ by setting $i^! \alpha = \sum n_i (X \cdot_Y V_i)$. If α' is rationally equivalent to α on Y', it must be shown that $i^! \alpha'$ is equal to $i^! \alpha$; the analogous statement with rational equivalence replaced by algebraic equivalence refines the "principle of continuity". A third important property is functoriality (or associativity) of this construction: if $i: X \to Y$, $j: Y \to Z$ are regular imbeddings, then $j \circ i$ is also, and $(j \circ i)^! = i^! \circ j^!$. The homomorphisms $i^!$ refine the Gysin homomorphisms $i^*: A_k Y \to A_{k-d} X$ which had been constructed by Verdier.

In addition, one needs formulas for these classes. Any cone C on a scheme W determines a *Segre class* $s(C)$ in the group of rational equivalence classes on W. In case $C = E$ is a vector bundle, this Segre class is dual to the inverse total Chern class of E: $s(E) = c(E)^{-1} \cap [W]$. In the situation of the basic construction, one then has the formula

$$X \cdot_Y V = c(N) \cap s(C) .$$

When the imbedding of W in V is also regular, this gives an *excess intersection formula* for $X \cdot_Y V$. By replacing V by the blow-up of V along W, one may always reduce to this situation. Another basic property expresses the compatibility of these classes with push-forward by proper morphisms. The value of allowing $V \to Y$ to be a morphism which is not an imbedding is evident in this reduction, as is the need to allow V to have arbitrary singularities.

The basic construction, properties, and formulas are based on joint work with R. MacPherson, and work of J.-L. Verdier. Originally these sources depended on the previously constructed "Chow rings" of cycle classes on non-singular quasi-projective varieties, as developed by Severi, B. Segre, Todd, Chevalley, Chow, Samuel, Weil, Grothendieck, et al. It was indicated in Fulton-MacPherson (1), however, how one could use the basic construction to develop intersection theory from scratch. This is the program carried out in Chapters 1–8 here. Note that in this program one also has no need for a preliminary study of intersection multiplicities. In the case of proper intersections, the intersection class is automatically a well-defined cycle, whose coefficients are then the intersection multiplicities; indeed, one sees readily that in this case the construction agrees with that of Samuel (1). Although the refined intersection classes respect variation in families of cycles, no moving lemma, or quasi-projective hypotheses, are needed. In a sense the approach is rather close to that of B. Segre (4), and relies on explicit deformations, and blowing up, rather than an abstract moving lemma. Ideas related to this point of view have also been published by H. Gillet, J. P. Jouanolou, J. King, A. T. Lascu, D. Mumford, J. P. Murre, and D. B. Scott. Work in intersection theory by S. L. Kleiman, D. Laksov, and R. Piene has particularly influenced this book.

Outline. The first chapter contains the definition of the group $A_k X$ of rational equivalence classes of algebraic k-cycles on an algebraic scheme X, and the verification that the natural definition of push-forward of cycles makes A_k a covariant functor for proper morphisms. In addition, a flat morphism of relative dimension n determines pull-back homomorphisms, raising dimensions by n. In the second chapter the basic construction is studied in the case of codimension one. This includes a construction of a first Chern class for line bundles. In Chapter 3 Chern classes $c_i(E)$ are constructed as homomorphisms $\alpha \to c_i(E) \cap \alpha$ from $A_k X$ to $A_{k-i} X$, for E a vector bundle on an algebraic scheme X. The expected formulas are proved for these Chern classes; they are used to prove that the pull-back from $A_k X$ to $A_{k+r} E$, $r = \operatorname{rank}(E)$, is an isomorphism. The first Chern class is also used to construct the Segre class of a cone, studied in the next chapter; this includes the notion of the multiplicity of a scheme along a subvariety. In Chapter 5 the deformation to the normal cone is constructed. This is a rational family of closed imbeddings containing a given imbedding $W \to V$ and the zero-section imbedding $W \to C$ in the normal cone C to W in V. The existence of such a deformation, together with the "principle of continuity", helps explain the key role of normal cones in the construction of intersection products. Chapter 6 contains the general construction and basic properties of the intersection products $X \cdot_Y V$ and classes $i^! \alpha$. This chapter also contains a new general blow-up formula for the pull-back of cycles by a monoidal transformation.

The rest of the book consists of largely independent applications of the first six chapters. The next two chapters consider the special cases of proper intersections (intersection multiplicities) and intersections on non-singular varieties. In Chapter 9 we prove a residual intersection theorem, more general than those previously available. With notation as in the basic construction, and Z a given closed subscheme of the intersection scheme W, this formula writes the intersection class $X \cdot_Y V$ as a sum of a class on Z and a class on a residual set R with $Z \cup R = W$. Following Laksov (3), this is applied to deduce a general double point formula.

Chapter 10 considers the variation of intersection classes in families, including a strong form of the principle of continuity. The decomposition of the normal cone into its irreducible components determines a decomposition of the intersection product. Chapter 11 includes R. Lazarsfeld's infinitesimal construction of intersection classes, and his proof that the decomposition constructed by decomposing the normal cone agrees with that obtained by a dynamic method, along lines suggested by Severi.

From our construction of intersection classes, various positivity or ampleness hypotheses on the normal bundle to X in Y can be expected to force corresponding positivity of intersection classes $X \cdot_Y V$. Such theorems, and applications to a refined Bézout's theorem, and inequalities for intersection multiplicities, are discussed in Chapter 12; this is joint work with Lazarsfeld. Similarly, since the construction is valid over an arbitrary field, and produces classes on the loci of interest, it can be used to prove the existence of rational solutions to algebraic equations (Chapter 13).

Formulas for degeneracy loci are among the most important applications of intersection theory. Our method, combined with ideas from Kempf-Laksov (1), gives refinements of the usual formulas, producing classes *on* the degeneracy loci in question, and valid on possibly singular varieties (Chapter 14). The classical Schubert calculus is deduced from these formulas. In Chapter 15 the geometry of the deformation to the normal bundle is used to give a short conceptual proof of the Grothendieck-Riemann-Roch theorem, as well as the formula for blowing up Chern classes. The basic algebra of correspondences is included in Chapter 16; our intersection formulas yield classical formulas of Pieri and Severi for the virtual number of fixed points of a correspondence, when the fixed point locus is infinite.

The *bivariant* language of Fulton-MacPherson (3) is introduced in Chapter 17. This codifies and strengthens the machinery of Chapters 1–8. Indeed, the book could be considerably shortened by an uncompromising use of this formalism throughout, but at the cost of its usefulness as a reference for those unfamiliar with it. In Chapter 18 it plays a key role in the analysis of MacPherson's graph construction, which is used to extend Riemann-Roch to singular quasi-projective varieties, as in Baum-Fulton-MacPherson (1) and Verdier (5). In addition, recent joint work with H. Gillet is included, which removes all quasi-projective hypotheses from these Riemann-Roch theorems.

Chapter 19 shows that, for complex varieties, the cycle map, from algebraic cycle classes to homology classes, is compatible with (refined) intersection products; included is a brief survey comparing rational, algebraic, homological and numerical equivalence of cycles on non-singular complex projective varieties. The final chapter sketches generalizations of the preceding chapters to schemes over Dedekind domains and other non-algebraic base schemes. Serre's intersection multiplicity, and Bloch's formula relating rational equivalence to higher K-theory, are also mentioned.

Appendix A contains the commutative algebra needed for Chapters 1–6, together with references for a few facts used later; this appendix can be consulted as required, and need not be read as prerequisite. Appendix B is a glossary of basic concepts and constructions needed from algebraic geometry; it is hoped that occasional use of Appendix B will help bridge gaps between the language of various introductory treatments of algebraic geometry and that used here. In addition a few special conventions adopted here are pointed out. Among these are the following, to be understood otherwise indicated: *schemes* are algebraic schemes over an arbitrary field *K*; *varieties* are irreducible and reduced schemes; *points* are closed points; *subvarieties* and *imbeddings* are assumed to be closed; *cycles* are algebraic cycles, i.e., integral linear combinations of subvarieties; *non-singular* varieties are smooth over *K*; *flat* morphisms are assumed to have some relative dimension; $P(E)$ denotes the projective bundle of *lines* in a bundle E.

A substantial portion of the book consists of *Examples* at the end of sections. As one would expect, these include illustrations and special cases of the theorems, and classical and modern applications. In addition, there are generalizations of the theorems, or counterexamples to possible generalizations. Some examples, such as a series on intersections of plane curves in the first chapter, are included primarily to motivate later developments. Unless otherwise indicated, the proofs of assertions in examples should be reasonably straight-forward from the preceding text. Hints are included in parentheses. References preceded by "cf." indicate where similar results may be found, although often with a different approach. References without "cf.", or with the more direct "see", indicate a closer relation of the example to the reference, which may be consulted for details. An unspoken assumption in computational examples is that the ground field is algebraically closed of characteristic zero; the interested reader may make the necessary modifications in positive characteristics.

At the end of each chapter a *Notes and References* section contains some historical remarks on material related to the chapter, and an attempt to attribute sources for the main ideas. Many other references may be found in the examples. Although it is hoped that an impression of the interesting history of intersection theory emerges from these notes and the examples, a thorough historical analysis is beyond the capacity of this book or its author. For similar reasons, only rarely do we discuss to what extent classical references meet modern standards of rigor. Other surveys are referred to for much of the closely related history of enumerative geometry, but we have tried to point out important contributions which are likely to be unfamiliar to modern readers. Both in the notes and the examples, emphasis is placed on classical topics, such as excess intersection formulas, which are closely related to the basic view-point of the text presented here. References are given by referring to the author, followed by a number in parentheses; however, Grothendieck-Dieudonné (1) is referred to by the familiar [EGA] and Berthelot-Grothendieck-Illusie et al. (1) by [SGA 6].

The bibliography is similarly only a sampling of the vast literature in intersection theory. Omission of topics or references which should be included may be attributed to lack of space, as usual, but is more likely due to incompetence of writer.

Chapter 1. Rational Equivalence

Summary

A cycle on an arbitrary algebraic variety (or scheme) X is a finite formal sum $\sum n_V[V]$ of (irreducible) subvarieties of X, with integer coefficients. A rational function r on any subvariety of X determines a cycle $[\operatorname{div}(r)]$. Cycles differing by a sum of such cycles are defined to be rationally equivalent. Alternatively, rational equivalence is generated by cycles of the form $[V(0)] - [V(\infty)]$ for subvarieties V of $X \times \mathbb{P}^1$ which project dominantly to $\mathbb{P}^1$. The group of rational equivalence classes on X is denoted $A_* X$.

For a proper morphism $f: X \to Y$, there is an induced push-forward of cycles. The fundamental theorem of this chapter states that rational equivalence pushes forward, so there is an induced homomorphism f_* from $A_* X$ to $A_* Y$, making A_* a covariant functor for proper morphisms.

For flat morphisms $f: X \to Y$ (of constant relative dimension) there are contravariant pull-back homomorphisms f^* from $A_* Y$ to $A_* X$. There is a useful exact sequence

$$A_* Y \to A_* X \to A_* (X - Y) \to 0$$

for a closed subscheme Y of X, and exterior products

$$A_* X \otimes A_* Y \xrightarrow{\times} A_* (X \times Y) .$$

The groups $A_* X$ will play a role analogous to homology groups in topology. In succeeding chapters it will be shown how geometric objects (vector bundles, regularly imbedded subschemes, ...) give rise to operations on these groups (Chern classes, intersection products, ...). Eventually corresponding contravariant, ring-valued functors A^* will be constructed, with cap-products from $A^* X \otimes A_* X$ to $A_* X$ and other properties familiar from topology. When X is non-singular, $A^* X \cong A_* X$; in the non-singular case, but not in general, $A_* X$ will have a ring structure. The actual relation of these groups to homology groups is discussed in Chapter 19.

1.1 Notation and Conventions

Until Chapter 20, by a *scheme* we shall mean an algebraic scheme over a field[1] (see Appendix B.1). A *variety* will be a reduced and irreducible scheme, and a

1 Except where exterior products (§ 1.10) occur, the ground field could be replaced by any local Artinian ring with no significant changes.

subvariety of a scheme will be a closed subscheme which is a variety. A *point* on a scheme will always be a closed point.

Affine n-space is denoted $\mathbb{A}^n$; projective n-space is $\mathbb{P}^n$.

The *local ring* of a scheme X along a subvariety V is denoted $\mathcal{O}_{V,X}$, its maximal ideal $\mathcal{M}_{V,X}$. The *field of rational functions* on a variety X is denoted $R(X)$; the non-zero elements of this field form the multiplicative group $R(X)^*$.

Little will be lost if a reader wishes to assume that all ambient schemes are varieties over algebraically closed fields in the sense of Serre (1). It is important, however, that arbitrary closed subschemes be allowed. In other words, the defining ideals must be remembered (cf. Example 1.1). For rationality questions (Chapter 13) it is useful, and no more difficult, to work over an arbitrary ground field.

It is particularly important that ambient varieties are allowed to have singularities: our constructions of intersection cycles, even on non-singular varieties, will involve blowing up along singular subschemes.

The role of subschemes can be seen already in the modern definition of intersection numbers for plane curves. Although the situation is considerably more complicated in higher dimensions, several important features of intersection theory can be seen in the plane curve examples in this chapter.

Example 1.1.1. If $f(x,y)$ and $g(x,y)$ are polynomials defining affine plane curves F and G over an algebraically closed field K, the intersection scheme Z is the subscheme of $\mathbb{A}^2$ defined by the ideal (f,g) in $K[x,y]$ generated by f and g. If $P=(a,b)$ is a point in the plane, the *intersection multiplicity* of F and G at P is defined to be

$$i(P, F\cdot G) = \dim_K \mathcal{O}_{P,Z} = \dim_K \mathcal{O}_{P,\mathbb{A}^2}/(f,g)\,.$$

This intersection number satisfies the following properties:

1) $$i(P, G\cdot F) = i(P, F\cdot G)\,.$$

2) $$i(P, (F_1+F_2)\cdot G) = i(P, F_1\cdot G) + i(P, F_2\cdot G)\,,$$

where F_1+F_2 is the curve defined by $f_1 f_2$, with f_i defining F_i.

3) $$i(P, F'\cdot G) = i(P, F\cdot G)\,,$$

if F' is defined by $f+g\,h$, some $h \in K[x,y]$.

4) $$i(P, F\cdot G) = 0 \quad \text{if} \quad P \notin F\cap G\,,$$

and $i(P, F\cdot G)=\infty$ if F and G have a common component through P. Otherwise $i(P, F\cdot G)$ is finite and positive.

5) $$i(P, F\cdot G) = 1 \quad \text{if} \quad f = x-a,\quad g = y-b\,,$$

or more generally if the Jacobian $\partial(f,g)/\partial(x,y)$ is not zero at P.

6) $$i(P, G\cdot H) \geqq \min(i(P, F\cdot G), i(P, F\cdot H))$$

if P is a simple point on F, and F has no common component with G or H through P (see Namba (1) 2.3.2).

For fixed P, properties 1)–5) characterize the intersection number (see Fulton (1) 3.3). A similar definition, with analogous properties, is valid when the plane is replaced by an arbitrary non-singular surface. These intersection multiplicities agree with the general definitions to be given later (cf. Example 7.1.10). In general, however, the multiplicities will not be determined by the intersection scheme alone.

1.2 Orders of Zeros and Poles

Let X be a variety, V a subvariety of X of codimension one. The local ring $A = \mathcal{O}_{V,X}$ is a one-dimensional local domain. Let $r \in R(X)^*$. We will define the *order* of vanishing of r along V, $\operatorname{ord}_V(r)$, which will be a homomorphism, i.e.,

$$(*) \qquad \operatorname{ord}_V(r\,s) = \operatorname{ord}_V(r) + \operatorname{ord}_V(s)$$

for $r, s \in R(X)^*$.

Any $r \in R(X)^*$ may be written as a ratio $r = a/b$, for $a, b, \in A$. By (*) we must define

$$\operatorname{ord}_V(r) = \operatorname{ord}_V(a) - \operatorname{ord}_V(b)\,.$$

Thus ord_V will be determined if $\operatorname{ord}_V(r)$ is defined for $r \in A$.

In case X is non-singular along V, i.e., A is a discrete valuation ring, then $r = u\,t^m$, for u a unit in A, t a generator for the maximal ideal of A, and m an integer. In this case one may set $\operatorname{ord}_V(r) = m$. When X is a curve over an algebraically closed field K, this is the same as setting

$$\operatorname{ord}_V(r) = \dim_K A/(r)\,.$$

The latter definition extends to singular curves, but not to higher dimensions, for then $A/(r)$ is not finite-dimensional over the ground field. The correct general definition for $r \in A$ is

$$\operatorname{ord}_V(r) = l_A(A/(r))$$

where l_A denotes the length of the A-module in parentheses. That this determines a well-defined homomorphism ord_V from $R(X)^*$ to $\mathbb{Z}$ is proved in Appendix A.3.

For a fixed $r \in R(X)^*$, there are only finitely many codimension one subvarieties V of X with $\operatorname{ord}_V(r) \neq 0$ (Appendix B.4.3)

Example 1.2.1. Let f and g be polynomials defining plane curves F and G, with f irreducible. Let $\bar{g}$ be the rational function on the curve F defined by the residue class of g in $K[x, y]/(f)$. If $P \in F$, then

$$i(P, F \cdot G) = \operatorname{ord}_P(\bar{g})\,.$$

Example 1.2.2. Let $f(x, y)$ and $g(x, y)$ define plane curves F and G, over an algebraically closed field K. Let $r(x) \in K[x]$ be the resultant of f and g with respect to the variable y. Write $f(x, y) = \sum_{i=0}^{d} f_i(x)\, y^i$. If $f_d(a) \neq 0$, then

$$\operatorname{ord}_a(r) = \sum_b i((a, b), F \cdot G).$$

In case the intersections of F and G have distinct x-coordinates, the intersection numbers are given by the order of vanishing of the resultant. This is one of the classical definitions of intersection number for plane curves (cf. Walker (1)). (The equality follows from Example A.2.1 and the fact that $K[x, y]/(f, g)$, being Artinian, is the direct product of its localizations $\mathcal{O}_{P,\mathbb{A}^2}/(f, g)$.)

Example 1.2.3. Let X be a variety, $\tilde{X} \to X$ the normalization of X in its function field. If $r \in R(X)^* = R(\tilde{X})^*$, then

$$\operatorname{ord}_V(r) = \sum \operatorname{ord}_{\tilde{V}}(r)\, [R(\tilde{V}) : R(V)],$$

where the sum is over all subvarieties $\tilde{V}$ of $\tilde{X}$ which map onto V, and $[R(\tilde{V}) : R(V)]$ denotes the degree of the field extension. (This follows from Example A.3.1.) The more familiar order function on normal varieties therefore determines the order function on arbitrary varieties.

Example 1.2.4. If $r \in \mathcal{O}_{V,X}$, then

$$\operatorname{ord}_V(r) \geqq \min \{ n \mid r \in \mathcal{M}_{V,X}^n \}.$$

This inequality is an equality if X is non-singular along V, but strict if $r \in \mathcal{M}_{V,X}$ and X is singular along V.

Example 1.2.5. Let f, g define plane curves F, G in the affine plane over an algebraically closed field K, and let $P = (0, 0)$.

(a) $$i(P, F \cdot G) = \dim_K (K[\![x, y]\!]/(f, g)),$$

where $K[\![x, y]\!]$ is the ring of formal power series (see Zariski-Samuel (1) VII, VIII for properties of formal and convergent power series).

(b) If f has only one branch at P, and $(x(t), y(t))$ is a power series parametrization of this branch, then $i(P, F \cdot G)$ is the order of vanishing of $g(x(t), y(t))$ at $t = 0$. (K$[\![x, y]\!]/(f, g)$ is imbedded in $K[\![t]\!]$, with finite dimensional cokernel; apply Lemmas A.2.1 and A.2.4.)

(c) If $K = \mathbb{C}$, then convergent power series may be used in place of formal power series in (a) and (b).

(d) The definition and properties of intersection multiplicities of Example 1.1 extend to (germs of) analytic curves, and to the formal case, i.e., to arbitrary f, g in $K[\![x, y]\!]$.

(e) If $K = \mathbb{C}$, and F and G have no intersections but P inside an ε-neighborhood of P, then for sufficiently small $\eta \neq 0$, $i(P, F \cdot G)$ is the number of intersections of the curves $F = \eta$ and $G = 0$ inside an $\varepsilon/2$-neighborhood of P. (Factoring g in $\mathbb{C}\{x, y\}$, one may assume g has one branch; then use (b).)

(f) (Zeuthen's rule). Let $u_1, \ldots, u_m$ (resp. $v_1, \ldots, v_n$) be the roots of $f(x, Y)$ resp. $(g(x, Y))$ in some extension of $\mathbb{C}[\![x]\!]$. Then

$$i(P, F \cdot G) = \sum_{i,j} \operatorname{ord}_x (u_i - v_j).$$

Here $u_i, v_j \in \mathbb{C}[[x^{1/N}]]$ for some N, $\mathrm{ord}_x(x^{1/N}) = 1/N$. If axes are chosen so that F and G have no common points on the y-axis except at P, the right side is the order of the resultant of f and g.

For more on intersection multiplicities for plane curves, see Example 12.4.2, C. Segre (1), Zeuthen (3), Walker (1), and Fulton (1).

1.3 Cycles and Rational Equivalence

Let X be an algebraic scheme. A *k-cycle* on X is a finite formal sum

$$\sum n_i[V_i]$$

where the V_i are k-dimensional subvarieties of X, and the n_i are integers. The group of k-cycles on X, denoted $Z_k X$, is the free abelian group on the k-dimensional subvarieties of X; to a subvariety V of X corresponds $[V]$ in $Z_k X$.

For any $(k+1)$-dimensional subvariety W of X, and any $r \in R(W)^*$, define a k-cycle $[\mathrm{div}(r)]$ on X by

$$[\mathrm{div}(r)] = \sum \mathrm{ord}_V(r)[V],$$

the sum over all codimension one subvarieties V of W; here ord_V is the order function on $R(W)^*$ defined by the local ring $\mathcal{O}_{V,W}$.

A k-cycle α is *rationally equivalent to zero*, written $\alpha \sim 0$, if there are a finite number of $(k+1)$-dimensional subvarieties W_i of X, and $r_i \in R(W_i)^*$, such that

$$\alpha = \sum [\mathrm{div}(r_i)].$$

Since $[\mathrm{div}(r^{-1})] = -[\mathrm{div}(r)]$, the cycles rationally equivalent to zero form a subgroup $\mathrm{Rat}_k X$ of $Z_k X$. The group of *k-cycles modulo rational equivalence* on X is the factor group

$$A_k X = Z_k X / \mathrm{Rat}_k X.$$

Define $Z_* X$ (resp. $A_* X$) to be the direct sum of the $Z_k X$ (resp. $A_k X$) for $k = 0, 1, \ldots, \dim(X)$. A *cycle* (resp. *cycle class*) on X is an element of $Z_* X$ (resp. $A_* X$). A more classical definition of $A_* X$ will be given in § 1.5.

If α is a class in $A_* X$, and k is an integer, we denote by

$$\{\alpha\}_k$$

the component of α in $A_k X$. Thus $\alpha = \sum_{k \geqq 0} \{\alpha\}_k$.

A cycle is *positive* if it is not zero, and each of its coefficients is a positive integer. A cycle class is positive if it can be represented by a positive cycle.

Example 1.3.1. (a) A scheme and its underlying reduced scheme have the same subvarieties, and therefore the groups of cycles and rational equivalence classes are canonically isomorphic:

$$A_k(X) \cong A_k(X_{\mathrm{red}}).$$

(b) If X is a disjoint union of schemes $X_1, \ldots, X_t$, then $Z_* X = \oplus Z_* X_i$ and

$$A_k X = \bigoplus_{i=1}^{t} A_k(X_i) .$$

(c) If X_1 and X_2 are closed subschemes of X, then there are exact sequences

$$A_k(X_1 \cap X_2) \to A_k X_1 \oplus A_k X_2 \to A_k(X_1 \cup X_2) \to 0 .$$

(See Example 1.8.1 for a generalization.)

Example 1.3.2. If X is n-dimensional, $A_n X = Z_n X$ is the free abelian group on the n-dimensional irreducible components of X. More generally, any two rationally equivalent cycles on X contain any irreducible component V of X with the same coefficient. (Indeed, a cycle of the form $[\mathrm{div}(r)]$, $r \in R(W)^*$, cannot include an irreducible component of X.) For any $\alpha \in A_* X$, and any irreducible component V of X, we define *the coefficient of V in α* to be the coefficient of $[V]$ in any cycle which represents α.

1.4 Push-forward of Cycles

Let $f: X \to Y$ be a proper morphism. For any subvariety V of X, the image $W = f(V)$ is then a (closed) subvariety of Y. There is an induced imbedding of $R(W)$ in $R(V)$, which is a finite field extension if W has the same dimension as V (Appendix B.2.2). Set

$$\deg(V/W) = \begin{cases} [R(V):R(W)] & \text{if} \quad \dim(W) = \dim(V) \\ 0 & \text{if} \quad \dim(W) < \dim(V) \end{cases}$$

where $[R(V):R(W)]$ denotes the degree of the field extension. Define

$$f_*[V] = \deg(V/W)[W] .$$

This extends linearly to a homomorphism

$$f_* : Z_k X \to Z_k Y .$$

These homomorphisms are functorial: if g is a proper morphism from Y to Z, then $(gf)_* = g_* f_*$, as follows from the multiplicativity of degrees of field extensions. In the complex case, if $\dim W = \dim V$, V is generically a covering of W with $\deg(V/W)$ sheets, and the push-forward agrees with the push-forward in topology (cf. § 19.1).

Theorem 1.4. *If $f: X \to Y$ is a proper morphism, and α is a k-cycle on X which is rationally equivalent to zero, then $f_* \alpha$ is rationally equivalent to zero on Y.*

There is therefore an induced homomorphism

$$f_*: A_k X \to A_k Y,$$

so that A_* is a covariant functor for proper morphisms.

Proof. We may assume $\alpha = [\mathrm{div}(r)]$, where r is a rational function on a subvariety of X. We may replace X by this subvariety, and we may replace Y by $f(X)$, so we may assume Y is a variety and f is surjective. The theorem then follows from the following more explicit proposition.

Proposition 1.4. *Let $f: X \to Y$ be a proper, surjective morphism of varieties, and let $r \in R(X)^*$. Then*

(a) $f_*[\mathrm{div}(r)] = 0$ *if* $\dim(Y) < \dim(X)$.

(b) $f_*[\mathrm{div}(r)] = [\mathrm{div}(N(r))]$ *if* $\dim(Y) = \dim(X)$.

In (b), $R(X)$ is a finite extension of $R(Y)$, and $N(r)$ is the *norm* of r, i.e., the determinant of the $R(Y)$-linear endomorphism of $R(X)$ given by multiplication by r.

Proof. Case 1: $Y = \mathrm{Spec}(K)$, K a field, $X = \mathbb{P}^1_K$. Then $R(X) = K(t)$, with $t = x_1/x_0$. Since the order functions are homomorphisms, we may assume r is an irreducible polynomial, of degree d, in $K[t]$. Then r generates a prime ideal p in $K[t]$ corresponding to a point P in X with $\mathrm{ord}_P(r) = 1$. The only other point along which r has non-zero order is the point $P_\infty = (0:1)$ at infinity, where $s = 1/t$ is a uniformizing parameter. Then $s^d r$ is a unit at P_∞, so $\mathrm{ord}_{P_\infty}(r) = -d$. Therefore

$$[\mathrm{div}(r)] = [P] - d[P_\infty].$$

Now $R(P) = K[t]/p$ is an extension of K of degree d, while $R(P_\infty) = K$. Therefore

$$f_*[\mathrm{div}(r)] = d[Y] - d[Y] = 0.$$

Case 2: f is finite. Let $K = R(Y)$, $L = R(X)$. Let W be a subvariety of Y of codimension one, $A = \mathcal{O}_{W,Y}$, $\mathcal{M} = \mathcal{M}_{W,Y}$. There is a domain B, finite over A, with quotient field L, $B \otimes_A K = L$, so that the subvarieties V_i of X mapping onto W correspond to the maximal ideals $\mathcal{M}_i$ of B, with $B_{\mathcal{M}_i} = \mathcal{O}_{V_i,X}$. (To see this, one may assume Y and X are affine, with coordinate rings Γ and Λ respectively; then A is the localization of Γ at the prime ideal corresponding to W, and $B = \Lambda \otimes_\Gamma A$.) To prove part (b) we must show that

$$\sum_i \mathrm{ord}_{V_i}(r) \cdot [R(V_i): R(W)] = \mathrm{ord}_W(N(r)).$$

Since N and the order functions are homomorphisms, it is enough to prove this when $r \in B$. By Lemmas A.2.3 and A.2.2 of Appendix A, the left side of this equation is $l_A(\mathrm{Coker}(\varphi))$, where φ is the endomorphism of B induced by multiplication by r. The norm $N(r)$ is by definition $\det(\varphi_K)$, where φ_K is the induced endomorphism of L. The required equality

$$l_A(\mathrm{Coker}(\varphi)) = \mathrm{ord}_V(\det(\varphi_K))$$

is a special case of Lemma A.3.

The general case of (b) can be proved in the same way, since there is always a B as in Case 2 (Appendix B.2.4). For a more elementary proof, one may let $\tilde{X} \to X$, $\tilde{Y} \to Y$ be the normalizations of X and Y in their function fields; the morphism f induces a morphism $\tilde{f}: \tilde{X} \to \tilde{Y}$. By functorially and the case proved for the (finite) normalization maps, we may assume X and Y are normal. If A is the local ring of W on Y, A is a discrete valuation ring. Let B be the integral closure of A in $L = R(X)$. By the valuative criterion for properness, for each prime ideal p_i in B, B_{p_i} dominates a local ring $\mathcal{O}_{V_i, X}$; the V_i thus obtained are distinct since proper maps are assumed to be separated. Since $\mathcal{O}_{V_i, X}$ is one-dimensional and normal, $\mathcal{O}_{V_i, X} = B_{p_i}$, and one concludes as in Case 2.

For the general case of (a), we may assume $\dim(Y) = \dim(X) - 1$. Let $K = R(Y)$. The coefficient of Y in $f_*[\operatorname{div}(r)]$ is

$$\sum \operatorname{ord}_V(r)\,[R(V):K]\,,$$

the sum over all codimension one subvarieties V of X which map onto Y. We may replace Y by $\operatorname{Spec}(K)$, and X by the base extension $X_K = X \times_Y \operatorname{Spec}(K)$, so we may assume X is a curve over $Y = \operatorname{Spec}(K)$. Let $h: \tilde{X} \to X$ be the normalization of X, and choose a finite morphism $g: \tilde{X} \to \mathbb{P}^1_K$. Let p the projection from $\mathbb{P}^1_K$ to Y, so $f \circ h = p \circ g$, and let $\tilde{r} \in R(\tilde{X})$ be the image of r by the isomorphism $R(X) \cong R(\tilde{X})$. By functoriality, and Case 2 for h,

$$f_*[\operatorname{div}(r)] = f_* h_*[\operatorname{div}(\tilde{r})] = p_* g_*[\operatorname{div}(\tilde{r})]\,.$$

The latter is zero by an application of Case 2 to g and Case 1 to p. $\square$

Definition 1.4. If X is a complete scheme, i.e., X is proper over $S = \operatorname{Spec}(K)$, K the ground field, and $\alpha = \sum_P n_P[P]$ is a zero-cycle on X, the *degree* of α, denoted $\deg(\alpha)$, or $\int_X \alpha$, is defined by

$$\deg(\alpha) = \int_X \alpha = \sum_P n_P[R(P):K]\,.$$

Equivalently, $\deg(\alpha) = p_*(\alpha)$, where p is the structure morphism from X to S, and $A_0 S = \mathbb{Z}[S]$ is identified with $\mathbb{Z}$. By the theorem, rationally equivalent cycles have the same degree. We extend the degree homomorphism to all of $A_* X$,

$$\int_X : A_* X \to \mathbb{Z}$$

by defining $\int_X \alpha = 0$ if $\alpha \in A_k X$, $k > 0$. For any morphism $f: X \to Y$ of complete schemes, and any $\alpha \in A_* X$,

$$\int_X \alpha = \int_Y f_*(\alpha)\,,$$

a special case of functoriality. We often write $\int$ in place of $\int_X$.

Convention 1.4. Let $Y_1, \ldots, Y_r$ be closed subschemes of a scheme X. Let Y be a closed subscheme of X which contains all the Y_i. Given $\alpha_i \in A_* Y_i$,

$i=1,\ldots,r$, and $\beta \in A_* Y$, we will usually write "$\beta = \sum_{i=1}^{r} \alpha_i$ in $A_* Y$" in place of the precise equation $\beta = \sum_{i=1}^{r} \varphi_{i*}(\alpha_i)$ where φ_i is the inclusion of Y_i in Y.

Example 1.4.1. Theorem 1.4 implies Bézout's theorem for plane curves over an algebraically closed field (see Chapters 8 and 12 for generalizations): if F and G are projective plane curves of degrees m, n, with no common components, then

$$\sum_{P \in \mathbb{P}^2} i(P, F \cdot G) = m\, n\,.$$

(One may assume F is irreducible. If G and G' both have degree n, then G/G' defines a rational function r on the curve F, and

$$\sum i(P, F \cdot G) - \sum i(P, F \cdot G') = \sum \operatorname{ord}_P(r) = 0\,,$$

the last equation by Theorem 1.4. Taking $G' = L^n$ for L linear, one is reduced to the case where G is linear. Similarly, one reduces to the case where F is also linear, where it is obvious.) Bézout's theorem may also be proved via resultants, using Example 1.2.2.

Example 1.4.2. The fact that proper morphisms are separated is crucial for the truth of the theorem. If X is constructed by identifying two copies of $\mathbb{P}^1_K$ except at $(1:0)$, and f is the projection from X to $\operatorname{Spec}(K)$, and $r = x_1/x_0$, then $f_*[\operatorname{div}(r)] \not\sim 0$.

Example 1.4.3. Let X be a non-singular projective curve of genus g over an algebraically closed field. Then $A_0 X$ is the Picard group $\operatorname{Pic}(X)$ of divisor classes on X. The kernel of the degree homomorphism is the Jacobian variety of X, an abelian variety of dimension g. If $g > 0$, $A_0 X$ is not finitely generated.

Example 1.4.4. Let X be an abelian variety over an algebraically closed field. If a zero-cycle $\sum n_i [P_i]$ on X is rationally equivalent to zero, then the sum $\sum n_i P_i$ is zero in X. This determines a homomorphism

$$S : A_0 X \to X\,.$$

More generally, if X is any non-singular variety, and $\varphi : X \to \operatorname{Alb}(X)$ is the universal mapping to the Albanese variety of X, then there is a homomorphism

$$S : A_0 X \to \operatorname{Alb}(X)$$

taking $\sum n_i [P_i]$ to $\sum n_i\, \varphi(P_i)$. (Let $\alpha = \sum n_i [P_i]$. By the definition of rational equivalence, there are non-singular curves C_j, finite morphisms $f_j : C_j \to X$ mapping C_j birationally onto their images, and $r_j \in R(C_j)^*$, so that $\alpha = \sum f_{j*}[\operatorname{div}(r_j)]$. Each f_j induces a morphism from $\operatorname{Alb}(C_j)$ to $\operatorname{Alb}(X)$, compatible with S. The conclusion follows by the identification of $\operatorname{Alb}(C_j)$ with the Jacobian of C_j. See Roĭtman (1) or Murthy-Swan (1) for details.)

1.5 Cycles of Subschemes

Let X be any scheme, and let $X_1, \ldots, X_t$ be the irreducible components of X. The local rings $\mathcal{O}_{X_i,X}$ are all zero-dimensional (Artinian). The *geometric multiplicity* m_i of X_i in X is defined to be the length of $\mathcal{O}_{X_i,X}$:

$$m_i = l_{\mathcal{O}_{X_i,X}}(\mathcal{O}_{X_i,X}) .$$

The *(fundamental) cycle* $[X]$ of X is the cycle

$$[X] = \sum_{i=1}^{t} m_i [X_i] .$$

This is regarded as an element of $Z_* X$. By abuse of notation we also write $[X]$ for its image in $A_* X$. If X is purely k-dimensional, i.e., $\dim X_i = k$ for all i, then $[X] \in Z_k X$. In this case $Z_k X = A_k X$ is the free abelian group on $[X_1], \ldots, [X_t]$.

If X is a closed subscheme of a scheme Y, then $Z_* X \subset Z_* Y$, and we write also $[X]$ for the image of $[X]$ in $Z_* Y$, and for its image in $A_* Y$.

Example 1.5.1. Let V be a variety of dimension $k+1$, and $f: V \to \mathbb{P}^1$ a dominant morphism. Let $0 = (1:0)$, $\infty = (0:1)$ be the zero and infinite points of $\mathbb{P}^1$. The inverse image schemes (cf. Appendix B.2.3) $f^{-1}(0)$ and $f^{-1}(\infty)$ are purely k-dimensional subschemes of V, and the cycle

$$[f^{-1}(0)] - [f^{-1}(\infty)]$$

is the cycle $[\operatorname{div}(f)]$ defined in § 1.2, where f also denotes the rational function in $R(V)$ determined by the morphism f. (This follows immediately from the definitions given in § 1.2, § 1.3, and § 1.5.)

1.6 Alternate Definition of Rational Equivalence

Let X be a scheme, and let $X \times \mathbb{P}^1$ be the Cartesian product of X with $\mathbb{P}^1$. Let p be the projection from $X \times \mathbb{P}^1$ to X. Let V be a $(k+1)$-dimensional subvariety of $X \times \mathbb{P}^1$ such that the projection to the second factor induces a dominant morphism f from V to $\mathbb{P}^1$. For any point P in $\mathbb{P}^1$ which is rational over the ground field, the scheme-theoretic fibre $f^{-1}(P)$ is a subscheme of $X \times \{P\}$, which p maps isomorphically onto a subscheme of X; we denote this subscheme by $V(P)$. Note in particular that $p_*[f^{-1}(P)] = [V(P)]$ in $Z_k X$.

The morphism $f: V \to \mathbb{P}^1$ determines a rational function $f \in R(V)^*$. From Example 1.5.1 it follows that

$$[f^{-1}(0)] - [f^{-1}(\infty)] = [\operatorname{div}(f)] ,$$

where $0 = (1:0)$ and $\infty = (0:1)$ are the usual zero and infinity points of $\mathbb{P}^1$.

Therefore

$$[V(0)]-[V(\infty)]=p_*[\operatorname{div}(f)],$$

which is rationally equivalent to zero on X by Theorem 1.4.

Proposition 1.6. *A cycle α in Z_kX is rationally equivalent to zero if and only if there are $(k+1)$-dimensional subvarieties $V_1,\dots,V_t$ of $X\times\mathbb{P}^1$, such that the projections from V_i to $\mathbb{P}^1$ are dominant, with*

$$\alpha=\sum_{i=1}^{t}[V_i(0)]-[V_i(\infty)]$$

in Z_kX.

Proof. Let $\alpha=[\operatorname{div}(r)]$, $r\in R(W)^*$, W a $(k+1)$-dimensional subvariety of X. Then r defines a rational mapping from W to $\mathbb{P}^1$, i.e., a morphism from some open $U\subset W$ to $\mathbb{P}^1$. Let V be the closure of the graph of this morphism in $X\times\mathbb{P}^1$. The projection p maps V birationally and properly onto W. Let f be the induced morphism from V to $\mathbb{P}^1$. Then

$$\begin{aligned}[\operatorname{div}(r)]&=p_*[\operatorname{div}(f)] \quad \text{by Proposition 1.4 (b)}\\ &=[V(0)]-[V(\infty)].\end{aligned}$$

The proposition follows from this and the preceding remarks. □

With more intersection theory at our disposal, we will see that two cycles are rationally equivalent if they are members of a family of cycles parametrized by any rational or unirational variety (cf. Example 10.1.7).

Example 1.6.1. A k-cycle α on a scheme X is rationally equivalent to zero if and only if there are a finite number of normal varieties V_i, with rational functions f_i on V_i determined by morphisms from V_i to $\mathbb{P}^1$, and proper morphisms p_i from V_i to X, with $\alpha=\sum p_{i*}([\operatorname{div}(f_i)])$. (Replace the V_i of Proposition 1.6 by their normalizations.)

Example 1.6.2. Say that a cycle $Z=\sum n_i[V_i]$ on $X\times\mathbb{P}^1$ projects dominantly to $\mathbb{P}^1$ if each variety V_i which appears with non-zero coefficient in Z projects dominantly to $\mathbb{P}^1$; in this case set $Z(0)=\sum n_i[V_i(0)]$, $Z(\infty)=\sum n_i[V_i(\infty)]$.

Two k-cycles α, α' on a scheme X are rationally equivalent if and only if there is a positive $(k+1)$-cycle Z on $X\times\mathbb{P}^1$ projecting dominantly to $\mathbb{P}^1$, and a positive k-cycle β on X, with

$$Z(0)=\alpha+\beta \quad\text{and}\quad Z(\infty)=\alpha'+\beta.$$

(If $\alpha-\alpha'=Z'(0)-Z'(\infty)$ for some positive Z', choose a positive cycle β so that the cycle $\gamma=\alpha-Z'(0)+\beta$ is positive. Write $\gamma=\sum[V_i]$, and set $Z=Z'+\sum[V_i\times\mathbb{P}^1]$.)

Example 1.6.3. Let X be a projective scheme over an algebraically closed field. Let S^nX be the nth symmetric product of X, whose points are identified with positive 0-cycles of degree n on X. Two 0-cycles α, α' are rationally equivalent if and only if there is a morphism $f\colon\mathbb{P}^1\to S^nX$, for some n, and a positive 0-cycle β on X with

$$f(0)=\alpha+\beta \quad\text{and}\quad f(\infty)=\alpha'+\beta.$$

(If X is a smooth curve, this follows from the existence of the universal 0-cycle on $X \times S^n X$; in general, $f: \mathbb{P}^1 \to S^n X$ factors through $S^n C$ for some smooth (possibly disconnected) curve C which maps finitely to X.)

This result may be generalized to k-cycles, $k > 0$, with the varieties $S^n X$ replaced by Chow varieties parametrizing positive k-cycles on X (cf. Samuel (3) Thm. 3).

Even if α and α' are positive, the criterion fails if β is omitted. (Let X be the blow-up of $C \times \mathbb{P}^1$ at $P \times 0$, C a non-singular, non-rational curve, $P \in C$; let $\alpha = [P \times \infty]$, and let α' be any point on the exceptional divisor except the one determined by the line $P \times \mathbb{P}^1$.)

Example 1.6.4. Let $\mathscr{C}$ be a category of algebraic schemes over a fixed field. Assume that all morphisms in $\mathscr{C}$ are proper, and that for any projective morphism from X' to X, if $X \in \mathscr{C}$, then $X' \in \mathscr{C}$. Let H be a covariant functor from $\mathscr{C}$ to the category of abelian groups. Assume that every variety V in $\mathscr{C}$ has a class $cl(V)$ in $H(V)$. Assume

(i) if $f: V \to W$ is a surjective morphism of varieties in $\mathscr{C}$, then

$$H(f)(cl(V)) = \deg(V/W)\, cl(W).$$

This determines a natural transformation $cl: Z_* \to H$ of covariant functors. Assume also

(ii) if X is a normal variety in $\mathscr{C}$ and $f: X \to \mathbb{P}^1$ is a dominant morphism, then

$$cl([f^{-1}(0)]) = cl([f^{-1}(\infty)]).$$

Then cl passes to rational equivalence, inducing a natural transformation $cl: A_* \to H$ of covariant functors. Thus A_* is the finest theory satisfying (i) and (ii).

Example 1.6.5. The Grothendieck group $K_o X$ of coherent sheaves on X has a filtration $F_* K_o X$, with $F_k K_o X$ generated by sheaves whose support has dimension at most k. If f is a proper morphism, the higher direct image functors induce a map from $K_o X$ to $K_o Y$ which preserves this filtration (cf. Example 15.1.5). The associated graded groups $\mathrm{Gr}_* K_o X$ therefore become covariant for proper morphisms. A reader familiar with this machinery may prefer to use $\mathrm{Gr}_* K_o$ in place of A_*. The homomorphism from $Z_* X$ to $\mathrm{Gr}_* K_o X$ which takes $[V]$ to the class of the structure sheaf $\mathscr{O}_V$ satisfies the conditions of Example 1.6.4. This gives a natural transformation $A_* \to \mathrm{Gr}_* K_o$. In Chapter 18 we will see that this becomes an isomorphism after tensoring with $\mathbb{Q}$.

Example 1.6.6. Let X be a complete scheme, and let

$$\tilde{A}_0(X) = \mathrm{Ker}(A_0(X) \xrightarrow{\deg} \mathbb{Z}).$$

If the ground field is algebraically closed, and X is irreducible, the $\tilde{A}_0(X)$ is a divisible group. ($\tilde{A}_0(X)$ is generated by zero-cycles of the form $f_*([P] - [Q])$, $f: C \to X$, C a non-singular projective curve, $P, Q \in C$. $\tilde{A}_0(C)$ is the Jacobian variety of C, and any abelian variety over an algebraically closed field is divisible (cf. Mumford (4) p. 62).)

1.7 Flat Pull-back of Cycles

Let $f: X \to Y$ be a flat morphism of relative dimension n (cf. Appendix B.2.5). The examples of primary importance for us will be:

(i) an open imbedding ($n = 0$).
(ii) the projection of a vector bundle or $\mathbb{A}^n$-bundle (cf. § 1.9), or a projective bundle, to its base.
(iii) the projection from a Cartesian product $X = Y \times Z$ to the first factor, where Z is a purely n-dimensional scheme.
(iv) any dominant morphism from an $(n+1)$-dimensional variety to a non-singular curve.

Convention. In this book, a *flat morphism* is always assumed to have relative dimension n for some integer n.

For such $f: X \to Y$, and any subvariety V of Y, set

$$f^*[V] = [f^{-1}(V)].$$

Here $f^{-1}(V)$ is the inverse image scheme (cf. Appendix B.2.3), a subscheme of X of pure dimension $\dim(V) + n$, and $[f^{-1}(V)]$ is its cycle (§ 1.5). This extends by linearity to *pull-back homomorphisms*

$$f^*: Z_k Y \to Z_{k+n} X.$$

Lemma 1.7.1. *If $f: X \to Y$ is flat, then for any subscheme Z of Y,*

$$f^*[Z] = [f^{-1}(Z)].$$

Proof. Let W be an irreducible component of $f^{-1}(Z)$, let V be the closure of $f(W)$. The first assertion of Lemma A.4.1 implies that V is an irreducible component of Z. The second assertion of Lemma A.4.1, applied to $A = \mathcal{O}_{V,Z}$, $B = \mathcal{O}_{W,f^{-1}Z}$, implies the required equality of multiplicities. □

It follows from this lemma that flat pull-backs are functorial: if $f: X \to Y$ and $g: Y \to Z$ are flat, then gf is flat, and $(gf)^* = f^* g^*$. For if V is a subvariety of Z, then

$$(gf)^*[V] = [(gf)^{-1}(V)] = [f^{-1}g^{-1}(V)] = f^*[g^{-1}(V)] = f^* g^* [V].$$

Proposition 1.7. *Let*

$$\begin{array}{ccc} X' & \xrightarrow{g'} & X \\ {\scriptstyle f'}\downarrow & & \downarrow{\scriptstyle f} \\ Y' & \xrightarrow[g]{} & Y \end{array}$$

be a fibre square, with g flat and f proper. Then g' is flat, f' is proper, and for all $\alpha \in Z_* X$,

$$f'_* g'^* \alpha = g^* f_* \alpha$$

in $Z_* Y'$.

Proof. Since flatness and properness are preserved by base change, we may assume X and Y are varieties, f is surjective, and $\alpha = [X]$. Let $f_*[X] = d[Y]$. We must show that $f'_*[X'] = d[Y']$. This is a local calculation involving local rings of irreducible components, so we may assume $X = \operatorname{Spec}(K)$, $Y = \operatorname{Spec}(L)$, with K, L fields, $Y' = \operatorname{Spec}(A)$, with A local Artinian, and $X' = \operatorname{Spec}(B)$, $B = A \otimes_K L$. Then the result follows from Lemma A.1.3. □

Theorem 1.7. *Let $f: X \to Y$ be a flat morphism of relative dimension n, and α a k-cycle on Y which is rationally equivalent to zero. Then $f^*\alpha$ is rationally equivalent to zero in $Z_{k+n}X$.*

There are therefore induced homomorphisms, the *flat pull-backs,*

$$f^*: A_k Y \to A_{k+n} X,$$

so that A_* becomes a contravariant functor for flat morphisms.

Proof. By Proposition 1.6 we may assume $\alpha = [V(0)] - [V(\infty)]$, where V is a subvariety of $Y \times \mathbb{P}^1$ and the projection g from V to $\mathbb{P}^1$ is dominant, hence flat. Let $W = (f \times 1)^{-1}(V)$, a closed subscheme of $X \times \mathbb{P}^1$, and let $h: W \to \mathbb{P}^1$ be the morphism induced by the projection to $\mathbb{P}^1$.

Let $p: X \times \mathbb{P}^1 \to X$, $q: Y \times \mathbb{P}^1 \to Y$ be the projections. Then

$$\begin{aligned} f^*\alpha &= f^* q_*([g^{-1}(0)] - [g^{-1}(\infty)]) \\ &= p_*(f \times 1)^*([g^{-1}(0)] - [g^{-1}(\infty)]) \end{aligned}$$

by Proposition 1.7, and this equals

$$p_*([h^{-1}(0)] - [h^{-1}(\infty)])$$

by Lemma 1.7.1. Let $W_1, \ldots, W_t$ be the irreducible components of W, h_i the restriction of h to W_i. Let $[W] = \sum m_i [W_i]$. Since

$$[h_i^{-1}(0)] - [h_i^{-1}(\infty)] = [\operatorname{div}(h_i)],$$

and p_* preserves cycles rationally equivalent to zero, it suffices to verify that

$$[h^{-1}(P)] = \sum m_i [h_i^{-1}(P)],$$

for $P = 0$ and $P = \infty$. This is a special case of the following general lemma.

Lemma 1.7.2. *Let X be a purely n-dimensional scheme, with irreducible components $X_1, \ldots, X_r$, and geometric multiplicities $m_1, \ldots, m_r$. Let D be an effective Cartier divisor on X, i.e., a closed subscheme of X whose ideal sheaf is locally generated by one non-zero-divisor. Let $D_i = D \cap X_i$ be the restriction of D to X_i. Then*

$$[D] = \sum_{i=1}^{r} m_i [D_i]$$

in $Z_{n-1}(X)$.

Proof. It must be checked that each codimension one subvariety V of X appears with the same multiplicity in both sides of the equation. Let A be the local ring of X along V, a a local equation for D in A. The minimal prime ideals p_i in A correspond to the irreducible components X_i of X which contain

V. The multiplicity m_i of $[X_i]$ in $[X]$ is the length $l_{A_{p_i}}(A_{p_i})$. The multiplicity of $[V]$ in $[D]$ is $l_A(A/aA)$. The multiplicity of $[V]$ in $[D_i]$ is $l_{A/p_i}(A/p_i + aA)$. The required equality

$$l_A(A/aA) = \sum m_i \, l_{A/p_i}(A/p_i + aA)$$

is given by Lemma A.2.7; the fact that a is a non-zero-divisor is used to know that $l_A(A/aA)$ is the multiplicity $e_A(a, A)$ of § A.2. □

Example 1.7.1. Theorem 1.7 may fail if f is not assumed to have constant relative dimension. Lemma 1.7.2 may fail if X is not pure-dimensional. (For example, let X be the subscheme of $\mathbb{A}^3$ defined by the ideal (zx, zy), and E the Cartier divisor defined by the function $z - x$.) In Fulton (2) § 1.5, Prop. 3, this assumption was mistakenly omitted.

Example 1.7.2. Lemma 1.7.2 is also valid for Cartier divisors which are not effective.

Example 1.7.3. Let D be an effective Cartier divisor on an n-dimensional scheme X. Let $[X]_n = \sum_{i=1}^{s} m_i [X_i]$ denote the n-dimensional component of the cycle $[X]$, and let $D_i = D \cap X_i$. Then (by the proof of Lemma 1.7.2)

$$[D]_{n-1} = \sum_{i=1}^{s} m_i [D_i] \, .$$

Example 1.7.4. Let $f: X' \to X$ be a finite and flat morphism; each point of X has affine neighborhood U such that the coordinate ring of $f^{-1}(U)$ is a finitely generated free module over the coordinate ring of U. One says that f has degree d if the rank of this module is d, for all such U. Then for all subvarieties V of X, $f_* f^* [V] = d[V]$ in $Z_*(X)$. The composite

$$A_* X \xrightarrow{f^*} A_* X' \xrightarrow{f_*} A_* X$$

is multiplication by d.

Example 1.7.5. If a subgroup $R_* X$ of $Z_* X$ is specified for all algebraic schemes X over a given field, which is preserved under proper push-forward and flat pull-back, and $R_* \mathbb{P}^1$ contains $[0] - [\infty]$, then $R_* X$ contains all cycles rationally equivalent to zero, for any X.

Example 1.7.6. Suppose a finite group G acts on a variety Y, with quotient variety $X = Y/G$. (Such X exists if Y is quasi-projective, cf. Mumford (4) p. 111.) Then there is a canonical isomorphism

$$A_* X_{\mathbb{Q}} = (A_* Y_{\mathbb{Q}})^G.$$

Here $A_* X_{\mathbb{Q}}$ denotes $A_* X \otimes_{\mathbb{Z}} \mathbb{Q}$; G acts on $A_* Y$ by covariance (§ 1.4), and $(A_* Y_{\mathbb{Q}})^G$ denotes the G-invariant subgroup of $A_* Y_{\mathbb{Q}}$. (Let $\pi: Y \to X$ be the quotient map. For any subvariety W of Y, let

$$I_W = \{g \in G \mid g|_W = \mathrm{id}_W\}$$

be the inertia group, and let

$$e_W = \text{card}(I_W)/\deg_i(W/V),$$

where $\deg_i(W/V)$ is the degree of inseparability of $R(W)$ over $R(V)$. For a subvariety V of X set

$$\pi^*[V] = \sum e_W[W],$$

the sum over all irreducible components W of $\pi^{-1}(V)$. This determines an isomorphism $Z_*X_{\mathbb{Q}} = (Z_*Y_{\mathbb{Q}})^G$, and $(A_*Y_{\mathbb{Q}})^G$ is the quotient of $(Z_*Y_{\mathbb{Q}})^G$ modulo the subspace generated by

$$\left\{\sum_{g \in G} g_*[\text{div}(r)] \,\middle|\, r \in R(W)^*, W \subset Y\right\}.)$$

Note that the composite

$$A_*X_{\mathbb{Q}} \xrightarrow{\pi^*} (A_*Y_{\mathbb{Q}})^G \hookrightarrow A_*Y_{\mathbb{Q}} \xrightarrow{\pi_*} A_*X_{\mathbb{Q}}$$

is multiplication by card(G).

1.8 An Exact Sequence

Proposition 1.8. *Let Y be a closed subscheme of a scheme X, and let $U = X - Y$. Let $i: Y \to X$, $j: U \to X$ be the inclusions. Then the sequence*

$$A_k Y \xrightarrow{i_*} A_k X \xrightarrow{j^*} A_k U \to 0$$

is exact for all k.

Proof. Since any subvariety V of U extends to a subvariety $\bar{V}$ of X, the sequence

$$Z_k Y \xrightarrow{i_*} Z_k X \xrightarrow{j^*} Z_k U \to 0$$

is exact. If $\alpha \in Z_k X$ and $j^*\alpha \sim 0$, then

$$j^*\alpha = \sum [\text{div}(r_i)]$$

for $r_i \in R(W_i)^*$, W_i subvarieties of U. Since $R(W_i) = R(\bar{W}_i)$, r_i corresponds to a rational function $\bar{r}_i$ on $\bar{W}_i$, and

$$j^*(\alpha - \sum [\text{div}(\bar{r}_i)]) = 0$$

in $Z_k U$. Therefore

$$\alpha - \sum [\text{div}(\bar{r}_i)] = i_*\beta$$

for some $\beta \in Z_k Y$, which implies the proposition. □

Example 1.8.1. Let

$$\begin{array}{ccc} Y' & \xrightarrow{j} & X' \\ {\scriptstyle q}\downarrow & & \downarrow{\scriptstyle p} \\ Y & \xrightarrow[i]{} & X \end{array}$$

be a fibre square, with i a closed imbedding, p proper, such that p induces an isomorphism of $X' - Y'$ onto $X - Y$. Then the sequence

$$A_k Y' \xrightarrow{a} A_k Y \oplus A_k X' \xrightarrow{b} A_k X \to 0$$

is exact, where $a(\alpha) = (q_* \alpha, -j_* \alpha)$, $b(\alpha, \beta) = i_* \alpha + p_* \beta$. (Use the definition of § 1.3 to show that $\operatorname{Ker}(j_*)$ surjects onto $\operatorname{Ker}(i_*)$.)

1.9 Affine Bundles

A scheme E, together with a morphism $p: E \to X$, is an *affine bundle* of rank n over X if X has an open covering by U_α, and there are isomorphisms

$$p^{-1}(U_\alpha) \cong U_\alpha \times \mathbb{A}^n$$

such that p restricted to $p^{-1}(U_\alpha)$ corresponds to the projection from $U_\alpha \times \mathbb{A}^n$ to U_α.

Proposition 1.9. *Let $p: E \to X$ be an affine bundle of rank n. Then the flat pull-back*

$$p^*: A_k X \to A_{k+n} E$$

is surjective for all k.

Proof. Choose a closed subscheme Y of X so that $U = X - Y$ is an affine open set over which E is trivial. There is a commutative diagram

$$\begin{array}{ccccccc} A_* Y & \longrightarrow & A_* X & \to & A_* U & \to & 0 \\ \downarrow & & \downarrow & & \downarrow & & \\ A_*(p^{-1} Y) & \to & A_* E & \to & A_*(p^{-1} U) & \to & 0 \end{array}$$

where the vertical maps are flat pull-backs, and the rows are exact by Proposition 1.8. By a diagram chase it suffices to prove the assertion for the restrictions of E to U and to Y. By Noetherian induction, i.e., repeating the process on Y, it suffices to prove it for $X = U$. Thus we may assume $E = X \times \mathbb{A}^n$. The projection factors

$$X \times \mathbb{A}^n \to X \times \mathbb{A}^{n-1} \to X,$$

so we may assume $n = 1$.

We must show that $[V]$ is in $p^* A_k X$ for any $(k+1)$-dimensional subvariety V of E. We may replace X by the closure of $p(V)$ (cf. Proposition 1.7), so we may assume X is a variety and p maps V dominantly to X. Let A be the coordinate ring of X, $K = R(X)$ the quotient field of A, and let q be the prime ideal in $A[t]$ corresponding to V. If $\dim X = k$, then $V = E$, so $V = p^*[X]$. So we may assume $\dim X = k+1$. Since V dominates X and $V \neq E$, the prime ideal $qK[t]$ is non-trivial; let $r \in K[t]$ generate $qK[t]$. Then

$$[V] - [\operatorname{div}(r)] = \sum n_i [V_i],$$

for some $(k+1)$-dimensional subvarieties V_i of E whose projections to X are not dominant. Therefore $V_i = p^{-1}(W_i)$, with $W_i = p(V_i)$, so

$$[V] = [\operatorname{div}(r)] + \sum n_i p^* [W_i],$$

as required. □

In particular, $A_k(\mathbb{A}^n)$ is zero for $k < n$, while $A_n(\mathbb{A}^n) = \mathbb{Z}$.

When E is a vector bundle over X, we will see in Chap. 3 that p^* is an isomorphism.

Example 1.9.1. Let X be a scheme with a "cellular decomposition", i.e., X has a filtration $X = X_n \supset X_{n-1} \supset \ldots \supset X_0 \supset X_{-1} = \emptyset$ by closed subschemes, with each $X_i - X_{i-1}$ a disjoint union of schemes U_{ij} isomorphic to affine spaces $\mathbb{A}^{n_{ij}}$. Then $A_* X$ is (finitely) generated by $\{[V_{ij}]\}$, where V_{ij} is the closure of U_{ij} in X. The Grassmann and flag varieties are examples (cf. Chap. 14).

Example 1.9.2. The conclusion of Proposition 1.9 holds for any locally trivial fibre bundle whose fibre is isomorphic to an open subscheme of an affine space $\mathbb{A}^n$ (cf. Grothendieck (1) § 6).

Example 1.9.3. Let L^k be a k-dimensional linear subspace of $\mathbb{P}^n$, $k = 0 \ldots n$.

(a) $A_k(\mathbb{P}^n)$ is generated by $[L^k]$. (Apply Proposition 1.8 with $X = \mathbb{P}^n$, $Y = L^{n-1}$, $U = \mathbb{A}^n$.)

(b) $A_k(\mathbb{P}^n) = \mathbb{Z}[L^k] = \mathbb{Z}$. (For $k = n-1$ this follows from the form of rational functions on $\mathbb{P}^n$. For $k < n-1$, if $d[L^k] = \sum n_i [\operatorname{div}(r_i)]$, $r_i \in R(V_i)^*$, let Z be the union of the V_i, and let $f: Z \to \mathbb{P}^{k+1}$ be projection from a linear $(n-k-2)$-dimensional subspace disjoint from Z; apply Theorem 1.4 to f.) A more general theorem will be proved in Chap. 3.

Example 1.9.4. Let H be a (reduced) hypersurface of degree d in $\mathbb{P}^n$. Then $[H] = d[L]$, for L a hyperplane, and

$$A_{n-1}(\mathbb{P}^n - H) = \mathbb{Z}/d\mathbb{Z}.$$

Example 1.9.5. (a) Let $f: X' \to X$ be a finite, birational morphism of n-dimensional varieties. For each codimension one subvariety V of X, let $d(V)$ be the greatest common divisor of the degrees of all field extensions $R(W)$ of $R(V)$, for all subvarieties W of X' such that $f(W) = V$. There is an exact sequence

$$A_{n-1} X' \to A_{n-1} X \to \bigoplus_V \mathbb{Z}/d(V)\mathbb{Z} \to 0.$$

(b) Let $X \subset \mathbb{A}^3$ be the Whitney umbrella: $x^2 = y z^2$. Then $A_1 X = \mathbb{Z}/2\mathbb{Z}$, generated by the class of the y-axis. (Apply (a) to $f\colon \mathbb{A}^2 \to X, f(s,t) = (st, s^2, t)$.)

(c) Let X be the quotient of $\mathbb{A}^2$ obtained by identifying $(s,0)$ and $(\mu s, 0)$ for all n^{th} roots of unity μ; equivalently,

$$X = \operatorname{Spec}(K[s^n, s\,t, s^2 t, \dots, s^{n-1} t, t]) .$$

Then $A_0 X = 0$, $A_1 X = \mathbb{Z}/n\,\mathbb{Z}$, $A_2 X = \mathbb{Z}$.

Example 1.9.6. (cf. Fulton (2) p. 166). Let $H(X) = A_*(X)_{\mathbb{Q}} = A_*(X) \otimes \mathbb{Q}$, regarded as a covariant functor from the category of complete schemes over a fixed field, to abelian groups. Suppose $T\colon H \to H$ is a natural transformation such that $T[\mathbb{P}^n] = [\mathbb{P}^n] + \beta_n$ for some class β_n of dimension $< n$ on $\mathbb{P}^n$, $n = 0, 1, 2, \dots$. Then T must be the identity. (Show first that $\beta_n = 0$, by choosing $f\colon \mathbb{P}^n \to \mathbb{P}^n$ with $f_*[\mathbb{P}^n] = d[\mathbb{P}^n]$, $f_*[L^k] = d_k[L^k]$, L^k a k-plane in $\mathbb{P}^n$, $k < n$, $d_k \neq d$; comparing $f_* T[\mathbb{P}^n]$ to $T f_*[\mathbb{P}^n]$ shows that the coefficient of $[L^k]$ in β_n must be zero. To show that $T[X] = [X]$ for all varieties X, by Chow's lemma ([EGA] II.5.6) one may assume X is projective. Replacing X by a variety that maps finitely to X, one may assume a finite group G acts on X with $X/G \cong \mathbb{P}^n$. Let $T[X] = [X] + \beta$. Applying the covariance with respect to the projection f from X to $\mathbb{P}^n$, and the preceding case for $\mathbb{P}^n$, one derives $f_*\beta = 0$. By the covariance with respect to the automorphisms in G, $\beta \in H(X)^G$. By Example 1.7.6, $\beta = 0$, as desired.)

Similarly on the category of all algebraic schemes over a fixed field, if $T\colon H \to H$ is covariant with respect to proper morphisms, and contravariant with respect to open imbeddings of quasi-projective schemes, and $T[\mathbb{P}^n] = [\mathbb{P}^n] + \beta_n$, $\dim(\beta_n) < n$, then T is the identity.

1.10 Exterior Products

For algebraic schemes X, Y over a field, $X \times Y$ denotes the Cartesian (fibre) product of X and Y over the ground field.

The *exterior product*

$$Z_k X \otimes Z_l Y \xrightarrow{\times} Z_{k+l}(X \times Y)$$

is defined by the formula

$$[V] \times [W] = [V \times W]$$

for V, W subvarieties of X, Y, and extending bilinearly to general cycles. (If the ground field is not algebraically closed, $V \times W$ may not be irreducible; its cycle $[V \times W]$ is defined by the prescription of § 1.5.)

Proposition 1.10. (a) *If $\alpha \sim 0$ or $\beta \sim 0$, then $\alpha \times \beta \sim 0$.*

(b) *Let $f\colon X' \to X$, $g\colon Y' \to Y$ be morphisms, $f \times g$ the induced morphism from $X' \times Y'$ to $X \times Y$.*

(i) *If f and g are proper, so is $f \times g$, and*

$$(f \times g)_* (\alpha \times \beta) = f_* \alpha \times g_* \beta$$

for all cycles α on X', β on Y'.

(ii) *If f and g are flat of relative dimensions m and n, then $f \times g$ is flat of relative dimension $m+n$, and*

$$(f \times g)^* (\alpha \times \beta) = f^* \alpha \times g^* \beta$$

for all cycles α on X, β on Y.

Proof. For (b), factoring $f \times g$ into $(f \times id_Y) \circ (id_{X'} \times g)$ reduces one to the easy cases where f or g is the identity; cf. Proposition 1.7 for (i). For (a), if $\alpha \sim 0$, we may assume $\beta = [W]$, W a subvariety of Y, and then assume $W = Y$ by (b) (i). In this case $\alpha \times \beta = p^*(\alpha)$, where $p: X \times W \to X$ is the projection, and the assertion is a special case of Theorem 1.7. □

It follows that there are *exterior products*

$$A_k X \otimes A_l Y \xrightarrow{\times} A_{k+l}(X \times Y)$$

satisfying the formulas of Proposition 1.10 (b).

Example 1.10.1. The exterior product is associative:

$$(\alpha \times \beta) \times \gamma = \alpha \times (\beta \times \gamma) \quad \text{for} \quad \alpha \in A_* X\,,\ \beta \in A_* Y\,,\ \gamma \in A_* Z\,.$$

Example 1.10.2. If X has a cellular decomposition as in Example 1.9.1, then, for all schemes Y, the product maps

$$\bigoplus_{k+l=m} A_k X \otimes A_l Y \xrightarrow{\times} A_m(X \times Y)$$

are surjective.

Notes and References

For divisors on a non-singular variety, rational equivalence coincides with linear equivalence, a subject which has long been central in algebraic geometry. For zero-cycles on a curve this is the study of the Jacobian variety. Much of the Italian school of algebraic geometry was devoted to the study of linear systems of curves on a surface.

The notion of rational equivalence for cycles of codimension greater than one was originated by Severi. The canonical divisor class had been a vital classical tool. In his seminal 1932 paper (Severi (6)), he revealed that – to use modern language – the second Chern class of a surface was not just a number, but was a well-defined rational equivalence of zero-cycles. Much of the subsequent work of Severi, as well as B. Segre, Eger and Todd, was devoted to developing the idea of rational equivalence for cycles of arbitrary dimension, and constructing "canonical" classes in all dimensions.

Crucial to these developments was the construction of an intersection product for two cycle classes on a non-singular variety. For example, intersecting canonical classes would lead to numerical invariants, generalizing the self-intersection number of a canonical divisor on a surface. These related problems will be discussed in later chapters.

Several definitions of rational equivalence were proposed by Severi. One difficulty at first was caused by the desire to phrase everything in terms of rational families of positive cycles (cf. Examples 1.6.2, 1.6.3). Evidently motivated by Lefschetz's ideas in topology, Todd (2) explicitly introduced the notion of the group of virtual cycles, and the subgroup of cycles rationally equivalent to zero; this simple artifact led to considerable clarification. Definitions generating rational equivalence by intersecting families of rationally equivalent divisors were also proposed. Discussions of these ideas may be found in Severi (14), (19), and Baldassarri (1). Attempts to reconcile competing definitions and develop a satisfactory theory of rational equivalence with intersections led to considerable debate among Severi, Van der Waerden, Samuel, and Weil (cf. Van der Waerden (5), (6)).

Weil (5), Samuel (3), and Chow (1) began a systematic study of equivalence relations on cycles, based on the new foundations of Weil (2). Many of Severi's intuitive geometric notions about families of cycles were replaced by a precise algebraic language of specialization, and a more axiomatic approach to equivalence relations on cycles was developed. The paper of Chow gave an accepted proof that rational equivalence classes (on a non-singular projective variety) determine a well-defined intersection class; his proof used a "moving lemma" which was based on a construction of Severi's.

In 1958 Chevalley's seminar focused on rational equivalence. In the notes (Chevalley (2)) the theory is developed from first principles, with no reference to, or even mention of, previous work on the subject; there the ring of rational equivalence classes was named the "Chow ring". The fact that a general foundational crisis was taking place in algebraic geometry helps to explain this break with the past. Indeed, rational equivalence and intersection theory were used as a major testing ground for the new foundations. It would be unfortunate, however, if Severi's pioneering work in this area were forgotten; and if incompleteness or the presence of errors are grounds for ignoring Severi's work, few of the subsequent papers on rational equivalence would survive.

Although most work on rational equivalence assumed a non-singular ambient variety, Chevalley (2) and Grothendieck (1) pointed out that the notion of rational equivalence, and several basic properties, can be extended to singular varieties. These ideas were worked out in the first part of Fulton (2), on which the present Chap. 1 is based. The proof that rational equivalence pushes forward follows Chevalley (2); the exact sequence of § 1.8 and Proposition 1.9 are found in Grothendieck (1). The direct proof (Example 1.2.2) of the agreement of modern and resultant definitions of intersection multiplicities for plane curves is apparently new. The alternative in Example 1.6.5 was proposed by S. Kleiman. R. Lazarsfeld suggested Example 1.9.5.

In spite of the formal analogy with homology groups, the groups A_*X are much more difficult to calculate. Some of the literature on this is discussed in Chap. 19. On the other hand, an equation in A_*X is much finer than the corresponding equation in homology. For zero-cycles on a curve, this difference amounts to knowing a point on the Jacobian; for an application of this principle, see Example 14.4.6. In addition, the exact sequence of § 1.8 may be more useful than the analogous long exact homology sequence. For applications to affine and projective surfaces, see Murthy-Swan (1). Collino (1) has calculated $A_*(X)$ when X is a symmetric product of a non-singular curve.

Chapter 2. Divisors

Summary

If D is a Cartier divisor on a scheme X, and α is a k-cycle on X, we construct an intersection class

$$D \cdot \alpha \in A_{k-1}(|D| \cap |\alpha|)$$

where $|D|$[1], $|\alpha|$ are the supports of D and α. For $\alpha = [V]$, V a subvariety, $D \cdot [V]$ is defined by one of two procedures: (i) if $V \not\subset |D|$, D restricts to a Cartier divisor on V, and $D \cdot [V]$ is defined to be the associated Weil divisor of this restriction; (ii) if $V \subset |D|$, the restriction of the line bundle $\mathcal{O}_X(D)$ to V is the line bundle of a well-defined linear equivalence class of Cartier divisors on V, and $D \cdot [V]$ is represented by the associated Weil divisor of any such Cartier divisor.

We prove that if α is rationally equivalent to zero on X, then $D \cdot \alpha$ is zero in $A_{k-1}(|D|)$; there are therefore induced homomorphisms

$$A_k X \to A_{k-1}(|D|) .$$

In the special but important case where D is the inverse image of a point for a morphism from X to a smooth curve, $D \cdot \alpha$ is the specialization of α; in this case (or whenever D is principal) $D \cdot \alpha$ can be well-defined as a cycle, setting $D \cdot [V] = 0$ if $V \subset D$. The above fact therefore includes the assertion that rational equivalence is preserved under specialization.

If D and D' are Cartier divisors on a scheme X, and α is *a* k-cycle on X, a crucial property is the commutative law

$$D \cdot (D' \cdot \alpha) = D' \cdot (D \cdot \alpha)$$

in $A_{k-2}(|D| \cap |D'| \cap |\alpha|)$. Consider, for example, the case where $f: X \to \mathbb{A}^2$ is a morphism, and D and D' are the inverse images of the two axes. One may specialize a cycle first to the part of X over the x-axis, and then specialize the resulting cycle to $f^{-1}(0)$; or one may first specialize over the y-axis, then over the origin. The resulting cycles one arrives at by these two routes may well be different[2], but the above says they are rationally equivalent.

---

1 This shorthand for $\mathrm{Supp}(D)$ should not be confused with a notation for complete linear systems, which do not occur in this chapter.

2 This corresponds to the fact that a family of cycles parametrized by a smooth parameter variety has a unique limiting cycle over a missing point when the parameter variety is a curve, but not when it has dimension two or more.

Both of the above facts follow from the identity (Theorem 2.4):

$$D \cdot [D'] = D' \cdot [D] \quad \text{in} \quad A_{n-2}(|D| \cap |D'|)$$

for Cartier divisors D, D' on an n-dimensional variety X, with $[D]$, $[D']$ their associated Weil divisors.

A Cartier divisor D on a scheme X determines a line bundle $L = \mathcal{O}_X(D)$ and a trivialization of L over $X - |D|$. Only the line bundle, the support, and the trivialization are needed to carry out the above intersection construction[3]. These concepts are formalized in the notation of a pseudo-divisor (§ 2.2); there is the added advantage that a pseudo-divisor, unlike the stricter notion of a Cartier divisor, pulls back under arbitrary morphisms.

Intersecting with divisors is used to construct homomorphisms

$$A_k X \to A_{k-1} X, \quad \alpha \to c_1(L) \cap \alpha,$$

for a line bundle L on X, and to construct Gysin homomorphisms

$$i^* : A_k X \to A_{k-1} D$$

when i is the inclusion of an effective Cartier divisor D in X. These operations will be generalized to higher codimension in subsequent chapters.

2.1 Cartier Divisors and Weil Divisors

Let X be an n-dimensional variety. A *Weil divisor* on X is an $(n-1)$-cycle on X. The Weil divisors form the group $Z_{n-1} X$ of § 1.3.

A *Cartier divisor* on X is defined by data (U_α, f_α), where the U_α form an open covering of X and the f_α are non-zero functions in $R(U_\alpha) = R(X)$, subject to the condition that f_α / f_β is a unit (i.e., a regular, nowhere vanishing function) on $U_\alpha \cap U_\beta$. The rational functions f_α are called *local equations* or D; they are determined up to multiplication by units on U_α (see Appendix B.4).

If D is a Cartier divisor on X, and V is a subvariety of X of codimension one, write

$$\operatorname{ord}_V D = \operatorname{ord}_V (f_\alpha)$$

---

3 Over the complex numbers, a model for these constructions is available from topology. A complex line bundle L on X has a first Chern class $c_1(L) \in H^2(X)$. If $L = \mathcal{O}_X(D)$, L is trivialized on $X - |D|$, so $c_1(L)$ comes from a class we may call $c(D)$ in $H^2(X, X - |D|)$ (cf. Example 19.2.6). For Y closed in X there are cap products (cf. § 19.1)

$$H^p(X, X - |D|) \otimes H_q Y \xrightarrow{\frown} H_{q-p}(Y \cap |D|).$$

In particular, capping with $c(D)$ gives homomorphisms from $H_{2k} Y$ to $H_{2k-2}(Y \cap |D|)$. If D, D' are Cartier divisors, then $c(D) \cup c(D') = c(D') \cup c(D)$ in $H^4(X, X - (|D| \cup |D'|))$, so

$$c(D) \cap (c(D') \cap \alpha) = c(D') \cap (c(D) \cap \alpha)$$

in $H_{2k-4}(Y \cap |D| \cap |D'|)$ for $\alpha \in H_{2k} Y$ – which motivates the commutativity law. Other properties such as the projection formula are also familiar in topology.

where ord_V is the order function on $R(X)$ defined by V in § 1.2, and f_α is a local equation for D on any affine open set U_α with $U_\alpha \cap V \neq \emptyset$; this is well-defined since f_α is well-defined up to units. Define the *associated Weil divisor* $[D]$ of D by setting

$$[D] = \sum \mathrm{ord}_V D\,[V]$$

the sum over all codimension one subvarieties V of X; as in § 1.2, there are only finitely many V with $\mathrm{ord}_V D \neq 0$ (Appendix B.4.3).

The Cartier divisors form an abelian group $\mathrm{Div}(X)$: if D and E are given by data (U_α, f_α) and (U_α, g_α), the sum $D+E$ is given by $(U_\alpha, f_\alpha g_\alpha)$. By the additivity of the order functions, the mapping $D \to [D]$ is a homomorphism

$$\mathrm{Div}(X) \to Z_{n-1}(X)\,.$$

Any f in $R(X)^*$ determines a *principal* Cartier divisor $\mathrm{div}(f)$, by taking all local equations equal to f. Note that the Weil divisor associated to $\mathrm{div}(f)$ is the cycle $[\mathrm{div}(f)]$ defined in § 1.3.

Two divisors D, D' are *linearly equivalent* if they differ by a principal divisor: $D' = D + \mathrm{div}(f)$. From the definition of rational equivalence, it follows that $[D]$ and $[D']$ are rationally equivalent cycles. If $\mathrm{Pic}(X)$ denotes the group of linear equivalence classes of Cartier divisors, there is an induced homomorphism

$$\mathrm{Pic}(X) \to A_{n-1}(X)\,.$$

This homomorphism is in general neither injective nor surjective (see Examples 2.1.1–2.1.3).

We shall see that Cartier divisors can be intersected with arbitrary cycles, corresponding to the fact that elements of $\mathrm{Pic}(X)$ define cohomology classes. Weil divisors in general do not have this ability – they determine homology classes (see Example 2.4.5).

The *support* of a Cartier divisor D, denoted by $\mathrm{Supp}(D)$, or $|D|$, is the union of all subvarieties Z of X such that a local equation for D in the local ring $\mathcal{O}_{Z,X}$ is not a unit. This is a closed algebraic subset of X.

On a general scheme X, an *effective Cartier divisor* is a subscheme which is locally defined by one equation, which is required to be a non-zero-divisor. The notion of Cartier divisor also extends to schemes which are not varieties (Appendix B.4), but this is not required for present purposes.

Example 2.1.1 (cf. [EGA]IV.21.6). If X is normal (resp. locally factorial) then $\mathrm{Div}(X) \to Z_{n-1}(X)$ and $\mathrm{Pic}(X) \to A_{n-1}(X)$ are injective (resp. isomorphisms). It follows for example that $\mathrm{Pic}(\mathbb{P}^n) \cong \mathbb{Z}$, with generator $\mathcal{O}(1)$ (cf. Example 1.9.3).

Example 2.1.2. Let X be the projective plane curve over $\mathbb{C}$ defined by the homogeneous equation $y^2 z = x^3$. Then $A_0 X \cong \mathbb{Z}$, and the homomorphism $\mathrm{Pic}(X) \to A_0(X)$ is surjective, with kernel the additive group $\mathbb{C}$. In case X is the curve $y^2 z = x^2 z + x^3$, the kernel is $\mathbb{C}^*$.

Example 2.1.3. Let X be the surface in $\mathbb{A}^3$ defined by the equation $z^2 = x\,y$. The line $V\colon x = y = 0$ (a generator for the cone) defined a Weil divisor which is not a Cartier divisor. In this case $\mathrm{Pic}(X) = 0$, and $A_1(X) = \mathbb{Z}/2\mathbb{Z}$.

Example 2.1.4. Let X be a projective scheme, and let L be an ample line bundle on X. For a $(k+1)$-dimensional subvariety V of X, and non-zero sections s_1, s_2 of the restriction of $L^{\otimes n}$ to V, with divisors of zeros $D_1, D_2, [D_1]$ is rationally equivalent to $[D_2]$. The group $\mathrm{Rat}_k X$ is generated by cycles $[D_1]-[D_2]$, as V, n, s_1, s_2 vary. (If $r \in R(V)^*$, there is, for large n, a section s_1 so that $D_1 + \operatorname{div}(r)$ is an effective divisor on V.)

2.2 Line Bundles and Pseudo-divisors

If D is a Cartier divisor on X and $f: X' \to X$ is a morphism, a pull-back Cartier divisor f^*D is defined only under certain assumptions (cf. [EGA]IV.21.4). If X' is a variety, for example, f^*D is defined by pulling back local equations for D provided $f(X') \not\subset |D|$, but no Cartier divisor pull-back is defined if $f(X') \subset |D|$. We will introduce a simple generalization of the notion of Cartier divisor, which will not have this defect, but will still carry enough information to determine intersection operations on cycle classes.

Definition 2.2.1. A *pseudo-divisor* on a scheme X is a triple (L, Z, s), where L is a line bundle on X, Z is a closed subset of X, and s is a nowhere vanishing section of L on $X-Z$ (equivalently, s is a trivialization of the restriction of L to $X-Z$). We call L the *line bundle*, Z the *support*, and s the *section*, of the pseudo-divisor. Data (L', Z', s') define the same pseudo-divisor if $Z=Z'$ and there is an isomorphism σ of L with L' such that the restriction of σ to $X-Z$ takes s to s'. Note that a pseudo-divisor with support X is simply an isomorphism class of line bundles on X.

Any Cartier divisor D on a scheme X determines a pseudo-divisor[4] $(\mathscr{O}_X(D), |D|, s_D)$ on X, where $\mathscr{O}_X(D)$ is the line bundle of D, $|D|$ is the support of D, and s_D is the canonical section of $\mathscr{O}_X(D)$ (Appendix B.4.5). We say that a Cartier divisor D *represents* a pseudo-divisor (L, Z, s) if $|D| \subset Z$, and there is an isomorphism from $\mathscr{O}_X(D)$ to L which, off Z, takes s_D to s. Note that we allow Z to be larger than $|D|$; for example, if $Z=X$, all linearly equivalent Cartier divisors represent the same pseudo-divisor.

A general pseudo-divisor will often be denoted by a single letter D, and we write $\mathscr{O}_X(D)$ for its line bundle, $|D|$ for its support, s_D for its section. This agrees with the notation for Cartier divisors, except that a Cartier divisor may have smaller support than a pseudo-divisor it represents.

Lemma 2.2. *If X is a variety, any pseudo-divisor (L, Z, s) on X is represented by some Cartier divisor D on X. Moreover,*

---

4 A Cartier divisor is a line bundle together with a "regular meromorphic" section, up to isomorphism (cf. [EGA]IV.21.1.4). More generally, a line bundle together with any "meromorphic" section ([EGA]IV.20.1) determines a pseudo-divisor in our sense. Both these definitions require conditions on the section over the support, for which we have no need.

(a) *If $Z \neq X$, D is uniquely determined.*
(b) *If $Z = X$, D is determined up to linear equivalence.*

Proof. Let $g_{\alpha\beta}$ be transition functions for L, for some affine open covering $\{U_\alpha\}$ of X. Fix one index α_0, and set $f_\alpha = g_{\alpha\alpha_0}$. Then $f_\alpha/f_\beta = g_{\alpha\beta}$, so the data (U_α, f_α) define a Cartier divisor D with $\mathcal{O}_X(D) \cong L$. In case $Z = X$, this gives the existence of D.

If $Z \neq X$, let $U = X - Z$. The section s is given by a collection of regular functions s_α on $U \cap U_\alpha$ such that $s_\alpha = g_{\alpha\beta} s_\beta$. (The functions f_α give the canonical section s_D.) Since $s_\alpha/f_\alpha = s_\beta/f_\beta$, there is a rational function $r \in R(X)^*$ with $r = s_\alpha/f_\alpha$ for all α. Set $D' = D + \operatorname{div}(r)$. The local equations for D' are $f'_\alpha = f_\alpha \cdot r = s_\alpha$, so the canonical section $s_{D'}$ corresponds to s. This proves the existence of D in case $Z \neq X$.

For the uniqueness, if D and D', with local equations f_α and f'_α, both determine (L, Z, s), then there is an $f \in R(X)^*$ with $f'_\alpha = f_\alpha f$ for all α. If $U \neq \emptyset$, and $s_{D'} = s_D$, f'_α and f_α must agree on $U \cap U_\alpha$, so $f = 1$ on U, i.e., $f = 1$ and $D = D'$. □

Definition 2.2.2. If D is a pseudo-divisor on an n-dimensional variety X, and $|D|$ is its support, define the *Weil divisor class*

$$[D] \in A_{n-1}(|D|)$$

of D as follows. Take a Cartier divisor which represents D, and let $[D]$ be the class in $A_{n-1}(|D|)$ of the associated Weil divisor. In case $|D| \neq X$, this Cartier divisor is unique (Lemma 2.1) and then $[D]$ is a well-defined $(n-1)$-cycle on $|D|$, as reflected in the fact that $Z_{n-1}(|D|)$ is $A_{n-1}(|D|)$. In case $|D| = X$, the Cartier divisor is only determined up to linear equivalence, but its associated Weil divisor is well-defined in $A_{n-1}(X)$ (§ 2.1).

If $D = (L, Z, s)$ and $D' = (L', Z', s')$ are pseudo-divisors on X, the *sum* $D + D'$ is the pseudo-divisor

$$D + D' = (L \otimes L', Z \cup Z', s \otimes s') .$$

(This agrees with the sum for Cartier divisors, except that the sum of two Cartier divisors may have smaller support than the union of their supports.) Similarly define

$$-D = (L^{-1}, Z, 1/s) .$$

For fixed $Z \subset X$ closed, the pseudo-divisors with support Z form an abelian group.

If $f: X' \to X$ is a morphism and $D = (L, Z, s)$ is a pseudo-divisor on X, the *pull-back* f^*D is the pseudo-divisor $(f^*L, f^{-1}(Z), f^*s)$ on X'. This pull-back is functorial, and agrees with pull-backs of representing Cartier divisors when those pull-backs are defined, and takes sums to sums.

Example 2.2.1. Let $\operatorname{Div}_Z X$ be the group of pseudo-divisors on X with support Z. If $f: X' \to X$ is a morphism, pull-back determines a homomorphism $f^*: \operatorname{Div}_Z X \to \operatorname{Div}_{Z'} X'$, $Z' = f^{-1}(Z)$. If X is an n-dimensional variety, the mapping $D \to [D]$ determines a homomorphism from $\operatorname{Div}_Z X$ to $A_{n-1} Z$.

2.3 Intersecting with Divisors

Definition 2.3. Let D be a pseudo-divisor on a scheme X, and let V be a k-dimensional subvariety of X. Define a class, denoted $D \cdot [V]$ or $D \cdot V$, in $A_{k-1}(|D| \cap V)$ as follows: Let j be the inclusion of V in X. The restriction (pull-back) $j^* D$ is a pseudo-divisor on V whose support is $|D| \cap V$. Define $D \cdot [V]$ to be the Weil divisor class (Definition 2.2.2) of $j^* D$:

$$D \cdot [V] = [j^* D].$$

When D is a Cartier divisor, this may be rephrased as follows: if $V \not\subset |D|$, D restricts to a Cartier divisor j^*D on V, and $D \cdot [V]$ is its associated Weil divisor; if $V \subset |D|$, $D \cdot [V]$ is the class in $A_{k-1}(V)$ represented by $[C]$, for any Cartier divisor C on V whose line bundle $\mathcal{O}_V(C)$ is isomorphic to $j^* \mathcal{O}_X(D)$.

In line with the convention in § 1.4, we will write $D \cdot [V]$ also for the image of the above class in $A_{k-1}(Y)$, for any closed subscheme Y of X which contains $|D| \cap V$.

For any k-cycle $\alpha = \sum n_V [V]$ on X, the *support* of α, written $|\alpha|$, is the union of the subvarieties V appearing with non-zero coefficient in α. For a pseudo-divisor D on X, each $D \cdot [V]$ is a class in $A_{k-1}(|D| \cap |\alpha|)$, and we define the *intersection class* $D \cdot \alpha$ in $A_{k-1}(|D| \cap |\alpha|)$ by setting

$$D \cdot \alpha = \sum_V n_V D \cdot [V].$$

As above, we also regard $D \cdot \alpha$ in $A_{k-1}(Y)$ for any $|D| \cap |\alpha| \subset Y \subset X$.

These intersection classes will be used for two important constructions:

(1) If $L = \mathcal{O}(D)$ is a line bundle on X, and $|D| = X$, $D \cdot \alpha$ will be $c_1(L) \cap \alpha$, the action of the first Chern class of L on α (§ 2.5).

(2) If D is an effective Cartier divisor on X, and i is the inclusion of D in X, $D \cdot \alpha$ will be the Gysin pull-back $i^*(\alpha)$ (§ 2.6).

Remark 2.3. In one important case, intersecting with D is defined on the cycle level. *If the restriction of $\mathcal{O}_X(D)$ to $|D|$ is a trivial line bundle*, then D determines a homomorphism

$$Z_k(X) \to Z_{k-1}(|D|),$$

also denoted $\alpha \to D \cdot \alpha$. As before, set $D \cdot [V] = [j^* D]$, with j the inclusion of V in X, if $V \not\subset |D|$, but set $D \cdot [V] = 0$ if $V \subset |D|$.

This condition holds when D is a principal Cartier divisor on X, or on a neighborhood of D in X. For example if a variety X is mapped dominantly to a curve C, and P is a simple point of C, then the inverse image $X(P)$ of P satisfies this condition. The resulting homomorphisms from $Z_k(X)$ to $Z_{k-1}(X(P))$ are called *specialization* homomorphisms (cf. § 10.1).

Proposition 2.3. (a) *If D is a pseudo-divisor on X, and α, α' are k-cycles on X, then*

$$D \cdot (\alpha + \alpha') = D \cdot \alpha + D \cdot \alpha'$$

in $A_{k-1}(|D| \cap (|\alpha| \cup |\alpha'|))$.

(b) *If D, D' are pseudo-divisors on X, and α is a k-cycle on X, then*

$$(D + D') \cdot \alpha = D \cdot \alpha + D' \cdot \alpha$$

in $A_{k-1}((|D| \bigcup |D'|) \bigcap |\alpha|)$.

(c) (Projection formula) *Let D be a pseudo-divisor on X, $f: X' \to X$ a proper morphism, α a k-cycle on X', and g the morphism from $f^{-1}(|D|) \bigcap |\alpha|$ to $|D| \bigcap f(|\alpha|)$ induced by f. Then*

$$g_*(f^*D \cdot \alpha) = D \cdot f_*(\alpha)$$

in $A_{k-1}(|D| \bigcap f(|\alpha|))$.

(d) *Let D be a pseudo-divisor on X, $f: X' \to X$ a flat morphism of relative dimension n, α a k-cycle on X, and g the induced morphism from $f^{-1}(|D| \bigcap |\alpha|)$ to $|D| \bigcap |\alpha|$. Then*

$$f^*D \cdot f^*\alpha = g^*(D \cdot \alpha)$$

in $A_{k+n-1}(f^{-1}(|D| \bigcap |\alpha|))$.

(e) *If D is a pseudo-divisor on X whose line bundle $\mathcal{O}_X(D)$ is trivial, and α is a k-cycle on X, then*

$$D \cdot \alpha = 0$$

in $A_{k-1}(|\alpha|)$.

Proof. (a) follows directly from the definition. For (b)−(e), there is therefore no loss of generality in assuming $\alpha = [V]$, V a variety. (b) follows from the fact that restricting to V and forming associated Weil divisor classes is compatible with sums. For (c), by functoriality of pull-back and push-forward we may assume $\alpha = [V]$, $V = X'$ and $f(V) = X$; D is represented by a Cartier divisor, which we also denote D, and the content of (c) is the identity of cycles on X:

$$f_*([f^*D]) = \deg(X'/X)[D].$$

This identity is a local assertion on X, so we may assume $D = \operatorname{div}(r)$ for some rational function on X. Then from Proposition 1.4, with $d = \deg(X'/X)$, one has

$$f_*[\operatorname{div}(f^*r)] = [\operatorname{div}(N(f^*r))] = \operatorname{div}(r^d) = d[\operatorname{div}(r)],$$

as required.

For assertion (d) we may also also assume $V = X$, so D is represented by a Cartier divisor. The identity to prove is that

$$[f^*D] = f^*[D]$$

as cycles on X'. This too is local on X, so we may assume D is the difference of two effective divisors. Since both sides are additive, it suffices to prove the identity when D is effective. This case is a special case of Lemma 1.7.1.

For (e), we may assume $\alpha = [V]$, $V = X$, and D is represented by a Cartier divisor on X. The assertion is then that $[D] = 0$ in $A_{k-1}(X)$ when D is principal, which we have seen (§ 2.1). □

Example 2.3.1. Let α be a k-cycle on X, β and l-cycle on Y, let D be a pseudo-divisor on X, and let p be the projection from $X \times Y$ to X. Then

$$(p^*D) \cdot (\alpha \times \beta) = (D \cdot \alpha) \times \beta$$

in $A_{k+l-1}((|D| \cap |\alpha|) \times |\beta|)$. (Reduce to the case $\beta = [Y]$, Y a variety, and apply Proposition 2.3 (d).)

2.4 Commutativity of Intersection Classes

If D and D' are Cartier divisors on a variety X, with associated Weil divisors $[D]$ and $[D']$, one may form the intersection classes $D \cdot [D']$ and $D' \cdot [D]$, both of which are well-defined classes in $A_*(|D| \cap |D'|)$. When D and D' *intersect properly,* i.e., no codimension one subvariety of X is contained in $|D| \cap |D'|$, the equality of these classes is quite straightforward. The "classical" method of moving D or D' to linearly equivalent divisors that intersect properly may be used – when such moving is possible – to show that the two classes agree in $A_*(|D'|)$ or $A_*(|D|)$, but this does not prove their equality in $A_*(|D| \cap |D'|)$, which we shall need. Instead we proceed by blowing up along subschemes of X to achieve the situation where D and D' are sums of divisors, each pair of which either intersect properly, or are equal (the other case where commutativity is obvious!).

Theorem 2.4. *Let D and D' be Cartier divisors on an n-dimensional variety X. Then*

$$D \cdot [D'] = D' \cdot [D]$$

in $A_{n-2}(|D| \cap |D'|)$.

Proof. Case 1: D and D' are effective and intersect properly. Let W be any codimension two subvariety of X, let $A = \mathcal{O}_{W,X}$, and let a, a' be local equations for D, D' in A. The subvarieties V of X of codimension one which contain W correspond to height one primes p in A. The coefficient of $[V]$ in $[D']$ is $l_{A_p}(A_p/a'A_p)$. The coefficient of $[W]$ in $D \cdot [V]$ is $l_{A/p}(A/p + aA)$. The coefficient of $[W]$ in $D \cdot [D']$ is therefore

$$\sum_p l_{A_p}(A_p/a'A_p) \cdot l_{A/p}(A/p + aA) \,.$$

By Lemma A.2.7 applied to the ring $A/a'A$, this coefficient is

$$e_A(a, A/a'A) \,.$$

By Lemma A.2.8.

$$e_A(a, A/a'A) = e_A(a', A/aA) \,,$$

which by the same argument is the coefficient of $[W]$ in $D' \cdot [D]$. (In short, the coefficient of $[W]$ in both sides is the multiplicity $e_A(a, a', A)$ discussed in Example A.5.2.) □

Some preparation is needed before proving the general case of the theorem. If D and D' are effective Cartier divisors on a variety X, define the *excess* of intersection $\varepsilon(D, D')$ by the formula

$$\varepsilon(D, D') = \max\{\operatorname{ord}_V(D) \cdot \operatorname{ord}_V(D') \mid \operatorname{codim}(V, X) = 1\},$$

the maximum over all codimension one subvarieties V of X. Thus D and D' meet properly precisely when $\varepsilon(D, D') = 0$.

Let $D \cap D'$ be the intersection scheme of D and D'. This is the subscheme of X which on an affine open set U is defined by the ideal (a, a'), where a and a' are local equations for D and D' in U. Let

$$\pi: \tilde{X} \to X$$

be the blow-up of X along $D \cap D'$, and let $E = \pi^{-1}(D \cap D')$ be the exceptional divisor. The local equations for $\pi^* D$ and $\pi^* D'$ are divisible by the local equations for E, so

$$\pi^* D = E + C, \quad \pi^* D' = E + C'$$

for effective Cartier divisors C, C' on $\tilde{X}$.

Lemma 2.4. *With the above notation,*

(a) *C and C' are disjoint.*

(b) *If $\varepsilon(D, D') > 0$, then $\varepsilon(C, E)$ and $\varepsilon(C' E)$ are strictly smaller than $\varepsilon(D, D')$.*

Proof. The assertions are local on X, so we assume $X = \operatorname{Spec}(A)$, and $D = \operatorname{div}(a)$, $D' = \operatorname{div}(a')$ are principal divisors. Then $\tilde{X} = \operatorname{Proj}(\oplus I^n)$, where $I = (a, a')$. The surjective graded homomorphism

$$A[S, T] \to \oplus I^n$$

given by sending S to a, T to a', determines a closed imbedding of $\tilde{X}$ in $\operatorname{Proj}(A[S, T]) = X \times \mathbb{P}^1$:

$$\begin{array}{ccc} \tilde{X} & \hookrightarrow & X \times \mathbb{P}^1 \\ & {}_{\pi}\searrow \quad \swarrow_{\mathrm{pr}_1} & \\ & X & \end{array}.$$

In fact, $\tilde{X}$ is contained in the subscheme of $X \times \mathbb{P}^1$ where $a'S - aT$ vanishes. Let $\mathscr{O}(1)$ be the pull-back of the standard line bundle on $\mathbb{P}^1$ to $\tilde{X}$, and let s, t be the sections of $\mathscr{O}(1)$ induced by S, T. Then C is the zero-scheme $Z(s)$ of s, and $C' = Z(t)$. Indeed, the equation $a' = (t/s)\,a$ shows that $\pi^* D$ agrees with E on the open set where $s \neq 0$, and the equation $a = (s/t)\,a'$ shows that $\pi^* D$ agrees with $E + Z(s)$ on the open set where $t \neq 0$. Since $Z(s) \cap Z(t) = \emptyset$, this proves that

$$\pi^* D = E + Z(s),$$

as desired, and (a) follows.

We also see from this description that $C \subset X \times \{0\}$ and $C' \subset X \times \{\infty\}$ map isomorphically by π into subschemes of D and D' respectively. Therefore if

$\tilde{V}$ is any codimension one subvariety of $\tilde{X}$ contained in $C \cap E$ or $C' \cap E$, then $V = \pi(\tilde{V})$ is a codimension one subvariety of X contained in $D \cap D'$. Since $[D] = \pi_* [E + C]$ (by Proposition 2.3 (c)),

$$\operatorname{ord}_V D \geqq \operatorname{ord}_{\tilde{V}} E + \operatorname{ord}_{\tilde{V}} C,$$

and similarly for D'. Suppose $\varepsilon(C, E) \geqq \varepsilon(D, D') > 0$, and $\tilde{V} \subset C \cap E$ is chosen with $\operatorname{ord}_{\tilde{V}} C \cdot \operatorname{ord}_{\tilde{V}} E = \varepsilon(C, E)$. Then

$$\begin{aligned}\varepsilon(D, D') &\geqq \operatorname{ord}_V D \cdot \operatorname{ord}_V D' \\ &\geqq (\operatorname{ord}_{\tilde{V}} E + \operatorname{ord}_{\tilde{V}} C)(\operatorname{ord}_{\tilde{V}} E + \operatorname{ord}_{\tilde{V}} C') \\ &\geqq (\operatorname{ord}_{\tilde{V}} E)^2 + \varepsilon(C, E);\end{aligned}$$

since $\operatorname{ord}_{\tilde{V}} E > 0$, this is a contradiction. □

To prove Theorem 2.4 we will also need the following fact:

(*) *If D, D' are Cartier divisors on X, $\pi : \tilde{X} \to X$ is a proper birational morphism of varieties, $\pi^* D = B \pm C$, $\pi^* D' = B' \pm C'$, for Cartier divisors B, C, B', C' on $\tilde{X}$ with $|B| \cup |C| \subset \pi^{-1}(|D|)$, $|B'| \cup |C'| \subset \pi^{-1}(|D'|)$, and the theorem holds for each of the pairs (B, B'), (B, C'), (C, B'), and (C, C') on $\tilde{X}$, then the theorem holds for (D, D') on X.*

Indeed, by Proposition 2.3, if g is the induced morphism from $\pi^{-1}(|D| \cap |D'|)$ to $|D| \cap |D'|$,

$$\begin{aligned}D \cdot [D'] &= g_* ((B \pm C) \cdot [B' \pm C']) \\ &= g_* (B \cdot [B'] \pm B \cdot [C'] \pm C \cdot [B'] + C \cdot [C']) \\ &= g_* (B' \cdot [B] \pm C' \cdot [B] \pm B' \cdot [C] + C' \cdot [C]) \\ &= g_* ((B' \pm C') \cdot [B \pm C]) \\ &= D' \cdot [D].\end{aligned}$$

We return to the proof of Theorem 2.4. *Case 2:* D and D' are effective. This is proved by induction on $\varepsilon(D, D')$, the start $\varepsilon = 0$ being Case 1 proved above. If $\varepsilon(D, D') > 0$, blow up X along $D \cap D'$ as in the lemma. By assertion (b) of the lemma, and the inductive hypothesis, the theorem holds for (E, C') and (C, E). The theorem is trivially true for (E, E), and it is true for (C, C') since $C \cap C' = \emptyset$. An application of (*) completes the proof.

Case 3: One of D, D' is effective. Suppose D' is effective. Let $\mathcal{J}$ be the ideal sheaf of denominators for D: on an open affine $U = \operatorname{Spec}(A)$ where D has local equation d, $\mathcal{J}$ is determined by the ideal

$$\{a \in A \mid a\, d \in A\}.$$

Let $\pi : \tilde{X} \to X$ be the blowup of X along the subscheme defined by $\mathcal{J}$, and let E be the exceptional divisor. Then

$$\pi^* D = C - E$$

for an effective divisor C on $\tilde{X}$. Since $|C| \cup |E| \subset \pi^{-1}(|D|)$, and Case 2 covers the pairs $(C, \pi^* D')$ and $(E, \pi^* D')$, an application of (*) completes the proof.

Case 4: D, D' arbitrary. Let $\pi: \tilde{X} \to X$ be the blowup along the denominators of D as in Case 3. Then the pairs $(C, \pi^* D')$ and $(E, \pi^* D')$ are covered by Case 3, and we conclude by another application of (*). □

Corollary 2.4.1. *Let D be a pseudo-divisor on a scheme X, α a k-cycle on X which is rationally equivalent to zero. Then*

$$D \cdot \alpha = 0$$

in $A_{k-1}(|D|)$.

Proof. If $\alpha = [\operatorname{div}(r)]$, $r \in R(V)^*$, V a subvariety of X, we may replace X by V, and D by a representing Cartier divisor. Then

$$D \cdot [\operatorname{div}(r)] = \operatorname{div}(r) \cdot [D] = 0$$

in $A_{k-1}(|D|)$, by Theorem 2.4 and Proposition 2.3 (e) respectively. □

Definition 2.4.1. If D is a pseudo-divisor on a scheme X, and Y is a closed subscheme of X, the mapping $\alpha \to D \cdot \alpha$ determines homomorphisms

$$Z_k Y \to A_{k-1}(|D| \cap Y).$$

By Corollary 2.4.1 (applied to the restriction of D to Y), $D \cdot \alpha = 0$ if $\alpha \sim 0$ on Y, so these homomorphisms pass to rational equivalence, defining homomorphisms

$$A_k Y \to A_{k-1}(|D| \cap Y),$$

also denoted $\alpha \to D \cdot \alpha$, and called *intersecting with D.*

Corollary 2.4.2. Let D and D' be pseudo-divisors on a scheme X. Then for any k-cycle α on X,

$$D \cdot (D' \cdot \alpha) = D' \cdot (D \cdot \alpha)$$

in $A_{k-2}(|D| \cap |D'| \cap |\alpha|)$.

Proof. Note that $D' \cdot \alpha \in A_{k-1}(|D'| \cap |\alpha|)$, and intersecting with D takes $A_{k-1}(|D'| \cap |\alpha|)$ to $A_{k-2}(|D| \cap |D'| \cap |\alpha|)$. Taking $\alpha = [V]$, and restricting D and D' to V, one is reduced to Theorem 2.4. □

Definition 2.4.2. Let $D_1, \ldots, D_n$ be pseudo-divisors on a scheme X. For any $\alpha \in Z_k X$, define $D_1 \cdot \ldots \cdot D_n \cdot \alpha$ in $A_{k-n}(|D_1| \cap \ldots \cap |D_n| \cap |\alpha|)$ by induction:

$$D_1 \cdot \ldots \cdot D_n \cdot \alpha = D_1 \cdot (D_2 \cdot \ldots \cdot D_n \cdot \alpha).$$

From Corollary 2.4.2 this is independent of the ordering of the D_i, and from Proposition 2.3 it is linear in each variable $D_1, \ldots, D_n$, and α. More generally, for any homogeneous polynomial $P(T_1, \ldots, T_n)$ of degree d with integer coefficients, and any closed subscheme Z of X containing $(|D_1| \cup \ldots \cup |D_n|) \cap |\alpha|$, the class

$$P(D_1, \ldots, D_n) \cdot \alpha \in A_{k-d}(Z)$$

is defined by adding terms in the preceding definition. This is additive in P and in α.

If $n=k$, and $Y=|D_1|\cap\ldots\cap|D_n|\cap|\alpha|$ is complete, we define an *intersection number* $(D_1\cdot\ldots\cdot D_n\cdot\alpha)_X$ by

$$(D_1\cdot\ldots\cdot D_n\cdot\alpha)_X=\int_Y D_1\cdot\ldots\cdot D_n\cdot\alpha\,.$$

Similarly if $Z=(|D_1|\cup\ldots\cup|D_n|)\cap|\alpha|$ is complete, $k=d$, and P is a polynomial as above, define

$$(P(D_1,\ldots,D_n)\cdot\alpha)_X=\int_Z P(D_1,\ldots,D_n)\cdot\alpha\,.$$

If V is a pure k-dimensional subscheme of X, we may write $P(D_1,\ldots,D_n)\cdot V$ in place of $P(D_1,\ldots,D_n)\cdot[V]$. If X is pure dimensional, we may abbreviate this further, writing simply $P(D_1,\ldots,D_n)$ in place of $P(D_1,\ldots,D_n)\cdot[X]$. For example, we may write $D^i\in A_{k-i}(|D|)$ in place of $D^i\cdot[X]$, if X is purely k-dimensional. Similarly if X is purely k-dimensional, and P is homogeneous of degree k, we write $(P(D_1,\ldots,D_n))_X$ in place of $(P(D_1,\ldots,D_n)\cdot[X])_X$.

Example 2.4.1. Let $\pi: X\to\mathbb{A}^2$ be the blow-up of $\mathbb{A}^2$ at the origin. Let D and D' be the inverse images of the x-axis and y-axis. Then D and D' are principal divisors on X, so $D\cdot[D']$ and $D'\cdot[D]$ are well-defined cycles on $D\cap D'=E$, the exceptional divisor (cf. Remark 2.3). These cycles are not equal, although they are rationally equivalent in $E\cong\mathbb{P}^1$.

Example 2.4.2. The operations of Definition 2.4.2 are compatible with proper push-forward and flat pull-back, as in Proposition 2.3 (c), (d). In particular, the construction is *local*, in the following sense. If U is an open subscheme of X which contains $|D_1|\cap\ldots\cap|D_n|\cap|\alpha|$, and D_i^U, α^U denote restrictions of D_i, α to U, then

$$D_1\cdot\ldots\cdot D_n\cdot\alpha=D_1^U\cdot\ldots\cdot D_n^U\cdot\alpha^U$$

in $A_*(|D_1|\cap\ldots\cap|D_n|\cap|\alpha|)$.

Example 2.4.3. If $f\colon X'\to X$ is a proper morphism, $D_1,\ldots,D_n$ are pseudo-divisors on X, P a polynomial of degree d, and α is a d-cycle on X' with $|D_1|\cap\ldots\cap|D_n|\cap f(|\alpha|)$ complete, then

$$(f^*D_1\cdot\ldots\cdot f^*D_n\cdot\alpha)_{X'}=(D_1\cdot\ldots\cdot D_n\cdot f_*\alpha)_X\,.$$

Example 2.4.4. Let V be an irreducible surface, P a singular point of V, and let $\pi: X\to V$ be a proper morphism, $E=\pi^{-1}(P)$. Assume that X is regular, that π maps $X-E$ isomorphically onto $V-P$, and E is connected. Then (cf. Mumford (1))

$$(D\cdot D)_X<0$$

for any effective, non-zero, divisor D on X which is supported in E. (To see this (cf. Deligne-Katz (1) X), let $E_1,\ldots,E_r$ be the irreducible components of E. Choose a rational function f on V, so that

$$\operatorname{div}(\pi^*f)=\sum_{i=1}^r m_iE_i+Z\,,$$

with $m_i > 0$, Z a divisor not containing any E_i, $(Z \cdot \quad {}_i)_X \geqq 0$ for all i, some $(Z \cdot E_i)_X > 0$. Set $D_i = m_i E_i$. Replacing D by a positive multiple, one may assume $D = \sum_{i=1}^{r} a_i D_i$. Then

$$\begin{aligned}(D \cdot D)_X &= \sum_i a_i \Big(D_i \cdot \sum_j a_j D_j\Big)_X = \sum_i a_i \Big(D_i \cdot \sum_j a_j D_j - a_i (\operatorname{div}(\pi^* f))\Big)_X \\ &= \sum_{i \neq j} a_i (a_j - a_i)(D_i \cdot D_j)_X - \sum a_i^2 (D_i \cdot Z)_X \\ &= -\sum_{i<j} (a_j - a_i)^2 (D_i \cdot D_j)_X - \sum a_i^2 (D_i \cdot Z)_X < 0 .\end{aligned}$$

Corollary 2.4.1 was used for the second equality.)

Example 2.4.5. Let $(x:y:z:t)$ be homogeneous on $\mathbb{P}^3$, and let X be the singular cone defined by the equation $z^2 = x\,y$. Let D be the Cartier divisor on X defined by the equation $x = 0$. Let l be the line $x = z = 0$, l' the line $y = z = 0$, P the point $(0:0:0:1)$. Then $[D] = 2[l]$ and $D \cdot [l'] = [P]$. It follows from Theorem 2.4 that there cannot be a Cartier divisor D' on X with $[D'] = [l']$, either as cycles or as classes in $A_1(X)$.

Example 2.4.6. In the construction in the proof of Lemma 2.4, the subscheme of $X \times \mathbb{P}^1$ defined by the equation $a' S = a T$ may not be reduced; in particular, it may not be equal to $\tilde{X}$. (Let $X = V(x\,w - y\,z) \subset \mathbb{A}^4$, $a = x$, $a' = y$.)

Example 2.4.7. Let $f\colon X \to C$ be a dominant morphism from a variety X to a curve C, P a simple point on C, $D = f^{-1}(P)$. Then D is an effective Cartier divisor on X, and $D^2 \cdot \alpha = 0$ in $A_{k-2}(|D| \cap |\alpha|)$ for any k-cycle α on X. (Shrink C, so D becomes principal.)

Example 2.4.8. Let $D_1, \ldots, D_n$ be effective divisors on an n-dimensional variety X, which intersect in a finite set. Assume that for each point P in the intersection, local equations for $D_1, \ldots, D_n$ form a regular sequence in $\mathcal{O}_{P,X}$. Then

$$(D_1 \cdot \ldots \cdot D_n)_X = \deg[D_1 \cap \ldots \cap D_n] .$$

This is the case whenever X is Cohen-Macaulay (Lemma A.7.1). For more on this intersection number, see Example 12.4.8, and Lomadze (1).

Example 2.4.9 (cf. Mumford (2), Deligne-Katz (1) X). Let X be a non-singular surface, D, E Cartier divisors on X, with $|D| \cap |E|$ complete, so that the intersection number $(D \cdot E)_X$ is defined.

(a) If D and E are effective and meet properly, then

$$(D \cdot E)_X = \deg[D \cap E] .$$

(b) If D is effective and complete, then

$$(D \cdot E)_X = \deg(\mathcal{O}(E)|_D) .$$

(c) If D and E are effective, then

$$(D \cdot E)_X = \chi(\mathcal{O}_D \otimes \mathcal{O}_E) - \chi(\operatorname{Tor}_1^X(\mathcal{O}_D, \mathcal{O}_E)) .$$

(d) If X is complete, then

$$(D \cdot E)_X = \chi(\mathcal{O}_X) - \chi(\mathcal{O}(-D)) - \chi(\mathcal{O}(-E)) + \chi(\mathcal{O}(-D-E)).$$

In (c) and (d), χ denotes the Euler characteristic (cf. § 15.1). (The formulas (b)–(d) are bilinear in D and E, so one is reduced to the case where D and E are effective and either meet properly or are equal.)

2.5 Chern Class of a Line Bundle

Let L be a line bundle on a scheme X. For any k-dimensional subvariety V of X, the restriction $L|_V$ of L to V is isomorphic to $\mathcal{O}_V(C)$ for some Cartier divisor C on V, determined up to linear equivalence (§ 2.3). The Weil divisor $[C]$ determines as well-defined element in $A_{k-1}(X)$, which we denote by $c_1(L) \cap [V]$:

$$c_1(L) \cap [V] = [C].$$

This is extended by linearity to define a homomorphism $\alpha \to c_1(L) \cap \alpha$ from $Z_k(X)$ to $A_{k-1}(X)$. If $L = \mathcal{O}_X(D)$ for a pseudo-divisor D on X, it follows from the definition of the intersection class (§ 2.3) that

$$c_1(\mathcal{O}_X(D)) \cap \alpha = D \cdot \alpha$$

in $A_{k-1}(X)$.

Proposition 2.5. (a) *If α is rationally equivalent to zero on X, then $c_1(L) \cap \alpha = 0$. There is therefore an induced homomorphism*

$$c_1(L) \cap _ : A_k X \to A_{k-1} X.$$

(b) (Commutativity). *If L, L' are line bundles on X, α a k-cycle on X, then*

$$c_1(L) \cap (c_1(L') \cap \alpha) = c_1(L') \cap (c_1(L) \cap \alpha)$$

in $A_{k-2}(X)$.

(c) (Projection formula). *If $f: X' \to X$ is a proper morphism, L a line bundle on X, α a k-cycle on X', then*

$$f_*(c_1(f^*L) \cap \alpha) = c_1(L) \cap f_*(\alpha)$$

in $A_{k-1}(X)$.

(d) (Flat pull-back). *If $f: X' \to X$ is flat of relative dimension n, L a line bundle on X, α a k-cycle on X, then*

$$c_1(f^*L) \cap f^*\alpha = f^*(c_1(L) \cap \alpha)$$

in $A_{k+n-1}(X')$.

(e) (Additivity). *If L, L' are line bundles on X, α a k-cycle on X, then*

$$c_1(L \otimes L') \cap \alpha = c_1(L) \cap \alpha + c_1(L') \cap \alpha$$

and

$$c_1(L^\vee) \cap \alpha = -c_1(L) \cap \alpha$$

in $A_{k-1} X$.

Proof. Since a line bundle on X determines a pseudo-divisor on X with support X, the assertions follow from the corresponding facts for pseudo-divisors, i.e., Corollaries 2.4.1, 2.4.2, and Proposition 2.3 (c), (d), and (b). □

Definition 2.5. From (a), (b), and (e) of the Proposition, it follows that arbitrary polynomials in Chern classes of line bundles act on A_*X. If $L_1, \ldots, L_n$ are line bundles on X, $\alpha \in A_k X$, and $P(T_1, \ldots, T_n)$ is a homogeneous polynomial of degree d with integer coefficients, then

$$P(c_1(L_1), \ldots, c_1(L_n)) \cap \alpha$$

is defined in $A_{k-d}(X)$. In particular, for a line bundle L on X, $\alpha \in A_k(X)$

$$c_1(L)^d \cap \alpha$$

is the element in $A_{k-d}X$ defined inductively by $c_1(L)^d \cap \alpha = c_1(L) \cap (c_1(L)^{d-1} \cap \alpha)$.

Example 2.5.1. Let L^k be a k-dimensional linear subspace of $\mathbb{P}^n$, $k = 0, \ldots, n$. Then

$$c_1(\mathcal{O}(1)) \cap [L^k] = [L^{k-1}],$$

for $k = 1, \ldots, n$. This gives another proof that $A_k \mathbb{P}^n = \mathbb{Z} \cdot [L^k] \cong \mathbb{Z}$ for $k = 0, \ldots, n$ (cf. Example 1.9.3 (b)).

The *degree* $\deg(\alpha)$ of a k-cycle α on $\mathbb{P}^n$ is defined to be the integer d such that $\alpha \sim d[L^k]$. Equivalently,

$$\deg(\alpha) = \int_{\mathbb{P}^n} c_1(\mathcal{O}(1))^k \cap \alpha .$$

Example 2.5.2. Let X be a closed subscheme of $\mathbb{P}^n$ of dimension $\leqq k$, and let $\Gamma(X)$ be the homogeneous coordinate ring of X.

(a) For t sufficiently large, the dimension of the tth graded piece $\Gamma(X)_t$ is a polynomial in t of degree $\leqq k$ called the Hilbert polynomial (cf. Hartshorne (5) I.7.5). Define $d_k(X)$ to be the coefficient of $t^k/k!$ in this polynomial.

(b) If $X_1, \ldots, X_r$ are the irreducible components of X, m_i the multiplicity of X_i in X, then $d_k(X) = \sum m_i d_k(X_i)$.

(c) If X is an irreducible variety, and H is a hypersurface of degree m not containing X, then $d_{k-1}(X \cap H) = m\, d_k(X)$. (Here $X \cap H$ is the scheme-theoretic intersection, so there are exact sequences

$$0 \to \Gamma(X)_{t-m} \to \Gamma(X)_t \to \Gamma(X \cap H)_t \to 0 .)$$

(d) For any purely k-dimensional subscheme X of $\mathbb{P}^n$,

$$d_k(X) = \deg[X] = \int_{\mathbb{P}^n} c_1(\mathcal{O}(1))^k \cap [X] .$$

(For X a subvariety, $c_1(\mathcal{O}(1)) \cap [X] = [X \cap H]$, where H is a hyperplane not containing X.)

Example 2.5.3. If α, β are cycles on X, Y, L a line bundle on X, p the projection from $X \times Y$ to X, then (cf. Example 2.3.1)

$$c_1(p^* L) \cap (\alpha \times \beta) = (c_1(L) \cap \alpha) \times \beta .$$

Example 2.5.4. The operation $c_1(L) \cap_$ is uniquely determined by properties (c), (e) of Proposition 2.5, and the normalization that

$$c_1(\mathcal{O}(D)) \cap [X] = [D]$$

for D an effective Cartier divisor on a normal variety X. (Argue as in the proof of Theorem 2.4, Case 3.)

Example 2.5.5. If D is an effective Cartier divisor on a variety X, the restriction of $\mathcal{O}_X(D)$ to D is the normal bundle $N_D X$, and

$$[D] = c_1(\mathcal{O}_X(D)) \cap [X] .$$

If L is another line bundle on X whose restriction to D is $N_D X$, however, $c_1(L) \cap [X]$ need not be represented by $[D]$; cf. Proposition 2.6 (c).

Example 2.5.6. *Plücker formulas.* Let C be a non-singular projective curve of genus g, and let $C(r) \subset C \times C$ be the subscheme defined by the ideal sheaf $\mathcal{I}^{r+1}$, where $\mathcal{I}$ is the ideal sheaf of the diagonal; let p and q be the first and second projections from $C(r)$ to C. For a line bundle L on C, the *bundle of principal parts* $P^r(L)$ is the sheaf on C defined by:

$$P^r(L) = p_* q^* L .$$

Then $P^0(L) = L$, and for $r > 0$ there is an exact sequence

$$0 \to (\Omega_C^1)^{\otimes r} \otimes L \to P^r(L) \to P^{r-1}(L) \to 0 .$$

Therefore $P^r(L)$ is locally free of rank $r+1$, and

$$c_1(\Lambda^{r+1} P^r(L)) = (r+1)\, c_1(L) + \binom{r+1}{2}\, c_1(\Omega_C^1) .$$

If $V \subset H^0(C, L)$ is a subspace, there are canonical homomorphisms of vector bundles on C,

$$\sigma : C \times V \to P^n(L) .$$

If $\dim V = r+1$, i.e., the linear system determined by V has dimension r, then $\det(\sigma)$ is a section of $\Lambda^{r+1} P^r(L)$, well-defined up to scalars. If $\det(\sigma) \neq 0$, its divisor of zeros, denoted δ_V, measures the osculation of the linear system. Then

$$\deg(\delta_V) = (r+1) \deg(L) + \binom{r+1}{2}(2g-2) .$$

For a sample of applications see Piene (1), Laksov (4), Eisenbud-Harris (1).

2.6 Gysin Map for Divisors

Let D be an effective Cartier divisor on a scheme X, and let $i : D \to X$ be the inclusion. There are *Gysin homomorphisms*

$$i^* : Z_k X \to A_{k-1} D$$

defined by the formula

$$i^*(\alpha) = D \cdot \alpha$$

where $D \cdot \alpha$ is the intersection class in $A_{k-1}D$ defined in § 2.3.

Proposition 2.6. (a) *If α is rationally equivalent to zero on X, then $i^*\alpha = 0$. There are therefore induced homomorphisms*

$$i^* : A_k X \to A_{k-1} D .$$

(b) *If α is a k-cycle on X, then*

$$i_* i^*(\alpha) = c_1(\mathcal{O}_X(D)) \cap \alpha .$$

(c) *If α is a k-cycle on D, then*

$$i^* i_*(\alpha) = c_1(N) \cap \alpha ,$$

where $N = i^*\mathcal{O}_X(D)$.

(d) *If X is purely n-dimensional, then*

$$i^*[X] = [D]$$

in $A_{n-1}D$.

(e) *If L is a line bundle on X, then*

$$i^*(c_1(L) \cap \alpha) = c_1(i^*L) \cap i^*(\alpha)$$

in $A_{k-2}(D)$ for any k-cycle α on X.

Proof. (a) and (e) are special cases of Corollary 2.4.1 and 2.4.2 respectively. (b) and (c) follow from the definitions: in both cases, both sides are represented by the intersection class $D \cdot \alpha$. (d) says that $[D] = D \cdot [X]$, which is a restatement of Lemma 1.7.2. □

Example 2.6.1. Let L be a line bundle on X, $p: L \to X$ the projection, $i: X \to L$ the imbedding of X in L by the zero section. Then $i^*(p^*\alpha) = \alpha$ for all $\alpha \in A_k X$. One concludes from Proposition 1.9 that

$$p^* : A_k X \to A_{k+1} L$$

is an isomorphism (see § 3.3 for generalizations).

Example 2.6.2. Let X be a closed subscheme of $\mathbb{P}^n$, and let X' be a cone over X in $\mathbb{P}^{n+1}$. Then $A_0(X') \cong \mathbb{Z}$, and $A_k(X') \cong A_{k-1}(X)$ for $k > 0$. (If P is the vertex of the cone, the complement of P in X' is a line bundle over X. Use Proposition 1.8 and Example 2.6.1.)

Example 2.6.3. (a) Let L be a line bundle on X, let $L - \{0\}$ be the complement of the zero section, and let η be the projection from $L - \{0\}$ to X. Then for all $k \geqq 0$, the sequence

$$A_{k+1} X \xrightarrow{c_1(L) \cap -} A_k X \overset{\eta^*}{\to} A_{k+1}(L - \{0\}) \to 0$$

is exact.

(b) Let X be a closed subscheme of $\mathbb{P}^n$, with canonical line bundle $\mathcal{O}(1)$. Let $V \subset \mathbb{A}^{n+1}$ be the affine cone over X. Then $A_0 V = 0$, and for $k > 0$ there is an exact sequence

$$A_k X \xrightarrow{c_1(\mathcal{O}(1)) \cap -} A_{k-1} X \to A_k V \to 0 .$$

Example 2.6.4. Pure-dimensionality is needed in Proposition 2.6 (d). Let $X = V(z\,x, z\,y) \subset \mathbb{P}^3$, D the Cartier divisor on X defined by the equation $z - x = 0$. Let i be the inclusion of D in X. Then (cf. Example 1.7.1)

$$D \cdot [X] = i^* [X] \neq [D], \quad c_1(\mathcal{O}(D)) \cap [X] \neq [D] .$$

Example 2.6.5. Let D be an effective Cartier divisor on a scheme X, i the inclusion of D in X. Let Y be a closed subscheme of X. Assume that Y is purely n-dimensional and $D \cap Y$ has dimension $n-1$. Let $V_1, \ldots, V_r$ be the (reduced) irreducible components of $D \cap Y$. Let A_i be the local ring of Y along V_i, and let a_i be a local equation for D in A_i. Then

$$i^* [Y] = D \cdot [Y] = \sum_{i=1}^{r} e_{A_i}(a_i, A_i) [V_i]$$

where $e_{A_i}(a_i, A_i)$ is the multiplicity defined in Appendix A.2. More generally, if one assumes only that $\dim Y \leqq n$ and $\dim D \cap Y \leqq n-1$, and $V_1, \ldots, V_r$ are the components of $D \cap Y$ of dimension $n-1$, then the right side of this equation gives a formula for $i^*([Y]_n)$, where $[Y]_n$ is the n-dimensional component of the cycle of Y. (These formulas follow from Lemma A.2.7.)

Example 2.6.6. Let X be a scheme, and let α be a k-cycle on $X \times \mathbb{P}^1$. Let i_0 and i_∞ be the imbeddings of X in $X \times \mathbb{P}^1$ at 0 and ∞ respectively. Then

$$i_0^* \alpha = i_\infty^* \alpha$$

in $A_{k-1} X$. (If $\alpha = [V]$, and the projection of V to $\mathbb{P}^1$ is not dominant, both sides are zero. If the projection is dominant, the equation says $[V(0)] = [V(\infty)]$.) More generally, for any $t \in \mathbb{P}^1$, rational over the ground field, if i_t is the imbedding of X in $X \times \mathbb{P}^1$ at t, then $i_t^* = i_0^* = i_\infty^*$.

Notes and References

Generalities on Cartier divisors and Weil divisors, including the material of § 2.1, may be found in [EGA] IV.21 (Many otherwise standard sources discuss one or both kinds of divisors only on normal or locally factorial varieties.)

The intersection class constructed in § 2.3 has not appeared before, either for Cartier divisors or the pseudo-divisors introduced here; its construction was implicit in Fulton-MacPherson (1), however. The statement and proof of the fundamental Theorem 2.4 follow Fulton (6). A special case of Corollary 2.4.1 – that rational equivalence specializes – had been conjectured by Grothendieck ([SGA 6] X.7) in the non-singular case; this case was proved in

Fulton (2), by a method similar to that used in Baum-Fulton-MacPherson (1) II.2.5, to prove a Riemann-Roch theorem. Conversely, having Theorem 2.4 will simplify the proof of the Riemann-Roch theorem (cf. § 18.1).

It has long been understood (cf. Hirzebruch (1), Grothendieck (1), (2)) that a good knowledge of first Chern classes for line bundles determines higher Chern classes for vector bundles. The development followed in the present text may be seen as a corresponding principle in intersection theory: a good knowledge of how divisors intersect determines intersections in arbitrary codimension. An important case of this principle was given by Verdier (5), who constructed Gysin maps in arbitrary codimension from the case of divisors.

Chapter 3. Vector Bundles and Chern Classes

Summary

We will construct, for any vector bundle E on a scheme X, Chern class operations

$$c_i(E) \cap _ : A_k X \to A_{k-i} X ,$$

satisfying properties expected from topology. From the special case of line bundles done in § 2.5, we first construct inverse Chern classes, or Segre classes, which are then inverted to produce Chern classes. The first Chern class operations are also used to describe $A_* E$ and $A_* P(E)$ in terms of $A_* X$.

Chern classes will be used later for one of the constructions of general intersection products. Although Chern classes are not absolutely needed for intersection theory, they are used in most applications. For the quickest route to intersection theory proper, the reader will need only Proposition 3.1 (a) and Theorem 3.3.

For vector bundles, Chern classes and Segre classes determine each other; Chern classes are preferred since they vanish beyond the rank of the bundle. We will see in the next chapter that for cones – "singular vector bundles" – there is a natural analogue of Segre classes, but not Chern classes. Segre classes for normal cones have other remarkable properties not shared by Chern classes (cf. § 4.2).

3.1 Segre Classes of Vector Bundles

Let E be a vector bundle of rank $e+1$ on an algebraic scheme X. Let $P = P(E)$ be the projective bundle of lines in E, $p = p_E$ the projection from P to X, and let $\mathcal{O}(1) = \mathcal{O}_E(1)$ denote the canonical line bundle on P, i.e., its dual $\mathcal{O}(-1)$ is the tautological subbundle of $p^* E$ (see Appendix B.5).

Define homomorphisms $\alpha \to s_i(E) \cap \alpha$ from $A_k X$ to $A_{k-i} X$ by the formula

$$s_i(E) \cap \alpha = p_* (c_1(\mathcal{O}(1))^{e+i} \cap p^* \alpha) .$$

Here p^* is the flat pull-back from $A_k X$ to $A_{k+e} P$ (§ 1.7), $c_1(\mathcal{O}(1))^{e+i} \cap _$ is the iterated first Chern class homomorphism from $A_{k+e} P$ to $A_{k-i} P$ (§ 2.5), and p_* is the push-forward from $A_{k-i} P$ to $A_{k-i} X$ (§ 1.4).

Proposition 3.1. (a) *For all* $\alpha \in A_k X$,

(i) $s_i(E) \cap \alpha = 0 \quad for \quad i < 0$;

(ii) $s_0(E) \cap \alpha = \alpha$.

(b) *If* E, F *are vector bundles on* X, $\alpha \in A_k X$, *then, for all* i, j,

$$s_i(E) \cap (s_j(F) \cap \alpha) = s_j(F) \cap (s_i(E) \cap \alpha).$$

(c) *If* $f: X' \to X$ *is proper,* E *a vector bundle on* X, $\alpha \in A_* X'$, *then, for all* i,

$$f_*(s_i(f^*E) \cap \alpha) = s_i(E) \cap f_*(\alpha).$$

(d) *If* $f: X' \to X$ *is flat,* E *a vector bundle on* X, $\alpha \in A_* X$, *then, for all* i,

$$s_i(f^*E) \cap f^*\alpha = f^*(s_i(E) \cap \alpha).$$

(e) *If* E *is a line bundle on* X, $\alpha \in A_* X$, *then*

$$s_1(E) \cap \alpha = -c_1(E) \cap \alpha.$$

Proof. We first prove (c) and (d). Given $f: X' \to X$, and a vector bundle E on X, there is a fibre square

$$\begin{array}{ccc} P(f^*E) & \xrightarrow{f'} & P(E) \\ {\scriptstyle p'}\downarrow & & \downarrow{\scriptstyle p} \\ X' & \xrightarrow{f} & X \end{array}$$

with $f'^* \mathcal{O}_E(1) = \mathcal{O}_{f^*E}(1)$. If f is proper, $\alpha \in A_* X'$, then

$$\begin{aligned} f_*(s_i(f^*E) \cap \alpha) &= f_* p'_*(c_1(\mathcal{O}_{f^*E}(1))^{e+i} \cap p'^*\alpha) \\ &= p_* f'_*(c_1(f'^*\mathcal{O}_E(1))^{e+i} \cap p'^*\alpha) && (\S\ 1.4) \\ &= p_*(c_1(\mathcal{O}_E(1))^{e+i} \cap f'_* p'^*\alpha) && \text{(Proposition 2.5 (c))} \\ &= p_*(c_1(\mathcal{O}_E(1))^{e+i} \cap p^* f_*\alpha) && \text{(Proposition 1.7)} \\ &= s_i(E) \cap f_*\alpha. \end{aligned}$$

The proof of (d) is similar, using the corresponding fact (Proposition 2.5 (d)) for line bundles, and is left to the reader.

To prove (a), we may assume $\alpha = [V]$, with V a k-dimensional subvariety of X. By (c), we may assume $X = V$. Then $A_{k-i} X$ is zero for $i < 0$, which proves (i). Also,

$$s_0(E) \cap \alpha = p_*(c_1(\mathcal{O}_E(1))^e \cap [P]) = m\,[X]$$

for some integer m. To check that $m = 1$, by (d) we may restrict to an open set of X, so we may assume E is trivial. Then $P(E) = X \times \mathbb{P}^e$, and $\mathcal{O}(1)$ has sections whose zero scheme is $X \times \mathbb{P}^{e-1}$. Then

$$c_1(\mathcal{O}(1)) \cap [X \times \mathbb{P}^e] = [X \times \mathbb{P}^{e-1}]$$

by definition of the Chern class of a line bundle. Repeating this e times shows that m must be 1.

To prove (b), form the fibre square

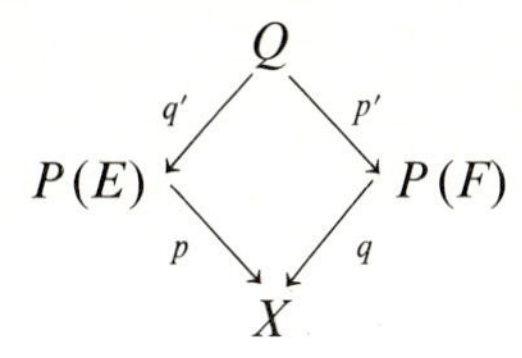

where p and q are the projections. Let $f+1$ be the rank of F. Then

$$\begin{aligned} s_i(E) \cap (s_j(F) \cap \alpha) &= p_*(c_1(\mathcal{O}_E(1))^{e+i} \cap p^*(q_*(c_1(\mathcal{O}_F(1))^{f+j} \cap q^*\alpha))) \\ &= p_*(c_1(\mathcal{O}_E(1))^{e+i} \cap q'_* p'^*(c_1(\mathcal{O}_F(1))^{f+j} \cap q^*\alpha)) \\ &\qquad \text{(Proposition 1.7)} \\ &= p_* q'_*(c_1(q'^*\mathcal{O}_E(1))^{e+i} \cap (c_1(p'^*\mathcal{O}_F(1))^{f+j} \cap p'^* q^*\alpha)) \\ &\qquad \text{Proposition 2.5 (c), (d))} \\ &= q_* p'_*(c_1(p'^*\mathcal{O}_F(1))^{f+j} \cap (c_1(q'^*\mathcal{O}_E(1))^{e+i} \cap q'^* p^*\alpha)) \end{aligned}$$

by Proposition 2.5 (b) and the functoriality of push-forward and pull-back. The steps may now be reverrsed, to arrive at

$$s_j(F) \cap (s_i(E) \cap \alpha) .$$

For (e), note that $P(E) = X$ and $\mathcal{O}_E(-1) = E$ in this case, so $\mathcal{O}_E(1) = E^\vee$, and

$$s_1(E) \cap \alpha = c_1(\mathcal{O}_E(1) \cap \alpha) = -c_1(E) \cap \alpha$$

by Proposition 2.5 (e). □

Corollary 3.1. *The flat pull-back*

$$p^*: A_k X \to A_{k+e}(P(E))$$

is a split monomorphism.

Proof. By (a) (ii), an inverse is $\beta \to p_*(c_1(\mathcal{O}_E(1))^e \cap \beta)$. □

For a geometric appearance of Segre classes, see Example 3.2.22.

Example 3.1.1. Let E be a vector bundle of rank $e+1$, L a line bundle. Then

$$s_p(E \otimes L) = \sum_{i=0}^{p} (-1)^{p-i} \binom{e+p}{e+i} s_i(E)\, c_1(L)^{p-i} .$$

(Identify $P(E)$ with $P(E \otimes L)$, with universal subbundle $\mathcal{O}_E(-1) \otimes p^* L$. Then

$$s_p(E \otimes L) \cap \alpha = p_*((c_1(\mathcal{O}_E(1)) - c_1(p^* L))^{e+p} \cap p^*\alpha) .)$$

3.2 Chern Classes

Let E be a vector bundle on a scheme X. Consider the formal power series

$$s_t(E) = \sum_{i=0}^{\infty} s_i(E)\, t^i = 1 + s_1(E)\, t + s_2(E)\, t^2 + \dots .$$

Define the *Chern polynomial*

$$c_t(E) = \sum_{i=0}^{\infty} c_i(E)\, t^i = 1 + c_1(E)\, t + c_2(E)\, t^2 + \dots$$

to be the inverse power series (which will be shown to be a polynomial):

$$c_t(E) = s_t(E)^{-1} .$$

Explicitly,

$$c_0(E) = 1\,, \quad c_1(E) = -\, s_1(E)\,,$$
$$c_2(E) = s_1(E)^2 - s_2(E), \dots$$
$$c_n(E) = -\, s_1(E)\, c_{n-1}(E) - s_2(E)\, c_{n-2}\, D - \dots - s_n(E)\,.$$

Here the $s_i(E)$ are regarded as endomorphisms of $A_* X$, with products denoting composition; since all such endomorphisms commute (Proposition 3.1 (b)), there is no ambiguity. We write, for $\alpha \in A_k X$,

$$c_i(E) \cap \alpha$$

for the element in $A_{k-i} X$ obtained by applying the endomorphism $c_i(E)$ to α.

The *total Chern class* $c(E)$ is the sum

$$c(E) = 1 + c_1(E) + \dots + c_r(E)\,, \qquad r = \operatorname{rank}(E).$$

In other words, $c(E) \cap \alpha = \sum_{i=0}^{r} c_i(E) \cap \alpha$ for all $\alpha \in A_* X$. Similarly, the *total Segre class* is

$$s(E) = 1 + s_1(E) + s_2(E) + \dots .$$

Although an infinite number of terms appear in this formel sum, when applied to $\alpha \in A_* X$, $s(E) \cap \alpha = \sum_{i \geqq 0} s_i(E) \cap \alpha$, only a finite number of non-zero terms appear.

Theorem 3.2. *The Chern classes satisfy the following properties:*

(a) (Vanishing) *For all bundles E on X, all $i >$* $\operatorname{rank}(E)$,

$$c_i(E) = 0\,.$$

(b) (Commutativity) *For all bundles E, F on X, integers i, j, and cycles α on X,*

$$c_i(E) \cap (c_j(F) \cap \alpha) = c_j(F) \cap (c_i(E) \cap \alpha)\,.$$

(c) (Projection formula) *Let E be a vector bundle on X, $f\colon X' \to X$ a proper morphism. Then*

$$f_*(c_i(f^*E) \cap \alpha) = c_i(E) \cap f_*(\alpha)$$

for all cycles α on X', all i.

(d) (Pull-back) *Let E be a vector bundle on X, $f: X' \to X$ a flat morphism. Then*

$$c_i(f^*E) \cap f^*\alpha = f^*(c_i(E) \cap \alpha)$$

for all cycles α on X, all i.

(e) (Whitney sum) *For any exact sequence*

$$0 \to E' \to E \to E'' \to 0$$

of vector bundles on X, then

$$c_t(E) = c_t(E') \cdot c_t(E''),$$

i.e.,

$$c_k(E) = \sum_{i+j=k} c_i(E')\, c_j(E'').$$

(f) (Normalization) *If E is a line bundle on a variety X, D a Cartier divisor on X with $\mathscr{O}(D) \cong E$, then*

$$c_1(E) \cap [X] = [D].$$

Note that from (c) and (f) it follows that the first Chern class for a line bundle defined here agrees with the definition given in § 2.5. Formula (f) will be generalized to bundles of arbitrary rank in Chap. 14.

Proof. Properties (b), (c), (d), and (f) follow directly from corresponding facts proved for Segre classes in Proposition 3.1. There are several proofs of (a) and (e). The following proof is based on the splitting construction, which will be useful later. Two shorter proofs are sketched in Examples 3.2.9 and 3.2.10.

Splitting construction. Given a finite collection $\mathscr{S}$ of vector bundles on a scheme X, there is a flat morphism $f: X' \to X$ such that

(1) $f^*: A_*X \to A_*X'$ is injective, and
(2) for each E in $\mathscr{S}$, f^*E has a filtration by subbundles

$$f^*E = E_r \supset E_{r-1} \supset \ldots \supset E_1 \supset E_0 = 0$$

with line bundle quotients $L_i = E_i/E_{i-1}$.

For one bundle E, f is constructed by induction n the rank of E. Let $P = P(E)$, $p: P \to X$ the projection. Then p^* is injective by Corollary 3.1, and p^*E has a subbundle $\mathscr{O}_E(-1)$ of rank one. If E' is the quotient bundle $p^*E/\mathscr{O}_E(-1)$, inductively we construct $q: X' \to P$ with q^* injective and q^*E' filtered. Then $f = p\,q$ has f^* injective, and an induced filtration on f^*E. The process may be repeated for any finite collection $\mathscr{S}$ of bundles.

To prove (a), it suffices by the splitting construction to prove that

$$c_t(E) = \prod_{i=1}^{r} (1 + c_1(L_i)\, t) \qquad (*)$$

when E has a filtration by subbundles

$$E = E_r \supset E_{r-1} \supset \ldots \supset E_0 = 0$$

with quotient line bundles $L_i = E_i/E_{i-1}$. For then if $f: X' \to X$ is as in the splitting construction,

$$f^*(c_i(E) \cap \alpha) = c_i(f^*E) \cap f^*\alpha = 0$$

for $i > r = \operatorname{rank}(E)$, and (a) follows since f^* is injective.

Lemma 3.2. *Assume that E is filtered as above, with line bundle quotients $L_1, \ldots, L_r$. Let s be a section E, and let Z be the closed subset of X where s vanishes. Then for any k-cycle α on X, there is a $(k-r)$-cycle β on Z with*

$$\prod_{i=1}^{r} c_1(L_i) \cap \alpha = \beta$$

in $A_{k-r}(X)$. In particular, if s is nowhere zero, then $\prod_{i=1}^{r} c_1(L_i) = 0$.

Proof. The section s determines a section $\bar{s}$ of the quotient bundle L_r. If Y is the zero-scheme of $\bar{s}$, then $(L_r, Y, \bar{s})$ determines a pseudo-divisor D_r on X (§ 2.2). Intersecting with D_r gives a class $D_r \cdot \alpha$ in $A_{k-1}(Y)$ such that

$$c_1(L_r) \cap \alpha = j_*(D_r \cdot \alpha),$$

where j is the inclusion of Y in X. By the projection formula (Proposition 2.5 (c)),

$$\prod_{i=1}^{r} c_1(L_i) \cap \alpha = j_*\left(\prod_{i=1}^{r-1} c_1(j^*L_i) \cap (D_r \cdot \alpha)\right).$$

The bundle j^*E_{r-1} has a section, induced by s, whose zero set is Z. By induction on r the term in parenthesis on the right side of the preceeding formula is represented by a cycle on Z, which concludes the proof. □

We return to the proof of (*). Let $p: P(E) \to X$ be the associated projective bundle. The universal subbundle $\mathcal{O}(-1)$ of p^*E corresponds to a trivial line subbundle of $p^*E \otimes \mathcal{O}(1)$, i.e., to a nowhere vanishing section of $p^*E \otimes \mathcal{O}(1)$. Since $p^*E \otimes \mathcal{O}(1)$ has a filtration with quotient line bundles $p^*L_i \otimes \mathcal{O}(1)$, Lemma 3.2 implies that

$$\prod_{i=1}^{r} c_1(p^*L_i \otimes \mathcal{O}(1)) = 0.$$

Let $\zeta = c_1(\mathcal{O}(1))$, and let σ_i (resp. $\tilde{\sigma}_i$) be the ith elementary symmetric function of $c_1(L_1), \ldots, c_1(L_r)$ (resp. $c_1(p^*L_1), \ldots, c_1(p^*L_r)$). Then $c_1(p^*L_i \otimes \mathcal{O}(1)) = c_1(p^*L_i) + \zeta$ by Proposition 2.5 (e), so that above displayed equation may be written

$$\zeta^r + \tilde{\sigma}_1 \zeta^{r-1} + \ldots + \tilde{\sigma}_r = 0.$$

Therefore, with $e = r - 1$,

$$\zeta^{e+i} + \tilde{\sigma}_1 \zeta^{e+i-1} + \ldots + \tilde{\sigma}_r \zeta^{i-1} = 0$$

for all $i \geqq 1$. It follows that for all $\alpha \in A_*X$,

$$p_*(\zeta^{e+i} \cap p^*\alpha) + p_*(\tilde{\sigma}_1 \zeta^{e+i-1} \cap p^*\alpha) + \ldots + p_*(\tilde{\sigma}_r \zeta^{i-1} \cap p^*\alpha) = 0.$$

From the definition of Segre classes, and the projection formula, this says

$$s_i(E) \cap \alpha + \sigma_1 s_{i-1}(E) \cap \alpha + \ldots + \sigma_r s_{i-r}(E) \cap \alpha = 0\,,$$

which means that

$$(1 + \sigma_1 t + \ldots + \sigma_r t^r)\, s_t(E) = 1\,,$$

and this is equivalent to (*).

The Whitney sum formula (e) also follows readily from (*). Given an exact sequence of vector bundles as in (e), apply the splitting construction to find $f: X' \to X$, with f^* injective, so that f^*E' filters with line bundle quotients L_i', and f^*E'' filters with line bundle quotients L_j''. Then f^*E has an induced filtration with line bundle quotients L_i' and L_j''. The formula

$$c_t(f^*E) = c_t(f^*E')\, c_t(f^*E'')$$

follows from (*), and the required formula on X follows from the injectivity of f^*. □

Remark 3.2.1. *Uniqueness.* The Chern classes are uniquely determined by properties (c), (d) and (f) of the theorem. In fact, formula (*) was proved using only these properties, and this determines the Chern classes by an application of the splitting construction, using (d) again.

Remark 3.2.2. *Notation and conventions.* By the commutative law (b), any polynomial in the Chern classes of vector bundles on X operates on A_*X. If

$$p = P(c_{i_1}(E_1), \ldots, c_{i_m}(E_m))$$

for vector bundles $E_1, \ldots, E_m$ on X, and $\alpha \in A_*X$, we write

$$p \cap \alpha = P(c_{i_1}(E_1), \ldots, c_{i_m}(E_m)) \cap \alpha$$

for the result of applying this polynomial to α. If p is isobaric of weight d, i.e., homogeneous of degree d, where $c_i(E)$ has weight i, and $\alpha \in A_k X$, then $p \cap \alpha \in A_{k-d} X$.

Notation is often abused by writing simply p in place of $p \cap [X]$. When X is non-singular, we will see in Chap. 8 that one may recover the action of p on any α from the class $p \cap [X]$, so there is no loss of information in this case. In any case, we may write

$$\int_X p = \int_X P(c_{i_1}(E_1), \ldots, c_{i_m}(E_m))$$

in place of $\int_X p \cap [X]$, the degree of the zero-dimensional component of $P \cap [X]$.

Another shorthand is quite useful. If p is a polynomial in Chern classes on X, as above, and a morphism $f: X' \to X$ is specified, then for any $\alpha \in A_*X'$, we may write $p \cap \alpha$ for the result of pulling back the bundles to X' and operating on α by the same polynomial in the pull-back bundles, i.e.,

$$p \cap \alpha = P(c_{i_1}(f^*E_1), \ldots, c_{i_m}(f^*E_m)) \cap \alpha\,.$$

If $g: X'' \to X'$ is a proper morphism, the projection formula (c) then reads simply

$$g_*(p \cap \alpha) = p \cap g_*(\alpha)$$

for $\alpha \in A_* X''$. Similarly, if g is flat,

$$g^*(p \cap \alpha) = p \cap g^*(\alpha)$$

for $\alpha \in A_* X'$.

In Chap. 17, we will define contravariant functors A^*, with cap products

$$A^i X \otimes A_j X \overset{\cap}{\to} A_{j-i} X .$$

Then a polynomial in Chern classes of bundles on X will be an element of $A^* X$. The convention introduced in the previous paragraph will become part of a general notation.

Remark 3.2.3. *Splitting principle.* Suppose one wants to prove a universal formula involving Chern classes of a finite number of vector bundles in a certain relation with each other. If the formula is true whenever the bundles have filtrations with line bundle quotients, and the relation among the bundles is preserved by flat pull-back, then the formula holds in general. This follows from the splitting construction and the pull-back property (d).

If the Chern polynomial of a bundle E of rank r is factored

$$c_t(E) = \prod_{i=1}^{r} (1 + \alpha_i t) ,$$

then $\alpha_1, \ldots, \alpha_r$ are called *Chern roots* of E. This factorization may be regarded as purely formal: the Chern classes of E are the elementary symmetric functions of $\alpha_1, \ldots, \alpha_r$. Or one may use the splitting construction: if $f^* E$ is filtered with line bundle quotients L_i, then

$$c_t(f^* E) = \prod_{i=1}^{r} (1 + \alpha_i t)$$

with $\alpha_i = c_1(L_i)$. Any symmetric polynomial in the Chern roots of E has a definite expression as a polynomial in the Chern classes of E. The Whitney sum formula says that for an exact sequence as in (d), Chern roots for E' and E'' taken together give Chern roots for E. We give some other applications.

(a) *Dual bundles.* The Chern classes of the dual bundle $E^\vee$ are given by the formula

$$c_i(E^\vee) = (-1)^i c_i(E) .$$

We have seen this when E is a line bundle (Proposition 2.5(e)). If E has a filtration with quotients L_i, then $E^\vee$ as a filtration with quotients $L_{r-i}^\vee$. Thus if $\alpha_1, \ldots, \alpha_r$ are Chern roots for E, $-\alpha_1, \ldots, -\alpha_r$ are Chern roots for $E^\vee$, which gives the displayed formula.

(b) *Tensor products.* The formula for Chern classes of a tensor product $E \otimes F$ has a simple expression in terms of Chern roots. If $\alpha_1, \ldots, \alpha_r$ are Chern roots for E, and $\beta_1, \ldots, \beta_s$ are Chern roots for F, then the sums

$$\alpha_i + \beta_j, \quad 1 \leqq i \leqq r, \quad 1 \leqq j \leqq s$$

are Chern roots for $E \otimes F$. This follows from the splitting principle and the corresponding result for line bundles (Proposition 2.5(e)). Thus $c_k(E \otimes F)$ is the kth elementary symmetric function of $\alpha_1 + \beta_1, \ldots, \alpha_r + \beta_s$, which may be written as a universal polynomial in the symmetric functions of $\alpha_1, \ldots, \alpha_r$ and $\beta_1, \ldots, \beta_s$, i.e., as a polynomial in the Chern classes of E and F. In general explicit formulas for these polynomials are rather complicated (see Example 14.5.2). In case $F = L$ is a line bundle, however, there is the useful formula

$$c_r(E \otimes L) = \sum_{i=0}^{r} c_1(L)^i c_{r-i}(E)$$

for the top Chern class. More generally,

$$c_t(E \otimes L) = \sum_{i=0}^{r} t^i c_t(L)^{r-i} c_i(E),$$

which follows from the identity

$$\prod_{i=1}^{r} ((1 + \beta t) + \alpha_i t) = \sum_{i=0}^{r} (1 + \beta t)^{r-i} c_i t^i$$

where c_i is the ith elementary symmetric function of $\alpha_1, \ldots, \alpha_r$. For a reformulation see Example 3.2.2.

(c) *Exterior powers.* Similarly, if $\alpha_1, \ldots, \alpha_r$ are the Chern roots of E, then

$$c_t(\Lambda^p E) = \prod_{i_1 < \ldots < i_p} (1 + (\alpha_{i_1} + \ldots + \alpha_{i_p}) t).$$

In particular

$$c_1(\Lambda^r E) = c_1(E).$$

For this one uses the fact that if there is an exact sequence

$$0 \to L \to E \to E' \to 0$$

with L a line bundle, then there are exact sequences

$$0 \to \Lambda^{p-1} E' \otimes L \to \Lambda^p E \to \Lambda^p E' \to 0$$

(cf. Hirzebruch (1) 4.13).

Remark 3.2.4. In the course of the proof of Theorem 3.2 the identity

$$\zeta^r + c_1(p^* E) \zeta^{r-1} + \ldots + c_r(p^* E) = 0$$

was proved. Here E is a bundle of rank r on a scheme X, p is the projection from $P(E)$ to X, and $\zeta = c_1(\mathcal{O}_E(1))$. Alternatively, this follows from the fact that $p^*(E) \otimes \mathcal{O}_E(1)$ has a nowhere vanishing section; then by the Whitney formula, $c_r(p^*(E) \otimes \mathcal{O}_E(1)) = 0$, which gives the displayed formula by Remark 3.2.3 (b).

Additional identities for Chern classes are given in Chap. 14, as well as in the examples.

Example 3.2.1. Let $p: P(E) \to X$ be as in § 3.1, and let $L_E = \mathcal{O}_E(-1)$. Then

$$s(E) \cap \alpha = p_*(s(L_E) \cap p^* \alpha)$$

for all $\alpha \in A_* X$, and $c(E) \cap \alpha = s(E)^{-1} \cap \alpha$.

Example 3.2.2. Let E be a vector bundle of rank r, L a line bundle. Then, for all $p \geqq 0$,

$$c_p(E \otimes L) = \sum_{i=0}^{p} \binom{r-i}{p-i} c_i(E)\, c_1(L)^{p-i}.$$

(Equate coefficients of t^p in the formula for $c_t(E \otimes L)$ in Remark 3.2.3.) Another equivalent formulation is

$$c_t(E \otimes L) = (1 + x\, t)^r c_\tau(E)$$

with $x = c_1(L)$ and $\tau = t/(1 + x\, t)$.

Example 3.2.3. The *Chern character* $\mathrm{ch}(E)$ of a bundle E is defined by the formula

$$\mathrm{ch}(E) = \sum_{i=1}^{r} \exp(\alpha_i)$$

where $\exp(x) = e^x = \sum_{n=0}^{\infty} x^n/n!$, and $\alpha_1, \ldots, \alpha_r$ are the Chern roots of E. The first few terms are

$$\mathrm{ch}(E) = r + c_1 + \frac{1}{2}(c_1^2 - 2c_2) + \frac{1}{6}(c_1^3 - 3c_1 c_2 + c_3)$$
$$+ \frac{1}{24}(c_1^4 - 4c_1^2 c_2 + 4c_1 c_3 + 2c_2^2 - 4c_4) + \ldots$$

where $c_i = c_i(E)$. Then n^{th} term is $p_n/n!$, where p_n is determined inductively by Newton's formula:

$$p_n - c_1 p_{n-1} + c_2 p_{n-2} - \ldots + (-1)^{n-1} c_{n-1} p_1 + (-1)^n n\, c_n = 0.$$

For any exact sequence of vector bundles as in Theorem 3.2 (e),

$$\mathrm{ch}(E) = \mathrm{ch}(E') + \mathrm{ch}(E''),$$

while for tensor products

$$\mathrm{ch}(E \otimes E') = \mathrm{ch}(E) \cdot \mathrm{ch}(E')$$

Example 3.2.4. The *Todd class* $\mathrm{td}(E)$ of a vector bundle E is defined by the formula

$$\mathrm{td}(E) = \prod_{i=1}^{r} Q(\alpha_i)$$

where

$$Q(x) = \frac{x}{1 - e^{-x}} = 1 + \frac{1}{2}x + \sum_{k=1}^{\infty} (-1)^{k-1} \frac{B_k}{(2k)!} x^{2k},$$

(B_k the Bernoulli numbers), and $\alpha_1, \ldots, \alpha_r$ are the Chern roots of E. The first few terms are

$$\mathrm{td}(E) = 1 + \frac{1}{2}c_1 + \frac{1}{12}(c_1^2 + c_2) + \frac{1}{24}(c_1 c_2)$$
$$+ \frac{1}{720}(-c_1^4 + 4c_1^2 c_2 + 3c_2^2 + c_1 c_3 - c_4) + \ldots .$$

For an exact sequence of vector bundles as in Theorem 3.2 (e),

$$\mathrm{td}(E) = \mathrm{td}(E') \cdot \mathrm{td}(E'') .$$

Example 3.2.5 (cf. Borel-Serre (1) Lemma 18). Let E be a vector bundle of rank r. Then

$$\sum_{p=0}^{r} (-1)^p \, \mathrm{ch}(\Lambda^p E^\vee) = c_r(E) \cdot \mathrm{td}(E)^{-1} .$$

(If $\alpha_1, \ldots, \alpha_r$ are Chern roots for E, the left side is

$$\sum_{p=0}^{r} (-1)^p \sum_{i_1 < \ldots < i_p} \exp(-\alpha_{i_1} - \ldots - \alpha_{i_p}) = \prod_{i=1}^{r} (1 - \exp(-\alpha_i))$$
$$= \alpha_1 \cdot \ldots \cdot \alpha_r \prod_{i=1}^{r} ((1 - \exp(-\alpha_i))/\alpha_i) ,$$

which equals the right side.)

Example 3.2.6. If $\alpha_1, \ldots, \alpha_r$ are Chern roots for E, then the sums $m_1 \alpha_1 + \ldots + m_r \alpha_r$, for all r-tuples of non-negative integers $m_1, \ldots, m_r$ adding to m, are the Chern roots for the mth symmetric power $\mathrm{Sym}^m(E)$ of E. For $m = 2$, explicit formulas are given in Example 14.5.1.

Example 3.2.7. (a) For bundles E, F on X write

$$c(F - E) = c(F)/c(E) = c(F)\, s(E)$$
$$= 1 + (c_1(F) - c_1(E)) + (c_2(F) - c_1(F)\, c_1(E) + c_1(E)^2 - c_2(E)) + \ldots$$

and let $c_k(F - E)$ be the kth term in this expansion. If rank $E = r$ and rank $F = s$, then

$$c_{s-r+1}(F - E) \cap \alpha = p_*(c_s(p^* F \otimes \mathcal{O}_E(1)) \cap p^* \alpha)$$

for all $\alpha \in A_* X$, where p projects $P(E)$ to X. Generalizations of this formula will be given in Chap. 14.

(b) For any element ξ in the Grothendieck group $K^\circ X$ of vector bundles on X (cf. § 15.1), there is a well-defined Chern class $c(\xi)$, and power series $c_t(\xi)$. If $\xi = \sum [F_j] - \sum [E_i]$, then

$$c(\xi) = \prod_j c(F_j) / \prod_i c(E_i) , \qquad c_t(\xi) = \prod_j c_t(F_j) / \prod_i c_t(E_i) .$$

It follows from the Whitney formula that these are well-defined on $K^\circ X$.

Example 3.2.8. Let X, Y be schemes, p and q the projections from $X \times Y$ to X and Y, E and F vector bundles on X and Y, $\alpha \in A_* X$, $\beta \in A_* Y$. Then (cf. Example 2.5.3)

$$(c_i(\mathrm{E}) \cap \alpha) \times \beta = c_i(p^* E) \cap (\alpha \times \beta)$$

and

$$(c(E) \cap \alpha) \times (c(F) \cap \beta) = c(p^* E \oplus q^* F) \cap (\alpha \times \beta) .$$

Example 3.2.9. The vanishing property (a) of Chern classes follows from the Whitney sum formula (e) and the splitting principle.

Example 3.2.10. There are other simple proofs of the Whitney sum formula (e):

Case 1: rank $E''=1$. It suffices to show that $s(E')\cap\alpha=c(E'')\,s(E)\cap\alpha$, for $\alpha=[V]$, $V=X$. Let $L_E=\mathcal{O}_E(-1)$. Consider the diagram

$$\begin{array}{ccc} P'=P(E') & \overset{i}{\hookrightarrow} & P(E)=P \\ & {\scriptstyle p'}\searrow \quad \swarrow{\scriptstyle p} & \\ & X & . \end{array}$$

On P, the composite $L_E\subset p^*E\to p^*E''$ determines a section of $p^*E''\otimes L_E^{\vee}$ whose zero-scheme is P'. So

$$i_*[P']=(c_1(p^*E'')-c_1(L_E))\cap[P]\,.$$

Now

$$\begin{aligned} c(E'')\,s(E)\cap\alpha &= p_*(c(p^*E'')\,s(L_E)\cap[P]) \\ &= p_*(s(L_E)\cap[P])+p_*(c_1(p^*E'')\,s(L_E)\cap[P])\,, \end{aligned}$$

and

$$\begin{aligned} s(E')\cap\alpha &= p_*i_*(s(i^*L_E)\cap[P'])=p_*(s(L_E)\cap i_*[P']) \\ &= p_*(s(L_E)\,c_1(p^*E'')\cap[P])-p_*(s(L_E)\,c_1(L_E)\cap[P])\,. \end{aligned}$$

Therefore, when rank$(E)>1$,

$$(c(E'')\,s(E)-s(E'))\cap\alpha=p_*(s(L_E)\,c(L_E)\cap[P])=p_*[P]=0\,.$$

General case. Let $Q=P(E''^{\vee})$, $q:Q\to X$ the projection, and let L be the universal line bundle quotient of q^*E''. Construct a commutative diagram of vector bundles on Q with exact rows and columns:

$$\begin{array}{ccccccccc} & & & & 0 & & 0 & & \\ & & & & \downarrow & & \downarrow & & \\ 0 & \to & q^*E' & \to & G & \to & F & \to & 0 \\ & & \| & & \downarrow & & \downarrow & & \\ 0 & \to & q^*E' & \to & q^*E & \to & q^*E'' & \to & 0 \\ & & & & \downarrow & & \downarrow & & \\ & & & & L & = & L & & \\ & & & & \downarrow & & \downarrow & & \\ & & & & 0 & & 0 & & \end{array}$$

Then $c(q^*E)=c(G)\,c(L)$, $c(q^*E'')=c(F)\,c(L)$, and $c(G)=c(q^*E')\,c(F)$, by Case 1 and induction. Combining these gives

$$c(q^*E)\cap q^*\alpha=c(q^*E')\,c(q^*E'')\cap q^*\alpha$$

for all $\alpha\in A_*X$, which suffices since q^* is one-to-one.

One may also deduce the general case from Case 1 by realizing the blow-up of $P(E)$ along $P(E')$ as a projective bundle $P(H)$ over $P(E'')$, where H fits into an exact sequence

$$0\to p''^*E'\to H\to L_{E''}\to 0\,.$$

One has a commutative diagram

$$\begin{array}{ccccc} & & P(H) & & \\ & \overset{\pi}{\swarrow} & & \overset{q}{\searrow} & \\ P(E) & & & & P(E'') \\ & \underset{p}{\searrow} & & \underset{p''}{\swarrow} & \\ & & X & & \end{array}$$

with $\pi^* L_E = L_H$. Then

$$\begin{aligned} s(E) \cap [X] &= p_*(s(L_E) \cap [P(E)]) = p_* \pi_*(s(L_H) \cap [P(H)]) \\ &= p''_*(s(H) \cap [P(E'')]) = p''_*(s(p''^* E')\, c(L_{E''}) \cap [P(E'')]) \\ &= s(E') \cap (s(E'') \cap [X]) \,. \end{aligned}$$

Example 3.2.11. From the exact sequence

$$0 \to \mathcal{O}_{\mathbb{P}^n} \to \mathcal{O}_{\mathbb{P}^n}(1)^{\oplus(n+1)} \to T_{\mathbb{P}^n} \to 0$$

on $\mathbb{P}^n$ (Appendix B.5.8),

$$c(T_{\mathbb{P}^n}) = (1 + H)^{n+1}$$

where $H = c_1(\mathcal{O}_{\mathbb{P}^n}(1))$ is the class of a hyperplane. More generally, if E is a vector bundle on a non-singular variety X, $p: P(E) \to X$ the projection, then

$$c_t(T_{P(E)}) = c_t(p^* T_X) \cdot c_t(p^* E \otimes \mathcal{O}_E(1)) = c_t(p^* T_X)\,(1 + x\,t)^r\, c_\tau(p^*(E)),$$

where $x = c_1(\mathcal{O}_E(1))$, $r = \operatorname{rank} E$, $\tau = t/(1 + x\,t)$. (From Appendix B.2.7 and B.5.8, there is an exact sequence

$$0 \to \mathcal{O}_{P(E)} \to p^* E \otimes \mathcal{O}_E(1) \to T_{P(E)} \to p^* T_X \to 0\,.)$$

Example 3.2.12. *Adjunction formula.* Let $i: X \to Y$ be a closed imbedding of codimension d of non-singular varieties with normal bundle N. From the exact sequence (Appendix B.7.2)

$$0 \to T_X \to i^* T_Y \to N \to 0$$

one has

$$c(T_X) = c(i^* T_Y)/c(N)\,.$$

If X is an intersection of divisors $D_1, \ldots, D_d$, then

$$c(N) = \prod_{i=1}^{d} (1 + c_1(i^* \mathcal{O}_Y(D_i)))\,.$$

From the projection formula one deduces

$$i_*(c(T_X) \cap [X]) = c(T_Y) \cdot \prod_{i=1}^{d} (D_i - D_i^2 + D_i^3 - \ldots)\,.$$

For example, if $Y = \mathbb{P}^m$, and $\deg D_i = n_i$, then

$$c(T_X) = (1 + h)^{m+1} \Big/ \prod_{i=1}^{d} (1 + n_i h)\,,$$

$h = c_1(i^*\mathcal{O}(1))$. In particular,

$$c_1(T_X) = \left(m + 1 - \sum_{i=1}^{d} n_i\right) h .$$

Example 3.2.13. Let X be a non-singular n-dimensional variety with tangent bundle T_x. The *total Chern class of* X, is $c(T_X) \cap [X]$. The *Euler characteristic* is $\int_X c_n(T_X)$. For example, if $n = 1$, the Euler characteristic is $2-2g$, where g is the *genus* of the curve X (see § 15 for other definitions). The *Todd class* of X is $\mathrm{td}(T_X)$. The *Todd genus* of X is $\int_X \mathrm{td}(T_X)$. When $n = 1$, the Todd genus is $1 - g$.

Example 3.2.14. If C is an effective Cartier divisor on a complete surface X, then

$$(C^2)_X = \int_C c_1(N)$$

where N is the normal bundle to C in X. If C and X are non-singular, and i is the inclusion of C in X, then $N = i^*T_X/T_C$, so

$$(C^2)_X = \int_C c_1(i^*T_X) - c_1(T_C) .$$

Equivalently,

$$(C \cdot (C + K))_X = 2g - 2 ,$$

where $K = - c_1(T_X)$ is the canonical divisor class of X, and g is the genus of C. In case X is a non-singular surface of degree d in $\mathbb{P}^3$,

$$(C^2)_X = 2g - 2 + (4 - d) \deg(C) .$$

Example 3.2.15. (a) Let X be an n-dimensional abelian variety, $i: X \to \mathbb{P}^m$ a closed imbedding, with normal bundle N. Then

$$c(N) = (1 + h)^{m+1}$$

where $h = c_1(i^*\mathcal{O}(1))$.

It follows that there can be no such imbedding if $m < 2n$. If $m = 2n$, such an imbedding is possible only if

$$\deg(X) = \binom{2n+1}{n} .$$

For example, if $n = 2$, $m = 4$, X must have degree 10; Horrocks and Mumford (1) have constructed such an example.

(b) Let i be the d-fold Veronese imbedding of $\mathbb{P}^n$ in $\mathbb{P}^m$, $m = \binom{n+d}{d} - 1$, $i^*\mathcal{O}_{\mathbb{P}^m}(1) = \mathcal{O}_{\mathbb{P}^n}(d)$. Then

$$c(N) = (1 + dh)^{m+1}/(1 + h)^{n+1} ,$$

where $h = c_1(\mathcal{O}_{\mathbb{P}^n}(1))$. For the Veronese surface $\mathbb{P}^2 \hookrightarrow \mathbb{P}^5$,

$$c(N) = 1 + 9h + 30h^2 ,$$
$$s(N) = 1 - 9h + 51h^2 .$$

(c) Let i be the Segre imbedding of $X = \mathbb{P}^{n_1} \times \ldots \times \mathbb{P}^{n_r}$ in $\mathbb{P}^m$, $m = \prod_{i=1}^{r} (n_i + 1) - 1$. Then

$$c(N) = \prod_{i=1}^{r} (1 + h_i)^{n_i+1} / (1 + h_1 + \ldots + h_r)^{m+1},$$

where $h_i = c_1(\mathrm{pr}_i^* \mathcal{O}_{\mathbb{P}^{n_i}}(1))$, pr_i the projection from X to the ith factor.

Example 3.2.16. Let E be a vector bundle of rank r on X, s a section of E, Z the zero-scheme of s.

(i) For any $\alpha \in A_k X$, there is a class β in $A_{k-r}(Z)$ whose image in $A_{k-r}(X)$ is $c_r(E) \cap \alpha$. In particular, if $Z = \emptyset$, $c_r(E) = 0$. (Use the splitting principle and Lemma 3.2.)

(ii) If X is purely n-dimensional, and s is a regular section of E, then Z is purely $(n-r)$-dimensional, and

$$c_r(E) \cap [X] = [Z].$$

These facts will be reproved and generalized in § 14.1. (cf. Grothendieck (2) § 5 in the projective non-singular case).

Example 3.2.17. Let F be a subbundle of a vector bundle E, with quotient bundle G. There is a regular section of $p^* G \otimes \mathcal{O}_E(1)$ on $P(E)$, whose zero-scheme is $P(F)$ (Appendix B.5.6). By the preceeding example,

$$[P(F)] = c_r(p^* G \otimes \mathcal{O}_E(1)) \cap [P(E)] = \sum_{i=0}^{r} c_1(\mathcal{O}_E(1))^i \cap p^*(c_{r-i}(G) \cap [X])$$

in $A_* P(E)$, where r is the rank of G. In the non-singular projective case this formula may be found in Ilori-Ingleton-Lascu (1).

Example 3.2.18. Let Y^m be a smooth subvariety of codimension d of a smooth variety X, with normal bundle N. Applying the preceding example to

$$\begin{array}{ccccc} \hat{Y} = P(T_Y) & \hookrightarrow & P(T_X|_Y) & \hookrightarrow & P(T_X) = \hat{X} \\ & & {\scriptstyle q}\downarrow & & {\scriptstyle p}\downarrow \\ & & Y & \overset{}{\underset{i}{\hookrightarrow}} & X \end{array}$$

one has

$$[\hat{Y}] = c_d(q^* N \otimes \mathcal{O}(1)) \cap [P(T_X|_Y)] = \sum_{i=0}^{d} \zeta^i \cap q^*(c_{d-i}(N) \cap [Y])$$

in $A_{2m-1}(P(T_X|_Y))$, with $\zeta = c_1(\mathcal{O}(1))$. Therefore,

$$[\hat{Y}] = \sum_{i=0}^{d} \zeta^i \cap p^* i_*(c_{d-i}(N) \cap [Y])$$

in $A_{2m-1}(\hat{X})$, a formula of D. B. Scott (1).

Example 3.2.19. Let X be a non-singular variety. Let $\hat{X} = P(T_X)$ be the projective bundle of tangent directions, $p : \hat{X} \to X$ the projection, $L \subset p^* T_X$ the universal line sub-bundle. Let $f : X \to \mathbb{P}^r$ be a morphism, i.e., a linear system without base points, and let $df\colon T_X \to f^*T_{\mathbb{P}^r}$ be the induced differential. On $\hat{X}$ there is a composite

$$L \subset p^*T_X \to p^*f^*T_{\mathbb{P}^r}.$$

The locus where this composite is zero is the set of tangent lines at points $x \in X$ which are tangent to every member of the linear system passing through x. The vanishing of the composite corresponds to the vanishing of the corresponding section of $p^*f^*T_{\mathbb{P}^r} \otimes L^\vee$, and the set is given a scheme structure Z as the zero scheme of this section. If codim $(Z, \hat{X}) = r$, then (cf. Example 3.16 and Lemma A.7.1)

$$\begin{aligned}[Z] &= c_r(p^*f^*T_{\mathbb{P}^r} \otimes L^\vee) \cap [\hat{X}] \\ &= \sum_{i=0}^{r} c_1(L^\vee)^{r-i} \cdot p^*(c_i(f^*T_{\mathbb{P}^r}) \cap [\hat{X}]) \\ &= \sum_{i=0}^{r} \binom{r+1}{i} \zeta^{r-i} p^*(h^i) \cap [\hat{X}],\end{aligned}$$

where $\zeta = c_1(L^\vee) = c_1(\mathcal{O}_{T_X}(1))$, and $h = c_1(f^*\mathcal{O}_{\mathbb{P}^r}(1))$ is the divisor class of any member of the linear system.

If a linear system has a base locus B, the above applies on $X - B$. If $\operatorname{codim}(B, X) \geqq r+1$, then the restriction from $A_k\hat{X}$ to $A_k(\hat{X} - p^{-1}(B))$ is an isomorphism for $k = \dim Z$, so the above formula holds for the closure of $[Z]$ in $A_*(\hat{X})$ (cf. Vainsencher (1)). When $r = 1$, this gives the formula

$$[Z] = \zeta + 2p^*(h)$$

which is useful for giving a geometric interpretation of ζ (cf. Example 16.2.4).

Example 3.2.20. *Ramification.* Let $f\colon X^n \to Y^n$ be a morphism of non-singular varieties. Let $R(f)$ be the subset of points of X where the induced map of tangent spaces is not an isomorphism. A scheme structure on $R(f)$ is given locally by the vanishing of the Jacobian determinant, i.e., $R(f)$ is the zero scheme of the map

$$\Lambda^n df\colon \Lambda^n T_X \to \Lambda^n f^*T_Y,$$

or the zero scheme of a section of the line bundle $\Lambda^n f^*T_Y \otimes \Lambda^n T_X^\vee$. If $R(f) \neq X$, then

$$[R(f)] = (c_1(f^*T_Y) - c_1(T_X)) \cap [X].$$

If $n = 1$, taking degrees of both sides yields the Riemann-Hurwitz formula

$$2g_X - 2 = \deg(f)(2g_Y - 2) + \deg R(f)$$

where g_X (resp. g_Y) is the genus of X (resp. Y). (See Examples 9.3.12 and 14.4.8 for generalizations.)

Example 3.2.21. *Dual varieties* (cf. Deligne-Katz (1) XVII).

(a) Let X^n be a non-singular subvariety of $\mathbb{P}^m$. Let $\check{\mathbb{P}}^m$ be the projective space of hyperplanes in $\mathbb{P}^m$, and let

$$\tilde{X} = \{(P, H) \in X \times \check{\mathbb{P}}^m \mid T_P X \subset H\}$$

where $T_P X$ is the tangent n-plane to X at P in $\mathbb{P}^m$. The projection $\pi: \tilde{X} \to X$ identifies $\tilde{X}$ with $P(N^\vee)$, where N is the normal bundle to X in $\mathbb{P}^m$. The image of $\tilde{X}$ by the projection $f: \tilde{X} \to \check{\mathbb{P}}^m$ is the *dual variety* to X, denoted $X^\vee$. Then

$$\deg f_*[\tilde{X}] = (-1)^n \int_X \frac{c(T_X)}{(1+h)^2}$$

where $h = c_1(\mathcal{O}_X(1))$. In particular, $X^\vee$ is a hypersurface if and only if the right side is non-zero; in this case

$$\deg(X^\vee) = \frac{(-1)^n}{\deg(\tilde{X}/X^\vee)} \int_X \frac{c(T_X)}{(1+h)^2}.$$

(The inclusion of $\tilde{X}$ in $X \times \check{\mathbb{P}}^m$ comes from the inclusion

$$N^\vee \subset (T_{\mathbb{P}^m})^\vee|_X \subset \mathcal{O}_X(-1)^{\oplus(m+1)},$$

so $f^*\mathcal{O}(1) = \mathcal{O}_{N^\vee}(1) \otimes \pi^*\mathcal{O}_X(-1)$. Therefore,

$$\begin{aligned}
\deg f_*[\tilde{X}] &= \int_{\tilde{X}} c_1(f^*\mathcal{O}(1))^{m-1} \cap [\tilde{X}] \\
&= \sum_{i=0}^{m-1} \binom{m-1}{i} (-1)^i \int_X c_1(\mathcal{O}_X(1))^i \cdot \pi_*(c_1(\mathcal{O}_{N^\vee}(1))^{m-1-i} \cap [\tilde{X}]) \\
&= \sum_{i=0}^{m-1} \binom{m-1}{i} (-1)^i \int_X h^i (-1)^{n-i} s_{n-i}(N) \\
&= (-1)^n \int_X (1+h)^{m-1} s(N).
\end{aligned}$$

And $c(T_X) = (1+h)^{m+1} s(N)$ by Examples 3.2.11, 3.2.12.)

(b) (cf. Kleiman (8) p. 364) Let X be a non-singular plane curve of degree d. Imbed X in $\mathbb{P}^m$, $m = \binom{r+2}{2} - 1$, by the r-fold Veronese imbedding of $\mathbb{P}^2$ in $\mathbb{P}^m$. The dual variety to X in $\mathbb{P}^m$ is the variety of plane curves of degree r which are tangent to X. Using (a), the degree of this variety is seen to be $d(d+2r-3)$.

Example 3.2.22. Regard $\check{\mathbb{P}}^3$ as the Grassmannian of 2-planes in $\mathbb{P}^3$, with universal rank 3 subbundle S and quotient line bundle Q. The variety Y of conics in $\mathbb{P}^3$ may be identified with the projective bundle of $\mathrm{Sym}^2 S^\vee$, or with $P(E)$, $E = \mathrm{Sym}^2 S^\vee \otimes Q^\vee$. With this description, the variety of conics meeting a given line is given by the vanishing of a section of $\mathcal{O}_E(1)$ (cf. Example 14.7.12). Thus the locus of conics meeting n given lines is represented by $c_1(\mathcal{O}_E(1))^n$. Computing the Segre class of E,

$$s(E) = 1 + 8h + 34h^2 + 92h^3$$

where $h = c_1(Q)$, one sees in particular that

$$\int_Y c_1(\mathcal{O}_E(1))^8 = \int_{\check{\mathbb{P}}^3} s_3(E) = 92,$$

the number of conics meeting 8 general lines. Similarly there are

$$\int_Y c_1(\mathcal{O}_E(1))^7 h = \int_{\check{\mathbb{P}}^3} s_2(E)\, h = 34$$

conics meeting 7 general lines, whose planes pass through a given general point, and $\int s_1(E)h^2 = 8$ conics meeting 6 general lines, whose planes contain a given general line (cf. Schubert (1) § 20).

3.3 Rational Equivalence on Bundles

Let E be a vector bundle of rank $r = e+1$ on a scheme X, with projection $\pi: E \to X$. Let $P(E)$ be the associated projective bundle, p the projection from $P(E)$ to X, and $\mathcal{O}(1)$ or $\mathcal{O}_E(1)$ the canonical line bundle on $P(E)$.

Theorem 3.3. (a) *The flat pull-back*

$$\pi^* : A_{k-r}X \to A_k E$$

is an isomorphism for all k.

(b) *Each element β in $A_k P(E)$ is uniquely expressible in the form*

$$\beta = \sum_{i=0}^{e} c_1(\mathcal{O}(1))^i \cap p^*\alpha_i,$$

for $\alpha_i \in A_{k-e+i}(X)$. Thus there are canonical isomorphisms

$$\bigoplus_{i=0}^{e} A_{k-e+i}(X) \cong A_k P(E).$$

Proof. The surjectivity of π^* was proved in § 1.9. Let θ_E be the canonical homomorphism from $\bigoplus_{i=0}^{e} A_*X$ to $A_*P(E)$ defined in (b):

$$\theta_E(\oplus \alpha_i) = \sum_{i=0}^{e} c_1(\mathcal{O}(1))^i \cap p^*\alpha_i.$$

To see that θ_E is surjective, the same Noetherian induction argument used in the proof of Proposition 1.9 reduces it to the case when E is trivial. By induction on the rank, it suffices to prove that θ_F is surjective when θ_E is known to be surjective, where $F = E \oplus 1$ is the direct sum of E and a trivial line bundle.

Let $P = P(E)$, $Q = P(F) = P(E \oplus 1)$, $q: Q \to X$ the projection. We have a commutative diagram

$$\begin{array}{ccccc} P & \xrightarrow{i} & Q & \xleftarrow{j} & E \\ & {\scriptstyle p}\searrow & \downarrow{\scriptstyle q} & \swarrow{\scriptstyle \pi} & \\ & & X & & \end{array}$$

identifying Q as the projective completion of E, P as the "hyperplane at infinity" (see Appendix B.5). By Proposition 1.8, the row in the following commutative diagram is exact:

$$\begin{array}{ccccccc} A_kP & \xrightarrow{i_*} & A_kQ & \xrightarrow{j^*} & A_kE & \to & 0 \\ & & {\scriptstyle q^*}\uparrow & \nearrow{\scriptstyle \pi^*} & & & \\ & & A_{k-r}X & & & & \end{array}.$$

Lemma 3.3. *For all* $\alpha \in A_* X$,

$$c_1(\mathcal{O}_F(1)) \cap q^*\alpha = i_* p^*\alpha .$$

Proof. It suffices to prove this for $\alpha = [V]$, V a subvariety of X. Since $\mathcal{O}_F(1)$ has a section vanishing precisely on P (Appendix B.5), the fact that $c_1(\mathcal{O}_F(1)) \cap [q^{-1} V] = [p^{-1} V]$ follows from the definition of Chern classes. □

Now if $\beta \in A_* Q$, write $j^*\beta = \pi^*\alpha$ for some α in $A_* X$. Then $\beta - q^*\alpha$ is in the kernel of j^*. Since $\operatorname{Ker}(j^*) = \operatorname{Im}(i_*)$, and inductively we are assuming θ_E is surjective,

$$\beta - q^*\alpha = i_*\left(\sum_{i=0}^{e} c_1(\mathcal{O}_E(1))^i \cap p^*\alpha_i\right)$$

for some $\alpha_i \in A_* X$. Since $i^*\mathcal{O}_F(1) = \mathcal{O}_E(1)$, the projection formula rewrites the right side as

$$\sum_{i=0}^{e} c_1(\mathcal{O}_F(1))^i \cap i_* p^*\alpha_i .$$

Now apply the lemma, obtaining

$$\beta = q^*\alpha + \sum_{i=0}^{e} c_1(\mathcal{O}_F(1))^{i+1} \cap q^*\alpha_i,$$

which shows that θ_F is surjective.

To show that the expression in (b) is unique, suppose there is a nontrivial relation

$$\beta = \sum_{i=0}^{e} c_1(\mathcal{O}(1))^i \cap p^*\alpha_i = 0 .$$

Let k be the largest integer with $\alpha_k \neq 0$. Then

$$p_*(c_1(\mathcal{O}(1))^{e-k} \cap \beta) = \alpha_k$$

by Proposition 3.1 (a), a contradiction.

Finally, to see that π^* is injective, let $F = E \oplus 1$, $Q = P(F)$, with other notation as before. If $\pi^*\alpha = 0$, $\alpha \neq 0$, then $j^* q^*\alpha = 0$, so

$$q^*\alpha = i_*\left(\sum_{i=0}^{e} c_1(\mathcal{O}_E(1))^i \cap p^*\alpha_i\right) = \sum_{i=0}^{e} c_1(\mathcal{O}_F(1))^{i+1} \cap q^*\alpha_i$$

using Lemma 3.3 again. But this contradicts the uniqueness assertion of (b) for the bundle $E \oplus 1$. □

Definition 3.3. Let $s = s_E$ denote the zero section of a vector bundle E; s is a morphism from X to E with $\pi \circ s = \mathrm{id}_X$. The result of Theorem 3.2 (a) allows us to define *Gysin homomorphisms*

$$s^* : A_k E \to A_{k-r} X ,$$

$r = \operatorname{rank} E$, by the formula

$$s^*(\beta) = (\pi^*)^{-1}(\beta) .$$

This Gysin homomorphism is an important intersection operation: given any subvariety of E, or k-cycle β on E, no matter how it meets the zero section, there is a well-defined cycle class $s^*(\beta)$ in $A_{k-r}X$. By the surjectivity of π^*, s^* is determined by the fact that $s^*[\pi^{-1}(V)]=[V]$ for all $V\subset X$, and the fact that s^* preserves rational equivalence.

This ability to intersect with zero sections of vector bundles will be the basis for the construction general Gysin homomorphisms and intersection products in later chapters.

Note that Theorem 3.3 was proved without using the higher Chern classes constructed in § 3.2. Only Chern classes of line bundles, and Proposition 3.1 (a) were used.

The following proposition gives a somewhat more constructive formula for s^*, using higher Chern classes.

Proposition 3.3. *Let $\beta\in A_kE$, and let $\bar{\beta}$ be any element of $A_k(P(E\oplus 1))$ which restricts to β in A_kE. Then*

$$s^*(\beta)=q_*(c_r(\xi)\cap\bar{\beta})$$

where q is the projection from $P(E\oplus 1)$ to X, and ξ is the universal rank r quotient bundle of $q^*(E\oplus 1)$.

If $\beta=\sum n_i[V_i]$, one may take $\bar{\beta}$ to be $\sum n_i[\bar{V}_i]$, $\bar{V}_i$ the closure of V_i in $P(E\oplus 1)$. Note that the composite $q^*1\subset q^*(E\oplus 1)\to\xi$ gives a section of ξ which vanishes precisely on the zero section $s(X)$.

Proof. Let $F=E\oplus 1$, $Q=P(F)$, i the inclusion of $P=P(E)$ in Q, j the inclusion of E in Q. We must show that

$$\pi^*q_*(c_r(\xi)\cap\bar{\beta})=j^*\bar{\beta}$$

for all $\bar{\beta}\in A^*Q$. By Theorem 3.3 (b) and Lemma 3.3, we may write

$$\bar{\beta}=q^*(\gamma)+i_*(\delta)$$

for some classes $\gamma\in A_*X$, $\delta\in A_*P$. Since $j^*q^*=\pi^*$ and $j^*i_*=0$, the required formula follows from the two formulas

(i) $q_*(c_r(\xi)\cap q^*\gamma)=\gamma$

(ii) $c_r(\xi)\cap i_*\delta=0$.

To prove (i), since ξ is a quotient of $q^*E\oplus 1$ by $\mathcal{O}_F(-1)$, the Whitney sum formula gives

$$c_r(\xi)=\sum_{i=0}^{r}c_1(\mathcal{O}_F(1))^i\,c_{r-i}(q^*E).$$

By Proposition 3.1 (a), (d),

$$q_*(c_r(\xi)\cap q^*\gamma)=q_*(c_1(\mathcal{O}_F(1))^r\cap q^*\gamma)=\gamma.$$

For (ii), since $i^*\xi$ has a nowhere vanishing section, i.e., a trivial subbundle, $c_r(i^*\xi)=0$ by the Whitney formula. Therefore,

$$c_r(\xi)\cap i_*\delta=i_*(c_r(i^*\xi)\cap\delta)=0. \quad\square$$

Example 3.3.1. When L is a line bundle on X, the zero section s imbeds X as a Cartier divisor on L. In this case the Gysin homomorphism of Definition

3.3 agrees with the Gysin homomorphism s^* defined in § 2.6. (It suffices to observe that the homomorphism of § 2.6 takes $[\pi^{-1}(V)]$ to $[V]$.)

Example 3.3.2. If s is the zero section of a bundle E of rank r on X, then

$$s^* s_*(\alpha) = c_r(E) \cap \alpha$$

for all $\alpha \in A_* X$. (If $\beta = s_*(\alpha)$, take $\bar{\beta} = \bar{s}_*(\alpha)$ in Proposition 3.3, where $\bar{s}: X \to Q$ is the section induced by s, i.e., $\bar{s} = j \circ s$.) This is a special case of the excess intersection formula (§ 6.3).

Example 3.3.3. If E has rank $e+1$, and η is the universal rank e quotient bundle of $p^* E$ on $P(E)$, then for all $\alpha \in A_* X$,

$$p_*(c_i(\eta) \cap p^* \alpha) = \begin{cases} \alpha & \text{if } i = e \\ 0 & \text{if } i \neq e. \end{cases}$$

(Argue as in the proof of (i) in the proof of Proposition 3.3, or as in the proof of Proposition 3.1 (a).)

Example 3.3.4. Let X be a non-singular subvariety of a non-singular n-dimensional variety Y, and let N be the normal bundle. Let $\tilde{Y}$ be the blow-up of Y along X, $\tilde{X} = P(N)$ the exceptional divisor, $\eta: \tilde{X} \to X$ the projection. Let $\tilde{X}^k$ denote the self-intersection of the divisor $\tilde{X}$ with itself k times:

$$\tilde{X}^k = \overbrace{\tilde{X} \cdot \ldots \cdot \tilde{X}}^{k} \cdot [\tilde{Y}],$$

which by the construction of § 2.4 is a well-defined class in $A_{n-k}(\tilde{X})$. Since $\mathcal{O}_{\tilde{Y}}(\tilde{X})$ restricts to $\mathcal{O}_N(-1)$ on $\tilde{X}$, we also have

$$\tilde{X}^k = (-1)^{k-1} c_1(\mathcal{O}_N(1))^{k-1} \cap [\tilde{X}].$$

It follows that

$$\sum_{k \geqq 1} (-1)^{k-1} \eta_*(\tilde{X}^k) = s(N) \cap [X]$$

in $A_* X$, where $s(N)$ is the total Segre class of N.

The left side of this formula was used by B. Segre (4) to construct invariants (or "covariants") of the imbedding of X in Y. For example, if $Y = X \times X$, and X is diagonally imbedded in Y, this class is the total inverse Chern class of T_X. Segre inverted these, much as in § 3.2, to give a new and intrinsic construction of the canonical classes of X, which had been constructed earlier by Severi, Segre, Eger and Todd, and are now known as Chern classes of X (up to sign).

From the results in Chapter 2, the classes $(-1)^{k-1} \eta_*(\tilde{X}^k)$ make sense for an arbitrary closed subscheme X of an arbitrary variety Y, $X \neq Y$. If X is regularly imbedded in Y, they are Segre classes of the normal bundle. In general they live only in $A_* X$, not as operators; they will play an important role in our construction and calculation of intersection products.

Note also that if Segre's construction is applied to the case $Y = E$ or $Y = P(E \oplus 1)$, with X imbedded by the zero-section, the Segre classes of an arbitrary bundle E result.

Example 3.3.5. Let E be a vector bundle of rank n on a scheme X, and let F be the flag manifold of complete flags in E, with projection $p: F \to X$, and universal flag of bundles $0 \subset E_1 \subset E_2 \subset \ldots \subset E_n = p^* E$ on F; F may be constructed by the splitting construction of § 3.2. Let $x_i = c_1(E_i/E_{i-1})$. Then every element of $A_* F$ can be written as a polynomial in $x_1, \ldots, x_n$, with coefficients in $A_*(X)$; this expression is unique modulo the relations

$$\sigma_i - p^* c_i(E) = 0, \quad 1 \leqq i \leqq n,$$

where σ_i is the ith elementary symmetric function of $x_1, \ldots, x_n$. (Factor p into $F \to P(E^\vee) \to X$ and induct; for details and a generalization to arbitrary flags, see Grothendieck (1) § 3. Example 3.2.17 also generalizes to flag bundles (cf. Ilori-Ingleton-Lascu (1).)

Example 3.3.6. The assertion of Example 2.6.6, that the Gysin map

$$i_t^* : A_k(X \times \mathbb{P}^1) \to A_{k-1}(X)$$

is independent of $t \in \mathbb{P}^1$, is also a consequence of Theorem 3.3 (b), and the fact that $i_t^* \mathscr{O}(1)$ is trivial.

Notes and References

Although the appearance of numerical invariants and canonical classes of varieties in algebraic geometry often anticipated their discovery in topology, the reverse is true for characteristic classes of vector bundles. Stiefel-Whitney classes in topology were constructed in the 1930's. Until the new foundations of Weil and Serre, which included flexable notions of abstract algebraic varieties, general vector bundles were not a natural object of study in algebraic geometry.

With those new foundations, Grothendieck (2) constructed Chern classes, in the rational equivalence ring, for an algebraic vector bundle on a non-singular quasi-projective variety. His method was to compute the intersection ring of the associated projective bundle as an extension of the intersection ring of the base, and to define the Chern classes by the formula of Remark 3.2.4. For this, intersection theory on such varieties had to be developed first. In Fulton (2), Chern classes were defined for vector bundles on singular quasi-projective varieties by using Grothendieck's theorem and the fact that such bundles are restrictions of bundles on suitable non-singular varieties.

The present treatment is considerably simpler, and requires no quasi-projective hypotheses. No intersection theory beyond the facts proved in Chapter 2 for divisors is needed. In fact, Grothendieck's procedure will be quite reversed, in that the results proved here about vector bundles will be used in later chapters to construct general intersection products.

The formula giving inverse Chern classes of E in terms of pushing forward powers of the first Chern class of the canonical line bundle on $P(E)$ appears in

Washnitzer (1); it has been rediscovered by almost everyone who has written about Chern classes. Although Segre (4) didn't work with bundles, he did construct higher codimension classes by a similar construction (cf. Example 3.3.4) which explains the naming of Segre classes of bundles. We shall see that Segre's fundamental construction has far wider application.

Theorems 3.2 and 3.3 appear in Grothendieck (1), (2) in the non-singular quasi-projective case. I. Vainsencher has also proved Theorem 3.3 in the singular quasi-projective case. Example 3.2.22 was suggested by J. Harris.

For the study of Chern classes in topology the reader is referred to Milnor-Stasheff (1) or Bott-Tu (1). The compatibility of the algebraic and topological constructions will be verified in Chapter 19.

Chapter 4. Cones and Segre Classes

Summary

If X is a proper subvariety of a variety Y, the Segre class $s(X, Y)$ of X in Y is the class in $A_* X$ defined as follows: let $C = C_X Y$ be the normal cone to X in Y, $P(C)$ the projectivized normal cone, p the projection from $P(C)$ to X. Then

$$s(X, Y) = \sum_{i \geqq 0} p_* (c_1(\mathcal{O}(1))^i \cap [P(C)]) .$$

When X is regularly imbedded in Y, $C = N$ is a vector bundle, and

$$s(X, Y) = c(N)^{-1} \cap [X] .$$

These Segre classes have a fundamental birational invariance: if $f: Y' \to Y$ is a birational proper morphism, and $X' = f^{-1}(X)$, then $s(X', Y')$ pushes forward to $s(X, Y)$.

The coefficient of $[X]$ in $s(X, Y)$ is the multiplicity of Y along X.

Segre classes will be used in one of our later constructions of intersection products, and in several intersection formulas.

This chapter contains the construction of Segre classes for general cones, and for general closed subschemes of a scheme. The birational invariance is a special case of a general proposition describing the behavior of Segre classes under proper push-forward and flat pull-back.

Segre classes arise naturally in many areas of algebraic geometry. Some of these occurrences are discussed in the examples and in the last two sections, which are not required for later chapters.

4.1 Segre Class of a Cone

Let C be a cone over a scheme X, i.e., $C = \operatorname{Spec}(S^{\bullet})$, where $S^{\bullet}$ is a sheaf of graded $\mathcal{O}_X$-algebras; we assume $\mathcal{O}_X \to S^0$ is surjective, S^1 is coherent, and $S^{\bullet}$ is generated by S^1. Let

$$P(C \oplus 1) = \operatorname{Proj}(S^{\bullet}[z])$$

be the projective completion of C, with projection $q: P(C \oplus 1) \to X$, and let $\mathcal{O}(1)$ be the canonical line bundle on $P(C \oplus 1)$. For more about cones, see Appendix B.5.1–5.4.

The *Segre class* of C, denoted $s(C)$, is the class in A_*X defined by the formula

$$s(C) = q_*\Big(\sum_{i \geqq 0} c_1(\mathcal{O}(1))^i \cap [P(C \oplus 1)]\Big).$$

Proposition 4.1. (a) *If E is a vector bundle on X, then*

$$s(E) = c(E)^{-1} \cap [X],$$

where $c(E) = 1 + c_1(E) + \ldots + c_r(E)$ is the total Chern class of X, $r =$ rank (E).

(b) *Let $C_1, \ldots, C_t$ be the irreducible components of C, m_i the geometric multiplicity of C_i in C. Then*

$$s(C) = \sum_{i=1}^{t} m_i s(C_i).$$

Proof. (a) Since $[P(E \oplus 1)] = q^*[X]$, the above definition of $s(E)$ agrees with the definition of $c(E \oplus 1)^{-1} \cap [X]$ in § 3.2. And $c(E \oplus 1) = c(E)$ by the Whitney sum formula.

(b) Since each C_i is open and dense in $P(C_i \oplus 1)$ (cf. Appendix B.5.3),

$$[P(C \oplus 1)] = \sum_{i=1}^{t} m_i [P(C_i \oplus 1)],$$

from which the assertion follows. Note that each $s(C_i)$ is a cycle class on the support of C_i, and the equality in (b) is understood in the sense of Convention 1.4. □

Example 4.1.1. For any cone C, $s(C \oplus 1) = s(C)$.

Example 4.1.2. Assume C is a purely n-dimensional cone, and $P(C_i)$ is not empty for each irreducible component C_i of C. Then

$$s(C) = \sum_{i \geqq 0} p_*(c_1(\mathcal{O}(1))^i \cap [P(C)]),$$

where $p: P(C) \to X$ is the projection. (Use Lemma 1.7.2 with Proposition 4.1 (b); by Appendix B.5.2, $c_1(\mathcal{O}(1)) \cap [P(C \oplus 1)] = [P(C)]$.)

Example 4.1.3. Although one looks for components of a normal cone $C_X Y$ over "special" subvarieties of X (cf. Example 4.2.2), such components need not be present. For example, if $Y = \mathbb{A}^{2n}$, and X is the union of two n-planes meeting transversally at the origin, then $C_X Y$ has two (reduced) irreducible components, and the fibre of $P(C_X Y)$ over the origin is isomorphic to $\mathbb{P}^{n-1} \times \mathbb{P}^{n-1}$.

Example 4.1.4. Let $Y = \mathbb{A}^{mn}$, the space of linear maps from $\mathbb{A}^m$ to $\mathbb{A}^n$, with coordinates (x_{ij}), $1 \leqq i \leqq n$, $1 \leqq j \leqq m$, with $m \geqq n$. Let $X \subset Y$ be the subspace of maps of rank $< n$ (cf. Lemma A.7.2), and let $C = C_X Y$. For each $(i) = (i_1, \ldots, i_n)$, $1 \leqq i_1 < \ldots < i_n \leqq m$, let $t_{(i)}$ be a variable, and let $\delta_{(i)}$ be the corresponding minor of (x_{ij}). Sending $t_{(i)}$ to $\delta_{(i)}$ imbeds $P(C)$ in $X \times \mathbb{P}^N$, $N = \binom{m}{n} - 1$. The graded ring of C is reduced. By the Plücker relations, $P(C)$ is contained in $X \times G$, where G is the Grassmannian of n-planes in $\mathbb{A}^m$. In fact,

$$P(C) = \{(\varphi, L) \mid \text{Image}(\varphi) \subset L\}.$$

Example 4.1.5. If C and D are cones on X defined by sheaves of algebras $S^{\bullet}$ and $T^{\bullet}$, the cone $C \times_X D$ is defined by $S^{\bullet} \otimes_{\mathcal{O}_X} T^{\bullet}$. If C is a cone and E is a vector bundle, we write $C \oplus E$ in place of $C \times_X E$. In this case

$$s(C \oplus E) = c(E)^{-1} \cap s(C).$$

(Reduce to the case where C is irreducible and $P(C) \neq \emptyset$. Let p be the projection from $P(C \oplus E)$ to X. There is a regular section of $p^*E \otimes \mathcal{O}(1)$ on $P(C \oplus E)$, whose zero-scheme may be identified with $P(C)$. By Example 3.2.16, if $r = \operatorname{rank}(E)$,

$$\begin{aligned}[P(C)] &= c_r(p^*E \otimes \mathcal{O}(1)) \cap [P(C \oplus E)] \\ &= \sum_i c_{r-i}(p^*E)\, c_1(\mathcal{O}(1))^i \cap [P(C \oplus E)].\end{aligned}$$

Therefore

$$\begin{aligned}s(C) &= \sum_{i,j} p_*(c_{r-i}(p^*E)\, c_1(\mathcal{O}(1))^{i+j} \cap [P(C \oplus E)]) \\ &= c(E) \cap s(C \oplus E).)\end{aligned}$$

Example 4.1.6. *Exact sequences of cones.* Consider cones $C = \operatorname{Spec}(S^{\bullet})$, $C' = \operatorname{Spec}(S'^{\bullet})$ on a scheme X, with $S^0 = S^{0\prime} = \mathcal{O}_X$. A *morphism* $\varphi: C \to C'$ is given by a homomorphism $\tilde{\varphi}: S'^{\bullet} \to S^{\bullet}$ of sheaves of graded $\mathcal{O}_X$-algebras. Let E be a vector bundle on X, $E = \operatorname{Spec}(\operatorname{Sym} \mathcal{E})$, $\mathcal{E}$ locally free. Given morphisms $\varphi: C \to C'$, $\psi: E \to C$, we call the sequence

$$(*) \qquad 0 \to E \overset{\psi}{\to} C \overset{\varphi}{\to} C' \to 0$$

exact if $\tilde{\varphi}: S'^{\bullet} \to S^{\bullet}$ is injective, $\tilde{\psi}: S^{\bullet} \to \operatorname{Sym} \mathcal{E}$ is surjective, and if, locally on X, there is a locally free subsheaf $\tilde{\mathcal{E}}$ of S^1 that maps onto $\mathcal{E}$ by $\tilde{\psi}$, such that the induced map

$$S'^{\bullet} \oplus_{\mathcal{O}_X} \operatorname{Sym}(\tilde{\mathcal{E}}) \to S^{\bullet}$$

is an isomorphism.

(a) Basic properties of this concept are: (i) The last condition in the definition is independent of the choice of $\tilde{\mathcal{E}}$. (ii) Exactness pulls back under flat base extensions $X' \to X$. (iii) If φ and ψ are given, and $(*)$ becomes exact under a faithfully flat base extension, then $(*)$ is exact. In particular, it suffices to verify exactness after base extension to $\operatorname{Spec}(\hat{\mathcal{O}}_{x,X})$ for all $x \in X$.

(b) If $\mathscr{C}$ is a cone over $X \times \mathbb{A}^1$ which is flat over $\mathbb{A}^1$. and C_t is the restriction of $\mathscr{C}$ to $X \times \{t\}$, $t = 0, 1$, then

$$s(C_0) = s(C_1) \in A_*(X).$$

(c) For an exact sequence $(*)$ of cones,

$$s(C') = c(E) \cap s(C) \in A_*(X).$$

(The proof of (a) is straightforward. (b) follows from the fact that $i_t^* s(\mathscr{C}) = s(C_t)$ for $t = 0, 1$, i_t the inclusion of X in $X \times \mathbb{A}^1$ at t; for a more general statement see Example 10.1.10. For (c), it suffices by (b) and Example 4.1.5 to show that there is a cone $\mathscr{C}$ on $X \times \mathbb{A}^1$, flat over $\mathbb{A}^1$, with $C_1 = C$ and $C_0 = C' \oplus E$. Let $p: X \times \mathbb{A}^1 \to X$ be the projection. Define $\mathscr{C}$ to be the closed subscheme of $p^*(C \times_X C')$ locally defined by the homogeneous ideal in $p^*(S^{\cdot} \otimes_{\mathcal{O}_X} S^{\cdot\prime})$ generated by

$$\{\tilde{\varphi}(s') \otimes 1 - T(1 \otimes s') \mid s' \in S^{1\prime}\}.$$

Using the local splitting of (*) one verifies that $\mathscr{C}$ is flat over $\mathbb{A}^1$.)

Example 4.1.7 (cf. Fulton-Johnson (1)). A coherent sheaf $\mathscr{F}$ on a scheme X determines a cone

$$C(\mathscr{F}) = \operatorname{Spec}(\operatorname{Sym}(\mathscr{F})).$$

One may define the Segre class of $\mathscr{F}$, $s(\mathscr{F})$, to be the Segre class of its cone $C(\mathscr{F})$. If

$$0 \to \mathscr{F}' \to \mathscr{F} \to \mathscr{E} \to 0$$

is an exact sequence of sheaves, with $\mathscr{E}$ locally free, then

$$s(\mathscr{F}') = c(E) \cap s(\mathscr{F})$$

where $E = \operatorname{Spec}(\operatorname{Sym}(\mathscr{E}))$ is the vector bundle whose sheaf of sections is $\mathscr{E}^{\vee}$. (The corresponding sequence of cones is exact in the sense of Example 4.1.6.)

Example 4.1.8. Let C be a purely k-dimensional closed subcone of a vector bundle E of rank r on X. Let s_E be the zero section of E, s_E^* the corresponding Gysin homomorphism (§ 3.3). Then

$$s_E^*([C]) = \{c(E) \cap s(C)\}_{k-r} = \sum_{i=0}^{r} c_i(E) \cap s_{k-r+i}(C)$$

in $A_{k-r}(X)$. (Use Proposition 3.3, with $\bar{\beta} = [P(C \oplus 1)]$.) See Proposition 6.1 for generalizations.

When X is non-singular and $E = T_X^{\vee}$ is the cotangent bundle, such cones, with $k = r$, appear as characteristic varieties of holonomic $\mathscr{D}$-modules, and the intersection with the zero section $s_E^*([C])$ is an important invariant (cf. Brylinski-Kashiwara (1)).

4.2 Segre Class of a Subscheme

Let X be a closed subscheme of a scheme Y. Let $C = C_X Y$ be the normal cone to X in Y:

$$C = \operatorname{Spec}\left(\sum_{n=0}^{\infty} \mathscr{I}^n/\mathscr{I}^{n+1}\right)$$

where $\mathscr{I}$ is the ideal sheaf defining X in Y. The *Segre class* of X in Y, denoted $s(X, Y)$, is defined to be the Segre class of the normal cone C:

$$s(X, Y) = s(C_X Y) \in A_* X.$$

In case X is regularly imbedded in Y, so the normal cone is a vector bundle, it follows from Proposition 4.1 that $s(X, Y)$ is the cap product of the total inverse Chern class of the normal bundle with $[X]$.

Lemma 4.2. *Let Y be a pure-dimensional scheme, $Y_1, \ldots, Y_r$ the irreducible components of Y, m_i the geometric multiplicity of Y_i in Y. If X is a closed subscheme of Y, and $X_i = X \cap Y_i$, then*

$$s(X, Y) = \sum m_i s(X_i, Y_i)$$

in $A_*(X)$.

Proof. Let $M_X Y$ denote the blow-up of $Y \times \mathbb{A}^1$ along $X \times \{0\}$. Since X is nowhere dense in $Y \times \mathbb{A}^1$, the varieties $M_{X_i} Y_i$ are the irreducible components of $M_X Y$, with multiplicities m_i:

$$[M_X Y] = \sum m_i [M_{X_i} Y_i].$$

The exceptional divisor in $M_X Y$ restricts to the exceptional divisor in $M_{X_i} Y_i$, so, by Lemma 1.7.2,

$$[P(C_X Y \oplus 1)] = \sum m_i [P(C_{X_i} Y_i \oplus 1)].$$

Capping by $\sum c_1(\mathcal{O}(1))^i$ and pushing forward to X gives the asserted equality. □

Proposition 4.2. *Let $f: Y' \to Y$ be a morphism of pure-dimensional schemes, $X \subset Y$ a closed subscheme, $X' = f^{-1}(X)$ the inverse image scheme, $g: X' \to X$ the induced morphism.*

(a) *If f is proper, Y irreducible, and f maps each irreducible component of Y' onto Y, then*

$$g_*(s(X', Y')) = \deg(Y'/Y)\, s(X, Y).$$

(b) *If f is flat, then*

$$g^*(s(X, Y)) = s(X', Y').$$

Proof. In (a), $\deg(Y'/Y)$ is defined to be $\sum_{i=1}^{r} m_i \deg(Y'_i/Y)$, where $Y'_1, \ldots, Y'_r$ are the irreducible components of Y', m_i the geometric multiplicity of Y'_i. By Lemma 4.2, we may assume Y' is also irreducible.

Let M be the blow-up of $Y \times \mathbb{A}^1$ along the subscheme $X \times \{0\}$. The exceptional divisor is $P(C \oplus 1)$. Similarly let M' be the blow-up of $Y' \times \mathbb{A}^1$ along $X' \times \{0\}$. There is an induced morphism F from M' to M, with $F^* P(C \oplus 1) = P(C' \oplus 1)$ as Cartier divisors. Let G be the induced morphism from $P(C' \oplus 1)$ to $P(C \oplus 1)$, and let q (resp. q') be the projection from $P(C \oplus 1)$ (resp. $P(C' \oplus 1)$) to X (resp. X'). If $\mathcal{O}(1)$ is the canonical line bundle on $P(C \oplus 1)$, $G^* \mathcal{O}(1)$ is the canonical line bundle on $P(C' \oplus 1)$.

In the situation of (a), $F_*[M'] = d[M]$, $d = \deg(Y'/Y)$, so

$$G_*[P(C' \oplus 1)] = d[P(C \oplus 1)]$$

by Proposition 2.3 (c). Therefore

$$\begin{aligned} g_* s(X', Y') &= g_* q'_* \left(\sum_i c_1 (G^* \mathcal{O}(1))^i \cap [P(C' \oplus 1)] \right) \\ &= q_* G_* \left(\sum_i c_1 (G^* \mathcal{O}(1))^i \cap [P(C' \oplus 1)] \right) \\ &= q_* \left(\sum_i c_1 (\mathcal{O}(1))^i \cap d[P(C \oplus 1)] \right) \\ &= d s(X, Y), \end{aligned}$$

making use of the projection formula (Proposition 2.5 (c)).

Similarly for (b),

$$\begin{aligned} g^* s(X, Y) &= g^* q_* \left(\sum_i c_1 (\mathcal{O}(1))^i \cap [P(C \oplus 1)] \right) \\ &= q'_* G^* \left(\sum_i c_i (\mathcal{O}(1))^i \cap [P(C \oplus 1)] \right) \\ &= q'_* \left(\sum_i c_i (G^* \mathcal{O}(1))^i \cap G^* [P(C \oplus 1)] \right) \\ &= s(X', Y'). \quad \square \end{aligned}$$

Corollary 4.2.1. *With the assumptions of Proposition* 4.2 (a), *if X' is regularly imbedded in Y', with normal bundle N', then*

$$g_*(c(N')^{-1} \cap [X']) = \deg(Y'/Y)\, s(X, Y).$$

If $X \subset Y$ is also regularly imbedded, with normal bundle N, then

$$g_*(c(N')^{-1} \cap [X']) = \deg(Y'/Y)(c(N)^{-1} \cap [X]).$$

Proof. Apply Proposition 4.1 (a). $\square$

Corollary 4.2.2. *Let X be a proper closed subscheme of a variety Y. Let $\tilde{Y}$ be the blow-up of Y along X, $\tilde{X} = P(C)$ the exceptional divisor, $\eta : \tilde{X} \to X$ the projection. Then*

$$\begin{aligned} s(X, Y) &= \sum_{k \geqq 1} (-1)^{k-1} \eta_*(\tilde{X}^k) \\ &= \sum_{i \geqq 0} \eta_*(c_1(\mathcal{O}(1))^i \cap [P(C)]). \end{aligned}$$

Proof. The self-intersection class $\tilde{X}^k$ is meant in the sense of Definition 2.4.2. The first formula follows from the fact that the normal bundle to $\tilde{X}$ in $\tilde{Y}$ is the restriction of $\mathcal{O}_{\tilde{Y}}(\tilde{X})$ to $\tilde{X}$, and the fact that multiplication by the first Chern class of the normal bundle is the same as intersecting with the divisor $\tilde{X}$ (Proposition 2.6 (c)). The second follows from the fact that, when $\tilde{X}$ is identified with $P(C)$, the normal bundle is dual to $\mathcal{O}(1)$; or one may apply Example 4.1.2. $\square$

Remarks. When f is birational, i.e. $\deg(Y'/Y) = 1$, Proposition 4.2(a) asserts the *birational invariance* of Segre classes: the Segre class of X' in Y' pushes forward to the Segre class of X in Y. In case $X \subset Y$ and $X' \subset Y'$ are regular imbeddings – for example, if all four schemes are non-singular – Corollary 4.2.1 gives a remarkable relation among the Chern classes of the normal bundles.

Consider the situation of Corollary 4.2.1 when X is a non-singular point on Y. A formula for the degree of f results:

$$\deg(Y'/Y) = \int_{X'} c(N')^{-1} \cap [X'] .$$

This can be used in a situation where the general fibre of d is difficult to describe, but a particular degenerate fibre is known very well. A procedure like this was carried out by R. Donagi and R. Smith (1) to calculate the degree of a Prym map.

The first formula of Corollary 4.2.1 is useful also when $X = P$ is a singular point on Y. We will see in the next section that $s(P, Y)$ is the multiplicity $e_P Y$ of Y at P. For example, if f is birational, one has

$$e_P(Y) = \int_{X'} c(N')^{-1} \cap [X'] .$$

This is essentially the procedure used by G. Kempf (2) to calculate the multiplicity of varieties of special divisors (cf. Example 4.3.2).

For $X \subset Y$, Y non-singular, the Segre class $s(X, Y)$, after correction by $c(T_Y)$, is independent of the imbedding (Example 4.2.6).

Example 4.2.1. Corollary 4.2.2 is valid for any purely n-dimensional scheme Y and any closed subscheme X of dimension less than n. This may also be deduced from the fact that $P(C)$ is a Cartier divisor on $P(C \oplus 1)$,

$$c_1(\mathscr{O}(1)) \cap [P(C \oplus 1)] = [P(C)] ,$$

and

$$q_*[P(C \oplus 1)] = 0 .$$

Example 4.2.2. Let A, B and D be effective Cartier divisors on a surface Y. Let $A' = A + D$, $B' = B + D$, and let X be the scheme-theoretic intersection of A' and B' on Y. Assume that A and B meet only at one point P, which is non-singular on Y, and that A and B meet transversally at P. Then

$$s(X, Y) = [D] + ([P] - D \cdot [D]) .$$

(To see this, let $f\colon \tilde{Y} \to Y$ be the blow-up of Y at P, E the exceptional divisor. Then $\tilde{X} = f^{-1}(X) = f^*D + E$. Therefore

$$\begin{aligned} s(X, Y) &= g_*[\tilde{X}] - g_*(\tilde{X} \cdot [\tilde{X}]) = [D] - g_*(f^*D \cdot [f^*D] + 2f^*D \cdot [E] + E \cdot [E]) \\ &= [D] - D \cdot [D] + [P] .) \end{aligned}$$

One may also calculate the normal cone to X in Y; it contains a component which lies over P.

Similarly if A and B have the same multiplicity m at P, and no common tangents at P, then

$$s(X, Y) = [D] + (m^2[P] - D \cdot [D]) .$$

For arbitrary intersections of A and B, the answer depends also on how A and B meet D (cf. Example 6.1.4).

Example 4.2.3. Under the hypotheses of Proposition 4.2(a), if X is regularly imbedded in Y with normal bundle N, then

$$g_*(c(g^*N) \cap s(X', Y')) = \deg(Y'/Y) \cdot [X].$$

Example 4.2.4. Lemma 4.2 may fail if Y is not pure-dimensional (cf. Example 1.7.1).

Example 4.2.5. Let X_i be a closed subscheme of a scheme Y_i for $i = 1, \ldots, r$. Then

$$s(X_1 \times \ldots \times X_r, Y_1 \times \ldots \times Y_r) = s(X_1, Y_1) \times \ldots \times s(X_r, Y_r).$$

(Reduce to the case where X_i is a Cartier divisor in Y_i, and use Example 3.2.8.)

Example 4.2.6. *Canonical classes of singular varieties.*

(a) Let X be a scheme which can be imbedded as a closed subscheme of a non-singular variety M. Then the class

$$c_*(X) = c(T_M|_X) \cap s(X, M)$$

in $A_*(X)$ is independent of the choice of imbedding. If X is a local complete intersection, then

$$c_*(X) = c(T_M|_X)\, c(N_X M)^{-1} \cap [X] = c(T_X) \cap [X].$$

Here $T_X = T_M|_X - N_X M$ is the virtual tangent bundle to X, a well-defined element of the Grothendieck group of vector bundles on X (cf. § 10.1 and Appendix B.7.6). (Two imbeddings are dominated by the diagonal, so it suffices to compare $i: X \to M$ with $j: X \to P$, M, P non-singular, when there is a smooth $\varrho: P \to M$ with $\varrho j = i$. By Example 4.1.6, it suffices to show that there is an exact sequence of cones

$$(*) \qquad 0 \to j^* T_\varrho \to C_j \to C_i \to 0$$

with C_i, C_j the normal cones to $X \subset M$, $X \subset P$, and T_ϱ the relative tangent bundle to ϱ. Let $Y = X \times_M P$, and consider the diagram

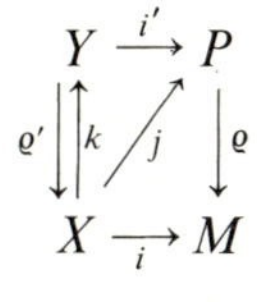

$$\begin{array}{ccc} Y & \xrightarrow{i'} & P \\ {\scriptstyle \varrho'}\downarrow\uparrow{\scriptstyle k} & \nearrow{\scriptstyle j} & \downarrow{\scriptstyle \varrho} \\ X & \xrightarrow[i]{} & M \end{array}$$

with i', ϱ' induced by i, ϱ, and $k = (1, j)$. Then k is a regular imbedding with $C_k = k^* T_{\varrho'} = j^* T_\varrho$. Since ϱ is flat, $C_{i'} = \varrho'^* C_i$, so $k^* C_{i'} = C_i$. The factoring of j into $i' \circ k$ determines morphisms

$$j^* T_\varrho = C_k \to C_j, \qquad C_j \to k^* C_{i'} = C_i.$$

To see that the resulting sequence (*) is exact, use Example 4.1.6(a) to reduce to the case where $M = \operatorname{Spec}(A)$, A a complete local ring and $P = \operatorname{Spec}(B)$, B a formal power series ring over A; in this case the local splitting can be verified directly.)

(b) If X is a plane curve of degree d, then

$$\deg(c_0(X)) = \int_X c_*(X) = 3d - d^2 = 2\chi(X, \mathcal{O}_X).$$

We know no simple general formula for $\deg c_0(X)$ for a general curve. The examples $X = V(x^2, xy, y^2)$ or $X = V(x^2, xy, z^2)$ in $\mathbb{P}^3$ show that $\deg(c_0(X))$ is not always $2\chi(X, \mathcal{O}_X)$.

(c) (cf. Fulton-Johnson (1)) If $X \subset M$ as in (a), and $\mathcal{N}_X M = \mathcal{I}/\mathcal{I}^2$ is the conormal sheaf to X in M, the class

$$c(T_M|_X) \cap s(\mathcal{N}_X M)$$

in $A_* X$ is also independent of the imbedding, where $s(\mathcal{N}_X M)$ is the Segre class of the sheaf $\mathcal{N}_X M$. (This follows similarly from Example 4.1.7.) This agrees with the canonical class in (a) for local complete intersections, but not in general.

Example 4.2.7. Let X be a closed subscheme of Y, E a vector bundle on Y, with Y imbedded in E by the zero section. Then

$$s(X, Y) = c(E|_X) \cap s(X, E).$$

(Reduce to the case where Y is a variety and X is a Cartier divisor on Y.)

Example 4.2.8. (a) If $X \overset{i}{\hookrightarrow} Y \overset{j}{\hookrightarrow} Z$, and j is a regular imbedding with normal bundle N, $s(X, Y)$ is not always equal to $c(i^*N) \cap s(X, Z)$. For example, let $Z = \mathbb{P}^2$, Y a curve, and X a singular point on Y.

(b) If $X \overset{i}{\hookrightarrow} Y \overset{j}{\hookrightarrow} Z$, and i imbeds X as a Cartier divisor on Y, with normal bundle N, $i^*s(Y, Z)$ is not always equal to $c(N) \cap s(X, Z)$. For example, let Z be the cone $x^2+y^2 = z^2$ in $\mathbb{A}^3$, Y the line $x = z - y = 0$, and X the point $(0, 0, 0)$.

Example 4.2.9. (a) *Local Euler obstruction.* Let X be an n-dimensional variety. Let $\nu: \tilde{X} \to X$ be a proper birational morphism such that there is a surjection of sheaves

$$\nu^* \Omega_X^1 \twoheadrightarrow \tilde{\Omega}$$

with $\tilde{\Omega}$ locally free of rank n on $\tilde{X}$. For example, $\tilde{X}$ could be the Nash blow-up of X. Let $\tilde{T} = \tilde{\Omega}^\vee$. For any point P in X, define the local Euler obstruction $\mathrm{Eu}_P X$ by the formula

$$\mathrm{Eu}_P X = \int_{\nu^{-1}(P)} c(\tilde{T}|_{\nu^{-1}(P)}) \cap s(\nu^{-1}(P), \tilde{X}).$$

This integer is independent of the choice of $\tilde{X}$. (Since two such $\tilde{X}$ are dominated by a third, this follows from the birational invariance of Segre classes.) This definition was made by Gonzalez-Springberg (1) and Verdier, who proved that it agrees with the original transcendental definition of MacPherson (1).

(b) *Mather Chern class.* With $\nu : \tilde{X} \to X$ as in (a), the Mather Chern class

$$c_M(X) = \nu_*(c(\tilde{T}) \cap [\tilde{X}])$$

in A_*X is independent of the choice of $\tilde{X}$. (Use the projection formula.)

4.3 Multiplicity Along a Subvariety

For an irreducible subvariety X of a variety Y, the coefficient of $[X]$ in the class $s(X, Y)$ is called the *multiplicity of Y along X*, or the *algebraic multiplicity of X on Y*, and is denoted $e_X Y$. If $\operatorname{codim}(X, Y) = n > 0$, then

$$\begin{aligned} e_X Y[X] &= q_*(c_1(\mathcal{O}(1))^n \cap [P(C \oplus 1)]) \\ &= p_*(c_1(\mathcal{O}(1))^{n-1} \cap [P(C)]) \\ &= (-1)^{n-1} p_*(\tilde{X}^n) . \end{aligned}$$

Here $C = C_X Y$, p and q are the projections from $P(C)$ and $P(C \oplus 1)$ to X, $\tilde{Y}$ is the blow-up of Y along X, with exceptional divisor $\tilde{X} = P(C)$. This definition is equivalent to the definition of the multiplicity of the local ring $\mathcal{O}_{X,Y}$ given by Samuel (1) (cf. Example 4.3.1).

If $X = P$ is a point, $C = C_P Y$ is the projective tangent cone to P in Y, and

$$e_P Y = \int_{P(C)} c_1(\mathcal{O}(1))^{n-1} \cap [P(C)] = \deg[P(C)] .$$

In this case $e_P Y$ is called the *multiplicity of Y at P*.

Example 4.3.1. Let A be the local ring of Y along X, $\mathcal{M}$ the maximal ideal of A, $A/\mathcal{M} = k$. Then

$$\dim_k \left(\sum_{i=1}^{t} \mathcal{M}^{i-1}/\mathcal{M}^i \right) = l_A(A/\mathcal{M}^t)$$

is a polynomial of degree $n = \operatorname{codim}(X, Y)$ in t for $t \gg 0$, whose leading term is $et^n/n!$, where $e = e_X Y$ (cf. Example 2.5.2).

Example 4.3.2. In this example C is a projective non-singular curve of genus g; $C^{(d)}$ is the dth symmetric product of C, parametrizing effective divisors of degree d on C; P_0 is a fixed point of C; J is the Jacobian of C; u_d is the morphism from $C^{(d)}$ to J which takes a divisor D to the divisor class of $D - dP_0$. The following facts are assumed: (i) the (scheme-theoretic) fibres of u_d are the linear systems $|D| \cong \mathbb{P}^r$; (ii) if $d > 2g - 2$, u_d makes $C^{(d)}$ a projective bundle over J; (iii) if $1 \leqq d \leqq g$, u_d maps $C^{(d)}$ birationally onto its image W_d.

(a) If $\deg D = d$, $\dim |D| = r$, then

$$s(|D|, C^{(d)}) = (1 + h)^{g-d+r} \cap [|D|]$$

where h is the first Chern class of the canonical line bundle on $|D| = \mathbb{P}^r$. (This follows from (ii) if d is large. For smaller d consider $C^{(d)} \subset C^{(d+s)}$ by

$E \to E + sP_0$; the normal bundle to this imbedding restricts to a bundle on $|D|$ whose Chern class is $(1+h)^s$.)

(b) From (a) and Proposition 4.2(a) follows the *Riemann-Kempf formula* (cf. Kempf (2): the multiplicity of W_d at $u_d(D)$ is

$$\binom{g-d+r}{r},$$

where $r = \dim |D|$.

R. Smith has shown how a similar procedure can be used to prove an assertion of Mumford (cf. Beauville (3)) that a theta divisor in the intermediate Jacobian of a non-singular cubic three-fold in $\mathbb{P}^4$ has a singular point of multiplicity 3. If F is the Fano variety of lines in the cubic three-fold, there is a morphism of degree 6 from $F \times F$ onto the theta divisor, so that the inverse image of the singular point is the diagonal in $F \times F$; calculating (Clemens-Griffiths (1), p. 326) that

$$\int_F s_2(T_F) = \int_F c_1(T_F)^2 - c_2(T_F) = 45 - 27 = 18,$$

it follows that the multiplicity is $18/6 = 3$.

Example 4.3.3. (a) (Schwarzenberger (1)) With the notation of the previous example, for $d > 2g-2$ there is a vector bundle E_d on J with $P(E_d) = C^{(d)}$, and whose canonical line bundle $\mathcal{O}_{E_d}(1)$ is the line bundle of the divisor $C^{(d-1)}$. This gives a geometric realization of the Segre classes of E_d:

$$s_i(E_d) \cap [J] = (u_d)_*[C^{(g-i)}] = [W_{g-i}].$$

Let δ be the involution of J which takes the class of a divisor D to the class of $K-(2g-2)P_0 - D$, with K a canonical divisor on C. There is an exact sequence

$$0 \to E_n \to M \to \delta^* E_n^\vee \to 0$$

for n large, with M a successive extension of trivial line bundles. Therefore

$$s_i(E_n) = c_i(\delta^* E_n^\vee) = (-1)^i c_i(\delta^* E_n),$$

so

$$c_i(E_n) \cap [J] = (-1)^i \delta_*(s_i(E_n) \cap [J]) = (-1)^i [\delta(W_{g-i})].$$

If w_i represents $[W_{g-i}]$, and u_i represents $[\delta(W_{g-i})]$, this explains Mattuck's formula

$$\left(\sum w_i\right)\left(\sum (-1)^i u_i\right) = 1.$$

(b) (cf. Mattuck (3)) The Chern class of the dth symmetric product $C^{(d)}$ of a non-singular projective curve C of genus g is given by the formula:

$$c_t(T_{C^{(d)}}) = (1+xt)^{d-g+1}/u_d^* \, w(t/(1+xt)).$$

Here x is the class of $C^{(d-1)}$, imbedded in $C^{(d)}$ by $D \to D + P_0$, and $w(\tau) = \sum w_i \tau^i$. (If $d > 2g-2$, $C^{(d)} = P(E_d)$, with $c_t(E_d) = w(t)^{-1}$, and Example 3.2.11 applies. If the formula is known for $C^{(d)}$, imbed $C^{(d-1)}$ in $C^{(d)}$ as above, with $1 + xt$ the Chern polynomial of the normal bundle, to deduce it for $C^{(d-1)}$.)

Example 4.3.4. For any closed subscheme X of a pure-dimensional scheme Y, and any irreducible component V of X, the *multiplicity of* Y *along* X *at* V, denoted $(e_X Y)_V$, may be defined as the coefficient of $[V]$ in the class $s(X, Y)$. If $V = X$, we write simply $e_X Y$. This multiplicity is the same as Samuel's multiplicity $e(q)$ of the primary ideal q determined by X in the local ring $A = \mathcal{O}_{V,Y}$, i.e., if $n = \dim(A) = \operatorname{codim}(V, Y)$, then

$$l(A/q^t) = e(q) \cdot t^n/n! + \text{lower terms}$$

for $t \gg 0$. In general, if $Y_1, \ldots, Y_r$ are the irreducible components of Y which contain V, with geometric multiplicities $m_1, \ldots, m_r$, then, by Lemma 4.2,

$$(e_X Y)_V = \sum_{i=1}^{r} m_i (e_{X \cap Y_i} Y_i)_V .$$

Example 4.3.5. Continue the notations of the preceding example.

(a) If the residue field $R(Z)$ of A is infinite, Samuel (1) has shown that there are $a_1, \ldots, a_n \in q$, generating an ideal $q' = (a_1, \ldots, a_n)$ with $\dim(A/q') = 0$ and

$$e(q') = e(q) .$$

If the ground field K is infinite, such a_i may be found among K-linear combinations of a given set of generators for q. (See Zariski-Samuel (1), Vol. II, p. 294 for a proof.)

(b) If $a_1, \ldots, a_n \in A$ generate an ideal q' with $\dim(A/q') = 0$, then

$$e(q') = e_A(a_1, \ldots, a_n)$$

where the right side is the alternating sum of the lengths of the Koszul complex defined by $a_1, \ldots, a_n$ (Appendix A.5). For the proof in this geometric context, see Example 7.1.2. For an algebraic proof, see Serre (4), p. IV-12.

(c) If A is a Cohen-Macaulay local ring, then

$$e(q) \geqq l(A/q) .$$

If K is infinite, equality holds if and only if q can be generated by a regular sequence of elements. (For $q' = (a_1, \ldots, a_n) \subset q$ as in (a), $a_1, \ldots, a_n$ is a regular sequence by Lemma A.7.1, so

$$e_A(a_1, \ldots, a_n) = l(A/q') \geqq l(A/q)$$

with equality when $q' = q$. If K is not infinite, make a base extension $A \otimes_K K'$ with K' infinite, e.g. $K' = K(T)$.)

(d) In this geometric context, $e(q) = (e_X Y)_V = 1$ if and only if A is regular and q is the maximal ideal of A. Assuming (a), this will follow from Proposition 7.2; for algebraic proofs see Samuel (1) or Nagata (2). This criterion is not valid for arbitrary Noetherian local rings A, even if A is a domain and q is the maximal ideal (Nagata (2) Appendix A.1).

Example 4.3.6. Let $f: Y' \to Y$ be a proper surjective homomorphism of irreducible varieties, X a closed subscheme of Y, $X' = f^{-1}(X)$. Let V be an irreducible component of X, and assume that each irreducible component V'

of $f^{-1}(V)$ has the same dimension as V. Then, by Proposition 4.2(a),

$$\deg(Y'/Y)(e_X Y)_V = \sum_{V'} \deg(V'/V)(e_{X'} Y')_{V'}$$

where the sum is over the irreducible components V' of $f^{-1}(V)$.

Example 4.3.7. Let $f\colon Y' \to Y$ be a proper surjective homomorphism of irreducible varieties. Let V' be a subvariety of Y', $V = f(V')$. If V' is an irreducible component of $f^{-1}(V)$, define the *ramification index* of f at V', $e_{V'}(f)$, to be the mulitplicity of Y' along $f^{-1}(V)$ at V' (cf. Example 4.3.4). If all irreducible components V' of $f^{-1}(V)$ have the same dimension as V, then

$$\deg(Y'/Y)\, e_V Y = \sum_{V'} \deg(V'/V)\, e_{V'}(f)\,.$$

For example, if Y is smooth over an algebraically closed field, and Q is a point of Y such that $f^{-1}(Q)$ is finite, then

$$\sum_{f(P)=Q} e_P(f) = \deg(Y'/Y)\,.$$

In particular, the sum of the ramification indices is independent of $Q \in Y$.

For a notion of ramification index related to the separable degree of f, see Gaffney-Lazarsfeld (1). For other geometric interpretations of multiplicity, see Example 12.4.5 and Mumford (5).

Example 4.3.8. Assume X_1 and X_2 are two subschemes of Y, each containing V as an irreducible component. Suppose there is a proper birational morphism $f\colon Y' \to Y$ so that $f^{-1}(X_1) = f^{-1}(X_2)$. Then $(e_{X_1} Y)_V = (e_{X_2} Y)_V$.

Example 4.3.9. Let X be a variety of dimension at least 2, P a simple point of X, $\pi\colon \tilde{X} \to X$ the blow-up of X at P, E the exceptional divisor, $\eta\colon E \to P$ the projection. Let D be an effective Cartier divisor on X, and let $\tilde{D}$ be the blow-up of D at P (the strict transform of D in $\tilde{X}$). If m is the least power of the maximal ideal of $\mathcal{O}_{P,X}$ which contains a local equation for D, then

$$\pi^* D = \tilde{D} + mE\,.$$

Therefore

$$\begin{aligned} 0 = D \cdot \pi_*(E^{n-1}) &= \eta_*(\tilde{D} \cdot E^{n-1}) + \eta_*(mE^n) \\ &= (-1)^{n-2} e_P(D) + (-1)^{n-1} m\, e_P(X) \end{aligned}$$

in $A_0 P = \mathbb{Z}$. This verifies the formula $m = e_P D$.

Example 4.3.10. If X_i is a subvariety of Y_i for $i = 1, \ldots, r$, then the multiplicity of $X_1 \times \ldots \times X_r$ in $Y_1 \times \ldots \times Y_r$ is the product of the multiplicities of X_i in Y_i. (Use Example 4.2.5.)

4.4 Linear Systems

If a subscheme is the base locus of a linear system, its Segre class is related to important invariants of the system.

Let L be a line bundle on an n-dimensional variety X, and let $V \subset H^0(X, L)$ be an $(r+1)$-dimensional space of sections of L. For $s \in V$, $s \neq 0$, let D_s be the zero-scheme of s, a Cartier divisor on X. Let

$$B = \bigcap_{s \in V - \{0\}} D_s$$

be the *base* of the linear system. Let $\pi : \tilde{X} \to X$ be the blow-up of X along B. There is a morphism

$$f : \tilde{X} \to P(V^\vee) = \mathbb{P}^r$$

extending the morphism $X - B \to P(V^\vee)$ which takes x to the hyperplane of divisors containing x. Indeed, if $\mathscr{I}$ is the ideal sheaf of B in X, there is a canonical surjection from the trivial bundle V_X to $\mathscr{I} \otimes L$. Therefore $\mathrm{Sym}(V_X \otimes L^{-1})$ surjects onto $\otimes \mathscr{I}^n$, which gives an imbedding of $\tilde{X}$ in

$$\mathrm{Proj}(\mathrm{Sym}(V_X \otimes L^{-1})) = \mathrm{Proj}(\mathrm{Sym}(V_X)) = X \times P(V^\vee).$$

The morphism f is the projection to $P(V^\vee)$. From this description it follows that

$$f^*\mathscr{O}(1) = \pi^*(L) \otimes \mathscr{O}(-E),$$

where E is the exceptional divisor on $\tilde{X}$.

Define $\deg_f \tilde{X}$ to be the degree of $f_*[\tilde{X}]$ as an n-dimensional cycle on $P(V^\vee)$, i.e.,

$$\begin{aligned} \deg \tilde{X} &= \deg(\tilde{X}/f(\tilde{X})) \cdot \int_{P(V^\vee)} c_1(\mathscr{O}(1))^n \cap [f(\tilde{X})] \\ &= \int_{\tilde{X}} c_1(f^*\mathscr{O}(1))^n. \end{aligned}$$

Proposition 4.4. *With the above notation,*

$$\deg_f \tilde{X} = \int_X c_1(L)^n - \int_B c(L)^n \cap s(B, X).$$

Proof. By the above description of $f^*\mathscr{O}(1)$,

$$\begin{aligned} \deg_f \tilde{X} &= \int_{\tilde{X}} (c_1(\pi^* L) - c_1(\mathscr{O}(E)))^n \\ &= \sum_{i=0}^{n} (-1)^i \binom{n}{i} \int_X c_1(L)^{n-i} \pi_*(c_1(\mathscr{O}(E))^i \cap [\tilde{X}]) \\ &= \int_X c_1(L)^n - \int_B (1 + c_1(L))^n \cap s(B, X), \end{aligned}$$

using the projection formula and Corollary 4.2.2. □

Example 4.4.1. Suppose that n divisors $D_1, \dots, D_n$ in the linear system cut out B scheme-theoretically, together with a finite set S disjoint from B. In this case $\deg_f \tilde{X}$ may be interpreted as the weighted number of points in S, and $\int_X c_1(L)^n$ is the total intersection number $(D_1 \cdot \ldots \cdot D_n)$. The remaining term $\int_B c(L)^n \cap s(B, X)$ therefore represents the contribution to this total intersection number which is carried by B. This is a special case of a general formula to be proved in Chap. 9.

Example 4.4.2. Let $B \subset \mathbb{P}^n$ be the rational normal curve. Let $V \subset H^0(\mathbb{P}^n, \mathcal{O}(2))$ be the linear system of quadrics passing through B. Then $\deg_f \tilde{\mathbb{P}}^n = 2^n - (n^2 - n + 2)$. If $n = 4$, $f(\tilde{\mathbb{P}}^4) \subset \mathbb{P}^5$ may be identified with the Grassmannian of lines in $\mathbb{P}^3$.

Example 4.4.3. Let X be an irreducible hypersurface in $\mathbb{P}^{n+1} = P(V)$ defined by an equation $F(X_0, \ldots, X_{n+1})$ of degree d. The *singular*, or *Jacobian* subscheme of X is the scheme J of zeros of the partial derivatives $\partial F/\partial X_0, \ldots, \partial F/\partial X_{n+1}$. If $\tilde{X}$ is the blow-up of X along J, we have morphisms

$$\begin{array}{ccc} & \tilde{X} & \\ {}^{\pi}\swarrow & & \searrow^{f} \\ X & & \mathbb{P}^{n+1} = P(V^{\vee}) . \end{array}$$

The image $f(\tilde{X})$ is called the *dual* variety of X and denoted $X^{\vee}$. In characteristic zero, biduality holds, i.e., $(X^{\vee})^{\vee} = X$ (cf. Kleiman (11)). In any case, we call $\deg f_*[\tilde{X}]$ the *degree of the dual* of X. Then

$$\deg f_*[\tilde{X}] = d(d-1)^n - \sum_{i=0}^{n} \binom{n}{i} (d-1)^i \deg(s_i)$$

where s_i is the i-dimensional component of $s(J, X)$. For example, if X is non-singular, its dual has degree $d(d-1)^n$ (cf. Example 3.2.21).

Example 4.4.4. Let $X \subset \mathbb{P}^2$ be an irreducible plane curve of degree d. Then the degree of the dual is

$$d(d-1) - \sum_{P} (e_J X)_P$$

where $(e_J X)_P$ is the multiplicity of X along J at P (cf. Example 4.3.4). Equivalently, if $\varrho: X' \to X$ is the normalization of X in its function field, and $J' = \varrho^{-1}(J)$, then the degree of the dual is

$$d(d-1) - \sum_{P'} \operatorname{ord}_{P'}(J') .$$

(Use Proposition 4.2(a).) For example, a node counts for 2, an ordinary m-fold point for $m(m-1)$, an ordinary cusp for 3; a higher cusp, of the form $y^p = x^q$, with $p < q$, and p, q relatively prime (and not divisible by the characteristic) counts for $(p-1)q$ (cf. Walker (1) IV.6). For an analogous result for surfaces, see Example 9.3.8.

Example 4.4.5. *Polar classes* (cf. Piene (3)). Let $X^n \subset \mathbb{P}^{n+1}$ be an irreducible hypersurface of degree d, over an algebraically closed ground field. For any $0 \leq k \leq n$, let $W_k \subset \mathbb{P}^{n+1}$ be a linear subspace of dimension $k-1$. The kth *polar locus* of X, with respect to W_k, denoted M_k, is defined to be the closure of

$$\{x \in X_{\text{reg}} \mid T_x X \supset W_k\} .$$

Here $T_x X$ denotes the imbedded tangent space to X at x. For general W_k, M_k has pure codimension k in X, and the class $[M_k] \in A_{n-k}(X)$ is independent of

W_k. Indeed, with $h = c_1(\mathcal{O}_X(1))$, $s_i = s_i(J, X)$ as in Example 4.4.3,

$$[M_k] = (d-1)^k h^k \cap [X] - \sum_{i=0}^{k-1} \binom{k}{i} (d-1)^i h^i \cap s_{n-k+i}.$$

The kth *class* of X, denoted ϱ_k, is defined to be the degree of $[M_k]$. Therefore

$$\varrho_k = d(d-1)^k - \sum_{i=0}^{k-1} \binom{k}{i} (d-1)^i \deg(s_{n-k+i}).$$

(Let $\tilde{X}$, π, f be as in Example 4.4.3, $L = \mathcal{O}_{P(V)}(d-1)$. Set

$$W^\vee = \{H \in P(V^\vee) \mid H \supset W_k\}.$$

Then $[M_k] = \pi_*[f^{-1}(W^\vee)] = \pi_*(c_1(f^*\mathcal{O}(1))^k \cap [\tilde{X}])$. The proof concludes as in Proposition 4.4.)

Note that $M_0 = X$. For $k = 1$, if F is an equation for X, and $W_1 = y = (y_0: \ldots : y_{n+1})$, let $F_y = \sum_{i=0}^{n+1} y_i \, \partial F/\partial X_i$. The hypersurface defined by $F_y = 0$ is the classical polar hypersurface of X with respect to y. A non-singular point $x \in X$ lies in M_1 if and only if $F_y(x) = 0$. For the kth polar locus, X_{reg} may be intersected with k polar hypersurfaces, choosing the points in general position.

For calculations of s_0 and s_1, and hence the polar classes, when X is a surface in $\mathbb{P}^3$ with ordinary singularities, see Example 9.3.7. Generalizations to higher codimension are discussed in Example 14.4.15.

Notes and References

The normal cone to a subvariety or subscheme, in the form of an associated graded ring, was used by Samuel (1) to define the multiplicity of a subvariety. Samuel used these multiplicities to define intersection numbers (cf. Chap. 7).

Normal cones were used explicitly by Verdier (5) for his construction of Gysin maps. They were implicit in MacPherson's graph construction (cf. Baum-Fulton-MacPherson (1)).

Segre classes of cones and general subvarieties and subschemes were defined in Fulton-MacPherson (1). At least in the non-singular case, these classes had been studied by B. Segre (4) (cf. Example 3.3.4 and Corollary 4.2.2 above), and more recently by J. King (3) and Lascu-Scott (1).

A sheaf-theoretic version of the multiplicity formula of § 4.3, together with results like Example 4.3.9, were given by Ramanujam (1).

Example 4.3.2, which gives a result of G. Kempf (2), was suggested by J. Harris, and Examples 4.1.3, 4.1.4, 4.2.6 and 4.4.2 by R. Lazarsfeld. Examples 4.1.6 and 4.2.6 were worked out jointly with A. Collino and R. MacPherson.

Chapter 5. Deformation to the Normal Cone

Summary

If X is a closed subscheme of Y, there is a family of imbeddings $X \hookrightarrow Y_t$, parametrized by $t \in \mathbb{P}^1$, such that for $t = 0$ (in fact for $t \neq \infty$) the imbedding is the given imbedding of X in Y, and for $t = \infty$ one has the zero section imbedding of X in the normal cone $C_X Y$. The existence of such a deformation, together with the "principle of continuity" that intersection products should vary nicely in families, explains the prominent role to be played by the normal cone in constructing intersection products.

5.1 The Deformation

Let X be a closed subscheme of a scheme Y, and let $C = C_X Y$ be the normal cone to X in Y. We will construct a scheme $M = M_X Y$, together with a closed imbedding of $X \times \mathbb{P}^1$ in M, and a flat morphism $\varrho : M \to \mathbb{P}^1$ so that

$$\begin{array}{ccc} X \times \mathbb{P}^1 & \hookrightarrow & M \\ {\scriptstyle pr} \searrow & & \swarrow {\scriptstyle \varrho} \\ & \mathbb{P}^1 & \end{array}$$

commutes, and such that:

(1) Over $\mathbb{P}^1 - \{\infty\} = \mathbb{A}^1$, $\varrho^{-1}(\mathbb{A}^1) = Y \times \mathbb{A}^1$ and the imbedding is the trivial imbedding:

$$X \times \mathbb{A}^1 \hookrightarrow Y \times \mathbb{A}^1 .$$

(2) Over ∞, the divisor $M_\infty = \varrho^{-1}(\infty)$ is the sum of two effective Cartier divisors:

$$M_\infty = P(C \oplus 1) + \tilde{Y}$$

where $\tilde{Y}$ is the blow-up of Y along X. The imbedding of $X = X \times \{\infty\}$ in M_∞ is the zero-section imbedding of X in C, followed by the canonical open imbedding of C in $P(C \oplus 1)$. The divisors $P(C \oplus 1)$ and $\tilde{Y}$ intersect in the scheme $P(C)$, which is imbedded as the hyperplane at infinity in $P(C \oplus 1)$, and as the exceptional divisor in $\tilde{Y}$.

In particular, the image of X in M_∞ is disjoint from $\tilde{Y}$. Letting $M^\circ = M^\circ_X Y$ be the complement of $\tilde{Y}$ in M, one has a family of imbeddings of X:

$$\begin{array}{ccc} X\times\mathbb{P}^1 & \hookrightarrow & M^\circ \\ {\scriptstyle pr}\searrow & & \swarrow{\scriptstyle \varrho^\circ} \\ & \mathbb{P}^1 & \end{array}$$

which deforms the given imbedding of X in Y to the zero-section imbedding of X in $C_X Y$.

To construct this deformation, let M be the blow-up of $Y\times\mathbb{P}^1$ along the subscheme $X\times\{\infty\}$. Since the normal cone to $X\times\{\infty\}$ in $Y\times\mathbb{P}^1$ is $C\oplus 1$, the exceptional divisor in this blow-up is $P(C\oplus 1)$.

From the sequence of imbeddings

$$X=X\times\{\infty\}\hookrightarrow X\times\mathbb{P}^1\hookrightarrow Y\times\mathbb{P}^1,$$

the blow-up of $X\times\mathbb{P}^1$ along $X\times\{\infty\}$ is imbedded as a closed subscheme of M (Appendix B.6.9); since $X\times\{\infty\}$ is a Cartier divisor on $X\times\mathbb{P}^1$, the blow-up of $X\times\mathbb{P}^1$ along $X\times\{\infty\}$ may be identified with $X\times\mathbb{P}^1$, so we have a closed imbedding

$$X\times\mathbb{P}^1\hookrightarrow M.$$

Similarly from

$$X\times\{\infty\}\hookrightarrow Y\times\{\infty\}\hookrightarrow Y\times\mathbb{P}^1$$

the blow-up $\tilde{Y}$ of Y along X is imbedded as a closed subscheme of M.

Since the projection from $Y\times\mathbb{P}^1$ to $\mathbb{P}^1$ is flat, the composite

$$\varrho: M\to\mathbb{P}^1$$

of the blow-down map from M to $Y\times\mathbb{P}^1$ followed by the projection to $\mathbb{P}^1$ is flat (Appendix B.6.7).

Since $M\to Y\times\mathbb{P}^1$ is an isomorphism away from $Y\times\{\infty\}$, assertion (1) is clear. The description (2) of $M_\infty=\varrho^{-1}(\infty)$ as the sum of Cartier divisors will follow from the explicit algebraic description given below; since we have $P(C\oplus 1)$ and $\tilde{Y}$ globally imbedded in M, it suffices to examine their structure locally on Y.

Assume $Y=\operatorname{Spec}(A)$, and X is defined by the ideal I in A. To study M near ∞, identify $\mathbb{P}^1-\{0\}$ with $\mathbb{A}^1=\operatorname{Spec}K[T]$, where K is the ground field. The blow-up of $Y\times\mathbb{A}^1$ along $X\times\{0\}$ is $\operatorname{Proj}(S^\bullet)$, with

$$S^n=(I,T)^n=I^n+I^{n-1}T+\ldots+AT^n+AT^{n-1}+\ldots.$$

$\operatorname{Proj}(S^\bullet)$ is covered by affine open sets $\operatorname{Spec}(S^\bullet_{(a)})$, where $S^\bullet_{(a)}$ is the ring of fractions

$$S^\bullet_{(a)}=\{s/a^n \mid s\in S^n\},$$

and a runs through a set of generators for the ideal (I,T) in $A[T]$. For $a\in I$, the exceptional divisor $P(C\oplus 1)$ is defined in $\operatorname{Spec}(S^\bullet_{(a)})$ by the equation $a/1$, $a\in S^0$, while $\tilde{Y}$ is defined by T/a; since $T=(a/1)\cdot(T/a)$, the description of M_∞ as the sum of $P(C\oplus 1)$ and $\tilde{Y}$ follows.

The complement of $\tilde{Y}$ in the blow-up of $Y\times\mathbb{A}^1$ along $X\times\{0\}$ is $\operatorname{Spec}(S^\bullet_{(T)})$, where

$$S^\bullet_{(T)}\cong\ldots\oplus I^nT^{-n}\oplus\ldots\oplus IT^{-1}\oplus A\oplus AT\oplus\ldots\oplus AT^n\oplus\ldots.$$

This is the ring studied by Rees (2), and Gerstenhaber (1). The canonical homomorphism from $A[T]$ to $S^{\cdot}_{(T)}$ becomes an isomorphism after localization at T, while

$$S^{\cdot}_{(T)}/TS^{\cdot}_{(T)} = \bigoplus_{n=0}^{\infty} I^n/I^{n+1}.$$

Remark 5.1.1. MacPherson's description of this deformation, as a special case of his graph construction, is particularly vivid. Assume E is a vector bundle on Y, and s is a section of E whose zero-scheme is X. (If Y is quasi-projective such E always exists, although its rank r may be larger than the codimension of X on Y.) For each scalar λ, the graph of λs is a line in $E \oplus 1$. This gives an imbedding

$$Y \times \mathbb{A}^1 \hookrightarrow P(E \oplus 1) \times \mathbb{P}^1,$$

$(y, \lambda) \to$ (graph of $\lambda s(y)$, $(1:\lambda)$). The deformation space $M_X Y$ is in fact the closure of $Y \times \mathbb{A}^1$ in this imbedding. We won't need this description; the construction given above is simpler because it relies only on standard properties of blowing up. The graph construction is more powerful, however, in that it generates to general vector bundle maps, or to complexes of vector bundles (cf. § 18.1).

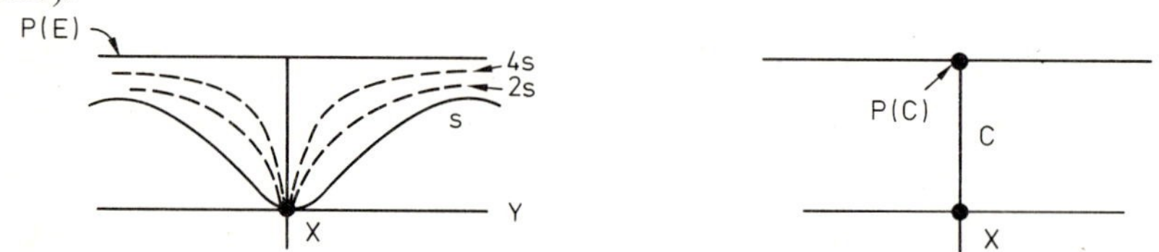

Remark 5.1.2. From the point of view of deformation theory of varieties, this construction would be called a deformation *from* the normal cone. This terminology seems inappropriate here, since we always start with the imbedding $X \hookrightarrow Y$. The alternative, "specialization" to the normal cone, is reserved for the associated homomorphisms of cycles or cycle classes (§ 5.2).

Example 5.1.1. Assume Y is purely n-dimensional. Since M is flat over $\mathbb{P}^1$, $[\varrho^{-1}(0)] \sim [\varrho^{-1}(\infty)]$, i.e.,

$$i_*[Y] = j_*[P(C \oplus 1)] + k_*[\tilde{Y}]$$

in $A_n(M)$. Here $i: Y \to M$ is the imbedding of Y at $t=0$, and j (resp. k) is the canonical imbedding of $P(C \oplus 1)$ (resp. $\tilde{Y}$) in M over $t=\infty$.

Example 5.1.2. Assume X is regularly imbedded in Y so $C_X Y = N$ is a vector bundle. The imbedding of X in $P(N \oplus 1)$ has several advantages over the given imbedding of X in Y. For example:

(i) There is a retraction from $P(N \oplus 1)$ to X.

(ii) There is a vector bundle ξ on $P(N \oplus 1)$ of rank equal to the codimension of X, with a regular section whose zero-scheme is X. Thus, on $P(N \oplus 1)$, X is represented by the top Chern class of the bundle ξ.

There is usually no such retraction or bundle for the given imbedding, even if Y is replaced by an open subscheme containing X. The deformation to the normal bundle may be thought of as an analogue in algebraic geometry of the tubular neighborhood construction in topology.

5.2 Specialization to the Normal Cone

Let X be a closed subscheme of a scheme Y, and let $C = C_X Y$ be the normal cone to X in Y. Define the *specialization* homomorphisms

$$\sigma : Z_k Y \to Z_k C$$

by the formula

$$\sigma[V] = [C_{V \cap X} V]$$

for any k-dimensional subvariety V of Y, and extending linearly to all k-cycles; note that $C_{V \cap X} V$ is a purely k-dimensional scheme (Appendix B.6.6), so it has a fundamental cycle $[C_{V \cap X} V]$ by § 1.5.

Proposition 5.2. *If a cycle $\alpha \in Z_k Y$ is rationally equivalent to zero on Y, then $\sigma(\alpha)$ is rationally equivalent to zero on C.*

Therefore σ passes to rational equivalence, defining *specialization homomorphisms*

$$\sigma : A_k Y \to A_k C .$$

Proof. Let $M^\circ = M_X^\circ Y$ be the deformation space constructed in § 5.1, i the inclusion of C in M°, j the inclusion of $Y \times \mathbb{A}^1$ in M. Consider the diagram

$$\begin{array}{ccccc} A_{k+1} C & \xrightarrow{i_*} A_{k+1} M^\circ \xrightarrow{j^*} & A_{k+1}(Y \times \mathbb{A}^1) & \to 0 \\ {\scriptstyle i^*}\downarrow & & \uparrow{\scriptstyle \mathrm{pr}^*} & \\ A_k C & \leftarrow\text{------} & A_k(Y) & \end{array}$$

The row is the exact sequence of § 1.8. The map i^* is the Gysin map for divisors; the composite $i^* i_*$ is zero since the normal bundle to C in M° is trivial (Proposition 2.6 (c)); in fact, C is a principal divisor on $M^\circ - \varrho^{-1}(0)$.

There is therefore an induced morphism from $\mathrm{Coker}(i_*) = A_{k+1}(Y \times \mathbb{A}^1)$ to $A_k C$, and therefore a morphism from $A_k Y$ to $A_k C$ obtained by composing with the flat pull-back from $A_k Y$ to $A_{k+1}(Y \times \mathbb{A}^1)$. To prove the proposition, it suffices to verify that this composite takes $[V]$ to $[C_{V \cap X} V]$. First, $\mathrm{pr}^*[V] = [V \times \mathbb{A}^1]$. The variety $M_{V \cap X}^\circ V$ is a closed subvariety of $M_X^\circ Y$ which restricts to $V \times \mathbb{A}^1$, so

$$j^*[M_{V \cap X}^\circ V] = \mathrm{pr}^*[V] .$$

The Cartier divisor $C = M^\circ \cap \varrho^{-1}(\infty)$ intersects $M_{V \cap X}^\circ V$ in $C_{V \cap X} V$, so

$$i^*[M_{V \cap X}^\circ V] = [C_{V \cap X} V] ,$$

which concludes the proof. □

The following example may provide a useful preview for the next few chapters.

Example 5.2.1. Let $i : X \to Y$ be a regular imbedding of codimension d, with normal bundle $N = N_X Y$. Define the *Gysin homomorphism*

$$i^* : A_k Y \to A_{k-d} X$$

to be the composite $i^* = s_N^* \circ \sigma$:

$$A_k Y \xrightarrow{\sigma} A_k N \xrightarrow{s_N^*} A_{k-d} X ,$$

where s_N^* is the Gysin homomorphism of Definition 3.3.

(a) If $d=1$ (resp. i is the zero section of a vector bundle) this Gysin homomorphism agrees with that defined in § 2.6 (resp. § 3.3).

(b) If Y is purely n-dimensional, $i^*[Y]=[X]$.

(c) For all $\alpha \in A_* X$, $i^* i_* (\alpha) = c_d(N) \cap \alpha$. (See § 6.3.)

(d) For any k-dimensional subvariety V of Y

$$i^*[V] = \{c(N) \cap s(V \cap X, V)\}_{k-d} .$$

(e) If X is an n-dimensional variety which is smooth over the ground field, the diagonal imbedding δ of X in $X \times X$ is a regular imbedding of codimension n. This defines an intersection product on $A_* X$:

$$A_p X \otimes A_q X \xrightarrow{\times} A_{p+q}(X \times X) \xrightarrow{\delta^*} A_{p+q-n}(X) .$$

(See Chap. 8.)

Notes and References

Deformation to the normal bundle, or cone, has an interesting history. It has appeared in at least three places:

(1) For a non-singular subvariety X of a non-singular quasi-projective variety Y, Mumford (unpublished, 1959), Jouanolou (2), Lascu and Scott (1), (2), and Lascu, Mumford and Scott (1) used the blow-up of $Y \times \mathbb{P}^1$ along $X \times \{\infty\}$ to prove important formulas in intersection theory: the self-intersection and key formulas, Riemann-Roch without denominators, and the formula for blowing up Chern classes.

(2) For an ideal I in a ring A, Gerstenhaber (1) deformed A to the associated graded ring $\oplus I^n/I^{n+1}$. The algebra for this deformation had previously appeared in Rees (2).

(3) For a regularly imbedded subscheme X of a variety Y, and a section s of a vector bundle on Y whose zero-scheme is X, MacPherson (cf. Baum-Fulton-MacPherson (1), and Example 18.1.7) deformed the graph of λs to $\lambda = \infty$, and identified the normal bundle to X in Y at ∞.

In (1) the deformation was not at first explicit. In fact, considerable simplification occurred when the rational equivalence between $[Y]$ at 0 and $[P(N \oplus 1)] + [\tilde{Y}]$ at ∞ was used (cf. Lascu and Scott (2)). The deformation in (2) had been used in algebraic geometry by Kleiman and Landolfi (1), Mumford, and others, but not in intersection theory. The graph construction (3) was used to solve problems of intersection theory. The identity of the three approaches was established in the 1974/75 seminars of Kleiman and Douady-Verdier (1).

When one realizes the role of normal cones in Samuel's construction of intersection multiplicities in the case of proper intersections, and the role of deformation to the normal bundle in proving excess intersection formulas, it becomes reasonable to expect general intersection products to be constructed using normal cones and bundles; this reasonable expectation, however, was apparently not formulated before these products were constructed.

Verdier (5) used deformation to the normal cone, together with the Gysin homomorphism for principal divisors constructed in Fulton (2), to construct specialization homomorphisms to the normal cone. He used these specialization homomorphisms to construct Gysin homomorphisms for regular imbeddings of arbitrary codimension. The present chapter follows Verdier's exposition closely. Except for (d) and (e) of Example 5.2.1, all the results appear in Verdier (5), at least in the quasi-projective case.

Chapter 6. Intersection Products

Summary

Given a regular imbedding $i: X \to Y$ of codimension d, a k-dimensional variety V, and a morphism $f: V \to Y$, an intersection product $X \cdot V$ is constructed in $A_{k-d}(W)$, $W = f^{-1}(X)$. Although the case of primary interest is when f is a closed imbedding, so $W = X \cap V$, there is significant benefit in allowing general morphisms f. Let $g: W \to X$ be the induced morphism. The normal cone $C_W V$ to W in V is a closed subcone of $g^* N_X Y$, of pure dimension k. We define $X \cdot V$ to be the result of intersecting the k-cycle $[C_W V]$ by the zero-section of $g^* N_X Y$:

$$X \cdot V = s^* [C_W V]$$

where $s: W \to g^* N_X Y$ is the zero-section, and s^* is the Gysin map constructed in Chapter 3. Alternatively $X \cdot V$ is the $(k-d)$-dimensional component of the class

$$c(g^* N_X Y) \cap s(W, V)$$

where $s(W, V)$ is the Segre class of W in V.

If the k-cycle $[C_W V]$ is written out as a sum $\sum m_i [C_i]$, with C_i irreducible, one has a corresponding decomposition $X \cdot V = \sum m_i \alpha_i$, with α_i a well-defined cycle-class on the support of C_i.

If the imbedding of W in V is regular of codimension d', then $E = g^* N_X Y / N_W V$ is the quotient bundle, there is an excess intersection formula

$$X \cdot V = c_{d-d'}(E) \cap [W].$$

Given $i: X \to Y$ as above, and a morphism $f: Y' \to Y$, form the fibre square

$$(*) \qquad \begin{array}{ccc} X' & \xrightarrow{j} & Y' \\ {\scriptstyle g}\downarrow & & \downarrow{\scriptstyle f} \\ X & \xrightarrow[i]{} & Y \end{array} .$$

There are refined Gysin homomorphisms

$$i^! : A_k Y' \to A_{k-d} X'$$

determined by the formula $i^![V] = X \cdot V$ for subvarieties V of Y'.

In this chapter the fundamental properties of these intersection operations are proved. After proving that $i^!$ is well-defined on rational equivalence classes, the most important of these properties are:

(i) Compatibility with proper push-forward (§ 6.2)
(ii) Compatibility with flat pull-back (§ 6.2)
(iii) Commutativity (§ 6.4)
(iv) Functoriality (§ 6.5).

For example, to calculate $X \cdot V$, by (i) it suffices to calculate $X \cdot V'$ for any V' mapping properly and birationally to V; one may blow up V along $V \cap W$ to reduce to a case where the excess intersection formula applies. A particular case of (ii) is the assertion that the intersection products restrict to open subschemes: one may often compute intersection products locally. An important case of commutativity asserts that intersections may be carried out before or after specialization in a family; this will include a strong version of the "principle of continuity" in Chapter 10.

When $Y' = Y$, $i^!$ determines the (ordinary) Gysin homomorphisms

$$i^* : A_k Y \to A_{k-d} X .$$

Functoriality (iv) refines the statement that $(j\,i)^* = i^* j^*$ for $i: X \to Y$, $j: Y \to Z$ regular embeddings.

More generally, if $f: X \to Y$ is a local complete intersection morphism, there are Gysin homomorphisms f^*, and refined homomorphisms $f^!$. These Gysin homomorphisms are used to describe the group $A_* \tilde{Y}$, when $\tilde{Y}$ is the blow-up of a scheme Y along a regularly imbedded subscheme. A new blow-up formula describes the Gysin map from $A_* Y$ to $A_* \tilde{Y}$ explicitly.

The rest of this book is based on this intersection product and the fundamental properties proved in § 6.1 – § 6.5. As in Chap. 2, the formal properties can be motivated from topology. As we shall see in Chap. 19, a regular imbedding $X \hookrightarrow Y$ of codimension d determines an orientation, or generalized Thom class, in $H^{2d}(Y, Y - X)$. The Gysin maps are the algebraic geometry versions of cap product by this orientation class, or with its pull-back to Y', if Y' maps to Y.

6.1 The Basic Construction

Let $i: X \to Y$ be a (closed) regular imbedding of codimension d, and denote the normal bundle by $N_X Y$. Let V be a purely k-dimensional scheme, and let $f: V \to Y$ be a morphism. Denote the inverse image scheme $f^{-1}(X)$ by W, and form the fibre square

$$\begin{array}{ccc} W & \xrightarrow{j} & V \\ {\scriptstyle g}\downarrow & & \downarrow{\scriptstyle f} \\ X & \xrightarrow[i]{} & Y \end{array} .$$

Let $N = g^* N_X Y$, a bundle of rank d on W, and let $\pi: N \to W$ be the projection. Since the ideal sheaf $\mathscr{I}$ of X in Y generates the ideal sheaf $\mathscr{J}$ of W in V, there is a surjection

$$\bigoplus_n f^*(\mathscr{I}^n/\mathscr{I}^{n+1}) \twoheadrightarrow \bigoplus_n \mathscr{J}^n/\mathscr{J}^{n+1}.$$

This determines a closed imbedding of the normal cone $C = C_W V$ as a subcone of the vector bundle N:

$$\begin{array}{ccc} C & \hookrightarrow & N \\ & \searrow \; \swarrow^{\pi} & \\ & W & \end{array}.$$

Since C is purely k-dimensional (Appendix B.6.6), it determines a k-cycle $[C]$ on N (§ 1.5). Let s be the zero section of the bundle N. Define the *intersection product of V by X on Y*, denoted $X \cdot V$ (or $X \cdot_Y V$, or $i^![V]$, see Example 6.2.1), to be the class on W obtained by "intersecting $[C]$ by the zero section of N". That is, set

$$X \cdot V = s^*[C] \tag{1}$$

in $A_{k-d}(W)$, where $s^*: A_k(N) \to A_{k-d} W$ is the Gysin homomorphism of Definition 3.3; equivalently, $X \cdot V$ is the unique class in $A_{k-d}(W)$ such that $\pi^*(X \cdot V) = [C]$ in $A_k(N)$.

Proposition 6.1. (a) *With the above notation,*

$$X \cdot V = \{c(N) \cap s(W, V)\}_{k-d}.$$

(b) *If ξ is the universal quotient bundle of rank d on $P(N \oplus 1)$, and q is the projection from $P(N \oplus 1)$ to W, then*

$$X \cdot V = q_*(c_d(\xi) \cap [P(C \oplus 1)]).$$

(c) *If $d = 1$, i.e. X is a Cartier divisor on X, V is a variety, and f is a closed imbedding, then $X \cdot V$ is the intersection class constructed in* § 2.3.

Proof. Since $P(C \oplus 1)$ restricts to C on N, (b) follows from Proposition 3.3. To prove (a), consider the universal exact sequence

$$0 \to \mathscr{O}(-1) \to q^* N \oplus 1 \to \xi \to 0$$

on $P(N \oplus 1)$. By the Whitney formula, $c(\xi)\, c(\mathscr{O}(-1)) = c(q^*N)$. Therefore

$$\begin{aligned} q_*(c_d(\xi) \cap [P(C \oplus 1)]) &= \{q_*(c(\xi) \cap [P(C \oplus 1)])\}_{k-d} \\ &= \{q_*(c(q^*N)\, s(\mathscr{O}(-1)) \cap [P(C \oplus 1)])\}_{k-d} \\ &= \{c(N) \cap q_*(s(\mathscr{O}(-1)) \cap [P(C \oplus 1)])\}_{k-d} \\ &= \{c(N) \cap s(C)\}_{k-d}, \end{aligned}$$

which proves (a).

For (c), if $V \subset X$, then $C = W = V$, so $s^*[C] = c_1(N) \cap [W]$ by Proposition 2.6(c) (cf. Example 3.3.1). If $V \not\subset X$, then W is the pull-back Cartier divisor f^*X, $C = N$, and $s^*[C] = [W] = [f^*X]$. These prescriptions agree with those of § 2.3. □

Definition 6.1.2. Let $C_1, \ldots, C_r$ be the irreducible components of the cone C, and let m_i be the geometric multiplicity of C_i in C. Thus $[C] = \sum_{i=1}^{r} m_i [C_i]$ is the cycle of C. Let $Z_i \subset W$ be the support of C_i, i.e.

$$Z_i = \pi(C_i) = q(P(C_i \oplus 1)),$$

a closed subvariety of W. The varieties $Z_1, \ldots, Z_r$ (which need not all be distinct) are called the *distinguished varieties* of the intersection of V by X. Let N_i be the restriction of N to Z_i. We have a commutative diagram

$$\begin{array}{ccc} C_i & \hookrightarrow & N_i \\ & \searrow \; \swarrow & \\ & Z_i & \end{array} .$$

Let s_i be the zero section of N_i, and set

$$\alpha_i = s_i^* [C_i]$$

in $A_{k-d}(Z_i)$. From the definition of s^* and s_i^*, it is clear that α_i maps to $s^*[C_i]$ in $A_{k-d} W$. Therefore

$$X \cdot V = \sum_{i=1}^{r} m_i \alpha_i$$

in $A_{k-d} W$. We call the equation $X \cdot V = \sum m_i \alpha_i$ the *canonical decomposition* of the intersection product.

If Z is a distinguished subvariety of W, the sum of those terms $m_i \alpha_i$ with $Z_i = Z$ is called the *equivalence* of Z for the intersection of V by X, or the *contribution* of Z to $X \cdot V$. It should be emphasized that a distinguished variety Z may have any dimension from $k-d$ to k, but that the contributions are always cycle classes of dimension $k-d$; only if $\dim Z = k-d$ is the equivalence of Z a multiple of $[Z]$ (cf. Chap. 7).

For any closed subset S of V, the *part of* $X \cdot V$ *supported on* S, denoted $(X \cdot V)^S$, is the class in $A_{k-d} S$ obtained by adding the equivalence of all distinguished varieties contained in S:

$$(X \cdot V)^S = \sum_{Z_i \subset S} m_i \alpha_i .$$

In the rest of this chapter we will study the classes $X \cdot V$, and show that they satisfy the formal properties one expects for intersection products. The individual terms in the canonical decompositions are more subtle, however. Examples and classical applications will appear later, particularly in Chapters 9 and 16. A canonical decomposition occurs naturally in the proof of a formula of Severi in Example 16.2.4. A dynamic interpretation of these equivalences will be given in Chap. 11.

Example 6.1.1. The formulas of Proposition 6.1 are also valid for the equivalences of the distinguished varieties:

$$\alpha_i = \{c(N_i) \cap s(C_i)\}_{k-d} = q_{i*}(c_d(\xi_i) \cap [P(C_i \oplus 1)])$$

where ξ_i is the universal bundle on $P(N_i \oplus 1)$, and q_i the projection to Z_i.

Example 6.1.2. Define divisors D_1, D_2 on $\mathbb{P}^2$ by $D_1 = 2A + B$, $D_2 = A + 2B$, where A and B are lines meeting in a point P. Let $X = D_1 \times D_2$, $Y = \mathbb{P}^2 \times \mathbb{P}^2$, $V = \mathbb{P}^2$, f the diagonal imbedding of V in Y. Then A, B, and P are the distinguished varieties for the intersection of V by X, and the corresponding canonical decomposition is

$$X \cdot V = \alpha + \beta + 3[P]$$

where α (resp. β) is any zero cycle of degree 3 on A (resp. B).

Example 6.1.3. Let $Y = \mathbb{P}^2$, X_1 the curve $xy = 0$, X_2 the curve $x = 0$, P the point $x = y = 0$. For the intersection product of X_2 by X_1, only X_2 is distinguished, but for the intersection product of X_1 by X_2 both X_2 and P are distinguished. (To see this, replace Y by the affine plane. If I is the ideal in $K[x, y]/(xy)$ generated by x, then

$$K[x, y, T]/(x, yT) \to \oplus\, I^n/I^{n+1},$$

sending T to x, is an isomorphism; thus the components of C are: the line $x = y = 0$, which maps to P, and the line $x = T = 0$, which maps to X_2.)

Similarly for the intersection product of the diagonal $\Delta_{\mathbb{P}^2}$ by $X_1 \times X_2$ in $\mathbb{P}^2 \times \mathbb{P}^2$, only X_2 is distinguished, but for the intersection of $X_1 \times X_2$ by $\Delta_{\mathbb{P}^2}$, X_2 and P are distinguished.

In each case, the canonical decompositions of the intersection classes may be calculated either directly, or by the dynamic interpretation of Chap. 11 (cf. Example 11.3.2).

Example 6.1.4 (cf. Examples 4.2.2 and 11.3.2). Let A, B and D be effective Cartier divisors on a non-singular surface X, with A and B assumed relatively prime. Let $A' = A + D$, $B' = B + D$. Consider the intersection of the diagonal $\Delta = X$ by $A' \times B'$ arising from the fibre square

$$\begin{array}{ccc} A' \cap B' & \to & X \\ \downarrow & & \downarrow \delta \\ A' \times B' & \to & X \times X\,. \end{array}$$

(a) The distinguished varieties for this intersection product are the irreducible components of D, and the points of $A \cap B$. (Identify the blow-up of X along $A \cap B$ with the blow-up of X along $A' \cap B'$.)

(b) If P is a point where A and B meet transversally, the equivalence of P for the intersection class is

$$(1 + \mathrm{ord}_P(D))\,[P]\,.$$

(c) Let E be an irreducible component of multiplicity m of D, and assume A and B meet transversally at any common point P on E. Then the equivalence of E for the intersection class is

$$m(A \cdot E + B \cdot E + D \cdot \mathrm{E}) - \sum_{P \in A \cap B} \mathrm{ord}_P(E)\,[P]\,.$$

(Use the blow-up of (a) to compute Segre classes.)

(d) When A and B are not transversal, the contributions may still be computed by successive blow-ups, but the results are more complicated. For example, if $X = \mathbb{P}^2$, and A, B, D are defined by polynomials $y^2 - xz$, $y^2 + xz$, and $y - \lambda x$ (for some $\lambda \in K$) respectively, and $P_0 = (0:0:1)$, the equivalence of P_0 is $3[P_0]$ if $\lambda \neq 0$, but $4[P_0]$ if $\lambda = 0$.

Example 6.1.5. With homogeneous coordinates (x, y, z, w, t) on $\mathbb{P}^4$, let

$$V_1 = V(z^3 - xy(y-2x), w)$$
$$V_2 = V(w^3 - yx(x-2y), z).$$

The distinguished varieties for the intersection product of the diagonal $\mathbb{P}^2$ by $V_1 \times V_2$ in $\mathbb{P}^2 \times \mathbb{P}^2$ are the lines $x = z = w = 0$, $y = z = w = 0$, and the point $x = y = z = w = 0$. Each contributes a zero-cycle of degree 3 to the intersection product.

Example 6.1.6. In the situation of Proposition 6.1 (a), $c(N) \cap s(W, V) = X \cdot V +$ lower terms, i.e., $\{c(N) \cap s(W, V)\}_i = 0$ for $i > k - d$.

Example 6.1.7. If the imbedding of W in V is a regular imbedding of codimension d', with normal bundle N', then

$$X \cdot V = c_{d-d'}(N/N') \cap [W].$$

(See § 6.3 for a generalization.)

Example 6.1.8. Let Q be the universal quotient bundle, of rank $d-1$, on $P(N)$. Assume that $\dim W \leqq k-1$. Then

$$X \cdot V = p_*(c_{d-1}(Q) \cap [P(C)])$$

where p is the projection from $P(N)$ to W. (Compare the proof of Proposition 6.1 (a) with Example 4.1.2.)

Example 6.1.9. *Uniqueness of intersection products.* The intersection product $X \cdot_Y V$ in $A_{k-d}(X)$, defined for any regular imbedding $X \hookrightarrow Y$ of codimension d, and any purely k-dimensional subscheme V of Y, is characterized by the following properties:

(i) ("normalization"). If Y is a vector bundle on X, $X \hookrightarrow Y$ the zero section imbedding, and $V = \pi^{-1}(W)$, where $\pi: Y \to X$ is the projection, then $X \cdot_Y V = [W]$.

(ii) ("continuity"). If $X \times \mathbb{P}^1 \to \mathscr{Y}$ is a family of regular imbeddings, $\mathscr{V}$ a subvariety of $\mathscr{Y}$, with $\mathscr{Y}$ and $\mathscr{V}$ flat over $\mathbb{P}^1$, then all the classes $X \cdot_{Y_t} V_t$ are equal. Here Y_t and V_t are the fibres of $\mathscr{Y}$ and $\mathscr{V}$ over rational points $t \in \mathbb{P}^1$.

(Apply (ii) to the deformation to the normal cones, i.e., $\mathscr{Y} = M_X^\circ Y$, $\mathscr{V} = M_{X \cap V}^\circ V$.)

6.2 Refined Gysin Homomorphisms

Let $i: X \to Y$ be a regular imbedding of codimension d, and let $f: Y' \to Y$ be a morphism. Form the fibre square

$$(*) \qquad \begin{array}{ccc} X' & \xrightarrow{j} & Y' \\ {\scriptstyle g}\downarrow & & \downarrow{\scriptstyle f} \\ X & \xrightarrow[i]{} & Y \end{array}$$

Define homomorphisms

$$i^! : Z_k Y' \to A_{k-d} X'$$

by the formula

$$i^!(\sum n_i [V_i]) = \sum n_i X \cdot V_i$$

where $X \cdot V_i$ is the intersection product constructed in the previous section. (Of course, $X \cdot V_i$ is constructed as a cycle class on $X' \cap V_i$; following our usual convention (§ 1.4) the same notation is used for its image in the larger scheme X'.)

To see that $i^!$ passes to rational equivalence, we give a variant of this definition. The normal cone $C' = C_{X'} Y'$ is a closed subcone of $N = g^* N_X Y$. Then $i^!$ is the composite

$$Z_k Y' \xrightarrow{\sigma} Z_k C' \to A_k N \xrightarrow{s^*} A_{k-d} X'$$

where σ is the specialization homomorphism of § 5.2, the second map is induced by the inclusion of C' in N, and s^* is the Gysin map for the zero-section s of X' in N (§ 3.3). By Proposition 5.2, σ passes to rational equivalence, so $i^!$ does also.

The induced homomorphisms

$$i^! : A_k Y' \to A_{k-d} X'$$

are called *refined Gysin homomorphisms.* We also may write $X \cdot_Y \alpha$ in place of $i^!(\alpha)$. If $Y' = Y$, $f = \mathrm{id}_Y$, these are called simply *Gysin homomorphisms,* and denoted i^* instead of $i^!$,

$$i^* : A_k Y \to A_{k-d} X .$$

A more precise notation for the refined Gysin homomorphism would be $i^*_{Y'}$ or i^*_f; we prefer to use the single notation $i^!$ for all these homomorphisms, taking care to specify where they act.

Theorem 6.2. *Consider a fibre diagram*

$$(*)\qquad \begin{array}{ccc} X'' & \xrightarrow{i''} & Y'' \\ {\scriptstyle q}\downarrow & & \downarrow{\scriptstyle p} \\ X' & \xrightarrow{i'} & Y' \\ {\scriptstyle g}\downarrow & & \downarrow{\scriptstyle f} \\ X & \xrightarrow[i]{} & Y \end{array}$$

with i a regular imbedding of codimension d.

(a) (Push-forward) *If p is proper, and* $\alpha \in A_k Y''$, *then*

$$i^! p_* (\alpha) = q_* (i^! \alpha)$$

in $A_{k-d} X'$.

(b) (Pull-back) *If p is flat of relative dimension n, and* $\alpha \in A_k Y'$, *then*

$$i^! p^* (\alpha) = q^* i^! (\alpha)$$

in $A_{k+n-d} X''$.

(c) (Compatibility) *If i' is also a regular imbedding of codimension d, and $\alpha \in A_k Y''$, then*

$$i^! \alpha = i'^! \alpha$$

in $A_{k-d} X''$.

Proof. (a) and (b) follow from the corresponding properties of Segre classes (Proposition 4.2). For (a), one may assume $\alpha = [V']$; let $V = p(V')$. Let $N = g^* N_X Y$. Then

$$\begin{aligned} i^! p_*[V'] &= \deg(V'/V)\,\{c(N) \cap s(X' \cap V, V)\}_{k-d} \\ &= \{c(N) \cap q_*(s(X'' \cap V', V'))\}_{k-d} \\ &= q_*\{c(q^*N) \cap s(X'' \cap V', V')\}_{k-d} \\ &= q_* i^! [V'] . \end{aligned}$$

The proof of (b) is similar and left to the reader. For (c), it suffices to observe that, when i' is a regular imbedding of the same codimension as i, $g^* N_X Y$ is the normal bundle to X' in Y'; indeed if $\mathscr{I}$ and $\mathscr{I}'$ are the ideal sheaves, the canonical epimorphism of $g^*(\mathscr{I}/\mathscr{I}^2)$ onto $\mathscr{I}'/\mathscr{I}'^2$ must be an isomorphism, since both sheaves are locally free of the same rank. □

Remark 6.2.1. If

$$\begin{array}{ccc} X' & \xrightarrow{i'} & Y' \\ {\scriptstyle g}\downarrow & & \downarrow{\scriptstyle f} \\ X & \xrightarrow[i]{} & Y \end{array}$$

is a fibre square, with i and i' regular imbeddings of codimensions d, an important case of (c) is the formula

$$i^!(\alpha) = i'^*(\alpha)$$

for all $\alpha \in A_* Y'$. If i' is not a regular imbedding, or if i' is a regular imbedding of codimension $< d$, then $i^!(\alpha)$ depends on i, not just on i' (cf. Theorem 6.3).

Remark 6.2.2. By the push-forward property (a), to calculate the intersection product $X \cdot V$, it suffices to calculate $X \cdot V'$ for any V' which maps properly and birationally onto V; for example, we may blow-up V along $X' \cap V$ to reduce to the case where $X' \cap V$ is a divisor in V (or $X' \cap V = V$). A simple formula for the intersection product in this case will be given in the next section. Note that even when one starts with subvarieties of Y, such reductions are possible only if one has intersection products for varieties *mapping* to Y.

An important special case of the pull-back property (b) is the case when Y'' is an open subscheme of Y'. For example, the part of an intersection product $i^! \alpha$ carried by a connected component of X' can be calculated by replacing Y' by any open neighborhood of the component.

Example 6.2.1. The intersection product of § 6.1 also determines the class $i^![V]$ for an arbitrary pure-dimensional scheme V, i.e.

$$i^![V] = X \cdot V$$

in $A_{k-d}(X')$. (To see this, if $C' = C_{X'}Y'$, and $W = V \cap X'$, it suffices to show that

$$\sigma[V] = [C_W V]$$

where $\sigma: A_k Y' \to A_k C'$ is the specialization homomorphism. From the construction of σ given in the proof of Proposition 5.2,

$$\sigma[V] = i^*[M^\circ_W V] .$$

If $[V] = \sum m_i [V_i]$ is the cycle of V, then, as in Lemma 4.2, $[M^\circ_W V] = \sum m_i [M^\circ_{V_i \cap W} V_i]$; an application of Proposition 2.6(d) completes the proof.) In particular, if Y is pure-dimensional, then

$$i^*[Y] = [X] .$$

Example 6.2.2. If V is not pure dimensional, the above must be modified. If M is the blow-up of $V \times \mathbb{P}^1$ along $W \times \infty$, $E = P(C \oplus 1)$ the exceptional divisor, $q: E \to W$ the projection, then

$$i^![V] = q_*(c_d(\xi) \cap E \cdot [M]) .$$

In this case, however, $E \cdot [M]$ is not necessarily the same as $[E]$ (cf. Example 2.6.4).

However, if $\dim V \leqq n$, and $[V]_n$ denotes the n-dimensional component of $[V]$, one always has the formula

$$i^!([V]_n) = s^*([C]_n) = q_*(c_d(\xi) \cap [E]_n)$$

where s is the zero-section of $g^* N_X Y$. (Apply Example 1.7.3 to $P(C \oplus 1) \hookrightarrow M_{X' \cap V} V$.)

Example 6.2.3. If E is a vector bundle of rank d on X, the zero-section s_E is a regular imbedding of codimension d, with normal bundle E, and the Gysin homomorphism $s_E^*: A_k E \to A_{k-d} X$ of this section agrees with the Gysin homomorphism constructed in Definition 3.3. (It suffices to check that $s_E^*[\pi^{-1} V] = [V]$, with $\pi: E \to X$ the projection, and this follows from Theorem 6.2(c).)

Example 6.2.4. Let (⁑) be a diagram as in Theorem 6.2(a). Let S be a closed subset of Y', $S' = p^{-1}(S)$, r the induced morphism from S' to S. Let V' be a subvariety of Y'', $V = p(V')$. Then

$$\deg(V'/V)(X \cdot V)^S = r_*((X \cdot V')^{S'}) .$$

The individual terms of the decomposition do not enjoy such a "birational invariance", however.

Example 6.2.5. Consider the diagram (*) of the beginning of this section. For any open subscheme Y'_o of Y', and any k-dimensional subvariety V of Y', the intersection class $X \cdot V$ in $A_{k-d} X'$ restricts to $X \cdot V_o$ in $A_{k-d}(X'_o)$, where $V_o = V \cap Y'_o$, $X'_o = X' \cap Y'_o$. In addition, for any closed subset S of Y', if $S_o = S \cap Y'_o$, the restriction

$$A_{k-d}(S) \to A_{k-d}(S_o)$$

takes $(X \cdot V)^S$ to $(X \cdot V_o)^{S_o}$. For example, if $\dim(S - S_o) < k - d$, this restriction is an isomorphism, so the part of $X \cdot V$ supported on S is determined by the part of $X \cdot V_o$ on S_o.

Example 6.2.6. Let δ be the diagonal imbedding of $\mathbb{P}^n$ in $\mathbb{P}^n \times \ldots \times \mathbb{P}^n$ (r factors). Let $[k]$ denote the generator of $A_k \mathbb{P}^n$ given by a k-plane in $\mathbb{P}^n$ (cf. Example 2.5.1). Then the Gysin homomorphism δ^* is determined by the formula $\delta^*([k_1] \times \ldots \times [k_r]) = [l]$ where $l = k_1 + \ldots + k_r - (r-1)n$. (Take the linear spaces in general position and apply Theorem 6.2(c)). For $V_1, \ldots, V_r$ closed subschemes of $\mathbb{P}^n$, with V_i of pure dimension k_i, it follows that $\delta^![V_1 \times \ldots \times V_r]$ is a cycle-class in $A_l(\bigcap V_i)$ whose degree is the product of the degrees of the V_i:

$$\deg \delta^* [V_1 \times \ldots \times V_r] = \prod_i \deg [V_i] .$$

(See Chapters 8 and 12 for more on Bézout's theorem.)

Example 6.2.7. Let E be a vector bundle of rank d on a scheme Y, s a regular section of E, $X = Z(s)$ the zero-scheme of s, i the (regular) imbedding of X in Y. Let α be a k-cycle on Y, $Y' = \operatorname{Supp}(\alpha)$, $X' = X \bigcap Y'$. Then

$$i^! \alpha = s_E^! (s_* \alpha) = s^! (s_E)_* (\alpha)$$

in $A_{k-d}(X')$, where s_E is the zero section of E. (Apply Theorem 6.2(c) to the diagram

$$(*) \qquad \begin{array}{ccc} X' & \rightarrow & Y' \\ \downarrow & & \downarrow \\ X & \xrightarrow{i} & Y \\ {\scriptstyle i}\downarrow & & \downarrow{\scriptstyle s} \\ X & \xrightarrow[s_E]{} & E \end{array}$$

and to the analogous diagram with s and s_E interchanged.)

Example 6.2.8. The results of this section extend with little change to the case when $i: X \rightarrow Y$ is a regular imbedding which is not assumed to be a closed imbedding. Let i be factored into a regular closed imbedding $i_o: X \rightarrow U$ followed by an open imbedding $U \subset Y$. Given a fibre square (*), define

$$i^!: A_k Y' \rightarrow A_{k-d} X'$$

to be the composite $A_k Y' \rightarrow A_k U' \rightarrow A_{k-d} X'$, where $U' = f^{-1}(U)$, the first map is the restriction homomorphism, and the second is $i_o^!$. This homomorphism $i^!$ is independent of the choice of U. (This is a special case of the Gysin map constructed in § 6.6.)

Example 6.2.9. The operations of intersection theory are compatible with field extension. For an algebraic scheme X over a field K, let X_L denote the scheme $X \otimes_K L$ over L. For a k-cycle $\alpha = \sum n_V [V]$ on X, let α_L be the k-cycle $\sum n_V [V_L]$ on X_L. This determines a homomorphism $\alpha \rightarrow \alpha_L$ from $A_k X$ to $A_k(X_L)$, which is compatible with proper push-forward, flat pull-back, Chern classes, and refined Gysin homomorphisms. (When L is a finite extension of

K, $\alpha \to \alpha_L$ is the flat pull-back for the projection $X_L \to X$, in which case the assertions have been proved in the text; the proofs for the general case are similar.)

6.3 Excess Intersection Formula

Consider a fibre diagram

$$(\ast\ast)\qquad \begin{array}{ccc} X'' & \longrightarrow & Y'' \\ {\scriptstyle q}\downarrow & & \downarrow{\scriptstyle p} \\ X' & \xrightarrow[i']{} & Y' \\ {\scriptstyle g}\downarrow & & \downarrow{\scriptstyle f} \\ X & \xrightarrow[i]{} & Y \end{array}$$

with i (resp. i') a regular imbedding of codimension d (resp. d') and normal bundle N (resp. N'). There is a canonical imbedding of N' in g^*N (see § 6.1). The quotient bundle

$$E = g^*N/N'$$

is a vector bundle of rank $e = d - d'$ on X'. We call E the *excess normal bundle* of the lower fibre square.

Theorem 6.3 (Excess Intersection Formula). *For any $\alpha \in A_k Y''$,*

$$i^!(\alpha) = c_e(q^*E) \cap i'^!(\alpha)$$

in $A_{k-d}(X'')$.

Proof. Let $Q' = P(q^*N' \oplus 1)$, $Q = P(q^*g^*N \oplus 1)$, and let ξ' and ξ be universal quotient bundles on Q' and Q. There is a canonical imbedding of Q' in Q, with the canonical line bundle on Q restricting to the canonical line bundle on Q'. There results an exact sequence of bundles

$$0 \to \xi' \to \xi|_{Q'} \to r^*(q^*E) \to 0$$

on Q; here r is the projection from Q' to X''. We may assume $\alpha = [V]$, V a subvariety of Y''. Set $P = P(C_{V \cap X''} V \oplus 1)$. Using the Whitney formula and the projection formula,

$$\begin{aligned} i^!(\alpha) &= r_*(c_d(\xi) \cap [P]) \\ &= c_e(q^*E) \cap r_*(c_{d'}(\xi') \cap [P]) \\ &= c_e(q^*E) \cap i'^!(\alpha) . \quad \square \end{aligned}$$

Corollary 6.3. *Let*

$$\begin{array}{ccc} X' & \xrightarrow{i'} & Y' \\ {\scriptstyle g}\downarrow & & \downarrow{\scriptstyle f} \\ X & \xrightarrow[i]{} & Y \end{array}$$

be a fibre square, with i a regular imbedding of codimension d and normal bundle N. Assume that i' is an isomorphism. Then

$$i^!(\alpha) = c_d(g^*N) \cap \alpha$$

for all $\alpha \in A_* Y'$. □

This includes the *self-intersection formula*

$$i^* i_*(\alpha) = c_d(N) \cap \alpha$$

for $\alpha \in A_*(X)$.

Remark 6.3. Given a diagram (*) as in § 6.2, and a class $\alpha \in A_k Y'$, if some connected component X'_o of X' is regularly imbedded in Y', then the excess intersection formula may be used for the part of $i^!(\alpha)$ supported on X'_o. As in Remark 6.2.2, this follows by restricting to open subschemes of Y'.

The fact that intersection products commute with Chern classes is a formal consequence of the properties proved so far.

Proposition 6.3. *Let $i: X \to Y$ be a regular imbedding of codimension d,*

$$(*) \qquad \begin{array}{ccc} X' & \xrightarrow{i'} & Y' \\ \downarrow & & \downarrow \\ X & \xrightarrow[i]{} & Y \end{array}$$

a fibre square, and let F be a vector bundle on Y'. Then for all $\alpha \in A_k(Y')$, and all $m \geqq 0$,

$$i^!(c_m(F) \cap \alpha) = c_m(i'^*F) \cap i^!(\alpha)$$

in $A_{k-d-m}(X')$.

Proof. Reduction Step. It suffices to find a proper morphism $h: \tilde{Y}' \to Y'$, and $\tilde{\alpha} \in A_k(\tilde{Y}')$ with $h_*(\tilde{\alpha}) = \alpha$, so that if we form the fibre square

$$\begin{array}{ccc} \tilde{X}' & \xrightarrow{\tilde{i}'} & \tilde{Y}' \\ {\scriptstyle h'}\downarrow & & \downarrow{\scriptstyle h} \\ X' & \xrightarrow[i']{} & Y' \end{array}$$

and set $\tilde{F} = h^*F$, then

$$i^!(c_m(\tilde{F}) \cap \tilde{\alpha}) = c_m(\tilde{i}'^*(\tilde{F})) \cap i^!(\tilde{\alpha})$$

in $A_{k-d-m}(\tilde{X}')$. This follows from the commutativity of Gysin homomorphisms and Chern class operations with push-forward (Theorem 6.2 (a) and Theorem 3.2 (c)):

$$\begin{aligned} i^!(c_m(F) \cap h_*\tilde{\alpha}) &= i^!(h_*(c_m(h^*F) \cap \tilde{\alpha})) = h'_* i^!(c_m(\tilde{F}) \cap \tilde{\alpha}) \\ &= h'_*(c_m(\tilde{i}'^*(\tilde{F})) \cap i^!\tilde{\alpha}) = c_m(i'^*F) \cap h'_* i^!(\tilde{\alpha}) \\ &= c_m(i'^*F) \cap i^!(h_*\tilde{\alpha}) . \end{aligned}$$

To prove the proposition, we first assume F is a line bundle, and $m = 1$. We may assume $\alpha = [V]$, V a k-dimensional subvarietiy of Y'. By the reduction step, we may replace Y' by the blow-up of V along $X' \cap V$. Thus we may assume Y' is a variety, $\alpha = [Y']$, and X' is either a Cartier divisor on Y' or $X' = Y'$. In case X' is a Cartier divisor, let E be the quotient of the normal bundles on X', as constructed in § 7.2. Then

$$\begin{aligned} i^!(c_1(F) \cap \alpha) &= c_{d-1}(E) \cap i'^!(c_1(F) \cap \alpha) && \text{(Theorem 6.3)} \\ &= c_{d-1}(E) \cap (c_1(i'^*F) \cap i'^!(\alpha)) && \text{(Proposition 2.6 (e))} \\ &= c_1(i'^*F) \cap (c_{d-1}(E) \cap i'^!(\alpha)) && \text{(Theorem 3.2 (b))} \\ &= c_1(i'^*F) \cap i^!(\alpha) && \text{(Theorem 6.3).} \end{aligned}$$

If $X' = Y'$ one uses similarly Corollary 6.3 with Theorem 3.2 (b).

For a general vector bundle F, and any m, let $F' = i'^* F$, and form the fibre square of projective bundles

$$\begin{array}{ccc} P(F') & \xrightarrow{\tilde{i}'} & P(F) \\ {\scriptstyle p'}\downarrow & & \downarrow{\scriptstyle p} \\ X' & \xrightarrow[i']{} & Y'. \end{array}$$

Let L be the canonical line bundle on $P(F)$, L' its restriction to $P(F')$. Since the Gysin homomorphisms commute with push-forward and pull-back (Theorem 6.2) and first Chern classes, we have

$$i^!(p_*(c_1(L)^{e+j} \cap p^*\alpha) = p'_*(c_1(L')^{e+j} \cap p'^*(i^!\alpha)),$$

or

$$i^!(s_j(F) \cap \alpha) = s_j(F') \cap i^!\alpha$$

for all α, j. Since the Chern classes are defined as polynomials in the Segre classes, the proposition follows. □

Example 6.3.1. Given a fibre square as in the beginning of this section, assume Y' is pure k-dimensional. Then

$$i^![Y'] = c_e(E) \cap [X']$$

in $A_{k-d}(X')$. (See Example 6.2.1.)

Example 6.3.2. If in a fibre square

$$\begin{array}{ccc} X' & \xrightarrow{j} & Y' \\ {\scriptstyle g}\downarrow & & \downarrow{\scriptstyle f} \\ X & \xrightarrow[i]{} & Y \end{array}$$

f, g, i and j are all regular imbeddings, then

$$g^*N_X Y/N_{X'} Y' \cong j^*N_{Y'} Y/N_{X'} X.$$

In other words, the excess normal bundle is independent of the orientation of the fibre square.

Example 6.3.3. Let X be a scheme. For any point $t \in \mathbb{P}^n$, rational over the ground field, let i_t be the imbedding of X in $X \times \mathbb{P}^n$ at t, i.e., $i_t(x) = (x, t)$. Let α

be a k-cycle on $X \times \mathbb{P}^n$. Then the classes $i_t^*(a)$ in $A_{k-n}(X)$ are independent of t. (By Theorem 3.3, write $\alpha = \sum_{j=0}^{n} \alpha_j \times [H^j]$, $\alpha_j \in A_{k-j}X$, H^j a j-plane in $\mathbb{P}^n$. Then $i_t^*(a) = \alpha_n$.)

Example 6.3.4. (a) If E is a vector bundle of rank d on a scheme Y, and s is a regular section of E, then the inclusion i of the zero-scheme $X = Z(s)$ in Y is a regular imbedding of codimension d, and $N_X Y$ is the restriction of E to X. If $f\colon Y' \to Y$ is a morphism, form the fibre square

$$(*) \qquad \begin{array}{ccc} X' & \xrightarrow{j} & Y' \\ {\scriptstyle g}\downarrow & & \downarrow{\scriptstyle f} \\ X & \xrightarrow[i]{} & Y. \end{array}$$

Then

$$j_* i^!(\alpha) = c_d(f^*E) \cap \alpha$$

for all α in A_*Y'. (Form the diagram ($\ast\!\ast$) as in Example 6.2.7; by Theorem 6.2(c), $i^!(\alpha) = s_E^!(\alpha)$. Taking $\alpha = [V]$, $V = Y'$, one has a fibre diagram

$$\begin{array}{ccc} X' & \xrightarrow{j} & Y' \\ \downarrow & & \downarrow{\scriptstyle f^*s} \\ Y' & \xrightarrow[s_{f^*E}]{} & f^*E \\ {\scriptstyle f}\downarrow & & \downarrow \\ Y & \xrightarrow[s_E]{} & E. \end{array}$$

Therefore $s_E^!(\alpha) = s_{f^*E}^!(\alpha)$, and one concludes by Example 6.2.3.)

Example 6.3.5. Let $0 \to E \to F \to G$ be an exact sequence of vector bundles on a scheme X. Let $\pi: F \to X$ be the projection, and let $g = \operatorname{rank} G$. Then

$$[E] = c_g(\pi^*G) \cap [F]$$

in A_*F.

Example 6.3.6. The refined Gysin homomorphism is uniquely determined by the excess intersection formula and the push-forward property. (If $\alpha = [V]$, $V \hookrightarrow Y'$, blow up V along $V \cap X'$.)

Example 6.3.7. Consider a fibre square

$$\begin{array}{ccc} X' & \xrightarrow{i'} & Y' \\ {\scriptstyle g}\downarrow & & \downarrow{\scriptstyle f} \\ X & \xrightarrow[i]{} & Y \end{array}$$

with i a regular imbedding of codimension d. Assume that $c_d(N_X Y) = 0$; for example, $N_X Y$ might have a nowhere vanishing section. Then there is a unique "specialization" map

$$\sigma: A_k(Y' - X') \to A_{k-d}(X')$$

which makes the diagram

$$\begin{array}{ccccc} A_k X' & \xrightarrow{i'_*} & A_k Y' & \xrightarrow{j'^*} & A_k(Y'-X') \to 0 \\ & & \downarrow{\scriptstyle i^!} & \swarrow{\scriptstyle \sigma} & \\ & & A_{k-d}X' & & \end{array}$$

commute, where j' is the inclusion of $Y'-X'$ in Y'. (The row is exact by Proposition 1.8, and $i^! i'_*(\alpha) = c_d(g^* N_X Y) \cap \alpha$ by Theorem 6.2 (a) and Corollary 6.3.)

In case $d=1$, and $\alpha = \sum n_i [V_i]$ is a cycle on $Y'-X'$, then

$$\sigma(\alpha) = \sum n_i [\bar{V}_i \bigcap X'],$$

where $\bar{V}_i$ is the closure of V_i in Y'.

6.4 Commutativity

In this section we prove that the refined Gysin homomorphisms defined in § 6.2 commute with each other. This will be done by blowing up to reduce to the case of divisors, which was proved in § 2.4.

Theorem 6.4. *Let $i: X \to Y$ be a regular imbedding of codimension d, $j: S \to T$ a regular imbedding of codimension e. Let Y' be a scheme, $f: Y' \to Y$, $g: Y' \to T$ two morphisms. Form the fibre diagram*

$$(\text{**}) \qquad \begin{array}{ccccc} X'' & \to & Y'' & \to & S \\ \downarrow & & \downarrow{\scriptstyle j'} & & \downarrow{\scriptstyle j} \\ X' & \xrightarrow[i']{} & Y' & \xrightarrow[g]{} & T \\ \downarrow & & \downarrow{\scriptstyle f} & & \\ X & \xrightarrow[i]{} & Y & & \end{array}$$

i.e., each of the three squares is a fibre square. Then for all $\alpha \in A_k Y'$,

$$j^! i^! (\alpha) = i^! j^! (\alpha)$$

in $A_{k-d-e} X''$.

Proof. Reduction Step. Let $\alpha \in A_k Y'$. Suppose $h: \tilde{Y}' \to Y'$ is a proper morphism, $\tilde{\alpha} \in A_k(\tilde{Y}')$, with $h_*(\tilde{\alpha}) = \alpha$. Form the fibre diagram

$$\begin{array}{ccccc} \tilde{X}'' & \to & \tilde{Y}'' & \to & S \\ \downarrow & & \downarrow & & \downarrow{\scriptstyle j} \\ \tilde{X}' & \to & \tilde{Y}' & \xrightarrow[\tilde{g}]{} & T \\ \downarrow & & \downarrow{\scriptstyle \tilde{f}} & & \\ X & \xrightarrow[i]{} & Y & & \end{array}$$

where $\tilde{f} = f \circ h$, $\tilde{g} = g \circ h$. If we prove that $j^! i^! (\tilde{\alpha}) = i^! j^! (\tilde{\alpha})$ in $A_{k-d-e}(\tilde{X}'')$, it follows from Theorem 6.2 (a) that $j^! i^! (\alpha) = i^! j^! (\alpha)$ in $A_{k-d-e}(X'')$. Indeed, if

$p: \tilde{X}' \to X', q: \tilde{Y}'' \to Y'', r: \tilde{X}'' \to X''$ are the morphisms induced by h, then

$$\begin{aligned} j^! i^! (\alpha) &= j^! i^! (h_* \tilde{\alpha}) = j^! p_* (i^! \tilde{\alpha}) = r_* j^! i^! (\tilde{\alpha}) = r_* i^! j^! (\tilde{\alpha}) = i^! q_* (j^! \tilde{\alpha}) \\ &= i^! j^! (h_* \tilde{\alpha}) = i^! j^! (\alpha) . \end{aligned}$$

By linearity, we may assume $\alpha = [V]$, V a subvariety of Y'. Applying the reduction step to the inclusion of V in Y' we may assume $V = Y'$. Let $h: \tilde{Y}' \to Y'$ be the blow-up of Y' along X'. By the reduction step again, we may assume either X' is a Cartier divisor on Y', or that $X' = Y'$. Similarly blowing Y' up along Y'', we may assume Y'' is a Cartier divisor on Y', or $Y'' = Y'$.

In case $X' = Y'$, let N' (resp. N'') be the pull-back to X' (resp. X'') of the normal bundle to X in Y. By Corollary 6.3 the homomorphisms $i^!$ are the Chern class operations of capping with $c_d(N')$ or $c_d(N'')$. Therefore

$$j^! i^! (\alpha) = j^! (c_d(N') \cap \alpha) = c_d(N'') \cap j^! (\alpha) = i^! j^! (\alpha)$$

by Proposition 6.3.

Thus we may assume X' and (by symmetry) Y'' are Cartier divisors on Y'. Let E (resp. F) be the excess normal bundle on X' (resp. Y'') constructed in § 6.3 for the square

$$\begin{array}{ccc} X' & \to & Y' \\ \downarrow & & \downarrow \\ X & \underset{i}{\to} & Y \end{array} \quad \left(\text{resp.} \begin{array}{ccc} Y'' & \to & Y' \\ \downarrow & & \downarrow \\ S & \underset{j}{\to} & T \end{array} \right).$$

Then by Theorem 6.3, and using the notation of Remark 3.3.2,

$$\begin{aligned} j^! i^! (\alpha) &= c_{e-1}(F) \cap j'^! (c_{d-1}(E) \cap i'^! (\alpha)) \\ &= c_{e-1}(F) \cap (c_{d-1}(E) \cap j'^! i'^! \alpha) \qquad \text{(Proposition 6.3)}. \end{aligned}$$

For the Cartier divisors X', Y'' on Y' we have the fundamental equation

$$Y'' \cdot [X'] = X' \cdot [Y'']$$

in $A_{k-2}(X'')$ (Theorem 2.4). Now

$$j'^! i'^! (\alpha) = j'^! [X'] = Y'' \cdot [X']$$

and

$$i'^! j'^! (\alpha) = i'^! [Y''] = X' \cdot [Y''] .$$

Therefore $i'^! j'^! (\alpha) = j'^! i'^! (\alpha)$, so

$$c_{e-1}(F) \cap (c_{d-1}(E) \cap j'^! i'^! \alpha) = c_{d-1}(E) \cap (c_{e-1}(F) \cap i'^! j'^! \alpha) ,$$

using the commutativity of Chern classes (Theorem 3.2 (b)). Reversing the previous argument, the right side is $i^! j^! (\alpha)$, which concludes the proof. □

6.5 Functoriality

We show that the refined Gysin homomorphisms for a composite of regular inbedding is the composite of the refined Gysin homomorphisms of the factors.

Theorem 6.5. *Consider a fibre diagram*

$$(**)\qquad \begin{array}{ccccc} X' & \xrightarrow{i'} & Y' & \xrightarrow{j'} & Z' \\ \downarrow h & & \downarrow g & & \downarrow f \\ X & \xrightarrow[i]{} & Y & \xrightarrow[j]{} & Z. \end{array}$$

If i (resp. j) is a regular imbedding of codimension d (resp. e), then $j\,i$ is a regular imbedding of codimension $d+e$, and for all $\alpha \in A_k Z'$,

$$(j\,i)^!(\alpha) = i^!(j^!(\alpha))$$

in $A_{k-d-e}X'$.

Proof. (For another proof, see Example 17.6.3.) The regularity of the imbedding $j\,i$ follows from Lemma A.5.2.

We first consider the case where $Z = E$ is a vector bundle over Y, $Z' = E'$ is the pull-back bundle g^*E, f is the natural map from g^*E to E, and j (resp. j') is the zero-section imbedding of Y in E (resp. Y' in E'). Let π (resp. π') be the bundle projection from E to Y (resp. E' to Y').

There is a canonical isomorphism of cones on X':

$$(1)\qquad C_{X'}E' \cong C_{X'}Y' \times_{X'} i'^*E'.$$

To see this, let $\mathscr{I}$ be the ideal sheaf defining X' in Y', and let $\mathscr{F}$ be the sheaf of sections of $(E')^\vee$, so that E' is $\operatorname{Spec}(S^\cdot)$, $S^\cdot = \operatorname{Sym}(\mathscr{F})$. The the ideal $\mathscr{J}$ of X' in E' is generated by $\mathscr{I} \subset S^0$ and by S^1. Then

$$\mathscr{J}^n = \mathscr{I}^n \oplus \mathscr{I}^{n-1} S^1 \oplus \ldots \oplus S^n \oplus S^{n+1} \oplus \ldots,$$

so

$$\mathscr{J}^n/\mathscr{J}^{n+1} = \mathscr{I}^n/\mathscr{I}^{n+1} \oplus (\mathscr{I}^{n-1}/\mathscr{I}^n \otimes S^1) \oplus \ldots \oplus (\mathscr{O}_{Y'}/\mathscr{I} \otimes S^n)$$

and

$$\oplus\, \mathscr{J}^n/\mathscr{J}^{n+1} = (\oplus\, \mathscr{I}^n/\mathscr{I}^{n+1}) \otimes_{\mathscr{O}_{X'}} (S^\cdot \otimes_{\mathscr{O}_{Y'}} \mathscr{O}_{X'}),$$

which proves (1).

In particular, the normal bundle $N_X E$ is a direct sum of $N_X Y$ and i^*E. Pulling back to X', this gives

$$(2)\qquad h^* N_X E = h^* N_X Y \oplus i'^* E'.$$

Let q be the projection from $h^*N_X E$ to $h^*N_X Y$, r the projection from $h^*N_X Y$ to X'; so $r\,q$ is the projection from $h^*N_X E$ to X'.

Assume also that Y' is an irreducible variety. From (1) and (2) it follows that

$$q^*[C_{X'}Y'] = [C_{X'}E']$$

in $A_*(E')$. From the construction of the refined Gysin homomorphism we have

$$[C_{X'}Y'] = r^* i^! [Y']$$

and

$$[C_{X'}E'] = (r\,q)^*(j\,i)^![E'] .$$

Since $(r\,q)^* = q^* r^*$ is one-to-one (Theorem 3.3(a)), the preceding three formulas give

$$(j\,i)^![E'] = i^![Y'] \tag{3}$$

in $A_*(X')$. It now follows for any Y', and any $\alpha \in A_k E'$, that

$$(j\,i)^!(\alpha) = i^! j^!(\alpha) . \tag{4}$$

For by Theorem 3.3(a), we may assume $\alpha = \pi'^*\beta$, and by linearity that $\beta = [V']$, V' an irreducible subvariety of Y'. By Theorem 6.2(a) we may replace Y' by V'. Then $\alpha = [E']$ and $j^!\alpha = [Y']$ by construction, and (4) therefore follows from (3).

To prove formula (4) in the general case, we may assume $\alpha = [Z']$, Z' an irreducible variety (Theorem 6.2(a)). Let $M = M_Y^\circ Z$ and $M' = M_{Y'}^\circ Z'$ be the deformation varieties constructed in § 5.1. For any $t \in \mathbb{P}^1$, rational over the ground field, let φ_t be the imbedding of $\{t\}$ in $\mathbb{P}^1$, and form the fibre diagram

$$\begin{array}{ccccc} X' \times \{t\} & \to & M'_t & \to & \{t\} \\ \downarrow & & \downarrow & & \downarrow \varphi_t \\ X' \times \mathbb{P}^1 & \xrightarrow{\varkappa'} & M' & \xrightarrow{\varrho'} & \mathbb{P}^1 \\ \downarrow & & \downarrow & & \\ X \times \mathbb{P}^1 & \xrightarrow[\varkappa]{} & M . & & \end{array} \tag{*}$$

Here $\varkappa$ is the composite of the inclusion $i \times \mathrm{id}$ of $X \times \mathbb{P}^1$ in $Y \times \mathbb{P}^1$ and the imbedding of $Y \times \mathbb{P}^1$ in M constructed in § 5.1. Since the fibres M'_t of M' over t in $\mathbb{P}^1$ are Cartier divisors on M',

$$\varphi_t^![M'] = [M_t^!] = \begin{cases} [Z'] & \text{if } t \neq \infty \\ [C_{Y'}Z'] & \text{if } t = \infty . \end{cases} \tag{5}$$

Consider the fibre diagrams

$$\begin{array}{ccc} X' \times \{t\} & \to & M'_t \\ \downarrow & & \downarrow \\ X \times \{t\} & \xrightarrow{\varkappa_t} & M_t \\ \downarrow & & \downarrow \\ X \times \mathbb{P}^1 & \xrightarrow{\varkappa} & M \end{array}$$

with $\varkappa_t = j\,i$ if $t \neq \infty$, and $\varkappa_\infty = \bar{j}\,i$ where $\bar{j}$ is the zero section imbedding of Y in $N_Y Z$. By Theorem 6.2(c),

$$\varkappa^! \varphi_t^! [M'] = \varkappa_t^! \varphi_t^! [M'] = \begin{cases} (j\,i)^![Z'] & \text{if } t \neq \infty \\ (\bar{j}\,i)^![C_{Y'}Z'] & \text{if } t = \infty . \end{cases} \tag{6}$$

But by the special case considered above,

$$(\bar{j}\,i)^![C_{Y'}Z'] = i^! \bar{j}^! [C_{Y'}Z'] = i^! j^! [Z'] , \tag{7}$$

the last equation from the construction of $j^!$. The conclusion then follows from (6) and (7), provided we know that $\varkappa^! \varphi_t^! [M']$ is independent of t. But by the fundamental commutativity result of § 6.4, applied to diagram (*),

$$\varkappa^! \varphi_t^! [M'] = \varphi_t^! \varkappa^! [M'] . \tag{8}$$

Let $\beta = \varkappa^! [M'] \in A_* (X' \times \mathbb{P}^1)$. The required claim follows from the fact that for any $\beta \in A_* (X' \times \mathbb{P}^1)$, the elements

$$\varphi_t^! (\beta) \in A_* (X' \times \{t\}) = A_* (X')$$

are independent of t (cf. Example 2.6.6 or 6.3.3). □

The following result is formally similar, but more straightforward to prove.

Proposition 6.5. Consider a fibre diagram

$$(**) \qquad \begin{array}{ccccc} X' & \xrightarrow{i'} & Y' & \xrightarrow{p'} & Z' \\ \downarrow h & & \downarrow g & & \downarrow f \\ X & \xrightarrow[i]{} & Y & \xrightarrow[p]{} & Z . \end{array}$$

(a) *Assume that i is a regular imbedding of codimension d, and that p and pi are flat of relative dimensions n and $n-d$. Then i' is a regular imbedding of codimension d, p' and $p' i'$ are flat, and for $\alpha \in A_k Z'$*

$$(p' i')^* (\alpha) = i'^* (p'^* \alpha) = i^! p'^* \alpha$$

in $A_{k+n-d} X'$.

(b) *Assume that i is a regular imbedding of codimension d, p is smooth of relative dimension n, and pi is a regular imbedding of codimension $d-n$. Then for all $\alpha \in A_k Z'$,*

$$(pi)^! (\alpha) = i^! (p'^* \alpha)$$

in $A_{k+n-d} X'$.

Proof. (a) The assertion that i' is a regular imbedding of codimension d follows from Lemma A.5.3. To prove the formula, we may, by our standard use of the projection formula, assume that $\alpha = [V]$, $V = Z'$, V a variety. Then $p'^* [V] = [Y']$, $(p' i')^* [V] = [X']$ by the definition of flat pull-back, and

$$i^! [Y'] = i'^* [Y'] = [X']$$

by Theorem 6.2 (c) and Example 6.2.1.

(b) Form the fibre square

$$\begin{array}{ccc} W & \xrightarrow{j} & Y \\ q \downarrow & & \downarrow p \\ X & \xrightarrow[pi]{} & Z & . \end{array}$$

Construct a corresponding square over this, induced by base extension from Z' to Z; denote corresponding schemes and morphisms in this square by corresponding letters with primes. The morphism i determines a section $s: X \to W$ with $q s = \mathrm{id}_X$. Since q is smooth, i is a regular embedding (Appendix B.7.3).

Therefore

$$\begin{aligned} i^! p'^* \alpha &= s^! j^! p'^* \alpha && \text{(Theorem 6.5)} \\ &= s^! q'^* (p\, i)^! \alpha && \text{(Theorem 6.2(b))} \\ &= (q' s')^* (p\, i)^! \alpha && \text{(Proposition 6.5(a))} \\ &= (p\, i)^! \alpha\,, \end{aligned}$$

since $q' s'$ is the identity on X'. □

Corollary 6.5. *Let E be a vector bundle of rank d on X, $\pi: E \to X$ the projection. Any section s of E is a regular imbedding, and*

$$s^*: A_k E \to A_{k-d} X$$

is the inverse isomorphism to π^*. In particular, s^* is independent of the choice of section s. If $Z(s)$ is the zero-scheme of s, and $\alpha \in A_k(E)$, the class $s^!(\alpha)$ in $A_{k-d}(Z(s))$ maps to $(\pi^*)^{-1}(\alpha)$ in $A_{k-d}(X)$.

Proof. By the proposition, $s^* \pi^* = (\pi s)^* = \mathrm{id}$; since π^* is surjective, both are isomorphisms. The last statement follows from Theorem 6.2(a). □

Example 6.5.1. (a) Let $i: V \to X$, $j: W \to X$ be regular imbeddings, and let α be a cycle on X. Then

$$i^! j^! \alpha = j^! i^! \alpha = (i \times j)^! (\alpha)$$

in $A_*(V \cap W \cap |\alpha|)$. Here $(i \times j)^!(\alpha)$ is constructed by intersecting α, diagonally imbedded in $X \times X$, by the subscheme $V \times W$. In case X is pure dimensional, and $\alpha = [X]$, this reads:

$$V \cdot W = W \cdot V = (V \times W) \cdot \Delta_X .$$

(Theorem 6.4 gives the first equality. For the second, form the diagram

$$\begin{array}{ccccc} & & V & \xrightarrow{\;i\;} & X \\ & & \downarrow & & \downarrow \delta \\ V \times W & \xrightarrow{1 \times j} & V \times X & \xrightarrow{i \times 1} & X \times X \\ \downarrow & & \downarrow & & \\ W & \xrightarrow{\;j\;} & X & & \end{array}$$

and apply Theorem 6.5 and Theorem 6.2(c).)

(b) The analogous formula is valid for more than two factors. For example, let $D_1, \ldots, D_r$ be effective Cartier divisors on a scheme X. Let i be the product imbedding of $D_1 \times \ldots \times D_r$ in $X \times \ldots \times X$. For any k-cycle α on X

$$i^!(\alpha) = D_1 \cdot \ldots \cdot D_r \cdot \alpha$$

in $A_{k-r}(D_1 \cap \ldots \cap D_r \cap |\alpha|)$. Here $D_1 \cdot \ldots \cdot D_r \cdot \alpha$ is constructed inductively by the process of § 2.2.

Example 6.5.2. Let $i_j: X_j \to Y_j$ be regular imbeddings of codimensions d_j, $j = 1, \ldots, r$. Let $f_j = Y_j' \to Y_j$ be morphisms, $\alpha_j \in A_{k_j}(Y_j')$. Then $i_1 \times \ldots \times i_r$ is a regular imbedding of $X_1 \times \ldots \times X_r$ in $Y_1 \times \ldots \times Y_r$, of codimension $\sum d_i$, and

$$(i_1 \times \ldots \times i_r)^! (\alpha_1 \times \ldots \times \alpha_r) = i_1^! (\alpha_1) \times \ldots \times i_r^! (\alpha_r)$$

in $A_{\sum(k_j-d_j)}(X'_1\times\ldots\times X'_r)$, $X'_i=X_i\times_{Y_i}Y'_i$. (One may assume $r=2$. Factoring $i_1\times i_2$ into $(1_{X_1}\times i_2)\circ(i_1\times 1_{Y_2})$, one may assume $i_2=1_{Y_2}$. Then the assertion follows from Theorem 6.2(c).)

Example 6.5.3. Commutativity was used in our proof of functoriality; conversely, using Cartesian products, functoriality implies commutativity. Given the situation of Theorem 6.4, form the fibre square

$$\begin{array}{ccc} X' & \longrightarrow & Y' \\ \downarrow & & \downarrow{\scriptstyle (f,g)} \\ X\times S & \xrightarrow[i\times j]{} & Y\times T \end{array} .$$

Then

$$j^! i^!(\alpha)=(i\times j)^!(\alpha)=i^! j^!(\alpha) .$$

(Factor $i\times j$ into $(i\times 1_T)(1_X\times j)$ and into $(1_Y\times j)(i\times 1_S)$.)

Example 6.5.4. In the situation of Proposition 6.5(b), one may prove more: if V is a subvariety of Z', and $W=p'^{-1}(V)$, then the intersection classes $X\cdot_Z V$ and $X\cdot_Y W$ have the same canonical decompositions. (There is a surjection from h^*N_XY onto h^*N_XZ (Appendix B.7.5), such that $C_{X'\cap W}W$ is the inverse image of $C_{X'\cap V}V$.)

6.6 Local Complete Intersection Morphisms

In this section, for simplicity[1], all schemes will be assumed to admit closed imbeddings into schemes which are smooth over the ground field. For example, all quasi-projective schemes are allowed. Any morphism $f:X\to Y$ then admits a factorization into a closed imbedding $i:X\to P$ followed by a smooth morphism $p:P\to Y$. For example, if j is a closed imbedding of X in a smooth M, one may take $P=Y\times M$, $i=(f,j)$, and p the projection from $Y\times M$ to Y.

A morphism $f:X\to Y$ is called a *local complete intersection* (l.c.i.) morphism of codimension d if f factors into a (closed) regular imbedding $i:X\to P$ of some (constant) codimension e, followed by a smooth morphism $p:P\to Y$ of (constant) relative dimension $d+e$. This notion is independent of the factorization of f into a closed imbedding followed by a smooth morphism (Appendix B.7.6).

For any l.c.i. morphism $f:X\to Y$ of codimension d, and any morphism $h:Y'\to Y$, form the fibre square

$$(*)\qquad \begin{array}{ccc} X' & \xrightarrow{f'} & Y' \\ {\scriptstyle h'}\downarrow & & \downarrow{\scriptstyle h} \\ X & \xrightarrow[f]{} & Y \end{array} .$$

1 What is actually needed for this treatment is that morphisms under consideration have factorizations into closed imbeddings followed by smooth morphisms, and that these may be chosen compatibly whenever a composite of morphisms is considered, as in the proof of Proposition 6.6(c).

Define a refined Gysin homomorphism

$$f^!: A_k Y' \to A_{k-d} X'$$

as follows; Factor f into $p \circ i$ as above, and form the fibre diagram

$$(**) \qquad \begin{array}{ccccc} X' & \xrightarrow{i'} & P' & \xrightarrow{p'} & Y' \\ {\scriptstyle h'}\downarrow & & \downarrow & & \downarrow{\scriptstyle h} \\ X & \xrightarrow[i]{} & P & \xrightarrow[p]{} & Y . \end{array}$$

Then p' is smooth, and we define

$$f^!(\alpha) = i^!(p'^* \alpha)$$

for $\alpha \in A_k Y'$. When $Y' = Y$, we write f^* for $f^!$.

Proposition 6.6. (a) *The definition of $f^!$ is independent of the factorization of f.*

(b) *If f is both a l.c.i. morphism and flat, then $f^! = f'^*$.*

(c) *The assertions stated in Theorems* 6.2, 6.4, *and* 6.5 *for regular imbeddings are valid for arbitrary l.c.i. morphisms. The excess intersection formula of Theorem 6.3 holds if the excess normal bundle E of a diagram* (*), *with f and f' l.c.i. morphisms, is defined to be*

$$E = h'^* N_X P / N_{X'} P'$$

with P, P' as in (**); *this definition of E is independent of choice of factorization.*

Proof. (a) If $X \xrightarrow{i_1} P_1 \xrightarrow{p_1} Y$ is another factorization of f, compare them both with the diagonal:

$$X \xrightarrow{(i,\, i_1)} P \times_Y P_1 \rightrightarrows \begin{array}{c} P_1 \xrightarrow{p_1} \\ P \xrightarrow{p} \end{array} Y .$$

The assertion then follows from Proposition 6.5 (b).

(b) follows from Proposition 6.5 (a).

(c) The extension of Theorems 6.2 and 6.4 are obvious, since the analogous statements are true for smooth (flat) morphisms as well as regular imbeddings. For the functoriality, if $f: X \to Y$ and $g: Y \to Z$ are l.c.i. morphisms, one chooses factorizations fitting into a commutative diagram

$$\begin{array}{ccccc} X & \to & P & \to & R \\ & {\scriptstyle f}\searrow & \downarrow & & \downarrow \\ & & Y & \to & Q \\ & & & {\scriptstyle g}\searrow & \downarrow \\ & & & & Z \end{array}$$

with the vertical morphisms smooth, the horizontal morphisms regular imbeddings, and the square a fibre square; for example; if $P = Y \times M$ as in the first paragraph of this section, one may take $R = Q \times M$. Then the functoriality $(gf)^! = f^! g^!$ follows from Theorem 6.5, using Theorem 6.2 (b) to go around the square.

To see that the excess bundle E is well-defined, by the diagonal construction used in (a) it suffices to compare factorizations $f=pi$ and $f=pqj$ where p, q are smooth, i, j are regular imbeddings, and $qj=i$. Consider the diagram

$$\begin{array}{ccccc} X' & \xrightarrow{j'} & Q' & \rightarrow & P' \\ {\scriptstyle h'}\downarrow & & \downarrow & & \downarrow \\ X & \xrightarrow[j]{} & Q & \xrightarrow[q]{} & P . \end{array}$$

There are exact sequences (Appendix B.7.5)

$$0 \rightarrow j^* T_{Q/P} \rightarrow N_X Q \rightarrow N_X P \rightarrow 0 ,$$

$$0 \rightarrow j'^* T_{Q'/P'} \rightarrow N_{X'} Q' \rightarrow N_{X'} P' \rightarrow 0 .$$

Since $T_{Q/P}$ pulls back to $T_{Q'/P'}$, there results a canonical isomorphism

$$h'^* (N_X Q)/N_{X'} Q' \cong h'^* N_X P/N_{X'} P' .$$

The excess formula then follows from Theorem 6.3 and Proposition 6.5(b). □

For more on l.c.i. morphisms see § 17.4.

6.7 Monoidal Transforms

Let X be a regularly imbedded subscheme of a scheme Y, of codimension d, with normal bundle N. Let $\tilde{Y}$ be the blow-up of Y along X, and let $\tilde{X} = P(N)$ be the exceptional divisor. We have a fibre square

$$(*) \qquad \begin{array}{ccc} \tilde{X} & \xrightarrow{j} & \tilde{Y} \\ {\scriptstyle g}\downarrow & & \downarrow{\scriptstyle f} \\ X & \xrightarrow[i]{} & Y . \end{array}$$

Since $N_{\tilde{X}} \tilde{Y} = \mathcal{O}_N(-1)$, the excess normal bundle E is the universal quotient bundle on $P(N)$:

$$E = g^* N/N_{\tilde{X}} \tilde{Y} = g^* N/\mathcal{O}_N(-1) .$$

We assume that Y can be imbedded in a non-singular scheme[2]. The same is then true for $\tilde{Y}$ (Appendix B.8.2). Then f is a l.c.i. morphism of relative dimension zero, in the sense of the preceding section. Indeed this is a local assertion, and the local case is proved in Lemma A.6.1.

Proposition 6.7. (a) (Key Formula). *For all* $x \in A_k X$,

$$f^* i_* (x) = j_* (c_{d-1} (E) \cap g^* x)$$

2 The precise assumption needed is that f factor into a closed imbedding followed by a smooth morphism.

in $A_k \tilde{Y}$.

(b) *For all* $y \in A_k Y$, $f_* f^* y = y$ *in* $A_k Y$.

(c) *If* $\tilde{x} \in A_k \tilde{X}$, *and* $g_* \tilde{x} = j^* j_* \tilde{x} = 0$, *then* $\tilde{x} = 0$.

(d) *If* $\tilde{y} \in A_k \tilde{Y}$, and $f_* \tilde{y} = j^* \tilde{y} = 0$, *then* $\tilde{y} = 0$.

(e) *There are split exact sequences*

$$0 \to A_k X \xrightarrow{\alpha} A_k \tilde{X} \oplus A_k Y \xrightarrow{\beta} A_k \tilde{Y} \to 0$$

with $\alpha(x) = (c_{d-1}(E) \cap g^* x, -i_* x)$, *and* $\beta(\tilde{x}, y) = j_* \tilde{x} + f^* y$. *A left inverse for* α *is given by* $(\tilde{x}, y) \to g_*(\tilde{x})$.

Proof. (a) By Theorem 6.2(a) and Theorem 6.3 (cf. Proposition 6.6),

$$f^* i_* x = j_* f^! x = j_*(c_{d-1}(E) \cap g^* x).$$

(b) One may assume $y = [V]$, for V a subvariety of Y. If $V \subset X$, then $y = i_* x$, $x = [V]$, and by (a),

$$f_* f^* y = f_* j_*(c_{d-1}(E) \cap g^* x)$$
$$= i_* g_*(c_{d-1}(E) \cap g^* x) = i_* x$$

(cf. Example 3.3.3). If $V \not\subset X$, let $\tilde{V} \subset \tilde{Y}$ be the blow-up of V along $V \cap X$. By Theorem 6.2(b),

$$f^*[V] = [\tilde{V}] + j_*(\tilde{x})$$

where $\tilde{x}$ is a class supported on $\tilde{V} \cap \tilde{X}$. Therefore

$$f_* f^*[V] = f_*[\tilde{V}] + f_* j_*(\tilde{x}) = [V] + i_* g_*(\tilde{x}).$$

But $g_*(\tilde{x})$ is a class supported on $V \cap X$, and $\dim(V \cap X) < k$, so $g_* \tilde{x} = 0$, as required.

(c) By Theorem 3.3(b), $\tilde{x} = \sum_{i=0}^{d-1} c_1(\mathcal{O}_N(1))^i \cap g^* x_i$ for some $x_i \in A_* X$. Then by Propositions 3.1(a) and 2.6(c)

$$0 = g_* \tilde{x} = x_{d-1},$$

and

$$0 = j^* j_* \tilde{x} = -\sum_{i=0}^{d-2} c_1(\mathcal{O}_N(1))^{i+1} \cap g^* x_i.$$

By the uniqueness assertion of Theorem 3.3(b), all $x_i = 0$, so $\tilde{x} = 0$.

We next verify that any $\tilde{y} \in A_k \tilde{Y}$ can be written in the form

$$\tilde{y} = f^* f_* \tilde{y} + j_*(\tilde{x})$$

for some $\tilde{x} \in A_k \tilde{X}$. Indeed, $\tilde{y} - f^* f_*(\tilde{y})$ restricts to zero on $\tilde{Y} - \tilde{X}$ (Theorem 6.2(b)), so it is the image of an element of $A_k \tilde{X}$ (Proposition 1.8).

(d) If $f_* \tilde{y} = j^* \tilde{y} = 0$, then by the preceding formula, $\tilde{y} = j_*(\tilde{x})$. Therefore $i_* g_* \tilde{x} = f_* \tilde{y} = 0$. Set

$$\tilde{x}' = \tilde{x} - c_{d-1}(E) \cap g^* g_* \tilde{x}.$$

By (a), $j_* \tilde{x}' = j_* \tilde{x} - f^* i_*(g_* \tilde{x}) = \tilde{y}$. But

$$g_*(\tilde{x}') = g_* \tilde{x} - g_*(c_{d-1}(E) \cap g^*(g_* \tilde{x})) = 0$$

(Example 3.3.3). Therefore, by (c), $\tilde{x}' = 0$, so $\tilde{y} = 0$.

(e) The surjectivity of β was verified before the proof of (d). That $\beta\alpha = 0$ is precisely (a). That the given map is a left inverse to α is Example 3.3.3. Finally, suppose

$$j_* \tilde{x} + f^* y = 0 .$$

By (b), $y = -f_* j_* \tilde{x} = -i_* g_* \tilde{x}$. Define $\tilde{x}'$ as in the proof of (d). Then $g_* \tilde{x}' = 0$, and

$$j_* \tilde{x}' = j_* \tilde{x} - f^* i_* (g_* \tilde{x}) = j_* \tilde{x} + f^* y = 0 .$$

By (c), $\tilde{x}' = 0$, i.e. $\tilde{x} = c_{d-1}(E) \cap g^*(g_* \tilde{x})$, so $(\tilde{x}, y) = \alpha(g_* \tilde{x})$, as required. □

Theorem 6.7 (Blow-up Formula). *Let V be a k-dimensional subvariety of Y, and let $\tilde{V} \subset \tilde{Y}$ be the proper transform of V, i.e. the blow-up of V along $V \cap X$. Then*

$$f^*[V] = [\tilde{V}] + j_* \{c(E) \cap g^* s(V \cap X, V)\}_k$$

in $A_k \tilde{Y}$.

Proof. If $V \subset X$, then $\tilde{V} = \emptyset$, $s(V \cap X, V) = [V]$, and the formula reduces to the key formula. So we assume $W = V \cap X \neq V$. By Proposition 6.7(d), it suffices to show that the two sides of the formula agree after applying f_*, and j^*.

When f_* is applied to the left side, one obtains $[V]$, by Proposition 6.7(b). Since $f_*[\tilde{V}] = [V]$, and $f_* j_* = i_* g_*$, to see that the two sides agree after applying f_* it suffices to note that

$$g_* \{c(E) \cap g^* s(W, V)\}_k = 0 .$$

By the projection formula, since $s(W, V)$ is a class on W, the left side of this equation is the image of a class on W. But $\dim W < k$, so the k-dimensional component of this class must be zero.

Now we apply j^* to each of the three terms in the blow-up formula:

$$\begin{aligned} j^* f^*[V] &= g^* i^*[V] = g^*(\{c(N) \cap s(W, V)\}_{k-d}) \\ &= \Big\{c(g^* N) \cap g^* g_* \Big(\sum_{i \geq 0} \zeta^i \cap [\tilde{W}]\Big)\Big\}_{k-1}, \end{aligned}$$

where $\zeta = c_1(\mathcal{O}_N(1))$ and $\tilde{W}$ is the exceptional divisor in $\tilde{V}$. For the second term,

$$j^*[\tilde{V}] = \tilde{X} \cdot \tilde{V} = [\tilde{W}] .$$

Since $c(E) = c(g^* N) \cdot c(\mathcal{O}_N(-1))^{-1}$ by the Whitney formula,

$$\begin{aligned} j^* j_*(\{c(E) \cap g^* s(W, V)\}_k &= -\{\zeta \cdot c(E) \cap g^*(s(W, V))\}_{k-1} \\ &= -\Big\{c(g^* N) \sum_{j \geq 1} \zeta^j \cap g^* g_* \Big(\sum_{i \geq 0} \zeta^i \cap [\tilde{W}]\Big)\Big\}_{k-1} . \end{aligned}$$

Setting $\beta = [\tilde{W}]$, $m = k-1$, and regarding the last three equations, it suffices to prove that

$$\Big\{c(g^* N) \sum_{j \geq 0} \zeta^j g^* g_* \Big(\sum_{i \geq 0} \zeta^i \cap \beta\Big)\Big\}_m = \beta$$

for all $\beta \in A_m \tilde{X}$. This is a formal identity, valid for any projective bundle. To verify it, by Theorem 3.3 we may assume $\beta = \zeta^q \cap g^* \alpha$, $q \leq d-1$, $\alpha \in A_* X$.

By Proposition 3.1,

$$g_*\Big(\sum_{i\geqq 0}\zeta^i\cap\beta\Big)=g_*\Big(\sum_{i\geqq 0}\zeta^i\cap g^*\alpha\Big)=s(N)\cap\alpha\,.$$

The left side of the identity is therefore

$$\Big\{\sum_{j\geqq 0}\zeta^j c(g^*N)\,s(g^*N)\cap g^*\alpha\Big\}_m=\Big\{\sum_{j\geqq 0}\zeta^j\cap g^*\alpha\Big\}_m=\beta\,,$$

as required. □

Corollary 6.7.1. *If $X=P$ is a point in Y, then*

$$f^*[V]=[\tilde{V}]+e_PV\,j_*[L]\,,$$

where L is a k-dimensional linear subspace of $E=\mathbb{P}_K^{d-1}$, K the residue field of $\mathcal{O}_{P,Y}$, and e_PV is the multiplicity of P on V (§ 4.3). □

Corollary 6.7.2. *If* $\dim V\cap X\leqq k-d$, *then*

$$f^*[V]=[\tilde{V}]\,.$$

Proof. In this case $g^*s(W,V)$ is supported on $g^{-1}(W)$, which has dimension $\leqq k-1$, so the other term is zero. □

Example 6.7.1. The key formula, and the more general blow-up formula, are also valid in refined versions: if $Y'\to Y$ is any morphism, and V is a k-dimensional subvariety of Y', then

$$f^![V]=[\tilde{V}]+j'_*\{c(E)\cap g'^*s(V\cap X',V)\}_k$$

in $A_k(\tilde{X}')$, where $X'=X\times_Y Y'$, $\tilde{X}'=\tilde{X}\times_Y Y'$ and the induced morphisms are denoted by primes. Likewise, $f'_*f^!y'=y'$ for any y' in A_*Y'. (The proofs are the same as those given in the absolute case. For a formalism which includes these generalities, see § 17.5.)

Notes and References

The construction and formulas for the general intersection product of § 6.1 were first given in Fulton-MacPherson (1)[3]. The properties of the refined Gysin homomorphisms were sketched in Fulton-MacPherson (3), where the formalism of bivariant theories provided a useful guide (cf. Chap. 17). Gysin homomorphisms for regular imbeddings (the case $Y'=Y$) had been constructed by Verdier (5).

The construction of refined intersections via specialization to normal cones that we use in § 6.2 follows Verdier closely. Our proof of functoriality (§ 6.5) is also modelled on Verdier's proof for unrefined Gysin maps.

---

3 To be precise, this paper relied on the intersection theory which had been developed previously for non-singular quasi-projective varieties.

The excess intersection formula (§ 6.3) has many precedents. In the non-singular case, the self-intersection formula had been proved by Mumford in 1959, the key formula by Jouanolou (1) § 4.1; see also Lascu-Mumford-Scott (1). A topological version of the general formula was given by Quillen (1); J. King (2) gave an analytic analogue. Illusie asked us if such a formula was known for rational equivalence. Such an excess intersection formula was given in Fulton-MacPherson (1) and by H. Gillet (unpublished).

The formula for intersection classes in terms of Segre classes of cones first appeared in Fulton-MacPherson (1). As mentioned in the notes to Chap. 4, such classes were constructed in many cases by B. Segre, who stressed the importance of blowing up to simplify problems in intersection theory.

The extension from regular imbeddings to l.c.i. morphisms follows the formalism of [SGA6]. Kleiman (12) has also developed and applied this extension.

In the case of smooth quasi-projective varieties, most of Proposition 6.7 was proved by Jouanoulou (1) § 9, by essentially the same calculations; the case of codimension 2 had been done by Samuel.

The blow-up formula of Theorem 6.7 is apparently new, even in the non-singular case.

Example 6.1.4 comes from R. Lazarsfeld.

Chapter 7. Intersection Multiplicities

Summary

As in Chap. 6, consider a fibre square

$$\begin{array}{ccc} W & \rightarrow & V \\ \downarrow & & \downarrow f \\ X & \underset{i}{\rightarrow} & Y \end{array}$$

with i a regular imbedding of codimension d, V a k-dimensional variety. If Z is an irreducible component of W of dimension $k-d$, the intersection multiplicity $i(Z, X\cdot V; Y)$ is defined to be the coefficient of Z in the intersection class $X\cdot V \in A_{k-d}(W)$. The intersection multiplicity is a positive integer, satisfying

$$i(Z, X\cdot V; Y) \leqq \text{length}(\mathscr{O}_{Z,W}) .$$

Examples show that this inequality may be strict; equality holds, however, if $\mathscr{O}_{Z,V}$ is a Cohen-Macaulay ring.

On the other hand, the criterion of multiplicity one asserts that $i(Z, X\cdot V; Y)$ is one precisely when $\mathscr{O}_{Z,V}$ is a regular local ring with maximal ideal generated by the ideal of X in Y.

The standard properties of intersection multiplicities, worked out in the examples, follow from the basic properties of the general intersection product which were proved in Chapter 6.

7.1 Proper Intersections

Consider, as in § 6.1, a fibre square

$$\begin{array}{ccc} W & \overset{j}{\rightarrow} & V \\ g\downarrow & & \downarrow f \\ X & \underset{i}{\rightarrow} & Y \end{array}$$

with i a regular imbedding of codimension d, V a purely k-dimensional scheme. Let

$$C = C_W V, \quad [C] = \sum_{i=1}^{r} m_i [C_i] ,$$

C_i the irreducible components of C, and let Z_i be the support of C_i; $Z_1, \ldots, Z_r$ are the distinguished varieties of the intersection.

Lemma 7.1. (a) *Every irreducible component of W is distinguished.*
(b) *For any distinguished variety Z,*

$$k-d \leqq \dim Z \leqq k .$$

Proof. (a) follows from the fact that the support of $C_W V$ is W, for any closed subscheme W of a scheme V. Since C_i is an irreducible subvariety of $g^* N_X Y$ which projects onto Z_i, if N_i is the restriction of $g^* N_X Y$ to Z_i, then

$$C_i \subset N_i .$$

Therefore

$$\dim Z_i \leqq \dim C_i \leqq \dim (N_i) = \dim Z_i + d .$$

Since C_i is k-dimensional (Appendix B.6.6), (b) follows. □

If $\dim Z_i = k-d$, the inclusion $C_i \subset N_i$ of irreducible k-dimensional varieties must be an isomorphism. In particular, the class α_i obtained by intersecting $[C_i]$ with the zero-section of N_i is just $[Z_i]$, and the equivalence of Z_i for the intersection is $m_i[Z_i]$.

Definition 7.1. An irreducible component Z of $W = f^{-1}(X)$ is a *proper* component of intersection of V by X if $\dim(Z) = k-d$. The *intersection multiplicity* of Z in $X \cdot V$, denoted

$$i(Z, X \cdot V; Y)$$

or simply $i(Z, X \cdot V)$, or $i(Z)$, is the coefficient of Z in the class $X \cdot V$ in $A_{k-d}(W)$. Equivalently, the equivalence of Z for the intersection class is

$$i(Z, X \cdot V; Y)[Z] .$$

If N_Z is the pull-back of $N_X Y$ to Z, then $i(Z, X \cdot V; Y)$ is the coefficient of N_Z in the cycle $[C]$ of the cone $C = C_W V$.

Let $A = \mathcal{O}_{Z,V}$ be the local ring of V along Z, and let $J \subset A$ be the ideal of W; A/J has finite length when Z is an irreducible component of W.

Proposition 7.1. *Assume Z is a proper component of W. Then*
(a) $1 \leqq i(Z, X \cdot V; Y) \leqq l(A/J)$, *where $l(A/J)$ is the length of A/J.*
(b) *If J is generated by a regular sequence of length d, then*

$$i(Z; X \cdot V; Y) = l(A/J) .$$

If A is Cohen-Macaulay (e.g. regular) the local equations for X in Y give a regular sequence generating J, and the equality in (b) holds.

Proof. Let $N = g^* N_X Y$. The restriction N_Z of N to Z is an irreducible component of N. Since N is a vector bundle over W, the coefficient of N_Z in the cycle $[N]$ is the same as the coefficient of Z in the cycle $[W]$, which is $l(A/J)$. Since C is a closed subscheme of N, the coefficient of any irreducible component of N is no larger in $[C]$ than it is in $[N]$ (Lemma A1.1). Since the coefficient of N_Z in $[C]$ is $i(Z, X \cdot V; Y)$, (a) follows.

If J is generated by a regular sequence of d elements, replacing V by an open subscheme which meets Z (which doesn't effect the intersection multiplicity, by Theorem 6.2(b)), we may assume the imbedding of W in V is regular of codimension d. Then C is a sub-bundle of N of rank d, so $C=N$, and the coefficients of N_Z in $[C]$ and in $[N]$ coincide.

The last assertion of the proposition follows from Lemma A.7.1. □

The inequalities in (a) may be strict, as shown by Macaulay (Example 7.1.5).

Example 7.1.1. Let $(e_W V)_z$ be the multiplicity of V along W at Z, as defined in Example 4.3.4. Then

$$i(Z, X \cdot V; Y) = (e_W V)_Z$$

i.e., the intersection multiplicity defined here agrees with Samuel's.

Example 7.1.2. Let $a_1, \ldots, a_d$ be the images in A of a regular sequence of elements defining X in Y (locally, in an open set which meets $f(Z)$). Then

$$i(Z, X \cdot V; Y) = \chi_A(\mathbf{a}) = e_A(a_1, \ldots, a_d)$$

where $\chi_A(\mathbf{a}) = \sum_{i=0}^{d} l_A(H_i(K_*(\mathbf{a})))$, with $K_*(\mathbf{a})$ the Koszul complex defined by $a_1, \ldots, a_d$ (Appendix A.5). Serre (4)IV.A3 showed generally that $\chi_A(\mathbf{a})$ gives Samuel's multiplicity. We sketch an alternate proof, by induction on d. If $d=1$ it says

$$\sum l_{A_p}(A_p) \cdot l_A(A/p + aA) = e_A(a, A)\,,$$

where the sum is over the minimal primes p of A; this is a special case of Lemma A.2.7. For the inductive step, localize so that one has a fibre diagram

$$\begin{array}{ccccc} W & \subset & W' & \subset & V \\ \downarrow & & \downarrow & & \downarrow \\ X & \subset & X' & \subset & Y\,, \end{array}$$

with $X' \subset Y$ and $X \subset X'$ regular imbeddings of codimension $d-1$ and 1, and local equations pulling back to $a_2, \ldots, a_d$ in A and $\bar{a}_1$ in $A/(a_2, \ldots, a_d)$ respectively; we may also assume the localization is sufficient so that Z is the only irreducible component of W, and all the irreducible components W'_i of W' contain Z, and therefore have dimension $k-d+1$. Let p_i be the prime ideal of A corresponding to W'_i. By induction

$$X' \cdot_Y V = \sum_i e_{A_{p_i}}(a_2, \ldots, a_d)\,[W'_i]\,.$$

By functoriality (§ 6.5), $X \cdot_Y V = X \cdot_{X'} (X' \cdot_Y V)$. Let $H_k = H_k(K_*(a_2, \ldots, a_d))$. Then

$$\begin{aligned} i(Z, X \cdot V; Y) &= \sum e_{A_{p_i}}(a_2, \ldots, a_d)\, l_A(A/p_i + a_1 A) \\ &= \sum_{i,k} (-1)^k l_{A_{p_i}}((H_k)_{p_i}) \cdot l_A(A/p_i + a_1 A) \\ &= \sum_k (-1)^k e_A(a_1, H_k) = e_A(a_1, \ldots, a_d) \end{aligned}$$

by Lemma A.2.7 and Example A.5.1. (Note that each H_k has support in $V(a_2, \ldots, a_d)$, so H_k is an $\bar{A}$-module for $\bar{A}=A/(a_2^m, \ldots, a_d^m)$, some $m>0$ – in fact $m=1$ will do; Lemma A.2.7 may be applied over the one-dimensional ring $\bar{A}$.)

Example 7.1.3. With the notation of the preceding example, the following are equivalent:

(i) $i(Z, X \cdot V; Y) = l(A/J)$.
(ii) J is generated by a regular sequence of length d.
(iii) $a_1, \ldots, a_d$ is a regular sequence in A.
(iv) $H_k(K_*(\mathbf{a})) = 0$ for all $k>0$.

In particular, $i=l$ if and only if A is Cohen-Macaulay. Algebraic proofs are given by Serre (4)IV. To prove directly that (i) implies (iii), the main point is to show that the equality of cycles $[C]=[N]$ implies that $C=N$, at least after replacing V by an open subset which meets Z. (Indeed, if A is an Artin local ring, and Q is a homogeneous ideal in $A[T_1, \ldots, T_d]$ whose localization at the minimal prime is zero, then $Q=0$.)

Example 7.1.4. Let

$$Y=\mathbb{A}^4, \quad X=V(x_1-x_3, x_2-x_4), \quad V=V(x_1x_3, x_1x_4, x_2x_3, x_2x_4).$$

Then V is purely 2-dimensional, $[V]=[V(x_1, x_2)]+[V(x_3, x_4)]$. The intersection number of the origin in $X \cdot V$ is 2, while $l(A/J)=3$ (cf. Hartshorne (5) p. 428.)

Example 7.1.5. Let $Y=\mathbb{A}^4$, $X=V(x_1, x_4)$, and let $V \subset \mathbb{A}^4$ be the image of the finite morphism φ from $\mathbb{A}^2$ to $\mathbb{A}^4$ given by

$$\varphi(s,t)=(s^4, s^3t, st^3, t^4).$$

The origin P is a proper component of the intersection of V by X.

(i) $I(V)=(x_1x_4-x_2x_3, x_1^2x_3-x_2^3, x_2x_4^2-x_3^3, x_2^2x_4-x_3^2x_1)$.
(ii) $[X \cap V]=5[P]$, $l(A/J)=5$.
(iii) $i(P, X \cdot V; Y)=4$.

(For (iii), note that $\varphi_*[\mathbb{A}^2]=4[V]$. Apply Theorem 6.2 (a) to the situation

$$\begin{array}{ccc} V(s^4, t^4) & \rightarrow & \mathbb{A}^2 \\ {\scriptstyle\psi}\downarrow & & \downarrow{\scriptstyle\varphi} \\ X \cap V & \rightarrow & V \\ \downarrow & & \downarrow \\ X & \underset{i}{\rightarrow} & Y. \end{array}$$

giving that $\psi_*(i^![\mathbb{A}^2])=i^!\varphi_*[\mathbb{A}^2]=4X \cdot V$. Since $\mathbb{A}^2$ is regular, Proposition 7.1 (b) gives

$$i(Q, X \cdot \mathbb{A}^2; Y)=l(K[s,t]/(s^4, t^4))=16.$$

with Q the origin in $\mathbb{A}^2$. Therefore $16=4i(P, X \cdot V; Y)$, as required.) Note that the kernel of multiplication by x_4 on A/x_1A has length 1, which accounts for the difference between (ii) and (iii).

Example 7.1.6. Without regularity assumptions, irreducible components Z of $X \cap V$ may have dimension smaller than $\dim V - \operatorname{codim}(X, Y)$. The standard example is $Y = V(x_1 x_4 - x_2 x_3) \subset \mathbb{A}^4$, $X = V(x_1, x_2)$, $V = V(x_3, x_4)$.

Example 7.1.7 (Commutativity). If $V \to Y$ is also a regular imbedding, then by Theorem 6.4,

$$i(Z, X \cdot V; Y) = i(Z, V \cdot X; Y) .$$

Example 7.1.8 (Associativity). Let $i: X \to Y$ factor into a composite $i': X \to X'$, $j: X' \to Y$ of regular imbeddings. Let $W'_1, \ldots, W'_r$ be the irreducible components of $f^{-1}(X')$ which contain Z. If Z is a proper component of W, then Z is a proper component of the intersection of each W'_h by X on X', each W'_h is a proper component of the intersection of V by X' on Y, and by Theorem 6.5,

$$i(Z, X \cdot V; Y) = \sum_{h=1}^{r} i(Z, X \cdot W'_h; X') \cdot i(W'_h, X' \cdot V; Y) .$$

Example 7.1.9 (Projection formula). Let $g: V' \to V$ be a proper surjective morphism of k-dimensional varieties, and let $Z_1, \ldots, Z_r$ be the irreducible components of $g^{-1}(Z)$. If each Z_j and Z have dimension $k-d$, then by Proposition 6.2 (a),

$$\deg(V'/V) \cdot i(Z, X \cdot V; Y) = \sum_{j=1}^{r} \deg(Z_j/Z) \cdot i(Z_j, X \cdot V'; Y) .$$

Example 7.1.10. (a) Let $D_1, \ldots, D_d$ be effective Cartier divisors in a k-dimensional variety V. An irreducible component Z of $\bigcap D_i$ of dimension $k-d$ is called a *proper* component, and the *intersection multiplicity*

$$i(Z, D_1 \cdot \ldots \cdot D_d; V)$$

is defined to be the intersection multiplicity of Z in the intersection of $V = \Delta_V$ by $D_1 \times \ldots \times D_d$:

$$\begin{array}{ccc} \bigcap D_i & \to & V \\ \downarrow & & \downarrow \delta \\ D_1 \times \ldots \times D_d & \to & V \times \ldots \times V . \end{array}$$

Equivalently (cf. Example 6.5.1), $i(Z, D_1 \cdot \ldots \cdot D_d; V)$ is the coefficient of Z in the intersection cycle $D_1 \cdot \ldots \cdot D_d$ in $A_{k-d}(\bigcap D_i)$ (Definition 2.4.2). If A is the local ring of V along Z, and a_i is a local equation for D_i in A, then

$$i(Z, D_1 \cdot \ldots \cdot D_d; V) = e_A(a_1, \ldots, a_d).$$

If A is Cohen-Macaulay, then

$$i(Z, D_1 \cdot \ldots \cdot D_d; V) = l_A(A/(a_1, \ldots, a_d)) .$$

For example, if $d = k$, and Z is a simple point on V, the intersection multiplicity is given by the length.

(b) Let $D_1, \ldots, D_d$ by hypersurfaces in $\mathbb{A}^d_K$ defined by polynomials $f_1, \ldots, f_d$, and assume $P = (0, \ldots, 0)$ is an isolated (i.e. proper) point of intersection of $\bigcap D_i$. Then

$$i(P, D_1 \cdot \ldots \cdot D_d; \mathbb{A}^d) = \dim_K(\mathcal{O}_{P,\mathbb{A}^d}/(f_1, \ldots, f_d))$$
$$= \dim_K(K[\![x_1, \ldots, x_d]\!]/(f_1, \ldots, f_d)) .$$

If $K = \mathbb{C}$, one may replace formal power series by convergent power series in the last formula. (Note that modules of finite length are not altered by completion, cf. Zariski-Samuel (1) VIII.2.)

Example 7.1.11. Let V be an n-dimensional variety, P a simple point on V, $\pi: \tilde{V} \to V$ the blow-up of V at P, E the exceptional divisor. For an effective Cartier divisor D on V, let $\tilde{D}$ be the blow-up of D along P, i.e. the proper transform of D in $\tilde{V}$. *If* $D_1, \ldots, D_n$ are divisors such that $\bigcap_i \tilde{D}_i \bigcap E$ is finite, then

$$i(P, D_1 \cdot \ldots \cdot D_n; V) = \prod_{i=1}^{n} e_P(D_i) + \sum_{Q \in E} i(Q; \tilde{D}_1 \cdot \ldots \cdot \tilde{D}_n; \tilde{V}) .$$

If $n = 2$, and D_1 and D_2 meet properly at P, it follows that the intersection multiplicity $i(P, D_1 \cdot D_2; X)$ is the sum of the products of multiplicities of D_1 and D_2 at all infinitely near points, a result of M. Noether. (By Example 4.3.9, $\pi^* D_i = \tilde{D}_i + e_P(D_i) E$. Write out the product of the $\tilde{D}_i$, and push forward to V.) Generalizations will be given in Example 12.4.8.

Example 7.1.12. The equivalence of a distinguished variety Z of the minimal dimension $k-d$ is always a positive multiple of $[Z]$. If $\dim Z > k-d$, the equivalence of Z may be represented by negative cycles. For example, if Y is the blow-up of a surface at a simple point, and $X = V = E$ is the exceptional divisor, then $Z = E$ is the only distinguished variety, and its equivalence is $-[P]$, P a point on E.

Example 7.1.13. Let E be a vector bundle of rank r on a purely n-dimensional scheme X, s a section of E, $Z(s)$ the zero-scheme of s:

$$(*) \qquad \begin{array}{ccc} Z(s) & \to & X \\ \downarrow & & \downarrow s \\ X & \underset{s_E}{\to} & E \end{array}$$

where s_E is the zero section. If Z is a proper component of $Z(s)$, i.e. $\dim Z = n - r$, then the intersection construction from (*) determines an intersection multiplicity $i(Z)$. If A is the local ring of X at Z, the stalk of E at Z is a free A-module with an induced section s_A, which determines a Koszul complex $\Lambda^*(s_A)$ (Definition A.5). Then

$$i(Z) = \chi_A(s_A) .$$

Example 7.1.14. Let $f: X \to C$ be a morphism from a smooth n-dimensional variety to a smooth curve C. The tangent map $df: T_X \to f^* T_C$ corresponds to a section s of $T_X \otimes f^* \Omega_C^1$. If $x \in X$ is an isolated zero of this section, the intersection multiplicity of x in the intersection of $s(X)$ by the zero section (as in the preceding example) is called the *multiplicity of x as a critical point* of f, and denoted $\mu_x(f)$. If f is given in local coordinates by a function

$f(z_1,\ldots,z_n)$, then

$$\mu_x(f) = l(\mathscr{O}_{x,X}/(\partial f/\partial z_1,\ldots,\partial f/\partial z_n)).$$

For a discussion of this multiplicity from an analytic and topological point of view, see Milnor (3) and Orlik (1), cf. Example 14.1.5.

Example 7.1.15. Let $f\colon Y'\to Y$ be a proper surjective morphism of varieties. Let X' be a subvariety of Y', $X=f(X')$; assume that $\dim X=\dim X'$, and X' is an irreducible component of $f^{-1}(X)$, and that X is regularly imbedded in Y. Then the ramification index of f at X' (Example 4.3.7) is given by an intersection multiplicity

$$e_{X'}(f) = i(X', X\cdot Y'; Y).$$

This applies in particular if f is finite and X is a simple point of Y.

Example 7.1.16. *Fractional intersection numbers on normal surfaces,* (cf. Mumford (1)II(b), Reeve (2)). Let $\pi\colon X\to V$ be a resolution of a singular point P on a surface V, $\pi^{-1}(P)=E_1\bigcup\ldots\bigcup E_r$ connected, as in Example 2.4.4. For an irreducible curve A on V, there are unique rational numbers $\lambda_1,\ldots,\lambda_r$ so that if $\tilde{A}$ is the proper transform of A on X,

$$(\tilde{A}\cdot E_i)_X+\sum_{j=1}^{r}\lambda_j(E_j\cdot E_i)_X=0$$

for all i. Set $A'=\tilde{A}+\sum\lambda_i E_i\in Z_1X_{\mathbb{Q}}=Z_1X\otimes_{\mathbb{Z}}\mathbb{Q}$. This extends to a homomorphism $\alpha\to\alpha'$ from Z_1V to $Z_1X_{\mathbb{Q}}$, satisfying

(i) $[D]'=[\pi^*D]$ for any Cartier divisor D on V.
(ii) If A is positive and contains P, then all the λ_i are positive.

(For (i), $(\pi^*D\cdot E_i)_X=(D\cdot\pi_*E_i)_V=0$. For (ii), with D_i, Z as in Example 2.4.4, let $A'=\tilde{A}+\sum\mu_jD_j$, with μ_i minimal among the μ_j, $\mu_i\leqq 0$. Then $0=A'\cdot D_i\geqq\sum_j\mu_j(D_j\cdot D_i)_X\geqq\mu_i(\sum_j(D_j\cdot D_i)_X=-\mu_i(Z\cdot D_i)_X\geqq 0$; the connectedness of E then implies that all μ_j are zero.)

For any two one-cycles A, B on V which meet only at P, set

$$j(P,A\cdot B)=(A'\cdot B')_X\in\mathbb{Q}.$$

(This is defined since $|A'|\bigcap|B'|\subset E$, which is complete.) This intersection number is symmetric and bilinear; it is non-negative if either A or B is positive, and positive if A and B are positive and pass through P. If $A=[D]$ is the Weil divisor of a Cartier divisor D, then

$$j(P,A\cdot B)=(D\cdot B)_V\in\mathbb{Z}.$$

This definition of multiplicity is independent of the resolution. (If $\varrho\colon\tilde{X}\to X$ blows up a point on X, ϱ^*A' is perpendicular to all exceptional components, and $(\varrho^*A'\cdot\varrho^*B')_{\tilde{X}}=(A'\cdot B')_X$.)

If X is a quadric cone with vertex P, and A and B are generating lines of the cone, then $j(P,A\cdot B)=1/2$.

Example 7.1.17. Let C be an irreducible curve on a scheme X, D an effective Cartier divisor on X, with C not contained in the support of D. Let

$f: C' \to X$ be a finite morphism which maps an irreducible curve C' birationally onto C. Then

$$i(P, D \cdot C; X) = \sum_{f(Q)=P} \operatorname{ord}_Q(f^* D) .$$

(Use Theorem 6.2(a).) For example if $C' = \mathbb{P}^1$, f^*D is given by a polynomial, and the intersection multiplicities are given by multiplicities of roots of this polynomial.

7.2 Criterion for Multiplicity One

Let Z be a proper component of the intersection of V by X on Y. Let $A = \mathcal{O}_{Z,V}$, J the ideal in A generated by the ideal of X in Y, and let m be the maximal ideal of A.

Proposition 7.2. *Assume that V is a variety. The following are equivalent:*

(i) $i(Z, X \cdot V; Y) = 1$.
(ii) *A is a regular local ring, and $J = m$.*

Recall that a d-dimensional local ring is regular if its maximal ideal has d generators, which necessarily form a regular sequence (Lemma A.6.2). Since J always has d generators, the regularity of A follows from the assertion that $J = m$.

Proof. The implication (ii) $\Rightarrow$ (i) is a special case of Proposition 7.1, since $l(A/m) = 1$. We prove (i) $\Rightarrow$ (ii) by induction on d. The assertions are unchanged if V and Y are replaced by open subschemes which meet Z and $f(Z)$ respectively. Therefore, we may assume that Z is the only irreducible component of W, that Y is affine, and X is defined in Y by a regular sequence in the coordinate ring of Y.

If $d = 1$, then $X \cdot V = [W]$. The coefficient of Z in $[W]$ is $l_A(A/J)$, which can be 1 only if $J = m$.

Let $d > 1$, and assume (i) $\Rightarrow$ (ii) for smaller d. Assume first that A is a normal domain. Let X' be the divisor on Y defined by the first of the equations defining X. Form the fibre diagram

$$\begin{array}{ccccc} W & \to & W' & \to & V \\ \downarrow & & \downarrow & & \downarrow \\ X & \underset{i'}{\to} & X' & \to & Y . \end{array}$$

Since $\dim W = k - d$, $W' \neq V$, so W' is a Cartier divisor on V and $X' \cdot_Y V = [W']$. By functoriality (Theorem 6.5),

$$i'^![W'] = i^![V] = [Z] .$$

In particular W' can have only one irreducible component which contains Z, and this component appears in the cycle $[W']$ with coefficient 1. In other words, $a_1 A$ has only one minimal prime ideal p containing it, and

$$l_{A_p}(A_p/a_1 A_p) = 1 .$$

Since A is a normal domain, p is the only prime ideal associated to $a_1 A$ (Lemma A.8.1), so $a_1 A$ is p-primary. Since $a_1 A_p = pA_p$, it follows that $a_1 A = p$. Therefore W' is a variety, and $A/a_1 A$ the local ring of Z on W'. By induction, the images of $a_2, \ldots, a_d$ in $A/a_1 A$ form a regular sequence generating $m/a_1 A$, so $a_1, \ldots, a_d$ form a regular sequence generating m.

Returning to the general case, it remains to show that A must be normal. Let $g: \tilde{V} \to V$ be the normalization of V in its function field. Let h be the induced morphism from $g^{-1}(W)$ to W. By Theorem 6.2(a), since g is proper and $g_*[\tilde{V}] = [V]$,

$$h_*(i^![\tilde{V}]) = i^![V] .$$

If $i^![\tilde{V}] = \sum_{i=1}^{r} m_i [\tilde{Z}_i]$, this gives

$$\sum_{i=1}^{r} m_i \deg(\tilde{Z}_i/Z) = 1 .$$

Therefore $r = m_1 = \deg(\tilde{Z}_1/Z) = 1$. The local ring A' of $\tilde{Z}_1$ in $\tilde{V}$ is the integral closure of A in its field of fractions. The case of (i) ⇒ (ii) proved above applies to $\tilde{V}$, so J generates the maximal ideal m' of A'; in particular, $mA' = m'$. Since $\deg(\tilde{Z}_1/Z) = 1$, the canonical map from A/m to A'/m' is an isomorphism. Therefore $A' = A + mA'$; since A' is finite over A, Nakayama's lemma implies that $A = A'$, so A is normal, as required. □

Example 7.2.1. It is not enough to assume that V is a pure-dimensional scheme in Proposition 7.2. For example, let $Y = \mathbb{A}^2$, $X = V(y)$, $V = V(xy, x^2)$. Then X and V meet properly at the origin, and the intersection multiplicity is 1, but the local ring of V at the origin is not regular. However, if all associated primes p in A have $\dim A/p = d$, then the intersection multiplicity is one only if A is regular and J is the maximal ideal. (Since $X \cdot V = X \cdot [V]$ (Example 6.2.1), A can have only one minimal prime p, and $l(A_p) = 1$; i.e. $p_p = 0$; since elements outside p are assumed to be non-zero divisors, $p = 0$, and Proposition 7.2 applies.)

Nagata has extended this to general local rings A whose completion is unmixed (cf. Nagata (20)40.6 and Huneke (1)). Nagata has given an example of a local Noetherian domain whose multiplicity is one without being regular (cf. Nagata (2) Appendix A1).

Notes and References

The problem of assigning a multiplicity to an isolated solution of n polynomial equations in n variables can be traced back near the beginnings of algebraic geometry, although clear statements did not appear until relatively recently.

Two points of view, which remain vital, can be found in the work of Newton and his contemporaries:

(1) The *dynamic* approach, where the multiplicity of a solution is the number of solutions near the given solution when the equations are varied. For example, a point of tangency of a line with a curve is a limit of intersections of nearby secant lines.

(2) The *static* approach, where the multiplicity is obtained without varying the given equations. For $n = 2$, Newton and Leibnitz showed how to eliminate one of the variables, obtaining a polynomial equation whose roots give the abscissas where the equations have common solutions, the multiplicity question is likewise reduced to the multiplicity of a root of a polynomial in one variable.

In 1822, Poncelet (1) made the dynamic point of view quite explicit with his "principle of continuity". Rules of this type were given for calculating intersection multiplicities, e.g. by Cayley (1), Halphen (1), Schubert (1), and Zeuthen (3). A useful summary of this era is given by Zeuthen and Pieri (1). We will discuss these principles in Chap. 11.

Elimination theory and the calculation of resultants also received considerable attention; the names of Euler, Bézout, Cayley, Sylvester, Kronecker, and Hilbert should at least be mentioned. This is discussed by Salmon (2) and B. Segre (8); cf. Example 8.4.13.

In 1915 Macaulay (1) gave a static definition in terms of the length of a ring modulo an ideal, and proved Bézout's theorem for n hypersurfaces in $\mathbb{P}^n$.

The intersection of more general varieties than hypersurfaces in n-space was taken up, from the dynamic point of view, by Severi, Van der Waerden, and Weil in the 1930's. In 1928 Van der Waerden (2), borrowing an example from Macaulay (Example 7.1.5 above), showed that the naive definition using length would not always work. Van der Waerden (3) also pointed out in 1930 that the Poincaré-Lefschetz intersection theory in topology includes a notion of intersection multiplicity for complex varieties, since they can be triangulated. Severi's treatments (cf. Severi (7) for a summary) were almost entirely geometric. Van der Waerden (1) and Weil (2) developed algebraic notions of specialization to make such geometric ideas rigorous, not relying on geometric intuition, and valid over general ground fields.

Chevalley (1), in 1945, gave an important new definition of intersection multiplicity in terms of completions of the local rings; his theory was therefore equally valid in the analytic or formal case. He also gave a criterion for multiplicity one, which includes that given in § 7.2. Samuel (1) gave the first definition valid for a general Noetherian local ring A. As in Examples 4.3.1 and 4.3.4, he defined a multiplicity $e_A(J)$ for an ideal J primary to the maximal ideal. Samuel proved many basic properties for this multiplicity, including its agreement with Chevalley's.

We can only mention a few of the very many subsequent treatises on multiplicities in general local rings. The books of Nagata (2), Northcott (2), and Kunz (1) may be consulted for this literature. Nagata proved that when J is generated by d elements, $d = \dim A$, then $e_A(J) = l(\hat{A}/J)$ precisely when A is Cohen-Macaulay (cf. Example 7.1.3). Nagata (1) also generalized Chevalley's criterion of multiplicity one: if $\hat{A}$ is unmixed, then $e_A(J) = 1$ if and only if A is regular and J is maximal. Nagata also gave an example to show that the criterion may fail for general local rings.

Lech (1) proved a remarkable asymptotic formula:

$$e_A(a_1, \ldots, a_d) = \lim_{\min(t_i) \to \infty} l(A/(a_1^{t_1}, \ldots, a_d^{t_d}))/t_1 \cdot \ldots \cdot t_d$$

which he used to prove the associativity formula for multiplicities (cf. Example 7.1.8).

In 1957 Serre (4) showed that $e_A(a_1, \ldots, a_d)$ is the alternating sum of the lengths of a Koszul complex (cf. Example 7.1.2); or an alternating sum of lengths of Tor modules, this definition, unlike previous algebraic definitions extends to intersections where neither factor is defined by a regular sequence. Several authors, beginning with Kleiman (5), have constructed other ideals J' in A so that $e_A(J) = l(A/J')$; one such is worked out and applied to Bézout's theorem by Vogel (1). Teissier (1), (2) has given some interesting new multiplicity formulas.

The definition of the intersection multiplicity $i(Z, X \cdot V)$ given in this chapter is also a length – the length of the local ring of the normal cone $C_{V \cap X} V$ at the component lying over Z. Since normal cones are constructed from associated graded rings $\oplus J^m/J^{m+1}$, it is not hard to see that this definition agrees with Samuel's (Example 7.1.1). This calculation of intersection multiplicities occurs implicitly in Verdier (5), and in Fulton-MacPherson (1), (2). Basic properties of intersection multiplicities, in this geometric context, follow from the properties proved for more general intersections in Chap. 6; other than the algebra in Appendix A, none of the previous multiplicity theory is required. The proof of the criterion of multiplicity one in § 7.2 is new, to our knowledge.

It should be emphasized that all of the above constructions of intersection multiplicities, with the notable exception of Serre's Tor definition, are valid only when one of the varieties being intersected is regularly imbedded in the ambient space. Intersection multiplicities for arbitrary varieties on a nonsingular variety are defined by reduction to the diagonal, as discussed in the next chapter.

Chapter 8. Intersections on Non-singular Varieties

Summary

If Y is a non-singular variety, the diagonal imbedding δ of Y in $Y \times Y$ is a regular imbedding. For $x, y \in A_* Y$, the product $x \cdot y \in A_* Y$ is defined by the formula

$$x \cdot y = \delta^*(x \times y) .$$

Setting $A^p Y = A_{n-p} Y$, $n = \dim Y$, this product makes $A^* Y$ into a commutative, graded, ring, with unit $[Y]$.

If $f: X \to Y$ is a morphism, with Y non-singular, the graph morphism γ_f from X to $X \times Y$ is a regular imbedding. For $x \in A_* X$, $y \in A^* Y$, define

$$x \cdot_f y = \gamma_f^* (x \times y) \qquad \in A_* X .$$

This product makes $A_* X$ into a graded module over $A^* Y$. If X is also non-singular, setting

$$f^*(y) = [X] \cdot_f y$$

defines a homomorphism $f^* : A^* Y \to A^* X$ of graded rings.

Using the refined operation $\gamma_f^!$ in place of γ_f^*, $x \cdot_f y$ has a canonical refinement in $A_*(|x| \cap f^{-1}(|y|))$. In particular, if V and W are subvarieties of a non-singular variety Y, the intersection class $V \cdot W$ is defined in $A_m(V \cap W)$, $m = \dim V + \dim W - \dim Y$. Any m-dimensional irreducible component Z of $V \cap W$ has a coefficient in $V \cdot W$, called the intersection multiplicity, and denoted $i(Z, V \cdot W; Y)$. The expected properties of these intersection products and multiplicities follow readily from the general properties proved in Chaps. 6 and 7.

Bézout's theorem, in its simplest form, states that $A^*(\mathbb{P}^n) \cong \mathbb{Z}[h]/(h^{n+1})$, where h is the class of a hyperplane. A deeper analysis of intersections on projective space will be given in Chap. 12.

8.1 Refined Intersections

A variety Y will be called *non-singular* if it is smooth over the given ground field. For our purposes, the important point (Appendix B.7.3) is that the diagonal imbedding

$$\delta : Y \to Y \times Y$$

is a regular imbedding of codimension n, $n = \dim\ Y$.

The (global) *intersection product* is the composite

$$A_k Y \otimes A_l Y \xrightarrow{\times} A_{k+l}(Y \times Y) \xrightarrow{\delta^*} A_{k+l-n} Y ,$$

where δ^* is the Gysin homomorphism (§ 6.2). We write $x \cdot y = \delta^*(x \times y)$ for $x, y \in A_*(Y)$.

More generally, if X is a scheme, Y a non-singular variety, $f: X \to Y$ a morphism, then the graph morphism

$$\gamma_f : X \to X \times Y ,$$

$\gamma_f(P) = (P, f(P))$, is a regular imbedding of codimension n, $n = \dim(Y)$. Define a *cap product*

$$A_i Y \otimes A_j X \to A_{i+j-n} X ,$$

denoted

$$y \otimes x \to f^*(y) \cap x ,$$

by the formula $f^*(y) \cap x = \gamma_f^*(x \times y)$. When f is the identity on Y, this is the previous product. This product is also denoted by $x \cdot_f y$. If X is also non-singular, we write $f^* y$ for $f^* y \cap [X]$.

By using the refined Gysin homomorphisms $\delta^!$ and $\gamma_f^!$ in place of δ^* and γ_f^*, these products can also be refined. If x and y are cycles on a non-singular variety Y, with supports $|x|$ and $|y|$, then $x \in A_*|x|$, $y \in A_*|y|$. Form the fibre square

$$\begin{array}{ccc} |x| \bigcap |y| & \to & |x| \times |y| \\ \downarrow & & \downarrow \\ Y & \to & Y \times Y \end{array} .$$

We have $\delta^!(x \times y) \in A_*(|x| \bigcap |y|)$. This product, also denoted $x \cdot y$, maps to the corresponding global product in $A_*(Y)$. If $x = [V]$, $y = [W]$, for V, W pure dimensional subschemes of Y of dimensions k, l, we write the refined product

$$V \cdot W = [V] \cdot [W] \in A_{k+l-n}(V \bigcap W) .$$

Recalling the procedure of Chap. 6.1, this product $V \cdot W$ is constructed as follows. The normal bundle to the diagonal imbedding of Y in $Y \times Y$ is the tangent bundle T_Y to Y. Let T be the restriction of T_Y to $V \bigcap W$, s the zero section of T. The normal cone $C = C_{V \bigcap W}(V \times W)$ is a $(k+l)$-dimensional subscheme of T, and

$$V \cdot W = s^*[C] ,$$

i.e., $V \cdot W$ is the intersection of the cycle of C with the zero section of T.

Similarly, if $f: X \to Y$, Y non-singular, x a cycle on X, y a cycle on Y, the cap product $f^* y \cap x$ has a canonical refinement in $A_*(|x| \bigcap f^{-1}(|y|))$, which we denote by $x \cdot_f y$:

$$x \cdot_f y = \gamma_f^!(x \times y) \in A_*(|x| \bigcap f^{-1}(|y|)) .$$

These products have the following common generalization.

Definition 8.1.1. Let $f: X \to Y$ be a morphism, with Y non-singular of dimension n. Let $p_X: X' \to X$, $p_Y: Y' \to Y$ be morphisms of schemes X', Y' to X and Y respectively, and let $x \in A_k X'$, $y \in A_l Y'$. Form the fibre square

$$\begin{array}{ccc} X' \times_Y Y' & \to & X' \times Y' \\ \downarrow & & \downarrow {\scriptstyle p_X \times p_Y} \\ X & \underset{\gamma_f}{\to} & X \times Y . \end{array}$$

Define

$$x \cdot_f y = \gamma_f^!(x \times y) \in A_{k+l-n}(X' \times_Y Y') ,$$

where $x \times y \in A_{k+l}(X' \times Y')$ is the exterior product (§ 1.10) and $\gamma_f^!$ is the refined Gysin homomorphism (§ 6.2). When $X' = X$, $Y' = Y$, these are the global products; when $X' = |x|$, $Y' = |y|$, the preceding refinements.

The following proposition proves the expected formal properties of these refined products. In this proposition, it is assumed that each named variety X, Y, Z, Y_i, comes equipped with a morphism $p_X: X' \to X$, $p_Y: Y' \to Y$, etc. and a class $x \in A_* X'$, $y \in A_* Y'$, etc.

Proposition 8.1.1. (a) (Associativity). *Let $X \xrightarrow{f} Y \xrightarrow{g} Z$, with Y and Z non-singular. Then*

$$x \cdot_f (y \cdot_g z) = (x \cdot_f y) \cdot_{gf} z$$

in $A_*(X' \times_Y Y' \times_Z Z')$.

(b) (Commutativity). *Let $f_i: X \to Y_i$, Y_i non-singular, $i = 1, 2$. Then*

$$(x \cdot_{f_1} y_1) \cdot_{f_2} y_2 = (x \cdot_{f_2} y_2) \cdot_{f_1} y_1$$

in $A_*(Y_1' \times_{Y_1} X' \times_{Y_2} Y_2')$.

(c) (Projection formula). *Let $X \xrightarrow{f} Y \xrightarrow{g} Z$, with Z non-singular. Let $f': X' \to Y'$ be a proper morphism such that $p_Y f' = f p_X$; let $f'' = f' \times_Z Z'$ be the base extension. Then*

$$f''_*(x \cdot_{gf} z) = f'_*(x) \cdot_g z$$

in $A_*(Y' \times_Z Z')$.

(d) (Compatibility). *Let $f: X \to Y$, with Y non-singular and let $g: V' \to Y'$ be a regular imbedding. Then*

$$g^!(x \cdot_f y) = x \cdot_f g^! y$$

in $A_*(X' \times_Y V')$.

Proof. For (a) consider the fibre square

$$\begin{array}{ccc} X & \xrightarrow{\gamma_{gf}} & X \times Z \\ {\scriptstyle \gamma_f} \downarrow & & \downarrow {\scriptstyle \gamma_f \times 1_Z} \\ X \times Y & \xrightarrow[1_X \times \gamma_g]{} & X \times Y \times Z . \end{array}$$

The canonical map from $X' \times Y' \times Z'$ to $X \times Y \times Z$ induces a fibre cube lying over this square. Then

$$\begin{aligned} x \cdot_f (y \cdot_g z) &= \gamma_f^!(x \times \gamma_g^!(y \times z)) \\ &= \gamma_f^!(1_X \times \gamma_g)^!(x \times y \times z) = (\gamma_f \times 1_Z)^!(1_X \times \gamma_g)^!(x \times y \times z) \end{aligned}$$

by two applications of Theorem 6.2 (c). Now by Theorem 6.4,

$$\begin{aligned}(\gamma_f \times 1_Z)^!(1_X \times \gamma_g)^!(x \times y \times z) &= (1_X \times \gamma_g)^!(\gamma_f \times 1_Z)^!(x \times y \times z) \\ &= \gamma_{gf}^!((x \cdot_f y) \times z) = (x \cdot_f y) \cdot_{gf} z \,,\end{aligned}$$

using Theorem 6.2 (c) again.

Similarly (b) follows by applying the commutativity theorem (§ 6.4) to the fibre square

$$\begin{array}{ccc} X & \xrightarrow{\gamma_{f_2}} & X \times Y_2 \\ {\scriptstyle \gamma_{f_1}}\downarrow & & \downarrow{\scriptstyle \gamma_{f_1} \times 1_{Y_2}} \\ Y_1 \times X & \xrightarrow[1_{Y_1} \times \gamma_{f_2}]{} & Y_1 \times X \times Y_2 \end{array}$$

and the class $y_1 \times x \times y_2$ in $A_*(Y_1' \times X' \times Y_2')$.

For (c), apply Theorem 6.2 (a) to the diagram

$$\begin{array}{ccc} X' \times_Z Z' & \rightarrow & X' \times Z' \\ {\scriptstyle f''}\downarrow & & \downarrow{\scriptstyle f' \times 1_{Z'}} \\ Y' \times_Z Z' & \rightarrow & Y' \times Z' \\ \downarrow & & \downarrow \\ Y & \xrightarrow[\gamma_g]{} & Y \times Z \,. \end{array}$$

This gives the formula $f'_*(x) \cdot_g z = f''_*(\gamma_g^!(x \times z))$. From the fibre square

$$\begin{array}{ccc} X & \xrightarrow{\gamma_{gf}} & X \times Z \\ {\scriptstyle f}\downarrow & & \downarrow{\scriptstyle f \times 1_Z} \\ Y & \xrightarrow[\gamma_g]{} & Y \times Z \,, \end{array}$$

Theorem 6.2 (c) gives

$$\gamma_g^!(x \times z) = \gamma_{gf}^!(x \times z) = x \cdot_{gf} z \,,$$

which concludes the proof of (c).

For (d), apply Theorem 6.4 to the diagram

$$\begin{array}{ccccc} X' \times_Y V' & \rightarrow & X' \times V' & \rightarrow & V' \\ \downarrow & & \downarrow & & \downarrow{\scriptstyle g} \\ X' \times_Y Y' & \rightarrow & X' \times Y' & \rightarrow & Y' \\ \downarrow & & \downarrow & & \\ X & \xrightarrow[\gamma_f]{} & X \times Y & & . \quad \square \end{array}$$

The following corollary follows from (d).

Corollary 8.1.1. *If Y is non-singular, and $j: V \rightarrow Y$ is a regular imbedding, and x is a cycle on Y, then*

$$x \cdot [V] = j^!(x)$$

in $A_*(|x| \cap V)$. □

Corollary 8.1.2. *If $f: X \rightarrow Y$, with X and Y non-singular, and $\Gamma_f \subset X \times Y$ is the graph of f, then for cycles x on X, y on Y,*

$$x \cdot_f y = (x \times y) \cdot [\Gamma_f]$$

in $A_*(|x| \cap f^{-1}(|y|))$. *In particular, for cycles x, y on a non-singular Y, $x \cdot y$ is the intersection product of $x \times y$ with the diagonal Δ_Y on $Y \times Y$.*

Proof. Apply Corollary 8.1.1 with $j = \gamma_f$ the imbedding of $X = \Gamma_f$ in $X \times Y$. □

Corollary 8.1.3. *Let $f: X \to Y$, Y non-singular, x a cycle on X. Then*

$$x \cdot_f [Y] = x .$$

Proof. By (c), one may assume $x = [X]$, X a variety. Then $x \cdot_f [Y] = \gamma_f^* [X \times Y] = [X]$ since γ_f is a regular imbedding. □

Definition 8.1.2. Let $f: X \to Y$ be a morphism from a purely m-dimensional scheme X to a non-singular n-dimensional variety Y. For any morphism $g: Y' \to Y$, define a refined Gysin homomorphism

$$f^!: A_k Y' \to A_{k+m-n} X'$$

with $X' = X \times_Y Y'$, by the formula

$$f^!(y) = [X] \cdot_f y .$$

Proposition 8.1.2. (a) *If f is also flat, then $f^!(y) = f'^*(y)$, where f' is the induced morphism from X' to Y'.*

(b) *If f is also a l.c.i. morphism, then $f^!$ agrees with the homomorphism constructed in § 6.6.*

Proof. (a) follows from Proposition 6.5 (a). For (*b*), if $i: X \to P$, $p: P \to Y$ factors f, with p smooth, i a regular imbedding, then we have a factorization

$$X \xrightarrow{\gamma_i} X \times P \xrightarrow{1 \times p} X \times Y \xrightarrow{\mathrm{pr}} Y$$

of f, and the conclusion follows from Proposition 6.5 (b). □

The *functoriality* of these refined Gysin homomorphisms follows from Proposition 8.1.1 (a), the *projection formula* from Proposition 8.1.1 (c), and the *commutativity* of these Gysin homomorphisms with l.c.i. Gysin homomorphisms from Proposition 8.1.1 (d). Similarly all other formal properties of refined Gysin homomorphisms given for l.c.i. morphisms in Chap. 6 are valid for morphisms to non-singular varieties. These results will all be subsumed in Chap. 17, so we do not write them out here.

Example 8.1.1. Let V be a closed subscheme of a non-singular variety Y such that the imbedding i of V in Y is a regular imbedding. Then

$$[V] \cdot y = i_* i^*(y) \quad \in A_* Y$$

for all $y \in A_*(Y)$. (Use Corollary 8.1.1.)

Example 8.1.2. Both classes in part (a) of Proposition 8.1.1 (a) are equal to $x \cdot_{(f, gf)} (y \times z)$, as well as to $(x \cdot_{gf} z) \cdot_f y$. (The morphism (f, gf) from X to $X \times Y \times Z$ is the composite of morphisms given in the proof of (a); use Theorem 6.5.)

Both classes in (b) are equal to $x \cdot_{(f_1, f_2)} (y_1 \times y_2)$.

Example 8.1.3. If in Proposition 8.1.1 (d) one assumes that g is flat instead of a regular imbedding, then

$$g'^*(x \cdot_f y) = x \cdot_f g^*(y)$$

where $g' : X' \times_Y V' \to X' \times_Y Y'$ is induced by g.

On the other hand, if g is assumed to be proper, and $v \in A_* V'$, then

$$g'_*(x \cdot_f v) = x \cdot_f g_*(v) .$$

Formula (d) is also valid for $g : V' \to Y'$ an arbitrary l.c.i. morphism as in § 6.6.

Example 8.1.4. If $f_i : X_i \to Y_i$ are morphisms, with Y_i non-singular, $i = 1, \ldots, r$, then

$$(x_1 \times \ldots \times x_r) \cdot_{(f_1 \times \ldots \times f_r)} (y_1 \times \ldots \times y_r) = (x_1 \cdot_{f_1} y_1) \times \ldots \times (x_r \cdot_{f_r} y_r)$$

in $A_*((X'_1 \times_{Y_1} Y'_1) \times \ldots \times (X'_r \times_{Y_r} Y'_r))$. (Use Example 6.5.2.)

Example 8.1.5. Let Y be non-singular, $V_1, \ldots, V_r$ regularly imbedded subschemes. Then the intersection product of $V_1 \times \ldots \times V_r$ by the diagonal $\Delta = Y$ in $Y \times \ldots \times Y$ is the same as the intersection product of Δ by $V_1 \times \ldots \times V_r$. (Use Theorem 6.4.)

Example 8.1.6. Let $f: X \to Y$, Y non-singular, and let E be a vector bundle on Y. Then, for all i,

$$x \cdot_f (c_i(E) \cap y) = (c_i(f^*E) \cap x) \cdot_f y$$

in $A_*(X' \times_Y Y')$. (Use Proposition 6.3.)

Example 8.1.7. *Projection formula.* Let $f: X \to Y$ be a proper morphism of non-singular varieties, and let x (resp. y) be a cycle on X (resp. Y). Then

$$f'_*(x \cdot_f y) = f_*(x) \cdot y$$

in $A_*(f(|x|) \cap |y|)$, where f' is the induced map from $|x| \cap f^{-1}(|y|)$ to $f(|x|) \cap |y|$. In particular,

$$f_*(f^*(y) \cap x) = y \cdot f_*(x)$$

in $A_*(Y)$. (Use Proposition 8.1.1 (c).)

Example 8.1.8. If $f: X \to Y$ is a morphism, with Y non-singular, there is a fibre square

$$\begin{array}{ccc} X & \xrightarrow{\gamma_f} & X \times Y \\ {\scriptstyle f}\downarrow & & \downarrow{\scriptstyle f \times 1_Y} \\ Y & \xrightarrow[\delta]{} & Y \times Y \end{array},$$

with δ and γ_f regular imbeddings of the same codimension. With x and y cycles as in Definition 8.1, it follows (Theorem 6.2 (c)) that

$$x \cdot_f y = \gamma_f^!(x \times y) = \delta^!(x \times y) .$$

In this sense all the intersections of this section are intersections with a diagonal.

Example 8.1.9. Let Y be non-singular, and let δ_r be the r-fold diagonal imbedding of Y in $Y \times \ldots \times Y$. for cycles $y_1, \ldots, y_r$ on Y, define

$$y_1 \cdot \ldots \cdot y_r = \delta_r^!(y_1 \times \ldots \times y_r)$$

in $A_*(|y_1| \cap \ldots \cap |y_r|)$. Then

$$y_1 \cdot \ldots \cdot y_r = y_1 \cdot (y_2 \cdot \ldots \cdot y_r)\,.$$

(See the proof of Proposition 8.1.1 (a).)

Example 8.1.10. Let X be a non-singular closed subvariety of a non-singular variety, i the inclusion of X in Y. Then for any cycles y_1, y_2 on Y,

$$[X] \cdot y_1 \cdot y_2 = i^!(y_1 \cdot y_2) = i^*(y_1) \cdot i^*(y_2)$$

in $A_*(X \cap |y_1| \cap |y_2|)$. (The intersection products in the first two formulas are taken on Y, the third on X.) In particular, if X is a hypersurface on Y, and V_1, V_2 are subvarieties of Y not contained in X, then

$$[X] \cdot [V_1] \cdot [V_2] = [X \cap V_1] \cdot [X \cap V_2]$$

in $A_*(X \cap V_1 \cap V_2)$, the first intersection taken on Y, the second on X.

Example 8.1.11. Let V, W be subvarieties of a non-singular, n-dimensional variety Y. If the diagonal imbedding of the intersection scheme $V \cap W$ into $V \times W$ is a regular imbedding of codimension n, then

$$V \cdot W = [V] \cdot [W] = [V \cap W]$$

in $A_*(V \cap W)$. This equation holds if the imbedding of $V \cap W$ in $V \times W$ is regular of codimension n on an open set of each component of $V \cap W$ (cf. Proposition 6.2 (b)). This is valid in particular if V and W are non-singular varieties meeting transversally at generic points of $V \cap W$.

Example 8.1.12. If X is an n-dimensional non-singular variety, and $\Delta \subset X \times X$ is the diagonal then (Corollary 8.1.1 and Corollary 6.3)

$$\Delta \cdot \Delta = c_n(T_X) \cap [X]\,.$$

In particular, the degree of $\Delta \cdot \Delta$ is the topological Euler characteristic $\int_X c_n(T_X)$. If $n = 1$, $\int \Delta \cdot \Delta = 2 - 2g$ (cf. Example 3.2.13).

Example 8.1.13. Consider a fibre square

$$\begin{array}{ccc} X' & \xrightarrow{f'} & Y' \\ {\scriptstyle h}\downarrow & & \downarrow{\scriptstyle g} \\ X & \xrightarrow[f]{} & Y \end{array}$$

with Y and Y' non-singular, and g a closed imbedding of codimension d and normal bundle N. Then for $x \in A_* X'$, $y \in A_* Y'$,

$$h_*(x) \cdot_f g_*(y) = h_*(c_d(f'^*N) \cap (x \cdot_{f'} y))$$

in $A_* X$. (Use Theorem 6.3.) This formula corrects Lemma 2.2 (4) of Fulton (2).

8.2 Intersection Multiplicities

Let Y be an n-dimensional non-singular variety. Let V and W be closed subschemes of Y of pure dimensions k and l. From the fibre square

$$\begin{array}{ccc} V\cap W & \to & V\times W \\ \downarrow & & \downarrow \\ \Delta_Y = Y & \xrightarrow{\delta} & Y\times Y \end{array}$$

and Lemma 7.1, it follows that every irreducible component Z of $V\cap W$ has dimension at least $k+l-n$. One says that Z is a *proper component* of the intersection of V and W, or that V and W *meet properly at* Z, if

$$\dim Z = k+l-n\,.$$

If Z is proper, the coefficient of Z in the intersection class $V\cdot W\in A_{k+l-n}(V\cap W)$ is called the *intersection multiplicity* of Z in $V\cdot W$, and denoted $i(Z, V\cdot W; Y)$. In other words,

$$i(Z, V\cdot W; Y) = i(Z, \Delta_Y\cdot(V\times W); Y\times Y)\,,$$

where the right side is defined in § 7.1. Setting $T=T_Y|_{V\cap W}$, and $C=C_{V\cap W}(V\times W)$, $i(Z, V\cdot W; Y)$ is the coefficient of $T|_Z$ in the cycle $[C]$.

If Y° is an open subscheme of Y which meets Z, and $V^\circ = V\cap Y^\circ$, $W^\circ = W\cap Y^\circ$, $Z^\circ = Z\cap Y^\circ$, then by Theorem 6.2 (c)

$$i(Z, V\cdot W; Y) = i(Z^\circ, V^\circ\cdot W^\circ; Y^\circ)\,.$$

If Y is a singular variety, but Z is not contained in the singular locus of Y, the preceding formula, with Y° the non-singular part of Y, defines $i(Z, V\cdot W; Y)$.

If every component of $V\cap W$ is proper, the intersection class $V\cdot W\in A_{k+l-n}(V\cap W)$ is a well-defined *cycle*

$$V\cdot W = \sum_Z i(Z, V\cdot W; Y)\cdot[Z]\,.$$

Proposition 8.2. *Assume Z is a proper component of $V\cap W$. Then*

(a) $1\leq i(Z, V\cdot W; Y) \leq l(\mathcal{O}_{Z,V\cap W})$.

(b) *If the local ring of $V\times W$ along Z is Cohen-Macaulay, then*

$$i(Z, V\cdot W; Y) = l(\mathcal{O}_{Z,V\cap W})\,.$$

(c) *If V and W are varieties, then $i(Z, V\cdot W; Y)=1$ if and only if the maximal ideal of $\mathcal{O}_{Z,Y}$ is the sum of the prime ideals of V and W; in this case $\mathcal{O}_{Z,V}$ and $\mathcal{O}_{Z,W}$ are regular.*

Proof. (a) and (b) follow from Proposition 7.1. By Proposition 7.2 (and Example 7.2.1 if $V\times W$ is not a variety), the intersection number is 1 precisely when there is an open subset U of $Y\times Y$, meeting Z, such that the diagonal Δ_Y intersects $U\cap(V\times W)$ scheme-theoretically in the (reduced) variety Z. Since $\Delta_Y\cap(V\times W)$ is scheme-theoretically isomorphic to $V\cap W$, this implies the equivalence stated in (c). To see that $\mathcal{O}_{Z,V}$ and $\mathcal{O}_{Z,W}$ are regular, let $A=\mathcal{O}_{Z,Y}$, and let p, q and m be the prime ideals of V, W and Z in A, and let $K = A/m$. From the exact sequence

$$0 \to (p+m^2)/m^2 \to m/m^2 \to m/(p+m^2) \to 0$$

and the identification of $m/(p+m^2)$ with $M_{V,Z}/(M_{V,Z})^2$, it follows that

$$\dim_K((p+m^2)/m^2) \leqq \dim(Y) - \dim(V),$$

with equality if and only if $\mathcal{O}_{V,Z}$ is regular (Lemma A.6.2); similarly for W. Since $p+q=m$,

$$(p+m^2)/m^2 + (q+m^2)/m^2 = m/m^2.$$

Since $(\dim(Y)-\dim(V)) + (\dim(Y)-\dim(W)) = \dim(Y)-\dim(Z)$, the spaces on the left must have the maximum dimensions, and the conclusion follows. □

Remark 8.2. If the ground field is algebraically closed, and Z is a (closed) point, the last displayed equation decomposes the cotangent space of Y at Z into a direct sum of the cotangent spaces of V and W at Z. The intersection number is one precisely when V and W are non-singular and meet transversally at Z. When $\dim(Z) > 0$, the condition says that V and W are generically non-singular along Z, and generically meet transversally along Z.

Example 8.2.1. If $V_1, \ldots, V_r$ are pure-dimensional closed subschemes of a non-singular variety Y, an irreducible component Z of $\bigcap V_i$ is a *proper component* if $\dim(Z) = \sum \dim(V_i) - (r-1)\cdot\dim(Y)$. The *intersection multiplicity*

$$i(Z, V_1 \cdot \ldots \cdot V_r; Y)$$

is the coefficient of Z in the class $V_1 \cdot \ldots \cdot V_r = \delta_r^!(V_1 \times \ldots \times V_r)$, where δ is the r-fold diagonal imbedding of Y in $Y \times \ldots \times Y$. Proposition 8.2 is valid for r factors.

Example 8.2.2. Let $V_1, \ldots, V_n$ be hypersurfaces in an n-dimensional variety Y. Assume: P is a (closed) point of Y, rational over the ground field; P is a non-singular point of Y and of each V_i; P is a component of $\bigcap V_i$; all of the V_i are tangent to each other at P. Then

$$i(P, V_1 \cdot \ldots \cdot V_n; Y) \geqq 2^{n-1}.$$

(Use Example 8.1.10.)

Example 8.2.3 (cf. Samuel (2)). Let K be a field of characteristic $p \neq 0$, with an element a in K with no p^{th} root in K. Let Y be the affine plane over K with coordinates x, y, and let $V = V(y)$, $W = V(y^2 - x^p + a)$, $Z = V(y, x^p - a)$. Then

(a) Z is a proper component of the intersection of V with W, and $i(Z, V \cdot W; Y) = 1$.

(b) If the ground field is extended to a field containing a pth root of a, the intersection still has one irreducible component, but the intersection number becomes p. (Note that $\mathcal{O}_{Z,W}$ is regular, but W is not smooth over $\operatorname{Spec}(K)$ at Z.)

Example 8.2.4. If Y, V, W, Z are as at the beginning of this section, and the imbedding of V in Y is a regular imbedding, then the intersection multiplicity $i(Z, V \cdot W; Y)$ defined in this section agrees with that defined in § 7.1. (Use Corollary 8.1.1, with $x = [W]$.) Similarly if $V_1, \ldots, V_r$ are regularly imbedded subschemes of a non-singular Y, and Z is a proper component of $\bigcap V_i$, then the intersection number $i(Z, V_1 \cdot \ldots \cdot V_r; Y)$ obtained by intersecting $V_1 \times \ldots \times V_r$ by the diagonal, is the same as that obtained by intersecting the diagonal by $V_1 \times \ldots \times V_r$ (cf. Example 8.1.5). In particular, if $V_1, \ldots, V_r$ are hypersurfaces in Y, the intersection number defined here agrees with that defined in Example 7.1.10.

Example 8.2.5. Let $f\colon Y' \to Y$ be a finite surjective morphism of non-singular varieties. Let V', W' be irreducible subvarieties of Y', $V = f(V')$, $W = f(W')$. Let Z be an irreducible component of $V \bigcap W$, and assume there is only one irreducible component Z' of $f^{-1}(Z)$ which is contained in either V' or W'. Assume that Z is a proper component of $V \bigcap W$, and that f is ètale at the generic point of Z'. Then V' and W' meet properly at Z', and

$$\deg(V'/V) \cdot \deg(W'/W) \cdot i(Z, V \cdot W; Y) = \deg(Z'/Z) \cdot i(Z', V' \cdot W'; Y').$$

(The assumptions imply that $(f \times f)^{-1}(\Delta_Y) \bigcap (V' \times W')$ is equal to $\Delta_{Y'} \bigcap (V' \times W')$ in a neighborhood of Z'. Therefore $i(Z', V' \cdot W'; Y')$ is the coefficient of Z' in the class $\delta^!(V' \times W')$ where δ is the diagonal imbedding of Δ_Y in $Y \times Y$ (cf. Theorem 6.2(c)). Then apply Theorem 6.2(a) to the diagram

$$\begin{array}{ccc} V' \times_Y W' & \to & V' \times W' \\ \downarrow & & \downarrow \\ V \bigcap W & \to & V \times W \\ \downarrow & & \downarrow \\ \Delta_Y & \xrightarrow[\delta]{} & Y \times Y \end{array})$$

Example 8.2.6. Let Y be a subvariety of $\mathbb{P}^N$, V, W subvarieties of Y, Z a proper component of $V \bigcap W$ which is not in the singular locus of Y. Then for a generic projection $f\colon Y \to \mathbb{P}^n$, with $n = \dim Y$,

$$i(Z, V \cdot W; Y) = i(f(Z), f(V) \cdot f(W); \mathbb{P}^n).$$

(The hypotheses of Example 8.2.5 are satisfied by a generic projection.) This was one of Severi's methods for reducing intersection multiplicities on general varieties to intersections on projective space (Severi (9) p. 203). A similar construction was used by Chevalley (1) in his algebraic definition of intersection multiplicities.

Example 8.2.7. The intersection number is given by the length as in (b) whenever V and W are Cohen-Macaulay schemes, i.e., all their local rings are Cohen-Macaulay. Indeed, the Cartesian product $V \times W$ is then Cohen-Macaulay ([EGA]IV.6.7.3).

8.3 Intersection Ring

For an n-dimensional, non-singular variety Y, set

$$A^p Y = A_{n-p} Y .$$

With this indexing by codimension, the product $x \otimes y \to x \cdot y$ of § 8.1 reads

$$A^p Y \otimes A^q Y \to A^{p+q} Y ,$$

i.e., the degrees add. Similarly if $f: X \to Y$, the cap product $y \otimes x \to f^*(y) \cap x$ reads

$$A^p Y \otimes A_q X \xrightarrow{\cap} A_{q-p} X .$$

If X is also non-singular the pull-back $y \to f^*(y)$ preserves degrees:

$$f^* : A^p Y \to A^p X .$$

Let $1 \in A^0 Y$ denote the class corresponding to $[Y]$ in $A_n Y$. Set $A^* Y = \bigoplus A^p Y$. We sometimes write $A(Y)$ in place of $A^* Y$ or $A_* Y$ when the grading is unimportant.

Proposition 8.3. (a) *If Y is non-singular, the intersection product makes $A^* Y$ into a commutative, associative ring with unit 1. The assignment*

$$Y \rightsquigarrow A^* Y$$

is a contravariant functor from non-singular varieties to rings.

(b) *If $f: X \to Y$ is a morphism from a scheme X to a non-singular variety Y, the cap product*

$$A^p Y \otimes A_q X \xrightarrow{\cap} A_{q-p} X$$

makes $A_* X$ into an $A^* Y$-module.

(c) *If $f: X \to Y$ is a proper morphism of non-singular varieties, then*

$$f_*(f^* y \cdot x) = y \cdot f_*(x)$$

for all classes x on X, y on Y.

Proof. The associativity and commutativity of $A^* Y$ follow from Proposition 8.1.1 (a) and (b), with $f_i = f = g$ the identity map of Y; that 1 is a unit follows from Corollary 8.1.3. The functoriality of the pull-back follows from Proposition 8.1.1 (a), with $x = [X]$, $y = [Y]$. The projection formula (c) is a special case of Proposition 8.1.1 (c), with $Y = Z$. Proposition 8.1.1 (a), with g the identity on $Y = Z$, gives the formula

$$f^*(y \cdot z) \cap x = f^* z \cap (f^* y \cap x)$$

for $x \in A_* X$, $y, z \in A^* Y$. This formula shows that $A_* X$ is an $A^* Y$-module; setting $x = [X]$, it shows that f^* preserves products. □

Remark 8.3. If Y is non-singular, and $y_1, \ldots, y_r$ are cycles on Y, then

$$y_1 \cdot \ldots \cdot y_r = \delta_r^* (y_1 \times \ldots \times y_r)$$

where δ_r is the diagonal imbedding of Y in $Y \times \ldots \times Y$ (r factors). This follows by writing $\delta_r = (\delta_{r-1} \times 1_Y) \circ \delta_2$ and applying Theorem 6.5 (or see Example 8.1.9).

The ring A^*Y is often called the *Chow ring* of Y; the notation $CH^*(Y)$ is also common.

Example 8.3.1. Let $f: X \to Y$ be a morphism, with Y non-singular, X pure dimensional. If f is flat, or a regular imbedding, or a l.c.i. morphism, then the cap product $f^*y \cap [X]$ agrees with the Gysin pull-back f^*y constructed in § 1.7, § 6.2, or § 6.6, respectively. (Use Proposition 8.1.2.)

Example 8.3.2. If $X \xrightarrow{f} Y \xrightarrow{g} Z$, with Y, Z non-singular, then (Proposition 8.1.1 (a))

$$(g f)^*(z) \cap x = f^*(g^*z) \cap x$$

for all $x \in A_* X$, $z \in A^*(Z)$.

Example 8.3.3. Let $P = P(c_{i_1}(E_1), \ldots, c_{i_r}(E_r))$ be a polynomial in Chern classes of vector bundles on a non-singular variety Y. Suppose that $P \cap [Y] = 0$ in $A_*(Y)$. Then for all morphisms $f: X \to Y$, and all $x \in A_* X$, $f^*(P) \cap x = 0$ in $A_* X$. (By Example 8.1.6, $f^*(P) \cap x = x \cdot_f (P \cap [Y])$.) There is therefore no loss of generality in identifying P with its image in $A_* Y$. This aspect of Poincaré duality will be formalized in Chap. 17. If P is homogeneous of isobaric degree m, we regard $P \in A^m Y$.

Example 8.3.4. Let E be a vector bundle of rank r on a non-singular variety Y, let $X = P(E)$, with projection $p: X \to Y$. Then

$$A^*X \cong A^*Y[\zeta]/(\zeta^r + c_1(E)\,\zeta^{r-1} + \ldots + c_r(E))$$

as graded rings. Here ζ corresponds to $c_1(\mathcal{O}_E(1))$. (This follows from Theorem 3.3 (b).)

Example 8.3.5. If $f: X \to Y$ is an isomorphism of non-singular varieties, with inverse f^{-1}, then

$$f_* = (f^{-1})^* = (f^*)^{-1}: A(Y) \to A(X)\,.$$

In particular, f_* preserves products. (By the projection formula, $f_*(f^*y) = y$.)

Example 8.3.6. Let Y be non-singular, $y_i \in A^*Y$, $U_i \subset Y$ open subschemes such that y_i restricts to 0 in A^*U_i, for $i = 1, \ldots, r$. Then $y_1 \cdot \ldots \cdot y_r \in A^*Y$ restricts to 0 in $A^*(U_1 \bigcup \ldots \bigcup U_r)$. In particular, if the U_i cover Y, then $y_1 \cdot \ldots \cdot y_r = 0$. (Let $Y'_i = Y - U_i$. By Proposition 1.8, y_i is represented by a cycle on Y'_i. The refined intersection product of § 8.1 then gives a class on $\bigcap Y'_i$ which represents $y_1 \cdot \ldots \cdot y_r$.)

Example 8.3.7. Let X and Y be non-singular varieties. Then the exterior product

$$A^*X \otimes A^*Y \xrightarrow{\times} A^*(X \times Y)$$

is a homomorphism of rings, preserving the grading. (Use Example 8.1.4.) If $Y = \mathbb{P}^n$, this homomorphism is an isomorphism.

Example 8.3.8. Let $X = \mathbb{P}^1 \times \mathbb{P}^1$, $i: X \to \mathbb{P}^3$ the Segre imbedding. The induced map from $A^1 \mathbb{P}^3 = \mathbb{Z}$ to $A^1(X) = \mathbb{Z} \oplus \mathbb{Z}$ takes 1 to $(1, 1)$. For any surface H in $\mathbb{P}^3$ of degree d, $i^*[H] = (d, d)$. Therefore no irreducible curve C on X, of bidegree (d, e) with $d \neq e$, can be – even set-theoretically – the intersection of X with any surface in $\mathbb{P}^3$.

Example 8.3.9. Let

$$\begin{array}{ccc} \tilde{X} & \xrightarrow{j} & \tilde{Y} \\ {\scriptstyle g}\downarrow & & \downarrow{\scriptstyle f} \\ X & \xrightarrow[i]{} & Y \end{array}$$

be a blow-up diagram as in § 6.7, with Y, X, and therefore $\tilde{Y}$, $\tilde{X}$ non-singular. The ring structure on $A(\tilde{Y})$ is determined by the following rules:

(i) $f^*y \cdot f^*y' = f^*(y \cdot y')$.

(ii) $j_*(\tilde{x}) \cdot j_*(\tilde{x}') = j_*(c_1(j^*\mathcal{O}_{\tilde{Y}}(\tilde{X})) \cdot \tilde{x} \cdot \tilde{x}')$.

(iii) $f^*y \cdot j_*(\tilde{x}) = j_*((g^*i^*y) \cdot \tilde{x})$.

Example 8.3.10. Let Y be a non-singular surface, and let $\tilde{Y} \xrightarrow{\pi} Y$ be the blow-up of Y at r points. Then

$$A_1 \tilde{Y} = A_1 Y \oplus \sum_{i=1}^{r} \mathbb{Z}[E_i]$$

with E_i the exceptional divisors, and $A_0 \tilde{Y} = A_0 Y$. The product is given by: $[E_i] \cdot [E_i] = -[P_i]$, $P_i \in E_i$; $[E_i] \cdot [E_j] = 0$ for $i \neq j$, $f^*[D] \cdot [E_i] = 0$, $f^*[D] \cdot f^*[D'] = f^*([D] \cdot [D'])$.

Example 8.3.11. Let V be a complete surface, $\pi: X \to V$ a resolution of singularities of V. Assume that $\pi^{-1}(P)$ is connected for all $P \in V$ (V normal, for example). For any irreducible curve A on V there is a unique one-cycle A^* supported on the exceptional locus of X, with rational coefficients, such that if $\tilde{A}$ is the proper transform of A, then

$$(\tilde{A} \cdot E_i)_X + (A^* \cdot E_i)_X = 0$$

for all irreducible components E_i of the exceptional locus (Example 7.1.16). Set $A' = \tilde{A} + A^*$. There is a unique homomorphism, denoted $\alpha \to \alpha'$, from $A_1 V$ to $A_1(X)_{\mathbb{Q}}$, which takes $[A]$ to $[A']$, for irreducible curves A. (Use Example 7.1.16 (i).). This determines a product

$$A_1 V \otimes A_1 V \to A_0(V)_{\mathbb{Q}}$$

by $\alpha \cdot \beta = \pi_*(\alpha' \cdot \beta')$, which is symmetric, bilinear, and independent of choice of X. For any Cartier divisor D on X, $[D] \cdot \beta$ is the image of the integral class $D \cdot \beta$ (defined in § 2.3) in $A_0 V$.

Example 8.3.12. If $X = Y/G$ is a quotient variety of a non-singular variety Y by a finite group G of automorphisms of Y, then $A_* X_{\mathbb{Q}}$ may also be made into a ring. Indeed, in this case one has an isomorphism (Example 1.7.6)

$$A_* X_{\mathbb{Q}} = (A_* Y_{\mathbb{Q}})^G,$$

so $A_*X_{\mathbb{Q}}$ is the ring of G-invariants of $A(Y)_{\mathbb{Q}}$. In fact, if V, W are subvarieties of X, one may construct a refined intersection class $V \cdot W$ in $A_m(V \cap W)_{\mathbb{Q}}$, $m = \dim V + \dim W - \dim X$; with the notation of Example 1.7.6,

$$V \cdot W = (1/\# G)\, \eta_*(\pi^*[V] \cdot \pi^*[W]),$$

where η is the projection from $\pi^{-1}(V \cap W)$ to $V \cap W$. In particular, for any m-dimensional component Z of $V \cap W$, one has a *rational* intersection number $i(Z, V \cdot W; X)$, the coefficient of Z in the class $V \cdot W$ (cf. Matsusaka (2), Briney (1)). Note that the product on X is determined so that $\pi^*(a \cdot b) = \pi^*(a) \cdot \pi^*(b)$, and $\pi_*((\pi^* a) \cdot c) = a \cdot \pi_*(c)$, for cycles a, b on X, c on Y. For a proof that these definitions are independent of the presentation of X as a quotient variety, see Examples 17.4.10 and 16.1.13.

Mumford (7) has constructed a ring structure on $A_*X_{\mathbb{Q}}$ when X is the moduli space of stable curves of genus g, in characteristic zero. He uses the fact that such X is locally (in the étale topology) a quotient of a non-singular variety by a finite group, and X is globally a quotient of a Cohen-Macaulay variety by a finite group, which dominates the local charts.

Example 8.3.13 (cf. Fulton (2) § 3). If X is a quasi-projective scheme, one may define a graded ring A^*X by

$$A^*X = \varinjlim A^*Y$$

where the limit is over all pairs (Y, f) with Y a non-singular quasi-projective scheme, f a morphism from X to Y. This is a contravariant functor from quasi-projective schemes to graded rings. There are (Example 8.3.2) cap products $A^pX \otimes A_qX \xrightarrow{\cap} A_{q-p}X$, with a projection formula $f_*(f^*x \cap x') = x \cap f_*x'$ for a proper morphism $f\colon X' \to X$.

Any vector bundle E on X has Chern classes $c_i(E)$ in A^iX, satisfying the formal properties of § 3.2, such that for any $f\colon X' \to X$, $x \in A_*X'$, $f^*c_i(E) \cap x'$ is the class $c_i(f^*E) \cap x'$ defined in § 3.2. (The essential points for this are the following facts (loc. cit. § 3.2): (i) There is a morphism $f\colon X \to Y$, Y non-singular, and a vector bundle $\tilde{E}$ on Y, with $f^*\tilde{E} \cong E$; (ii) If $f'\colon X \to Y'$, $\tilde{E}'$ is another, there is a non-singular Z, $g\colon X \to Z$, $h\colon Z \to Y$, $h'\colon Z \to Y$ with $hg = f$, $h'g = f'$, and $h^*\tilde{E} \cong h'^*\tilde{E}'$; (iii) Any exact sequence of vector bundles on X pulls back from an exact sequence on some non-singular Y.)

More generally, for any scheme X which admits a closed imbedding in a non-singular scheme, one may define A^*X to be $\varinjlim A^*Y$, the limit over all $X \to Y$, Y non-singular. One has the same properties as in the quasi-projective case. For (i), one uses Lemma 18.2.

Another "cohomology theory" to pair with A_* is discussed in Chap. 17.

Example 8.3.14. *Ruled varieties* (cf. B. Levi (1)). Let Y be a non-singular projective curve, let E be a vector bundle of rank d on Y, and let $X = P(E)$, $p\colon X \to Y$ the projection. Let V be a vector space of dimension $m+1$, $P(V) \cong \mathbb{P}^m$, and let $f\colon X \to P(V)$ be a morphism, with $f^*\mathcal{O}_V(1) = \mathcal{O}_E(1) \otimes p^*L$ for some line bundle L on Y; the fibres X_y of p are imbedded as linear subspaces of $P(V)$. Set $\zeta = c_1(\mathcal{O}_E(1))$, $h = c_1(f^*\mathcal{O}_V(1))$. For any subvariety W

of X of dimension w set

$$n(W) = \int_X h^w \cap [W] = \deg(W/f(W)) \cdot \deg(f(W)).$$

(In characteristic zero, $\deg(W/f(W))$ is the number of rulings, i.e., fibres X_y of p, which pass through a general point of $f(W)$.) Set

$$k(W) = \int_X \zeta^{w-1}[X_y] \cdot [W] = \int_X h^{w-1} \cdot p^*[y] \cdot [W].$$

($k(W)$ is the degree of $X_y \cap W$ as a subscheme of $X_y = \mathbb{P}^{d-1}$, for generic $y \in Y$.) Then

$$[W] = k(W)\, h^{d-w} + p^*(\alpha)\, h^{d-w-1},$$

where α is a zero-cycle on Y of degree $n(W) - k(W) \cdot N$, with $N = n(X)$. (Use Theorem 3.3(b).) If $W_1, \ldots, W_r$ are subvarieties of X with $\sum \operatorname{codim}(W_i) = d$, then

$$\deg(W_1 \cdot \ldots \cdot W_r) = \sum_{i=1}^{r} n_i \Big(\prod_{j \neq i} k_j\Big) - (r-1)\, N \prod_{i=1}^{r} k_i,$$

where $n_i = n(W_i)$, $k_i = k(W_i)$, $N = n(X)$.

Let Y be a non-singular curve of genus g in $P(V) = \mathbb{P}^m$, and let X be the ruled surface of tangent lines to Y, i.e. $X = P(E)$, where E is the "principal parts" bundle, constructed to make the following diagram have exact columns and rows:

$$\begin{array}{ccccccc}
 & & 0 & & 0 & & \\
 & & \downarrow & & \downarrow & & \\
0 \to \mathcal{O} \to & & E & \to & T_Y & \to 0 \\
\| & & \downarrow & & \downarrow & & \\
0 \to \mathcal{O} \to & & V \otimes \mathcal{O}_Y(1) & \to & T_{\mathbb{P}^m}|_Y & \to 0 \\
 & & \downarrow & & \downarrow & & \\
 & & N_Y \mathbb{P}^m & = & N_Y \mathbb{P}^m & & \\
 & & \downarrow & & \downarrow & & \\
 & & 0 & & 0 & &
\end{array}$$

In this case $N = n(X) = 2g - 2 + 2\deg(Y)$. If $m = 2$, N is the degree of the dual curve.

For more on ruled surfaces, see Beauville (2) III.

8.4 Bézout's Theorem (Classical Version)

Intersections on projective space $\mathbb{P}^n$ are particularly simple, as well as important for applications.

We saw in § 3.3 that $A_k(\mathbb{P}^n) \cong \mathbb{Z}$, with generator $[L^k]$, L^k a k-dimensional linear subspace of $\mathbb{P}^n$, $k = 0, 1, \ldots, n$. For any k-cycle α on $\mathbb{P}^n$, the *degree* of α, $\deg(\alpha)$, is defined to be the integer such that

$$\alpha = \deg(\alpha) \cdot [L^k]$$

in $A_k \mathbb{P}^n$. Equivalently, $\deg(\alpha) = \int_{\mathbb{P}^n} c_1(\mathcal{O}(1))^k \cap \alpha$. This follows from the fact that $c_1(\mathcal{O}(1))^k \cap [L_k] = 1$ (Proposition 3.1 (a) (ii)).

Write $A^d \mathbb{P}^n = A_{n-d} \mathbb{P}^n$.

Proposition 8.4 (Bézout's Theorem). *Let* $\alpha_i \in A^{d_i} \mathbb{P}^n$, $i = 1, \ldots, r$. *If* $d_1 + \ldots + d_r \leqq n$, *then*

$$\deg(\alpha_1 \cdot \ldots \cdot \alpha_r) = \deg(\alpha_1) \cdot \ldots \cdot \deg(\alpha_r).$$

Proof. From the preceding considerations, the ring homomorphism from $\mathbb{Z}[h]/(h^{n+1})$ to $A^* \mathbb{P}^n$ which takes h to $c_1(\mathcal{O}(1)) \cap [\mathbb{P}^n] = [L^{n-1}]$ is an isomorphism, from which the proposition follows. (For a more direct proof, see Example 6.2.6.) □

If $V_1, \ldots, V_r$ are pure-dimensional subschemes of $\mathbb{P}^n$ which meet properly, i.e. the irreducible components $Z_1, \ldots, Z_t$ of $\bigcap V_i$ all have codimension equal to $\sum \operatorname{codim}(V_i, \mathbb{P}^n)$, then

$$V_1 \cdot \ldots \cdot V_r = \sum_{j=1}^{t} i(Z_j, V_1 \cdot \ldots \cdot V_r; \mathbb{P}^n) \cdot [Z_j],$$

where $i(Z_j, V_1 \cdot \ldots \cdot V_r; \mathbb{P}^n)$ is the intersection multiplicity (cf. Example 8.2.1). In this case Bézout's theorem says that

$$\sum_{j=1}^{t} i(Z_j, V_1 \cdot \ldots \cdot V_r; \mathbb{P}^n) \cdot \deg(Z_j) = \prod_{i=1}^{r} \deg(V_i). \tag{1}$$

For example, if H is a hypersurface, and V is a subvariety of $\mathbb{P}^n$ not contained in H, then

$$H \cdot V = [H \cap V] = \sum_{j=1}^{t} m_j [Z_j]$$

with

$$\sum m_j \cdot \deg(Z_j) = \deg(H) \cdot \deg(V). \tag{2}$$

Another important case is the one considered originally by Bézout: Let $H_1, \ldots, H_n$ be hypersurfaces in $\mathbb{P}^n_K$ with only a finite number of common points. For each $P \in H_i$, let $\mathcal{O}_P(\bigcap H_i) = \mathcal{O}_{P,\mathbb{P}^n}/(h_1, \ldots, h_n)$ be the local ring of $\mathbb{P}^n$ at P modulo the ideal generated by local equations h_i for H_i at P (i.e., the local ring of the scheme $\bigcap H_i$ at P). Then

$$\sum_P \dim_K(\mathcal{O}_P(\bigcap H_i)) = \prod_{j=1}^{n} \deg(H_j). \tag{3}$$

Indeed, in this case each H_i is Cohen-Macaulay, so by Proposition 8.2 (b)

$$i(P, H_1 \cdot \ldots \cdot H_n; \mathbb{P}^n) = l(\mathcal{O}_P(\bigcap H_i)).$$

For a (closed) point P, $\deg P = [R(P):K]$, and

$$\dim_K \mathcal{O}_P(\bigcap H_i) = l(\mathcal{O}_P(\bigcap H_i)) \cdot [R(P):K]$$

by Lemma A.1.3. Thus (3) follows from (1).

Example 8.4.1. Let Y be an n-dimensional variety, smooth over a field K. If $V_1, \ldots, V_r$ are pure-dimensional subschemes of Y which meet properly in a finite set of points, then

$$\sum_P i(P, V_1 \cdot \ldots \cdot V_r; Y) \cdot [R(P):K] = \int_Y [V_1] \cdot \ldots \cdot [V_r].$$

if $H_1, \ldots, H_n$ are hypersurfaces meeting properly, then

$$\sum_P \dim_K(\mathscr{O}_P(\bigcap H_i)) = \int_Y [H_1] \cdot \ldots \cdot [H_r].$$

Example 8.4.2. (a) If s (resp. t) is the class of a hyperplane on $\mathbb{P}^n$ (resp. $\mathbb{P}^m$), then (Example 8.3.7)

$$A^*(\mathbb{P}^n \times \mathbb{P}^m) = \mathbb{Z}[s, t]/(s^{n+1}, t^{m+1}).$$

(b) If $H_1, \ldots, H_{n+m}$ are hypersurfaces in $\mathbb{P}^n \times \mathbb{P}^m$, and H_i has bidegree (a_i, b_i) (i.e., $[H_i] = a_i \cdot s + b_i \cdot t$), then

$$\int [H_1] \cdot \ldots \cdot [H_{n+m}] = \sum a_{i_1} \cdot \ldots \cdot a_{i_n} b_{j_1} \cdot \ldots \cdot b_{j_m},$$

where the sum is over all permutations $(i_1, \ldots, i_n, j_1, \ldots, j_m)$ of $(1, \ldots, n+m)$ with $i_1 < i_2 < \ldots < i_n$ and $j_1 < \ldots < j_m$.

(c) if Δ is the diagonal in $\mathbb{P}^n \times \mathbb{P}^n$, then

$$[\Delta] = \sum_{i=0}^{n} s^i t^{n-i}$$

in $A^n(\mathbb{P}^n \times \mathbb{P}^n)$. (Write $[\Delta] = \sum a_i s^i t^{n-i}$, and intersect both sides with $[L \times M]$, where L and M are linear spaces of complementary dimensions meeting transversally at a point. Alternatively, the composite of the canonical maps

$$\mathrm{pr}_1^* \mathscr{O}(-1) \hookrightarrow \mathscr{O}^{\oplus(n+1)} \hookrightarrow \mathrm{pr}_2^*(T_{\mathbb{P}^n}(-1))$$

corresponds to a section of $\mathrm{pr}_2^*(T_{\mathbb{P}^n}) \otimes \mathscr{O}(1, -1)$ whose zero-scheme is Δ; the top Chern class of this bundle is $\sum s^i t^{n-i}$.)

Example 8.4.3. (a) Let $\varphi: \mathbb{P}^n \times \mathbb{P}^m \to \mathbb{P}^N$ be the Segre imbedding, $N = nm + n + m$. If u is the hyperplane class on $\mathbb{P}^N$, then $\varphi^* u = s + t$. The degree of the image of φ is $\binom{n+m}{n}$.

(b) If $\psi: \mathbb{P}^n \to \mathbb{P}^N$ is the m-fold Veronese imbedding, $N = \binom{n+m}{n} - 1$, and s, u are hyperplane classes on $\mathbb{P}^n$ and $\mathbb{P}^N$, then $\psi^* u = m \cdot s$. If V is a k-dimensional subvariety of $\mathbb{P}^n$ of degree d, then $\psi(V)$ has degree $d \cdot m^k$.

Example 8.4.4. Let V be a subvariety of $\mathbb{P}^n \times \mathbb{P}^m$ of dimension k, so

$$[V] = \sum_{i+j=k} a_{ij} s^{n-i} t^{m-j};$$

the a_{ij} are the *bidegrees* of V. Define a variety $V' \subset \mathbb{P}^{n+m+1}$ as follows. Geometrically,

$$V' = \{(\lambda x_0 : \lambda x_1 : \ldots : \lambda x_n : \mu y_0 : \ldots : \mu y_m) \in \mathbb{P}^{n+m+1} \mid (x) \times (y) \in V, (\lambda : \mu) \in \mathbb{P}^1\}.$$

Algebraically, if I is the bihomogeneous ideal in $K[X_0, \ldots, X_n, Y_0, \ldots, Y_m]$ defining V, then the ideal of V' is generated by those elements in I which are homogeneous in all variables. Then V' is a variety of dimension $k+1$, and

$$\deg V' = \sum_{i+j=k} a_{ij}$$

(cf. Van der Waerden (7)).

Example 8.4.5. Let V, W be irreducible subvarieties of $\mathbb{P}^n$ of dimensions k, l respectively. Let $J(V, W) \subset \mathbb{P}^{2n+1}$ be the *ruled join* of V and W, i.e. let $\mathbb{P}^n_1$ (resp. $\mathbb{P}^n_2$) be the linear subspace of $\mathbb{P}^{2n+1}$ where the last (resp. the first) $n+1$ coordinates vanish; regard $V \subset \mathbb{P}^n_1$, $W \subset \mathbb{P}^n_2$, and define $J(V, W)$ be the union of all lines in $\mathbb{P}^{2n+1}$ from points of V to points of W. In the notation of the previous example, $J(V, W) = (V \times W)'$. If p, q are the ideals of V and W, then the ideal of $J(V, W)$ is the ideal in $K[X, Y]$ generated by all $f(X)$, $f \in p$ and $g(Y)$, $g \in q$.

Let L be the linear subspace of $\mathbb{P}^{2n+1}$ defined by

$$L = \{(x_0 : \ldots : x_n : y_0 : \ldots : y_n) \,|\, x_i = y_i \quad \text{for} \quad 0 \leqq i \leqq n\}.$$

The imbedding $i: \mathbb{P}^n \to \mathbb{P}^{2n+1}$, which takes $(x_0 : \ldots : x_n)$ to $(x_0 : \ldots : x_n : x_0 : \ldots : x_n)$, maps $\mathbb{P}^n$ isomorphically onto L, and determines an isomorphism of schemes:

$$V \cap W \cong L \cap J(V, W).$$

Moreover, if $\mathbb{P}^{2n+1}_0 = \mathbb{P}^{2n+1} - (\mathbb{P}^n_1 \cup \mathbb{P}^n_2)$, the canonical projection p from $\mathbb{P}^{2n+1}_0$ to $\mathbb{P}^n \times \mathbb{P}^n$ maps L isomorphically onto the diagonal Δ, and maps $L \cap J(V, W)$ to $\Delta \cap (V \times W)$. This projection p is smooth – in fact an $(\mathbb{A}^1 - \{0\})$-bundle. The claim is that

$$V \cdot W = L \cdot J(V, W)$$

in $A_{k+l-n}(V \cap W)$. To prove this consider the diagram

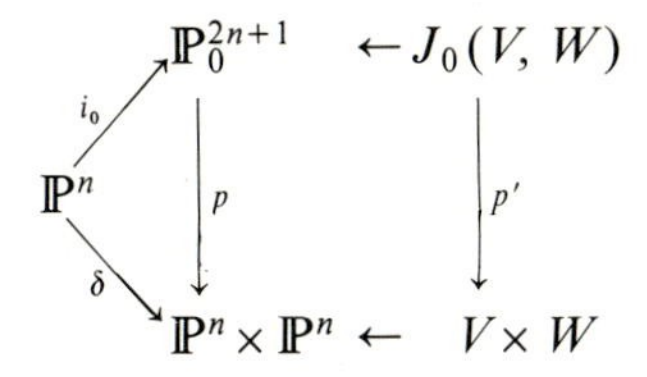

where $J_0(V, W) = J(V, W) \cap \mathbb{P}^{2n+1}_0$. By the definition of $V \cdot W$, and Proposition 6.5 (b),

$$V \cdot W = \delta^![V \times W] = i_0^! p'^*[V \times W] = i_0^![J_0(V, W)].$$

And $i_0^![J_0(V, W)] = i^![J(V, W)] = L \cdot J(V, W)$ by Theorem 6.2 (b) and Corollary 8.1.1.

In fact, by Example 6.5.4, the canonical decomposition of the intersection class $\Delta \cdot (V \times W)$ is the same as the canonical decomposition of the intersection class $L \cdot J(V, W)$. Note in particular that $\deg J(V, W) = \deg(V) \cdot \deg(W)$, as stated in Example 8.4.4.

A similar discussion is valid for the intersection of r varieties in $\mathbb{P}^n$; the ruled join is a subvariety of $\mathbb{P}^{r(n+1)-1}$. In summary, all intersection products on projective space can be realized by an intersection of a subvariety by a linear space (cf. Gaeta (1)).

Example 8.4.6. Let $V_1, \ldots, V_r$ be subvarieties of $\mathbb{P}^n$. Let $Z_1, \ldots, Z_t$ be the irreducible components of $V_1 \cap \ldots \cap V_r$. Then

$$\sum_{i=1}^{t} \deg(Z_i) \leqq \prod_{j=1}^{r} \deg(V_j).$$

In particular, the number of irreducible components of V_j is at most the product of the degrees. (By the preceding example, one may assume $r = 2$ and V_1 is a linear subspace. Write V_1 as an intersection of hyperplanes; by induction one may assume V_1 is a hyperplane. Then either $V_1 \supset V_2$, or $V_1 \cdot V_2 = \sum m_i [Z_i]$, from which the inequality is clear.) A refinement of this inequality will be discussed in § 12.3. For an application, see Bass-Connell-Wright (1) p. 293.

A typical classical application of Bézout's theorem is a proof that an irreducible subvariety $X \subset \mathbb{P}^n$ of degree d, not contained in a hyperplane, satisfies

$$\dim X + d \geqq n + 1.$$

Indeed, taking generic linear sections, one is reduced to the case where X is a curve. There is a hyperplane through any n points, which contradicts Bézout's theorem if $n > d$ and the n points are on the curve.

Example 8.4.7. Let X be a projective scheme. Let $V_1, \ldots, V_r$ be subvarieties of X, and let d_i be the degree of V_i with respect to some imbedding of X in $\mathbb{P}^N$. If $V_1 \cap \ldots \cap V_r$ is finite, then

$$\operatorname{card}(V_1 \cap \ldots \cap V_r) \leqq d_1 \cdot \ldots \cdot d_r.$$

(Use Example 8.4.6.) Note that this upper bound depends only on the (numerical) equivalence classes of the $[V_i]$.

Example 8.4.8. Let L be a linear subspace of $\mathbb{P}^n$, V a subvariety of $\mathbb{P}^n$, Z a proper component of $L \cap V$. If $\dim Z = k$, then for a generic linear subspace $M \subset \mathbb{P}^n$ of codimension k, and a point P on $Z \cap M$, and $L' = L \cap M$, then P is a proper component of $L' \cap V$ and

$$i(Z, L \cdot V; \mathbb{P}^n) = i(P, L' \cdot V; \mathbb{P}^n).$$

(Choose M transversal to Z at P and apply associativity of intersection products.) Together with Examples 8.4.5 and 8.2.6, this shows how intersection multiplicities on arbitrary non-singular varieties are determined by intersection multiplicities of varieties of complementary dimension in $\mathbb{P}^n$, with one factor a linear subspace.

Example 8.4.9. Let $V_1, \ldots, V_r$ be subvarieties of $\mathbb{P}^n$ with

$$m = \sum_{i=1}^{r} \dim(V_i) - (r-1)\,n \geqq 0.$$

(a) The intersection $\prod_{i=1}^{r} V_i$ cannot be empty.

(b) Assume that each V_i has odd degree. Then $V_1 \cap \ldots \cap V_r$ must contain an m-dimensional variety of odd degree. (The intersection class $V_1 \cdot \ldots \cdot V_r$ is represented by a cycle of odd degree *on* $\bigcap V_i$.) In particular, for at least one irreducible component Z of $\bigcap V_i$, the restriction of $\mathscr{O}(1)$ to Z is not the square of any line bundle.

Example 8.4.10 (cf. Fulton-MacPherson (2)). Given $V, W \subset \mathbb{P}^n$, with $\dim V + \dim W = \mathbb{P}^n$, there is no way to assign integers to the irreducible components of $V \cap W$ so that the sum is the Bézout number, at least if one requires the assignment to be preserved by automorphisms of $\mathbb{P}^n$. For example, let $n = 4$,

$$V = V(x_3^3 - x_1 x_2 (x_2 - 2x_1), x_4)$$
$$W = V(x_4^3 - x_2 x_1 (x_1 - 2x_2), x_3)\,.$$

Then $V \cap W$ is the union of two lines; the involution interchanging x_1 and x_2, and x_3 and x_4, takes V to W and interchanges the two lines; thus each would have to be assigned the same number, but the Bézout number 9 is odd. (In fact, the point of intersection of the two lines is a distinguished variety for the intersection of $V \times W$ by Δ.)

Example 8.4.11. Let V be a subvariety of $\mathbb{P}^n_K$ of dimension d, with K infinite. The there is a linear space $L \subset \mathbb{P}^n$ of codimension d which meets V properly, and

$$\deg(V) = \sum_P i(P, L \cdot V; \mathbb{P}^n) \cdot [R(P):K]\,.$$

(Choose, inductively, d hyperplanes $H_1, \ldots, H_d$ so that H_i meets all components of $V \cap H_1 \cap \ldots \cap H_{i-1}$ properly.)

Example 8.4.12. *A Bertini theorem.* Let X be an n-dimensional subvariety of $\mathbb{P}^m$, over an algebraically closed field. Let $0 < k < m$, and let G be the Grassmannian of k-planes in $\mathbb{P}^m$ (cf. § 14.7).

(a) There is a non-empty open set $U \subset G$ such that for all L in U,

$$\dim(X \cap L) = n + k - m$$

if $n + k - m \geqq 0$, or $X \cap L = \emptyset$ if $n + k - m < 0$. (Set

$$I = \{(x, L) \in X \times G \mid x \in L\}\,.$$

The projection from I to X is smooth of relative dimension $k(m-k)$, so $\dim(I) - \dim(G) = n + k - m$. The generic fibre of $I \to G$ therefore has the indicated dimension.)

(b) Let X_0 be the smooth locus of X. For $x \in X_0$, let $T_x \subset \mathbb{P}^m$ be the projective n-plane tangent to X at x. If $n + k \geqq m$, there is a non-empty open set $U \subset G$ such that every L in U meets X_0 transversally, i.e.

$$\dim(T_x \cap L) = n + k - m$$

for all $x \in X_0 \cap L$. (Set

$$J_l = \{(x, L) \in X_0 \times G \mid x \in L,\ \dim(T_x \cap L) = n + k - m + l\}\,,$$

$l = 1, 2, \ldots$. The projection from J_l to X_0 is smooth, and a dimension count shows that $\dim J_l < \dim G$.)

(c) Set $k = m - n$. There is an open set $U \subset G$ such that every $L \in U$ meets X transversally in $\deg(X)$ points. (Apply (b) to X_0, (a) to the components of $X - X_0$. More generally for X, Y subvarieties of $\mathbb{P}^m$ of complementary dimension, X meets $\sigma(Y)$ transversally, in $\deg(X)\deg(Y)$ points, for generic $\sigma \in \operatorname{Aut}(\mathbb{P}^m)$ (Appendix B.9).

Example 8.4.13. *Resultants.* Fix positive integers $d_1, \dots, d_n$. The hypersurfaces of degree d_i in $\mathbb{P}^n$ are parametrized by $\mathbb{P}^{m_i}$, $m_i = \binom{d_i + n}{n} - 1$. Let $x_0, \dots, x_n$ be homogeneous coordinates on $\mathbb{P}^n$, $t^i_{(j)}$ homogeneous coordinates on $\mathbb{P}^{m_i}$, so that $\mathcal{F}_i = \sum t^i_{(j)} x^{(j)}$ defines the universal hypersurface in $\mathbb{P}^n \times \mathbb{P}^{m_i}$. Set $T = \mathbb{P}^{m_1} \times \dots \times \mathbb{P}^{m_n}$, and define $\mathcal{Y} \subset \mathbb{P}^n \times T$ by

$$\mathcal{Y} = \{(x, t^1, \dots, t^n) \mid \mathcal{F}_i(t^i, x) = 0 \quad \text{for} \quad i = 1, \dots, n\}.$$

By projecting to $\mathbb{P}^n$, one verifies that $\mathcal{Y}$ is a smooth irreducible subvariety of $\mathbb{P}^n \times T$ of codimension n. The fibre $\mathcal{Y}_t$ of $\mathcal{Y}$ over t in T is the intersection of the corresponding hypersurfaces. Set

$$\begin{aligned} L &= \{(x) \in \mathbb{P}^n \mid x_0 = x_1 = 0\}, \\ T^0 &= \{t \in T \mid \mathcal{Y}_t \cap L = \emptyset\}, \\ \mathcal{Y}^0 &= \mathcal{Y} \cap (\mathbb{P}^n \times T^0), \end{aligned}$$

so T^0 is a non-empty open subvariety of T, $\mathcal{Y}^0$ is open in $\mathcal{Y}$, and closed in $(\mathbb{P}^n - L) \times T^0$. Let p be the projection from $\mathbb{P}^n - L$ to $\mathbb{P}^1$, $p(x) = (x_0 : x_1)$. Let

$$\mathcal{R}^0 = (p \times 1)(\mathcal{Y}^0) \subset \mathbb{P}^1 \times T^0.$$

Since $\mathcal{Y}^0$ is proper over T^0, the induced morphism f from $\mathcal{Y}^0$ to $\mathcal{R}^0$ is proper, so $\mathcal{R}^0$ is a closed subvariety of $\mathbb{P}^1 \times T^0$. In fact, f maps $\mathcal{Y}^0$ birationally onto $\mathcal{R}^0$, i.e.

$$f_*[\mathcal{Y}^0] = [\mathcal{R}^0].$$

Indeed, if T^{00} consists of those $t \in T^0$ for which the corresponding hypersurfaces meet in $d_1 \cdot \ldots \cdot d_n$ points with distinct projections to $\mathbb{P}^1$, f restricts to an isomorphism over T^{00}. In particular $\mathcal{R}^0$ is a hypersurface in $\mathbb{P}^1 \times T^0$. Let

$$R = R(x_0, x_1; t^1, \dots, t^n)$$

be the equation, unique up to scalars, for the closure of $\mathcal{R}^0$ in $\mathbb{P}^1 \times T$; R is called the *resultant*. For any particular hypersurfaces $F_1, \dots, F_n$ of degrees $d_1, \dots, d_n$, the resultant $R(F_1, \dots, F_n)$ is obtained by substituting the corresponding coefficients $t^1, \dots, t^n$ into R. The resultant is characterized, up to multiplication by polynomials in $t^1, \dots, t^n$, by being homogeneous of degree $d_1 \cdot \ldots \cdot d_n$ in x_0, x_1 (cf. below) and vanishing at points $((x_0 : x_1), (t^1, \dots, t^n))$, in any algebraic closure of the ground field, where the $F_i(t^i, x)$ have common zeros not on L. (For some of the many classical methods for calculating resultants, see Salmon (2) pp. 66–98, and Van der Waerden (4) § XI. For a modern discussion see Jouanolou (3).)

Theorem. *Let $V_i \subset \mathbb{P}^n$ be the hypersurface defined by equation F_i, $i = 1, \dots, n$. Assume $\bigcap V_i \bigcap L = \emptyset$, and $\bigcap V_i$ is finite. Then, for any $Q \in \mathbb{P}^1$,*

$$(*) \qquad \operatorname{ord}_Q(R(F_1, \dots, F_n)) = \sum_{\substack{P \in \mathbb{P}^n - L \\ p(P) = Q}} i(P, V_1 \cdot \ldots \cdot V_n; \mathbb{P}^n).$$

To prove this, let $t: \operatorname{Spec}(K) \to T$ be the point corresponding to $F_1, \ldots, F_n$. Let $\mathscr{V}_i^0$ be the subscheme of $(\mathbb{P}^n - L) \times T^0$ defined by $F_i(x, t^i)$. Form the fibre diagram:

$$\begin{array}{ccccccc}
\bigcap V_i & \to & V_1^0 \times \ldots \times V_n^0 & \to & \mathbb{P}^1 & \to & \operatorname{Spec}(K) \\
\downarrow & & \downarrow & & \downarrow & & \downarrow t \\
\mathscr{V}^0 & \to & \mathscr{V}_1^0 \times \ldots \times \mathscr{V}_n^0 & \to & \mathbb{P}^1 \times T^0 & \to & T^0 \\
\downarrow & & \downarrow & & & & \\
\mathbb{P}^n & \underset{\delta}{\to} & \mathbb{P}^n \times \ldots \times \mathbb{P}^n . & & & &
\end{array}$$

By Theorem 6.4, and Theorem 6.2 (b), (c),

$$\begin{aligned} t^![\mathscr{V}^0] &= t^! \delta^! [\mathscr{V}_1^0 \times \ldots \times \mathscr{V}_n^0] = \delta^! t^! [\mathscr{V}_1^0 \times \ldots \times \mathscr{V}_n^0] \\ &= \delta^! [V_1^0 \times \ldots \times V_n^0] = [V_1] \cdot \ldots \cdot [V_n] \end{aligned}$$

in $A_0(\bigcap V_i)$. Let f' be the morphism from $\bigcap V_i$ to $\mathbb{P}^1$ induced by p. By Theorem 6.2 (a),

$$f'_*([V_1] \cdot \ldots \cdot [V_n]) = f'_* t^! [\mathscr{V}^0] = t^! f_* [\mathscr{V}^0] = t^! [\mathscr{R}^0] .$$

Since $R(F_1, \ldots, F_n) \neq 0$, $t^! [\mathscr{R}^0]$ is the cycle determined by the divisor of $R(F_1, \ldots, F_n)$. The equality of these cycles in $A_0(f'(\bigcap V_i))$ gives (*).

If one does not assume $\bigcap V_i$ finite, but only that $\bigcap V_i$ is contained in a finite number of fibres of p, the same proof shows that $\operatorname{ord}_Q(R(F_1, \ldots, F_n))$ is the total contribution of the intersection class $V_1 \cdot \ldots \cdot V_n$ which is supported on $\bigcap V_i \bigcap p^{-1}(Q)$.

Adding formula (*) over $Q \in \mathbb{P}^1$, for any $V_1, \ldots, V_n$ which meet properly (off L) shows that R has degree $d_1 \cdot \ldots \cdot d_n$ in x_0, x_1.

Notes and References

The procedure of *reduction to the diagonal*, i.e., of intersecting two cycles on a non-singular variety by intersecting their exterior product with the diagonal, has played an important role in intersection theory. One may detect its presence in the nineteenth century theory of correspondences (cf. Pieri (1)). Apparently Weil (2) in 1946 was the first to use this principle in modern geometry, although Lefschetz (3) had made extensive use of cycles on product manifolds.

Following Lefschetz's model in topology, and ideas of Severi in algebraic geometry, the construction of an intersection ring A^*Y for a non-singular projective variety has usually proceeded in two separate steps: (1) a theory of intersection multiplicities was developed (see the notes to Chap. 7), so that properly intersecting cycles had a well-defined intersection cycle; (2) one showed that two cycle classes have representatives which meet properly, and that the resulting product is well defined up to rational equivalence. This is the approach followed by Samuel (3), Chow (1), and Chevalley (2). Earlier descriptions of A^*Y along these lines were made by Severi (9), Todd (2), (6), and Segre (4), among others.

In the present version, sketched in Fulton-MacPherson (1), the intersection product of two cycles on a non-singular variety is constructed directly, in one step, as a well-defined rational equivalence class on the intersection of the supports of the two cycles. In addition to its simplicity, a primary advantage of this approach is that it leads to useful formulas for intersection classes in cases where the intersection is not proper. There is also the benefit that the construction works equally well on non-projective varieties; the construction of the rational equivalence ring for regular algebraic schemes has also been carried out using higher K-theory (see Chap. 20).

In the projective case, for varieties V, W of complementary dimension, Murre (1) showed how to assign an intersection number to each connected component of $V \cap W$, thus answering a question of Weil. His method, based on the moving lemma, would apparently also lead to the class $V \cdot W$ in $A_*(V \cap W)$ that we have constructed in § 8.1, when the ambient variety is projective.

The original theorem of Bézout (1) concerned the number of solutions of n polynomials in n variables; precedents can be found in the work of Jacques Bernoulli, Euler, Braikenridge, and Maclaurin (cf. Berzolari (2) and Zeuthen-Pieri (1)). Now its name is attached to a number of theorems concerning intersections of arbitrary of cycles on projective space – and often to more general situations whenever an intersection ring of a variety is explicitly computed. Poncelet's approach to Bézout's theorem was to deform the varieties to be intersected until they are unions of linear spaces, using his principle of conservation of number to know this didn't change the number of solutions. This approach, suitably justified, is the one followed in this chapter; indeed, once the intersection product is known to be well-defined on rational equivalence classes, Bézout's theorem is evident.

The first modern algebraic – and perhaps the first fully complete – treatment of Bézout's original theorem was given by Macaulay (1). For the relation to resultants, see Example 8.4.13. A discussion of Bézout's theorem emphasizing its algebraic aspects is given by Vogel (1); his methods also yield the conclusion of Example 8.4.6.

For intersections on $\mathbb{P}^n$, a further reduction beyond reduction to the diagonal is possible; the intersection of V and W in $\mathbb{P}^n$ is equivalent to the intersection of the *ruled join* of V and W by a linear space in $\mathbb{P}^{2n+1}$ (Example 8.4.5). This construction was made by Gaeta (1) to compare the Chow point of a product cycle with Chow points of the factors. An algebraic analogue has been used by Boda and Vogel (1). The geometric construction given in Example 8.4.5 we learned from Deligne; this construction has been useful in the study of the topology of projective varieties (cf. Fulton-Lazarsfeld (1)). The result of Example 8.4.6 was discovered and proved with MacPherson and Lazarsfeld, in answer to a question of Kleiman.

The intersection products $x \cdot_f y$ of § 8.1 generalize those of Serre (4) V.C7, who defined such products and stated similar formal properties in the case of proper intersections.

Chapter 9. Excess and Residual Intersections

Summary

If $X \hookrightarrow Y$ is a regular imbedding, $V \subset Y$ a subvariety, we have constructed (§ 6.1) an intersection product $X \cdot V$ in $A_m(X \cap V)$, where $m = \dim V - \operatorname{codim}(X, Y)$. If a closed subscheme Z of $X \cap V$ is given, the basic problem of residual intersections is to write $X \cdot V$ as the sum of a class on Z and a class on a "residual set" R. There is a canonical choice for the class on Z, namely

$$\{c(N) \cap s(Z, V)\}_m$$

where N is the restriction to Z of $N_X Y$, and $s(Z, V)$ is the Segre class. Our problem is therefore to compute this class on Z, and to construct and compute a residual intersection class $\mathbb{R}$ in $A_m(R)$, for an appropriate closed set R such that $Z \cup R = X \cap V$, with

$$X \cdot V = \{c(N) \cap s(Z, V)\}_m + \mathbb{R}\,.$$

If $m = 0$, and R is a finite set, knowing $X \cdot V$ and $\{c(N) \cap s(Z, V)\}_0$ gives a formula for the weighted number of points of R. This is the basis for applications of the excess intersection formula to enumerative geometry.

In case Z is a (scheme-theoretic) connected component of $X \cap V$, R is the union of the other connected components; since $A_*(X \cap V) = A_*(Z) \oplus A_*(R)$, the above decomposition is part of the construction of Chap. 6. Computations, applications, and a few of the many classical examples are considered in § 9.1.

The general case is considered in § 9.2. In the main theorem Z is assumed to be a Cartier divisor on V; in this case there is a natural scheme structure on the residual set, which can be used to construct $\mathbb{R}$. If Z is arbitrary, one blows up V along Z to reduce to the divisor case.

An important and typical application of the residual intersection theorem is to the formula for the double point cycle class of a morphism, which is given in § 9.3.

9.1 Equivalence of a Connected Component

Let Y be a scheme, $X_i \hookrightarrow Y$ regularly imbedded subschemes, $1 \leqq i \leqq r$, and V a k-dimensional subvariety of Y. The intersection product

$$X_1 \cdot \ldots \cdot X_r \cdot V$$

is a class in $A_m(\bigcap X_i \bigcap V)$, $m = \dim V - \sum_{i=1}^{r} \operatorname{codim}(X_i, Y)$. It is constructed by the procedure of § 6.1, applied to the diagram

$$\begin{array}{ccc} \bigcap X_i \bigcap V & \hookrightarrow & V \\ \downarrow & & \downarrow \delta \\ X_1 \times \ldots \times X_r & \hookrightarrow & Y \times \ldots \times Y. \end{array}$$

If Z is a connected component of $\bigcap X_i \bigcap V$, we write

$$(X_1 \cdot \ldots \cdot X_r \cdot V)^Z \in A_m(Z)$$

for the part of $X_1 \cdot \ldots \cdot X_r \cdot V$ supported on Z, and call it the *equivalence* of Z for the intersection $X_1 \cdot \ldots \cdot X_r \cdot V$.

Proposition 9.1.1 *Let N_i be the restriction of $N_{X_i} Y$ to Z. Then*

$$(X_1 \cdot \ldots \cdot X_r \cdot V)^Z = \left\{ \prod_{i=1}^{r} c(N_i) \cap s(Z, V) \right\}_m.$$

If Z is regularly imbedded in V, with normal bundle $N_Z V$, then

$$(X_1 \cdot \ldots \cdot X_r \cdot V)^Z = \left\{ \prod_{i=1}^{r} c(N_i) \cdot c(N_Z V)^{-1} \cap [Z] \right\}_m.$$

If V and Z are non-singular, then

$$(X_1 \cdot \ldots \cdot X_r \cdot V)^Z = \left\{ \prod_{i=1}^{r} c(N_i)\, c(T_V|_Z)^{-1} c(T_Z) \cap [Z] \right\}_m.$$

Proof. The first assertion follows from Proposition 6.1 (a) and the Whitney sum formula (Theorem 3.2 (e)). The second follows from Proposition 4.1 (a). The last uses the identification of $N_Z V$ with the quotient of tangent bundles $T_V|_Z / T_Z$ (Appendix B.7.2) and the Whitney sum formula. □

The global intersection class $X_1 \cdot \ldots \cdot X_r \cdot V$ is always the sum of the equivalences for the connected components of $\bigcap X_i \bigcap V$. The following case suffices for many enumerative applications.

Proposition 9.1.2. *Suppose $m = 0$, and $\bigcap X_i \bigcap V$ consists of a connected component Z together with a finite set S. Then the degree of $X_1 \cdot \ldots \cdot X_r \cdot V$ is*

$$\deg((X_1 \cdot \ldots \cdot X_r \cdot V)^Z) + \sum_{P \in S} i(P, X_1 \cdot \ldots \cdot X_r \cdot V); Y) \cdot [R(P):K],$$

where $[R(P):K]$ denotes the degree of the residue field of P over the ground field. □

In enumerative applications, $\deg(X_1 \cdot \ldots \cdot X_r \cdot V)$ is known for global reasons, e.g. Bézout's theorem. Knowing the equivalence of Z then predicts the (weighted) number of residual points. Following the classical terminology, when $m = 0$, we also call the degree of $(X_1 \cdot \ldots \cdot X_r \cdot V)^Z$ the *equivalence* of Z in the intersection product. If $V = Y$, we write $(X_1 \cdot \ldots \cdot X_r)^Z$ in place of $(X_1 \cdot \ldots \cdot X_r \cdot V)^Z$.

Notation 9.1. For an arbitrary imbedding $Z \subset Y$, and any k, $s(Z, Y)_k$ denotes the k-dimensional component of the total Segre class $s(Z, Y)$. We regard $s(Z, Y)_k$ as an element of $A_k Z$, or of $A_k Z'$ for any closed subscheme Z' of Y which contains Z.

Example 9.1.1. *Let* $X_1, \ldots, X_r$ be Cartier divisors in an r-dimensional variety Y. Suppose a connected component Z of $\bigcap X_i$ is a curve. Then

$$(X_1 \cdot \ldots \cdot X_r)^Z = \sum_{i=1}^{r} X_i \cdot s(Z, Y)_1 + s(Z, Y)_0 .$$

If Z is a scheme-theoretic complete intersection of Cartier divisors $D_1, \ldots, D_{r-1}$, then

$$s(Z, Y) = [Z] - \sum_{j=1}^{r-1} D_j \cdot [Z] .$$

Therefore

$$(X_1 \cdot \ldots \cdot X_r)^Z = (X_1 + \ldots + X_r - D_1 - \ldots - D_{r-1}) \cdot D_1 \cdot \ldots \cdot D_{r-1} .$$

For example, if $Y = \mathbb{P}^r$, and $\deg X_i = n_i$, $\deg D_j = d_j$, then the equivalence of Z is

(i) $$(n_1 + \ldots + n_r - d_1 - \ldots - d_{r-1}) \cdot d_1 \cdot \ldots \cdot d_{r-1} .$$

On the other hand, if Z and Y are non-singular, then

$$s(Z, Y) = [Z] + c_1(T_Z) \cap [Z] - c_1(T_Y|_Z) \cap [Z] .$$

Therefore

$$(X_1 \cdot \ldots \cdot X_r)^Z = \left(\sum_{i=1}^{r} X_i - c_1(T_Y) \right) \cdot Z + c_1(T_Z) \cap [Z] .$$

For example, if $Y = \mathbb{P}^r$, and $\deg X_i = n_i$, the equivalence of Z is

(ii) $$(n_1 + \ldots + n_r - (r+1)) \cdot \deg Z + 2 - 2g$$

where g is the genus of Z.

Note that the equivalence of Z need not be a multiple of the degree of Z.

Example 9.1.2. A non-singular curve of genus 1 and degree 5 in $\mathbb{P}^3$ can never be the scheme-theoretic intersection of three surfaces in $\mathbb{P}^3$ (such a curve cannot lie on a quadric, say by Example 15.2.2. But there are no solutions to the equation

$$n_1 n_2 n_3 = 5(n_1 + n_2 + n_3 - 4)$$

in integers $n_i \geqq 3$.)

Since a linear system of degree 5 and dimension 3 on a curve of genus 1 is not complete, such curves are not linearly normal. The existence of such a curve which was not the intersection of three surfaces was posed by Peskine and Szpiro (1) Problem 7.3. Different examples have been given by Rao (1).

Many other curves which are not the complete intersection of three surfaces can be constructed by the same method. For any degree and genus, equation (ii) of the preceding example severely restricts the possibilities for the degrees of the three surfaces which cut it out.

Example 9.1.3. The twisted cubic curve in $\mathbb{P}^3$ may be written as the intersection of three quadrics. If three surfaces have a twisted cubic curve as their scheme-theoretic intersection, the surfaces must all be quadrics. (If the degrees are n_1, n_2, n_3, then

$$n_1 n_2 n_3 = 3(n_1 + n_2 + n_3 - 4) + 2 .$$

The only solution to this equation in integers $\geqq 2$ is $n_1 = n_2 = n_3 = 2$.)

Similarly, if $n > 3$, the rational normal curve in $\mathbb{P}^n$ cannot be written as the scheme-theoretic intersection of any n hypersurfaces. This should be contrasted with the following facts, valid over an algebraically closed ground field:

(i) Any algebraic set in $\mathbb{P}^n$ is the set-theoretic intersection of n hypersurfaces (Eisenbud-Evans (1)).

(ii) Any subscheme of $\mathbb{P}^n$ which is locally a complete intersection is the scheme-theoretic intersection of $n+1$ hypersurfaces. (Choose d so large that $\mathscr{I} \otimes \mathscr{O}(d)$ is generated by its sections, where $\mathscr{I}$ is the ideal of the subscheme; apply the construction of § 4.4 to this linear system, obtaining a morphism f from $\widetilde{\mathbb{P}}^n$ to $\mathbb{P}^r$. If $H_1, \ldots, H_{n+1}$ are hyperplanes in $\mathbb{P}^r$ so that $\bigcap H_i \bigcap f(\widetilde{\mathbb{P}}^n) = \emptyset$, then the corresponding hypersurfaces in the linear system cut out the subscheme.)

Another general fact explains why the equivalence of a connected component is studied mainly for zero-cycles:

(iii) If $X_1, \ldots, X_r, V$ are subvarieties of $\mathbb{P}^n$, and $m = \dim V - \sum \operatorname{codim}(X_i) > 0$, then $\bigcap X_i \bigcap V$ is always connected (Fulton-Hansen (1)).

Example 9.1.4. (a) If three surfaces in $\mathbb{P}^3$, of degrees n_1, n_2, n_3, contain a line as a connected component, the equivalence of this line is $n_1 + n_2 + n_3 - 2$. The three quadrics $x_0 x_2 = x_1^2$, $x_1 x_3 = x_2^2$, $x_0 x_1 = x_2 x_3$ meet in the line $x_1 = x_2 = 0$ and in 4 points outside this line.

(b) The equivalence of a non-singular rational quartic curve in the intersection of four quadrics in $\mathbb{P}^4$ is 14, leaving two points of intersection outside the curve.

Example 9.1.5. Let $X_1, \ldots, X_r$ be hypersurfaces in $\mathbb{P}^r$, $n_i = \deg(X_i)$, and assume a connected component of $\bigcap X_i$ is a non-singular surface Z. Then the equivalence of Z in $X_1 \cdot \ldots \cdot X_r$ is

$$\deg(c_2) + (n - r - 1) \cdot \deg(c_1) + \left(N - n(r+1) + \binom{r+2}{2}\right) \cdot \deg(Z) ,$$

where $n = \sum_{i=1}^{r} n_i$, $N = \sum_{i<j} n_i n_j$, and $c_i = c_i(T_Z)$.

Example 9.1.6 (cf. Severi (2), James (1), Semple and Roth (1) p. 228). Let $X_1, \ldots, X_r$ be divisors on a non-singular r-dimensional variety Y. Suppose a non-singular curve Z is a set-theoretic connected component of $\bigcap X_i$, but that the multiplicity of Z on X_i is m_i. Let $\pi : \tilde{Y} \to Y$ be the blow-up of Y along Z, E the exceptional divisor, and write

$$\pi^* X_i = m_i E + X_i' .$$

Assume that $X'_1 \bigcap \ldots \bigcap X'_r \bigcap E = \emptyset$. Set $m = m_1 \cdot \ldots \cdot m_r$, $\hat{m}_i = m/m_i$. Then

$$(*) \qquad (X_1 \cdot \ldots \cdot X_r)^Z = \sum_{i=1}^{r} \hat{m}_i X_i \cdot Z + m(c_1(T_Z) - c_1(T_Y|_Z)).$$

If $Y = \mathbb{P}^r$, and $n_i = \deg X_i$, $g = \text{genus}(Z)$, then the equivalence of Z is

$$\left(\sum_{i=1}^{r} \hat{m}_i n_i - m(r+1)\right) \cdot \deg(Z) + m \cdot (2 - 2g).$$

(Let η be the projection from E to Z. One may shrink Y so that $\bigcap X'_i = \emptyset$. Then

$$\begin{aligned} 0 &= \eta_*((\pi^* X_1 - m_1(E)) \cdot \ldots \cdot (\pi^* X_r - m_r E)) \\ &= (X_1 \cdot \ldots \cdot X_r)^Z - \sum_{i=1}^{r} \hat{m}_i X_i \cdot s(Z, Y)_1 - m\, s(Z, Y)_0.) \end{aligned}$$

It is necessary to make the above assumption that the X_i do not have infinitely near intersections over Z, i.e., that $\bigcap X'_i \bigcap E = \emptyset$. In case all $m_i = 1$, this corresponds to our previous assumption that the X_i cut out Z scheme-theoretically. For example, if $Y = \mathbb{P}^3$, and X_1, X_2, X_3 are defined by equations

$$x_2 = 0, \quad x_1^2 = x_0 x_2, \quad x_1 x_3 = x_0 x_2,$$

and Z is the line $x_1 = x_2 = 0$, then Z is the set-theoretic intersection of the X_i, and $m_i = 1$ for all i, but

$$(\sum n_i - 4) \cdot \deg(Z) + 2 - 2g = 1 + 2 = 3,$$

although $\deg((X_1 \cdot X_2 \cdot X_3)^Z) = n_1 n_2 n_3 = 4$.

If $X'_1 \bigcap \ldots X'_r \bigcap E \neq \emptyset$, one must add the image of $(X'_1 \cdot \ldots \cdot X'_r)^E$ to the right side of (*). In the above example, the X'_i meet transversally at one point.

Example 9.1.7 (cf. B. Segre (1) for $\dim Y = 3$, Todd (3) for $\dim Y = 4$). If $Y, X_1, \ldots, X_r$, and Z are non-singular, and Z is a curve, and $m = 0$, then

$$(X_1 \cdot \ldots \cdot X_r)^Z = \sum_{i=1}^{r} (K_{X_i} \cdot Z)_{X_i} - (r-1)(K_Y \cdot Z)_Y - K_Z.$$

(Here $K_V = -c_1(T_V)$ is the canonical divisor class, and for $Z \subset V$, $(K_V \cdot Z)_V = -c_1(T_V|_Z) \cap [Z]$.)

If Y is a 4-fold, and X_1, X_2 are surfaces, the equivalence of Z is

$$(K_{X_1} \cdot Z)_{X_1} + (K_{X_2} \cdot Z)_{X_2} - (K_Y \cdot Z)_Y - K_Z.$$

For example, if $Y = \mathbb{P}^4$, X_1 is the projection of the Veronese surface, X_2 is a plane, and Z a conic, then the equivalence of Z is 3; there is therefore one residual point outside Z.

If there are a finite number of points which are singular on Z, X_1, and X_2, such that the proper transforms in the blow-up $\tilde{Y}$ of Y at these points become non-singular, one may use the formula in $\tilde{Y}$ to deduce the equivalence on Y. For an example where X_1, X_2 are surfaces with improper nodes, see Todd (3).

Example 9.1.8. The plane conics are parametrized by $\mathbb{P}^5$. The conics tangent to a given line l form a hypersurface H_l in $\mathbb{P}^5$ of degree 2. If 5 lines

$l_1, \ldots, l_5$ are given, no three passing through any point, then the Veronese surface Z of double lines ($Z \cong \mathbb{P}^2$) is a scheme-theoretic component of $\bigcap H_{l_i}$. Therefore the equivalence of Z for $H_{l_1} \cdot \ldots \cdot H_{l_5}$ is

$$\int_Z (1+4h)^5 \cdot (1-9h+51h^2) = 160 - 180 + 51 = 31$$

where h is the class of a line in $Z = \mathbb{P}^2$ (cf. Example 3.2.15 (b)). Therefore there is one residual point, which represents the unique conic tangent to the given five lines.

Example 9.1.9 (Fulton-MacPherson (2)). The conics tangent to a given conic D form a hypersurface H_D of degree 6 in $\mathbb{P}^5$. Let $\pi: \tilde{\mathbb{P}}^5 \to \mathbb{P}^5$ be the blow-up of $\mathbb{P}^5$ along the Veronese Z, with exceptional divisor E. Then $\pi^* H_D = 2E + G_D$ for some divisor G_D. If $D_1, \ldots, D_5$ are given conics, such that (i) no three pass through any point, and (ii) there is no line with two points on it such that each D_i is either tangent to the line or passes through one of the points, then $\bigcap G_{D_i} \bigcap E = \emptyset$. The equivalence of Z for the intersection $H_{D_1} \cdot \ldots \cdot H_{D_5}$ is therefore

$$\int_Z (1+6(2h))^5 (2^3 - 2^4 \cdot 9h + 2^5 \cdot 51 h^2) = 4512\,.$$

Assume in addition: (iii) no two of the five conics are tangent, and (iv) the pairs of lines, each tangent to two of the conics, do not intersect on the fifth conic. Then all the points of $\bigcap H_{D_i}$ outside Z are isolated points which represent non-singular conics C, and the intersection number of the hypersurfaces at the point corresponding to C is $i(C)$, where

$$i(C) = \prod_{i=1}^{5} (4 - \#(C \cap D_i))\,.$$

Since the total intersection number is 6^5 (by Bézout),

$$\sum_C i(C) = 6^5 - 4512 = 3264\,.$$

In particular, if the D_i are chosen in general enough position so that all $i(C) = 1$, then there are 3264 non-singular conics tangent to the 5 given conics. (These calculations hold in all characteristics but 2. In characteristic 2 the H_D are cubics, and a simpler calculation shows that the number 3264 should be replaced by 51. See Example 10.4.3 or Vainsencher (3) for a different approach in characteristic 2.) For generalizations, see § 10.4.

Example 9.1.10. The class $(X_1 \cdot \ldots \cdot X_r)^Z$ in $A_* Z$ depends only on the rational equivalence classes of $X_1, \ldots, X_r$ in $A_* Y$: if X_i' is rationally equivalent to X_i, and Z is also a connected component of $\bigcap X_i'$, then

$$(X_1' \cdot \ldots \cdot X_r')^Z = (X_1 \cdot \ldots \cdot X_r)^Z\,.$$

However, contrary to an assertion of Segre (4) p. 20, the class may change – even numerically – if Z is replaced by a rationally equivalent variety. For example, if $Y = \mathbb{P}^3$, Z is a twisted cubic, and Z' is a plane elliptic curve, and surfaces X_i, X_i' are taken of large degree d_i, then

$$(X_1 \cdot X_2 \cdot X_3)^Z - (X_1' \cdot X_2' \cdot X_3')^{Z'} = 2(g(Z') - g(Z)) = 2\,.$$

Example 9.1.11. Let Y be an r-dimensional variety, $X_1, \ldots, X_{r-1}$ divisors on Y, such that $X_1 \bigcap \ldots \bigcap X_{r-1}$ is one-dimensional. Let Z be a curve which is contained in $X_1 \bigcap \ldots \bigcap X_{r-1}$. The number i of intersections of Z with the rest of $X_1 \bigcap \ldots \bigcap X_{r-1}$ is given by the formula

$$i = \deg\{c(N) \cap s(Z, Y)\}_0$$

where $N = \oplus N_{X_i} Y|_Z$. Note that $\{c(N) \cap s(Z, Y)\}_1$ represents the contribution of Z to the intersection product; it is interesting to see a geometric interpretation for a lower dimensional term. To justify this interpretation, let $\pi: \tilde{Y} \to Y$ be the blow-up of Y along Z, E the exceptional divisor, $\eta: E \to Z$ the projection. Write

$$\pi^* X_i = E + X_i'.$$

Then $\deg(X_1' \cdot \ldots \cdot X_{r-1}' \cdot E)$ is a reasonable geometric definition for the number i, and

$$\eta_*((\pi^* X_1 - E) \cdot \ldots \cdot (\pi^* X_{r-1} - E) \cdot E) = \sum_{i=1}^{r-1} X_i \cdot s(Z, Y)_1 + s(Z, Y)_0,$$

which gives the formula stated.

If $Y = \mathbb{P}^r$, $\deg X_i = n_i$, and Z is non-singular of degree e and genus g, then

$$i = \left(\sum_{i=1}^{r-1} n_i - (r+1)\right) \cdot e + 2 - 2g.$$

Example 9.1.12. Suppose $r-1$ hypersurfaces in $\mathbb{P}^r$, of degrees $n_1, \ldots, n_{r-1}$, intersect in a union of two non-singular curves Z and Z' which intersect transversally (i.e., with distinct tangents) at i points. If e, g (resp. e', g') denote the degree and genus of Z (resp. Z'), then

(i) $$e + e' = n_1 \cdot \ldots \cdot n_{r-1}$$

(ii) $$g' - g = \tfrac{1}{2}\left(\sum_{i=1}^{r-1} n_i - (r+1)\right) \cdot (e' - e)$$

(iii) $$i = \left(\sum_{i-1}^{r-1} n_i - (r+1)\right) \cdot e + 2 - 2g.$$

Thus from $n_1, \ldots, n_{r-1}$, e, and g one may calculate e', g' and i. (Bézout's theorem gives (i), the preceding example gives (iii). Writing (iii) with the roles of Z and Z' reversed gives an equation equivalent to (ii).)

For example, if Z is a twisted cubic contained in two quadrics, Z' will be a line meeting Z in two points. If Z is a curve of genus 1 and degree 5 contained in two cubic surfaces, this predicts that Z' will be a rational curve of degree 4 meeting Z in 10 points.

Example 9.1.13 (Le Barz (5)). Let $\mathrm{Hilb}^k \mathbb{P}^N$ be the Hilbert scheme of k-tuples of points in $\mathbb{P}^N$, $Al^k \mathbb{P}^N$ the subvariety consisting of k-tuples which are aligned. For any subvariety X of $\mathbb{P}^N$, the intersection

$$\mathrm{Hilb}^k X \bigcap Al^k \mathbb{P}^N$$

corresponds to the k-fold multisecant lines to X. However, any line L contained in X gives an excess component of this intersection. For example, suppose X is a surface in $\mathbb{P}^4$ containing a finite number of lines, and L is a line in X not meeting any singular point of X. Then the equivalence of L for the above intersection is $\binom{7+m}{6}$ where m is the self-intersection number of L on X. We refer to the papers of Le Barz for details and other examples.

Example 9.1.14. *Herbert's multiple point formula.* Let $f: X \to Y$ be a proper immersion of non-singular varieties. Assume that f is *completely regular*, i.e., for any distinct points $x_i \in X$ with the same image $y \in Y$, the images of the tangent spaces $T_{x_i}X$ are in general position in $T_y Y$. Let

$$Y_k = \{y \in Y \mid \# f^{-1}(y) \geqq k\},$$

and $X_k = f^{-1}(Y_k)$, with their reduced structures. Then

$$(*) \qquad [X_k] = f^*[Y_{k-1}] - c_d(\nu_f) \cap [X_{k-1}]$$

in $A_{m-dk}(X)$, with $m = \dim Y$, $m - d = \dim X$, and $\nu_f = f^* T_Y / T_X$. Therefore

$$[X_k] = \sum_{j=0}^{k-1} (-1)^j c_d(\nu_f)^j \cap f^*[Y_{k-1-j}].$$

These formulas are due to Herbert (1). Ronga (3) deduces (*) from the excess intersection formula. Let

$$\hat{X}_k = \{(x_1, \ldots, x_k) \in X \times \ldots \times X \mid x_i \neq x_j \ \forall i \neq j, f(x_1) = \ldots = f(x_k)\}.$$

Let $\bar{X}_k = \hat{X}_k / S_k$, $X'_k = \hat{X}_k / S_{k-1}$, the symmetric group S_{k-1} acting on the last $k-1$ factors. There are canonical proper immersions $f_k: X'_k \to X$ with $(f_k)_*[X'_k] = [X_k]$ and $g_k: \bar{X}_k \to Y$ with $(g_k)_*[\bar{X}_k] = [Y_k]$. There is a fibre square

$$\begin{array}{ccc} X'_k \amalg X'_{k-1} & \rightarrow & \bar{X}_k \\ {\scriptstyle f_k \amalg f_{k-1}} \downarrow & & \downarrow {\scriptstyle g_k} \\ X & \xrightarrow[f]{} & Y \end{array}.$$

From the excess intersection formula (Proposition 6.5),

$$f^*(g_k)_*[\bar{X}_k] = (f_k)_*[X'_k] + (f_{k-1})_*(c_d(E) \cap [X'_{k-1}]),$$

where E is the excess normal bundle. In fact, E is the pull-back of ν_f to X'_{k-1}, and (*) results.

For a general discussion of multiple point formulas, including some cases where ramification may be present, see Kleiman (8), (12), and Ran (2).

9.2 Residual Intersection Theorem

Definition 9.2.1. Let $D \subset W \subset V$ be closed imbeddings of schemes. Assume that D is a Cartier divisor on V. There is a closed subscheme R of W, called the *residual scheme to D in W* (with respect to V), such that

$$W = D \bigcup R$$

and, moreover, the ideal sheaves on Y are related by

$$\mathscr{I}(W) = \mathscr{I}(D) \cdot \mathscr{I}(R).$$

Indeed, the inclusion $D \subset W$ means that $\mathscr{I}(W) \subset \mathscr{I}(D)$, so that every local equation for W is uniquely divisible by a local equation for D; the quotients give local equations for R.

Proposition 9.2. *Let $D \subset W \subset V$ be closed imbeddings, with V a k-dimensional variety, and D a Cartier divisor on V. Let R be the residual scheme to D in W. Then, for all m,*

$$s(W,V)_m = s(D,V)_m + \sum_{j=0}^{k-m} \binom{k-m}{j} (-D)^j \cdot s(R,V)_{m+j}$$

in $A_m(W)$.

Proof. Assume first that $W \neq V$. Let $\pi: \tilde{V} \to V$ be the blow-up of V along R. Let $\tilde{W} = \pi^{-1}(W)$, $\tilde{R} = \pi^{-1}(R)$, $\tilde{D} = \pi^{-1}(D) = \pi^*(D)$, so that $\tilde{W} = \tilde{D} + \tilde{R}$ as Cartier divisors on $\tilde{Y}$. Let η be the induced morphism from $\tilde{W}$ to W. Set $d = k - m$. By the definition of Segre classes (cf. Corollary 4.2.2) and the projection formula (Proposition 2.3 (c))

$$\begin{aligned} s(W,V)_m &= (-1)^{d-1} \eta_*((\tilde{D} + \tilde{R})^d) \\ &= (-1)^{d-1} \eta_*(\tilde{D}^d) + \sum_{j=0}^{d-1} (-1)^{d-1} \binom{d}{j} \eta_*(\pi^* D^j \cdot \tilde{R}^{d-j}) \\ &= s(D,V)_m + \sum_{j=0}^{d-1} (-1)^j \binom{d}{j} D^j \cdot s(R,V)_{m+j}, \end{aligned}$$

which is the required equation since $s(R,V)_{m+d} = 0$. If $W = V$, the required equation reads

$$\{[V]\}_m = s(D,V)_m + (-D)^{k-m} \cdot [V],$$

which amounts to the definition of $s(D,V)$. □

Theorem 9.2 (Residual Intersection Theorem). *Consider a diagram*

$$\begin{array}{ccccc} & & R & & \\ & & \downarrow b & & \\ D & \xrightarrow{a} & W & \xrightarrow{j} & V \\ & & \downarrow g & & \downarrow f \\ & & X & \xrightarrow[i]{} & Y \end{array}$$

with the square a fibre square, i, j, a, b closed imbeddings, and V a k-dimensional variety. Assume:

(i) *i is a regular imbedding of codimension d;*
(ii) *j a imbeds D as a Cartier divisor on V;*
(iii) *R is the residual scheme to D in W.*

Let $N = g^* N_X Y$, *and* $\mathscr{O}(-D) = j^* \mathscr{O}_V(-D)$. *Define the* residual intersection class $\mathbb{R}$ *in* $A_{k-d}(R)$ *by the formula*

$$(*)\qquad \mathbb{R} = \{c(N \otimes \mathscr{O}(-D)) \cap s(R, V)\}_{k-d}.$$

Then

$$X \cdot V = \{c(N) \cap s(D, V)\}_{k-d} + \mathbb{R}$$

in $A_{k-d} W$.

Proof. Set $r = k - d$. By Proposition 6.1 (a),

$$(1)\qquad X \cdot V = \{c(N) \cap s(W, V)\}_r = \sum_{i=0}^{d} c_i(N) \cap s(W, V)_{r+i}.$$

By definition,

$$(2)\qquad \{c(N) \cap s(D, V)\}_{k-d} = \sum_{i=0}^{d} c_i(N) \cap s(D, V)_{r+i},$$

and, by Example 3.2.2,

$$\begin{aligned}(3)\qquad \{c(N \otimes \mathscr{O}(-D)) \cap s(R, V)\}_r &= \sum_{p=0}^{d} c_p(N \otimes \mathscr{O}(-D)) \cap s(R, V)_{r+p}\\ &= \sum_{p=0}^{d} \sum_{i=0}^{p} \binom{d-i}{d-p} c_i(N)\, c_1(\mathscr{O}(-D))^{p-i} \cap s(R, V)_{r+p}\\ &= \sum_{i=0}^{d} c_i(N) \left(\sum_{j=0}^{d-i} \binom{d-i}{j} (-D)^j \cap s(R, V)_{r+i+j} \right).\end{aligned}$$

From Proposition 9.2, for $m = r + i$, it follows that the right side of equation (1) is the sum of the right sides of (2) and (3), as required. □

Notation 9.2. For a power series P in Chern classes of vector bundles, denote by $(P)_m$ the term of P of isobaric weight m.

Corollary 9.2.1. *In addition to the assumptions of the theorem, assume that* R *is regularly imbedded in* V *of codimension* d'. *Then*

$$X \cdot V = c(i/j\,a) \cap [D] + c(i/j\,b; D) \cap [R],$$

where

$$c(i/j\,a) = (c(N)/c(\mathscr{O}(D)))_{d-1},$$

$$c(i/j\,b; D) = (c(N \otimes \mathscr{O}(-D))/c(N_R V))_{d-d'}.$$

In particular, if $d' = d$, *then*

$$X \cdot V = c(i/j\,a) \cap [D] + [R].$$

Proof. Since $s(R, V) = c(N_R V)^{-1} \cap [R]$,

$$\{c(N \otimes \mathscr{O}(-D)) \cap s(R, V)\}_{k-d} = \{c(N \otimes \mathscr{O}(-D)) \cdot c(N_R V)^{-1} \cap [R]\}_{k-d},$$

from which the corollary follows. □

The residual scheme R is always locally defined in V by d equations – the quotients of local equations for X in Y by an equation for D in V. In particular, if V is Cohen-Macaulay, and $\operatorname{codim}(R, V) \geqq d$, then R is regularly imbedded in V of codimension d (Lemma A7.1), and the last formula in the corollary applies. If V is not Cohen-Macaulay, the theorem gives the following result.

Corollary 9.2.2. *In the situation of the theorem, assume that* $\operatorname{codim}(R, V) = d$. *Let* $R_1, \ldots, R_t$ *be the irreducible components of* R, *and let* $e_i = (e_R V)_{R_i}$ *be the algebraic multiplicity of* V *along* R *at* R_i (§ 4.3). *Then*

$$X \cdot V = \{c(N) \cap s(D, V)\}_{k-d} + \sum_{i=1}^{t} e_i [R_i] . \quad \square$$

Definition 9.2.2. Let $Z \subset W \subset V$ be closed imbeddings, with V a variety, and $Z \neq V$. The *residual set* to Z in W is defined as follows. Let $\pi: \tilde{V} \to V$ be the blow-up of V along Z. Let $D = \pi^{-1}(Z)$ be the exceptional divisor, $\tilde{W} = \pi^{-1}(W)$. Then $D \subset \tilde{W} \subset \tilde{V}$, so the residual scheme $\tilde{R}$ to D in $\pi^{-1}(W)$ is defined. Define the residual set R to be the image of $\tilde{R}$ in V: $R = \pi(\tilde{R})$. (R may be given the image scheme structure, but this will play no role here.) Consider a diagram

$$\begin{array}{ccccc} & & R & & \\ & & \downarrow b & & \\ Z & \xrightarrow{a} & W & \xrightarrow{j} & V \\ & & \downarrow g & & \downarrow f \\ & & X & \xrightarrow[i]{} & Y \end{array}$$

satisfying the conditions of Theorem 9.2, but without assuming Z is a Cartier divisor on V. Define the *residual intersection class* $\mathbb{R}$ in $A_{k-d}R$ by the formula

$$\mathbb{R} = \eta_* \{c(N \otimes \mathcal{O}(-D)) \cap s(\tilde{R}, \tilde{V})\}_{k-d}$$

where η is the induced morphism from $\tilde{R}$ to R. In other words, $\mathbb{R}$ is the image of the residual intersection class on $\tilde{R}$ defined in Theorem 9.2.

Corollary 9.2.3 (Residual Intersection Formula). *With the above assumptions,*

$$X \cdot V = \{c(N) \cap s(Z, V)\}_{k-d} + \mathbb{R} .$$

Proof. By Theorem 6.2(a), $X \cdot V$ is the push-forward of $X \cdot \tilde{V}$. The corollary follows from the theorem, applied to $\tilde{V}$, and the identity

$$c(N) \cap s(Z, V) = \eta_* (c(\eta^* N) \cap s(D, \tilde{V})) ,$$

which follows from the projection formula and Proposition 4.2(a). $\square$

A more general residual intersection theorem together with a geometric explanation for the terms involved, will be given in § 17.6.

Example 9.2.1. If Z is a connected component of W, the residual scheme R is the union of the other components, and

$$\mathbb{R} = \{c(N) \cap s(R, V)\}_{k-d}$$

is the contribution of R to the intersection product $X \cdot V$.

Example 9.2.2. In the situation of Theorem 9.2, let E be the restriction of $N \otimes \mathcal{O}(-D)$ to R. The normal cone $C_R V$ is canonically imbedded as a sub-cone of E. If s_E is the zero-section of E, then

$$\mathbb{R} = \{c(N \otimes \mathcal{O}(-D)) \cap s(R, V)\}_{k-d} = s_E^*[C_R V].$$

Thus the residual class $\mathbb{R}$ has a canonical decomposition: $\mathbb{R} = \sum m_i s_E^*[C_i]$, where C_i are the irreducible components of $C_R V$, m_i is the geometric multiplicity of C_i in $C_R V$. (For the first statement, the equation $\mathcal{I}(X) \cdot \mathcal{O}_V = \mathcal{I}(D) \cdot \mathcal{I}(R)$ determines surjections

$$h^* \mathcal{I}(X)^n / \mathcal{I}(X)^{n+1} \twoheadrightarrow \mathcal{I}(D)^n \otimes (\mathcal{I}(R)^n / \mathcal{I}(R)^{n+1}),$$

where $h = g b$ is the morphism from R to X. Therefore

$$\bigoplus_n h^* \mathcal{I}(X)^n / \mathcal{I}(X)^{n+1} \otimes (\mathcal{O}_V(D)|_R)^{\otimes n} \twoheadrightarrow \bigoplus_n \mathcal{I}(R)^n / \mathcal{I}(R)^{n+1}$$

which corresponds to an imbedding of $C_R V$ in $h^* N \otimes \mathcal{O}_V(-D)|_R$, as asserted. The second statement follows from Proposition 6.1 (a).)

Example 9.2.3. In Corollary 9.2.1, the class $c(i/j\,b; D)$ is the top Chern class of the bundle $E/N_R V$, where E is the restriction of $N \otimes \mathcal{O}(-D)$ to R (cf. Example 9.2.2).

Example 9.2.4. Let A, B, D be effective divisors on a non-singular surface X, with A and B relatively prime. Let $A' = A + D$, $B' = B + D$. Then the residual scheme to D in $A' \cap B'$ (with respect to X) is the scheme $A \cap B$. For the intersection of the diagonal Δ_X in $X \times X$ by $A' \times B'$, the residual intersection theorem says that

$$\begin{aligned}(A' \times B') \cdot \Delta_X &= \{c(\mathcal{O}(A') \oplus \mathcal{O}(B')) \cap (D - D \cdot D)\}_0 + [A \cap B] \\ &= (A' \cdot D + B' \cdot D - D \cdot D) + A \cdot B .\end{aligned}$$

Note that this decomposition of the intersection class differs from the canonical decomposition using distinguished components (cf. Example 6.1.4).

Example 9.2.5. Let $V = \mathbb{P}^2$, $Y = \mathbb{P}^4$, $X = \mathbb{P}^2$ the plane in $\mathbb{P}^4$ defined by the vanishing of the first two coordinates. Let f be the morphism from V to Y given by

$$f(x_0 : x_1 : x_2) = (x_0^4 : x_0^3 x_1 : x_0 x_1^3 : x_1^4 : x_2^4) .$$

Then $W = f^{-1}(X)$ is the scheme $x_0^4 = x_0^3 x_1 = 0$. If $D \subset V$ is the triple line defined by $x_0^3 = 0$, then the reduced point $R = (0:0:1)$ is the residual scheme to D in W. Therefore

$$\begin{aligned}X \cdot V &= \{c(N) \cap s(D, \mathbb{P}^2)\}_0 + [R] \\ &= \{(1 + 4h)^2 (1 + 3h)^{-1} \cdot 3[h]\}_0 + [R] \\ &= 15[P] + [R] = 16[P] .\end{aligned}$$

where P is a point of D. Note that the image $f(V)$ is a subvariety of $\mathbb{P}^4$ of degree 4 which meets X geometrically in a triple line; since $f_*(X \cdot V) = X \cdot f_*[V] = 4X \cdot [f(V)]$, the above equation is compatible with Bézout's theorem.

Example 9.2.6. Let $X = \mathbb{A}^1$, $Y = \mathbb{A}^2$, $f: X \to Y$ by $f(t) = (a(t), b(t))$. Then $(f \times f)^{-1}(\Delta_Y)$ contains Δ_X, and the residual scheme R is the scheme defined by

$a_\delta(t_1, t_2)$ and $b_\delta(t_1, t_2)$, where

$$a(t_2) - a(t_1) = (t_2 - t_1)\, a_\delta(t_1, t_2)\,,$$

and similarly for b_δ. The scheme-theoretic intersection of R with Δ_X is defined by the derivatives $a'(t)$ and $b'(t)$. Off Δ_X, R is the double point set, while $R \cap \Delta_X$ measures ramification. (See Example 9.3.12 for a generalization.)

Example 9.2.7. Let Y be a non-singular variety, and let $Z \subset Y$ be a union of t curves $Z_1, \ldots, Z_t$, such that no point lies on three of these curves, and if $P \in Z_i \cap Z_j$, $i \neq j$, then P is simple on Z_i and Z_j with distinct tangent lines. Then

$$s(Z, Y) = \sum_{i=1}^{t} s(Z_i, Y) + \sum_{i<j} [Z_i \cap Z_j]\,.$$

(Blow up Y at $\bigcup_{i<j} (Z_i \cap Z_j)$, and use Proposition 9.2.)

If $X_1, \ldots, X_r$ are divisors on Y, $\dim Y = r$, and Z is a connected component of $\bigcap X_i$, and each Z_i is non-singular, then

$$(X_1 \cdot \ldots \cdot X_r)^Z = \left(\sum_{i=1}^{r} X_i - c_1(T_Y)\right) \cdot Z + \sum_{i=1}^{r} c_1(T_{Z_i}) + \sum_{i<j} [Z_i \cap Z_j]\,.$$

If $Y = \mathbb{P}^r$, $n_i = \deg X_i$, $g_i = \operatorname{genus}(Z_i)$, and N is the total number of points in more than one Z_i, then the equivalence of Z is

$$\left(\sum_{i=1}^{r} n_i - (r+1)\right) \cdot \deg(Z) + \sum_{i=1}^{r} (2 - 2g_i) + N\,.$$

(cf. Semple and Roth (1) IV.8.3).

Example 9.2.8. For any $Z \subset W \subset V$, one may define a residual *scheme* R to Z in W by defining the ideal sheaf $\mathscr{I}(R)$ on affine open sets of V to be the ideal of functions which multiply all functions in $\mathscr{I}(Z)$ into elements of $\mathscr{I}(W)$, i.e. (cf. Peskine and Szpiro (1)),

$$\mathscr{I}(R)/\mathscr{I}(Z) = \mathscr{H}om_{\mathscr{O}_V}(\mathscr{O}_Z, \mathscr{O}_W)\,.$$

This agrees with the description in the text if Z is a Cartier divisor on V. The residual class $\mathbb{R}$ constructed in Definition 9.2.2 is a class in $A_{k-d} R$. It would be valuable to have conditions which would imply that $\mathbb{R}$ is the cycle $[R]$ associated with this scheme structure, or to compute $\mathbb{R}$ using this scheme structure.

9.3 Double Point Formula

Let $f\colon X^n \to Y^m$ be a morphism of non-singular varieties of the indicated dimensions, with X complete. Then $(f \times f)^{-1}(\Delta_Y)$ contains Δ_X, and the residual set is the set of *double point pairs*, which we denote by $D'(f)$. Let $X \tilde{\times} X$ be the blow up of $X \times X$ along Δ_X, π the projection from $X \tilde{\times} X$ to $X \times X$, $E = P(T_X)$

the exceptional divisor. Let $F=(f\times f)\circ\pi$ be the induced morphism from $X\tilde{\times}X$ to $Y\times Y$. The *double point scheme* $\tilde{D}(f)$ is defined to be the residual scheme to E in $F^{-1}(\Delta_Y)$. We have a diagram as in § 9.2:

$$\begin{array}{ccccc} & & \tilde{D}(f) & & \\ & & \downarrow & & \\ P(T_X)=E & \to & F^{-1}(\Delta_Y) & \to & X\tilde{\times}X \\ & & \downarrow & & \downarrow F \\ & & \Delta_Y & \underset{\delta}{\to} & Y\times Y. \end{array}$$

One may verify (cf. Fulton-Laksov (1) Prop. 4) that the points of $\tilde{D}(f)$ are those point pairs (x_1, x_2) with $x_1 \neq x_2$ and $f(x_1)=f(x_2)$, together with those tangent directions in $P(T_X)$ on which the induced tangent map vanishes. By definition, $D'(f)=\pi(\tilde{D}(f))$. We define the *double point set* $D(f)\subset X$ to be the image of $D'(f)$ by the first (or second) projection from $X\times X$ to X. Let η be the induced morphism from $\tilde{D}(f)$ to $D(f)$.

Define $\tilde{\mathbb{D}}(f)\in A_{2n-m}(\tilde{D}(f))$ to be the residual intersection class, defined by formula (*) of Theorem 9.2. If $\tilde{D}(f)$ has the expected codimension m in $X\tilde{\times}X$, then $\tilde{\mathbb{D}}(f)=[\tilde{D}(f)]$. Define the *double point class* $\mathbb{D}(f)$ in $A_{2n-m}(D(f))$ by

$$\mathbb{D}(f)=\eta_*(\tilde{\mathbb{D}}(f)).$$

Theorem 9.3 (Double Point Formula). *With the above notation,*

$$\mathbb{D}(f)=f^*f_*[X]-(c(f^*T_Y)\,c(T_X)^{-1})_{m-n}\cap[X]$$

in $A_{2n-m}X=A^{m-n}X$.

Proof. By Corollary 9.2.3, since $N_{\Delta_Y}(Y\times Y)=T_Y$,

$$\Delta_Y\cdot(X\times X)=\{c(f^*T_Y)\cap s(\Delta_X, X\times X)\}_{2n-m}+\pi_*[\tilde{\mathbb{D}}(f)]$$

in $A_{2n-m}(X\times X)$. Let p be the first projection from $X\times X$ to X, and apply p_* to this equation. The first term on the right projects to

$$\{c(f^*T_Y)\cdot c(T_X)^{-1}\cap[X]\}_{2n-m},$$

the second to $\mathbb{D}(f)$. It suffices to show that

$$p_*(\Delta_Y\cdot(X\times X))=f^*f_*[X].$$

To see this, recall that $f^*f_*[X]=\gamma_f^*([X]\times f_*[X])$, where γ_f is the graph imbedding of X in $X\times Y$. Consider the fibre diagram

$$\begin{array}{ccc} X\times_Y X & \to & X\times X \\ p'\downarrow & & \downarrow 1\times f \\ X & \underset{\gamma_f}{\to} & X\times Y \\ f\downarrow & & \downarrow f\times 1 \\ Y & \underset{\delta}{\to} & Y\times Y. \end{array}$$

By Theorem 6.2 (a) and (c),

$$\gamma_f^*([X]\times f_*[X])=p'_*(\gamma_f^![X\times X])=p'_*(\delta^![X\times X])=p_*(\Delta_Y\cdot(X\times X)).\quad\square$$

Example 9.3.1. If f is a closed imbedding, then $\mathbb{D}(f) = 0$, and the formula reduces to the self-intersection formula. From the self-intersection formula one knew that the right side of the double point formula restricts to zero on $X - D(f)$, so it comes from some class on $D(f)$. The residual construction produces such a class $\mathbb{D}(f)$ explicitly.

Example 9.3.2. If $n = 1$, and $m = 2$, and f maps X birationally onto its image $\bar{X}$ in Y, then

$$\mathbb{D}(f) = [\mathscr{C}]$$

where $\mathscr{C}$ is the subscheme of X whose ideal sheaf is the conductor of X over $\bar{X}$. This may be proved inductively by comparing both sides for $f: X \to Y$ and $f': X \to Y'$, where Y' is the blow-up of Y at a singular point of Y (see Fulton (6) § 4 for details). The analogous formula, for $f: X^n \to Y^{n+1}$ finite and birational onto its image, is still open, although it follows from Theorem 9.3 or Kleiman (8) that the two cycles are rationally equivalent on X. An example of Artin and Nagata (1) (cf. Fulton (6) § 2.4) shows that the conductor and $\mathbb{D}(f)$ may disagree if $m > n + 1$.

Example 9.3.3. Let $C \subset Y = \mathbb{P}^2$ be a nodal curve, and let X be the normalization of C. The double point formula for the induced morphism $f: X \to Y$ gives the classical formula

$$2d = n^2 - 3n + 2 - 2g$$

relating the number d of nodes, the degree n of C, and the genus g of X.

(b) The mapping $df: T_X \to f^* T_{\mathbb{P}^2}$ is injective, with quotient bundle N_f. In fact,

$$N_f \cong f^* \mathscr{O}_{\mathbb{P}^2}(n) \otimes \mathscr{O}_X(-\mathbb{D})$$

where $\mathbb{D} = \mathbb{D}(f)$ is the double point divisor on X.

(c) The nodes on C impose independent conditions on curves of degree $k \geqq n - 3$. (It is enough to prove this for $k = n - 3$. The vector space L curves of degree $n - 3$ passing through the nodes is a subspace of

$$H^0(X, N_f \otimes f^* \mathscr{O}(-3)) = H^0(X, K_X)\,.$$

Clearly $\dim L \geqq (n-1)(n-2)/2 - d$, and this integer is $g = \dim H^0(X, K_X)$ by (a).)

(d) The linear system cut out on X cut out by curves of degree $k \geqq n - 3$ passing through the nodes of C is complete. (The dimension of the space of such curves is

$$(k+1)(k+2)/2 - d - (k-n+1)(k-n+2)/2\,,$$

which is $k\,n - 2d + 1 - g$, the dimension predicted by Riemann-Roch (Example 15.2.1). By (b), $f^* \mathscr{O}_{\mathbb{P}^2}(k) \otimes \mathscr{O}(-\mathbb{D})$ is non-special.)

Example 9.3.4. Let $f: X^n \to Y^{n+1}$ be a finite morphism, with X and Y non-singular, and assume that the sets

$$A = \{x \in X \mid df_x : T_x X \to T_{f(x)} Y \text{ is not injective}\},$$

$$B = \{x \in X \mid \exists\, x' \neq x'', \text{ distinct from } x, \text{ with } f(x') = f(x'') = f(x)\}$$

have dimensions at most $n-2$. Let $D(f)$ be the double point set, endowed with its reduced scheme structure. Then

$$\mathbb{D}(f)=[D(f)].$$

(Let $X_o=X-(f^{-1}(f(A))\cup B)$. It suffices to show $\mathbb{D}(f)$ and $[D(f)]$ agree on X_0. Over X_0, $\tilde{D}(f)$ is smooth and maps isomorphically to $D(f)$.)

Example 9.3.5. Let $\bar{X}^n$ be a non-singular hypersurface in a non-singular variety Y^{n+1}, $n>1$. Let X be the blow-up of $\bar{X}$ along some non-singular variety, and let D be the exceptional divisor. Then D is the double point set $D(f)$ for the induced morphism f from X to Y, and D has the expected codimension. However $\tilde{D}(f)$ does not have the correct codimension, and in fact

$$\mathbb{D}(f)=-[D(f)]$$

is a negative cycle.

Example 9.3.6 (cf. Johnson (1) and Fulton-Laksov (1)). With $f\colon X^n\to Y^m$ as in this section, but with X singular, the same procedure produces a double point class $\mathbb{D}(f)$ in $A_{2n-m}(D(f))$, and then

$$\mathbb{D}(f)=f^*f_*[X]-\{c(f^*T_Y)\cap s(\Delta_X, X\times X)\}_{2n-m}$$

in $A_{2n-m}X$.

Example 9.3.7. Let $f\colon X^2\to Y^3$ be a morphism of non-singular complete varieties over an algebraically closed field K of characteristic $\neq 2$. Assume that f maps X finitely and birationally onto its image $\bar{X}\subset\mathbb{P}^3$, and assume that the singularities are all *ordinary*, i.e., the singular locus of $\bar{X}$ is a curve $\bar{D}$, with a finite number t of *triple points* $\bar{Q}_1,\ldots,\bar{Q}_t$, and a finite number v of *pinch points* $\bar{P}_1,\ldots,\bar{P}_v$. The completion of the local ring of $\bar{X}$ is isomorphic to:

$K[[x_1,x_2,x_3]]/(x_1x_2)$	at a general point of $\bar{D}$
$K[[x_1,x_2,x_3]]/(x_1x_2x_3)$	at a triple point $\bar{Q}_i$
$K[[x_1,x_2,x_3]]/(x_2^2-x_1^2x_3)$	at a pinch point $\bar{P}_j$.

The curve $\bar{D}$ is non-singular except for the triple points. The double curve $D\subset X$ is non-singular except at the $3t$ triple points $Q_1,\ldots,Q_{3t}$ lying over the triple points of $\bar{D}$; each Q_i is a node on D. There is one point P_i on D mapping to each $\bar{P}_i$. The induced map $\bar{f}\colon D\to\bar{D}$ is generically two to one, with simple ramification points at the P_i. (A local analytic equation for f at a pinch point is given by $f(t_1,t_2)=(t_1,t_1t_2,t_2^2)$.)

Let $\bar{J}$ be the singular subscheme of $\bar{X}$: if $g(x_1,x_2,x_3)$ is a local equation for $\bar{X}$ in Y, then $\bar{J}$ is defined by the vanishing of the three partial derivatives of g; more intrinsically, the ideal of $\bar{J}$ is the second Fitting ideal of $\Omega^1_{\bar{X}}$ (cf. Piene (4)). Let $J=f^{-1}(\bar{J})$. Then J contains D, and the residual scheme to D in J is the reduced scheme of pinch points $\{P_1,\ldots,P_v\}$. From Proposition 9.2, one has

(a) $$s(J,X)=[D]+(-D\cdot D+\sum_{j=1}^{v}[P_j]).$$

From the double point formula (cf. Example 9.3.2),

(b) $$[D]=f^*[\bar{X}]-c_1(f^*T_Y)+c_1(T_X)\,.$$

In addition,

(c) $$f^*[\bar{D}]=f^*[\bar{X}]\cdot D-D\cdot D+\sum_{i=1}^{3t}[Q_i]-\sum_{j=1}^{v}[P_j]\,.$$

(To prove this, first consider the case when there are no triple points. Then $\bar{D}$, D are non-singular, $f^{-1}(\bar{D})=D$, and by the excess intersection formula (and Corollary 8.1.1),

$$\begin{aligned}f^*[\bar{D}]&=c_1(\bar{f}^*N_{\bar{D}}Y/N_D X)\cap[D]\\&=(c_1(f^*T_Y)-c_1(f^*T_{\bar{D}})-c_1(T_X)+c_1(T_D))\cap[D]\\&=(c_1(f^*T_Y)-c_1(T_X))\cap[D]-\sum_{j=1}^{v}[P_j]\,,\end{aligned}$$

the last step from the ramification formula (Example 3.2.20 or Example 9.3.12). Combine this with (b) to get (c). For the general case, let $\pi: X'\to X$ (resp. $\varrho: Y'\to Y$) be the blow-up of X at the Q_i (resp. the $\bar{Q}_i$), and apply the previous case to the induced map f' from X' to Y', noting that

$$f^*[y]=\pi_*\pi^*f^*[y]=\pi_*f'^*\varrho^*[y]$$

for $y=[\bar{D}]$ or $[\bar{X}]$. Use Proposition 6.7 to compute ϱ^*y.) It follows that

(d) $$s(J,X)=[D]+\left(f^*[\bar{D}]-f^*[\bar{X}]\cdot D-\sum_{i=1}^{3t}[Q_i]+2\sum_{j=1}^{v}[P_j]\right).$$

Therefore, by Proposition 4.1 (a),

(e) $$s(\bar{J},\bar{X})=2[\bar{D}]+\left(-\bar{X}\cdot\bar{D}-3\sum_{i=1}^{t}[\bar{Q}_i]+2\sum_{j=1}^{v}[\bar{P}_j]\right).$$

Example 9.3.8. Let $f: X\to\bar{X}\subset Y$ be as in the preceding example, but with $Y=\mathbb{P}^3$. By Example 4.4.3, the degree of the dual of $\bar{X}$ is

$$d(d-1)^2-\deg s(\bar{J},\bar{X})_0-2(d-1)\deg s(\bar{J},\bar{X})_1$$

where d is the degree of $\bar{X}$. By Example 9.3.7, one derives the classical formula

$$d(d-1)^2+(4-3d)\,e+3t-2v$$

for the degree of the dual, where e is the degree of the double curve $\bar{D}$, and t, v are the number of triple points and pinch points on $\bar{X}$. (For another proof, generalizations, history, and examples, see Piene (2).)

Example 9.3.9. In the preceding example, if the double curve of $\bar{X}$ is a line, there are $2d-2$ pinch points on the line, and the degree of the dual is

$$d(d-1)^2-(7d-12)$$

(cf. Salmon (1) § 20).

Example 9.3.10. Let $\bar{X}$ be the surface in $\mathbb{P}^3$ with equation $x_1^2 x_3 = x_2^2 x_0$. The double curve is a line with two pinch points, so the degree of the dual is

$$3 \cdot 2^2 - 5 - 4 = 3 .$$

In fact, the dual of $\bar{X}$ is isomorphic to $\bar{X}$.

Example 9.3.11 (cf. Salmon (1)). Assume the ground field is algebraically closed of characteristic zero. Let $\bar{X} \subset \mathbb{P}^3$ be an irreducible surface with equation $F(x_0, x_1, x_2, x_3) = 0$, of degree d. For a point $P = (\lambda_0 : \lambda_1 : \lambda_2 : \lambda_3) \in \mathbb{P}^3$, the *polar surface of* $\bar{X}$ with respect to P is the surface V_P of degree $d-1$ with equation

$$\sum_{i=0}^{3} \lambda_i \cdot \partial F / \partial X_i = 0 .$$

A point Q is in $\bar{X} \cap V_P$ if and only if the (projective) tangent space to $\bar{X}$ at Q contains P. If P_1, P_2 are two general points of $\mathbb{P}^3$, then

$$\bar{X} \cap V_{P_1} \cap V_{P_2} = S \cup T$$

where S is the singular locus of V, and T is a finite set of non-singular points Q such that the tangent plane to V at Q contains the line through P_1 and P_2. Thus the cardinality of T is the degree of the dual surface. If $\bar{X}$ has ordinary singularities, as in the preceding example, the equivalence of the singular curve for the intersection $\bar{X} \cdot V_{P_1} \cdot V_{P_2}$ is

$$(3d - 4)\, e - 3t + 2v$$

where e is the degree of the curve, t the number of triple points, v the number of pinch points.

Example 9.3.12 *Ramification Formula.* Let $f \colon X^n \to Y^m$ be as in this section, and let $\tilde{R}(f) = \tilde{D}(f) \cap E$. Define $\tilde{\mathbb{R}}(f) \in A_{2n-m-1}(\tilde{R}(f))$ by the formula

$$\tilde{\mathbb{R}}(f) = E \cdot \tilde{\mathbb{D}}(f) .$$

Let $R(f)$ be the image of $\tilde{R}(f)$ in X, and let $\mathbb{R}(f) \in A_{2n-m-1}(R(f))$ be the push-forward of $\tilde{\mathbb{R}}(f)$; $\mathbb{R}(f)$ is the *ramification class.* Then

$$\mathbb{R}(f) = (c(f^* T_Y)\, c\, (T_X)^{-1})_{m-n+1}$$

in $A_{2n-m-1}(X)$. (By Theorem 9.2 one has

$$\Delta_Y \cdot X \tilde{\times} X = \sum_{i \geq 1} (-1)^{i-1}\, c_{m-i}(\bar{F}^* T_Y)\, E^i + \tilde{\mathbb{D}}(f)$$

where $\bar{F}$ is the induced morphism from E to Y. Intersect both sides by the divisor E and push down to X. If $n > 1$, then $\pi_*(\Delta_Y \cdot E) = \Delta_Y \cdot \pi_* E = 0$, and the formula

$$0 = -\{c(f^* T_Y) \cap s(\Delta_X, X \times X)\}_{2n-m-1} + \mathbb{R}(f)$$

results. If $n = m = 1$, $E = \Delta_X$, and $\Delta_Y \cdot E = c_1(f^* T_Y)$, $E \cdot E = c_1(T_X)$, and the formula reads

$$c_1(f^* T_Y) = c_1(T_X) + \mathbb{R}(f)$$

(cf. Example 3.2.30).) As in Example 9.3.6, this extends to the case when X may be singular.

Example 9.3.13 (Johnson (1)). Let X^n be a subvariety of $\mathbb{P}^N$, and let $f_m: X^n \to \mathbb{P}^m$ be a general linear projection. Let $\mathbb{D}(f_m)$ and $\mathbb{R}(f_m)$ be the double point and ramification classes of f, and let $H = c_1(\mathcal{O}_X(1))$ be a hyperplane section. Then

$$(*) \qquad H \cdot \mathbb{D}(f_m) - \mathbb{D}(f_{m+1}) = \mathbb{R}(f_m)$$

in $A_{2n-m-1}(X)$. (This is a formal calculation, using Theorem 9.3 and Example 9.3.12.) If $N \leqq 2n$, and a projection to some $\mathbb{P}^m$ is unramified, it follows that the projection is an imbedding. This remarkable discovery sparked the development of a general theory, which includes a proof that an arbitrary morphism $X^n \to \mathbb{P}^m$ with X^n projective and irreducible, and $m < 2n$, cannot be unramified unless it is an imbedding (Fulton-Hansen (1)). For example, the number ν of pinch points in Example 9.3.8 must be positive unless the surface is non-singular.

J. Hansen (1) has recently given a geometric explanation for the identity (*).

Example 9.3.14. With $f: X^n \to Y^m$ as in this section let $\bar{D}(f) = f(D(f))$ be the image in Y of the double point set. There is a class $\bar{\mathbb{D}}(f) \in A_{2n-m}(\bar{D}(f))$ so that the push-forward of $\mathbb{D}(f)$ by the canonical map from $D(f)$ to $\bar{D}(f)$ is $2\bar{\mathbb{D}}(f)$.

This may be seen by an alternative construction of $\tilde{\mathbb{D}}(f)$, following Ran (2). For a non-singular V, let $V^{[2]}$ denote the blow-up of $V \times V$ along Δ_V. Let $Z = X \times Y$. There are imbeddings of $X^{[2]}$ and $X^{[2]} \times Y$ in $Z^{[2]}$ induced by the graph imbedding of X in $X \times Y$ and by the diagonal imbedding of Y. Then $\tilde{\mathbb{D}}(f)$ is the intersection class of $X^{[2]}$ and $X^{[2]} \times Y$ on $Z^{[2]}$. (If $\tilde{f}$ is the projection from $Z^{[2]}$ to $Y \times Y$, then $X^{[2]} \times Y$ is the residual scheme to the exceptional divisor of $Z^{[2]}$ in $\tilde{f}^{-1}(\Delta_Y)$; this follows from Appendix B.6.10, applied to the inclusions $X \times Y \to X \times X \times Y \to Z \times Z$ induced by the diagonal imbeddings of X and Y. The fact that $\tilde{\mathbb{D}}(f) = i^![X^{[2]}]$, where i is the imbedding of $X^{[2]} \times Y$ in $Z^{[2]}$, follows from the construction of $\tilde{\mathbb{D}}(f)$; Theorem 17.6 gives a general formula which implies this.)

For V non-singular, let $V^{(2)}$ be the quotient of $V^{[2]}$ by the involution which reverses the factors, i.e., $V^{(2)}$ is the Hilbert scheme of length 2 subschemes of V. As above, one has imbeddings of $X^{(2)}$ and $X^{(2)} \times Y$ in $Z^{(2)}$, and one may construct their intersection class, denoted $\tilde{\tilde{\mathbb{D}}}(f)$. Since the quotient mapping from $Z^{[2]}$ to $Z^{(2)}$ has degree 2, $\tilde{\mathbb{D}}(f)$ pushes forward to $2\tilde{\tilde{\mathbb{D}}}(f)$ (by Theorem 6.2). If one defines $\bar{\mathbb{D}}(f)$ to be the push-forward of $\tilde{\tilde{\mathbb{D}}}(f)$ by the projection from $X^{(2)} \times Y$ to Y, then $\bar{\mathbb{D}}(f)$ is the required class on $\bar{D}(f)$.

For a similar approach to higher multiple point formulas, see Ran (2).

Notes and References

We point out only a few of the landmarks in the extensive literature on residual intersections and multiple point formulas. The interested reader may find many additional references, and hundreds of examples, in the cited works.

To our knowledge, the first excess intersection problem was formulated in 1847 by G. Salmon (1), although precedents can be found in Jacobi (1). To find the degree of the dual of a surface with a multiple curve, arguing as in Example 9.3.11, he was led to the general problem of finding the "equivalence" of a curve for three surfaces which contain it, so that the equivalence, plus the number of intersections of the surfaces outside the curve, totals the product of the degrees of the surfaces. These ideas are developed further in Salmon-Fiedler (1).

More general formulas of this kind were given by Cayley (2), (3), who also showed that many enumerative problems – such as the problem of how many curves in a family are tangent to a collection of given curves – can be formulated as problems of excess intersection (cf. Example 9.1.9, and § 10.4). Cayley also pointed out that these excess intersection problems can be very difficult.

The impressive accomplishments of the great enumerative geometers such as Chasles, Schubert, and Zeuthen, did not follow the ideas of Cayley. In modern terms, we can see that what they did often amounts to doing intersection theory on suitable blow-ups of the original spaces; in such blow-ups the proper transforms of the hypersurfaces corresponding to given geometric conditions will meet properly. In fact, however, these parameter spaces were not mentioned explicitly, and the intersection theory was done in a purely formal, symbolic manner. It is, nevertheless, difficult to reconcile the clarity and precision of Cayley's formulation with the debate still raging half a century later over such basic foundational problems as the principle of continuity. It should be noted that there was a concurrent dichotomy between the synthetic and the analytic approaches to geometry, cf. Chasles (4), p. 1168.

Residual intersection problems were considered by several others in the nineteenth century, notably Noether, Pieri, Caporali, and Bertini. Quite general intersections in projective space were considered by Severi (2). B. Segre (1) and Todd (3) generalized to the case of general ambient varieties of dimension 3 and 4, and extended Severi's results from numerical results to equalities modulo rational equivalence. Todd (6) gave a version of a residual intersection theorem in higher dimensions, and Segre (4) gave general formulas for equivalences. Many applications may be found in the books of Salmon (3), Enriques–Chisini (1), Semple and Roth (1), and Baker (1), (2).

The classical approach to the problem may be illustrated by the original case considered by Salmon: to calculate the equivalence of a curve which is a component of the intersection of three surfaces in $\mathbb{P}^3$. Two of the surfaces contain the original curve Z together with another curve Z'. First the degree and genus of Z', and the number i of intersections of Z with Z', were determined from the invariants of Z and the surfaces (cf. Example 9.1.12). The number i was then subtracted from the total (Bézout) number of intersections of the third surface with Z', to obtain the number of intersections of the three surfaces outside Z.

Such inductive proofs make a number of implicit assumptions. For example, it is apparently assumed that the residual curve Z' is non-singular, or at least reduced, and meets Z transversally. If not, one must count the points of

$Z \bigcap Z'$ with multiplicities; but then the equality of different calculations of the number i, on which the proof is based, is not at all obvious. Similar problems are met in many of the applications of intersection theory by classical geometers.

The problem encountered in this discussion, to compare properties of Z' with those of Z, is interesting in its own right. It has received considerable attention under the name of linkage, or liaison. For an approach to this problem using homological algebra, see Peskine and Szpiro (1), and Rao (1). Other properties of residual schemes are proved by Artin and Nagata (1). Hironaka (1) has used residual constructions for smoothing cycles.

The use of the modern excess intersection formula avoids the problems of general position arising in classical inductive proofs. The approach followed here is related to the method of Segre (4); it confirms his conjecture that the general equivalence $(X_1 \cdot \ldots \cdot X_r)^Z$ can be calculated purely in terms of the "covariants" (Segre classes) of the varieties involved (but see Example 9.1.10).

Double point formulas also have a long history, going back to Clebsch's formula for the genus of a plane curve (cf. Example 9.3.2). Severi (2) gave general numerical formulas for projections to projective spaces, which were rediscovered by Holme (1) and Peters and Simonis (1). Todd (6) gave a formula for the rational equivalence class of double point cycles, based on his residual intersection formula. The extension to morphisms to general non-singular varieties, and the simple approach followed here, was made possible by the residual construction of the double point scheme made by Laksov (3).

Extensions to singular varieties have been made by Holme, Johnson, and Fulton and Laksov (see Example 9.3.6). Kleiman (12) has a version of the residual intersection theorem which he uses to prove formulas for triple points and higher multiple loci (cf. Example 17.6.2). We refer to the article of Kleiman (8) for more on the history of multiple point formulas. Ran (2) has recently given a new application of the residual intersection formula to prove secant and multiple point formulas.

Most of the applications to classical problems are newly worked out in the examples of § 9.1, although the case of conics tangent to five conics (Example 9.1.9) was included in Fulton-MacPherson (2), and the application to multi-secants (Example 9.1.13) is from Le Barz. Example 9.3.3 was suggested by R. Lazarsfeld, and Example 9.3.14 by Z. Ran. The application to curves in $\mathbb{P}^n$ which cannot be the scheme-theoretic intersection of n hypersurfaces (Examples 9.1.2, 9.1.3) is, apparently, new.

The residual intersection theorem in § 9.2 is based on the construction of Laksov (3). The original theorem made rather stringent requirements on the residual scheme, which have gradually been removed as technique in intersection theory has improved (cf. Fulton (6), Fulton-Laksov (1), Kleiman (12), Fulton-MacPherson (3)). The present Theorem 9.2 is the first to make no assumptions on the residual scheme. Corollary 9.2.1 is also new when the residual scheme has larger dimension than predicted; Corollaries 9.2.2 and 9.2.3 have not appeared before.

Similarly, Theorem 9.3 is the first to allow the double point locus to have arbitrary dimension. The proof follows Fulton-Laksov (1), and is based on Laksov's fundamental construction.

The calculation of Segre classes of singular subschemes in Example 9.3.7 is apparently new; numerical formulas, when $Y = \mathbb{P}^3$, had been given by Piene (2).

The construction and simple proof of the ramification formula in Example 9.3.12 is also new. It generalizes the formula of Johnson (1) to arbitrary morphisms, without assumption on the ramification set.

It should be remarked that there have occasionally been alternate proposals for assigning numbers to components of intersections which have larger than the expected dimension. Severi (7) has discussed this when the expected dimension is negative, Samuel (1) has given a definition using his algebraic multiplicity. These definitions, whatever their virtues, all violate the principle of continuity. They are therefore quite a different sort of notion than that considered in classical intersection theory, and in this text.

Chapter 10. Families of Algebraic Cycles

Summary

If T is a non-singular curve, and $p: \mathscr{Y} \to T$ is a morphism, any $(k+1)$-cycle $\alpha = \sum n_i[\mathscr{V}_i]$ on $\mathscr{Y}$ determines an algebraic family of k-cycles α_t on the fibres $Y_t = p^{-1}(t)$:

$$\alpha_t = \sum_{\mathscr{V}_i \not\subset Y_t} n_i[(V_i)_t] .$$

Rationally equivalent $(k+1)$-cycles on $\mathscr{Y}$ determine rationally equivalent k-cycles in each fibre. The basic operations of intersection theory preserve algebraic families. For example, if $\mathscr{Y}$ is smooth over T, and $\{\alpha_t\}$ and $\{\beta_t\}$ are algebraic families of cycles, then the intersection products $\alpha_t \cdot \beta_t$ also vary in an algebraic family. These facts are consequences of the general theorems of Chap. 6, and the recognition of α_t as the image of α by the refined Gysin homomorphism constructed from the diagram

$$\begin{array}{ccc} Y_t & \to & \mathscr{Y} \\ \downarrow & & \downarrow p \\ \{t\} & \to & T . \end{array}$$

In this formulation, T may be replaced by any variety of arbitrary dimension, with t a regular point of T. This provides a simple method for studying algebraic equivalence.

The principle of continuity, or conservation of number, has two parts. First, in an algebraic family of zero-cycles, on a scheme which is proper over the parameter space, all the cycles have the same degree. Second, as mentioned above, the operations of intersection theory preserve algebraic families.

Refined intersection theory yields an improvement over classical formulations of this principle. For example, the ambient variety need not be complete; all that is necessary is that the locus of intersections is proper over the parameter space. This is useful for applications to enumerative geometry, when the ambient space is a space of non-degenerate geometric figures. In the last section an example of this kind is worked out: the formula for the number of curves in an r-dimensional family of plane curves which are tangent to r given plane curves in general position, in terms of the characteristics of the family, and the degrees and classes of the given curves.

10.1 Families of Cycle Classes

In this chapter T will denote an irreducible variety of dimension $m > 0$. The notation "$t \in T$" will be used to denote a regular, closed point of T (Appendix B.1). By abuse of notation we write $\{t\}$ in place of $\operatorname{Spec}(\varkappa(t))$, where $\varkappa(t)$ is the residue field of the local ring of T at the point, and we denote by

$$t : \{t\} \to T$$

the canonical inclusion of $\operatorname{Spec}(\varkappa(t))$ in T. The assumption that the point is regular means that t is a regular imbedding of codimension m.

Script letters will be used to denote schemes over T, with corresponding Latin letters, subscripted by t, denoting the fibre over $t \in T$. If $p : \mathscr{Y} \to T$ is given, then

$$Y_t = p^{-1}(t) ;$$

Y_t is regarded as an algebraic scheme over the ground field $\varkappa(t)$.

Any $(k+m)$-cycle α on $\mathscr{Y}$, or more generally any rational equivalence class $\alpha \in A_{k+m}\mathscr{Y}$ determines a *family of k-cycle classes* $\alpha_t \in A_k(Y_t)$, for all $t \in T$, by the formula

$$\alpha_t = t^!(\alpha)$$

where $t^! : A_{k+m}\mathscr{Y} \to A_k Y_t$ is the refined Gysin homomorphism defined from the fibre square

$$\begin{array}{ccc} Y_t & \to & \mathscr{Y} \\ {\scriptstyle p_t}\downarrow & & \downarrow{\scriptstyle p} \\ \{t\} & \underset{t}{\to} & T \end{array}$$

by the construction of § 6.2. We may say that α_t is the *specialization* of α at t.

More precisely, if $\alpha = [\mathscr{V}]$ where $\mathscr{V}$ is a subscheme of $\mathscr{Y}$ of pure dimension $k+m$, then

$$\alpha_t = [\mathscr{V}]_t = \{s(V_t, \mathscr{V})\}_m$$

where $V_t = \mathscr{V} \cap Y_t$, and $s(V_t, \mathscr{V})$ is the Segre class of V_t in $\mathscr{V}$. (This follows from Proposition 6.1 (a) and the fact that the normal bundle to $\{t\}$ in T is trivial.) In particular, if V_t is k-dimensional, then $[\mathscr{V}]_t$ is a well-defined positive k-cycle supported on V_t (cf. Example 10.1.1). On the other hand, if $\mathscr{V} \subset Y_t$, then $[\mathscr{V}]_t = 0$.

If T is a curve, and α is a $(k+1)$-cycle on $\mathscr{Y}$, then each α_t is well-defined as a k-cycle on Y_t. For if $\alpha = \sum n_i [\mathscr{V}_i]$, with $\mathscr{V}_i$ a variety, then we define

$$\alpha_t = \sum_{\mathscr{V}_i \not\subset Y_t} n_i [(V_i)_t] .$$

Thus all components of α which do not map dominantly to T are simply discarded. In case $\dim T > 1$, a general $(k+m)$-cycle $\alpha = \sum n_i [\mathscr{V}_i]$ on $\mathscr{Y}$ will not determine a k-cycle on Y_t, unless one assumes that $\dim (V_i)_t = k$ for all i.

If $f : \mathscr{X} \to \mathscr{Y}$ is a morphism, with $\mathscr{Y} \to T$ as above, we denote by

$$f_t : X_t \to Y_t$$

the induced morphism on fibres over $t \in T$.

The next proposition states that this specialization in algebraic families is compatible with the main operations of intersection theory.

Proposition 10.1 (a) *If $f: \mathscr{X} \to \mathscr{Y}$ is proper, and α is a $(k+m)$-cycle on $\mathscr{X}$, then*

$$f_{t*}(\alpha_t) = (f_*(\alpha))_t$$

in $A_k(Y_t)$.

(b) *If $f: \mathscr{X} \to \mathscr{Y}$ is flat of relative dimension n, and α is a $(k+m)$-cycle on $\mathscr{Y}$, then*

$$f_t^*(\alpha_t) = (f^*(\alpha))_t$$

in $A_{k+n}(X_t)$.

(c) *If $i: \mathscr{X} \to \mathscr{Y}$ is a regular imbedding of codimension d, such that $i_t: X_t \to Y_t$ is also a regular imbedding of codimension d, $f: \mathscr{V} \to \mathscr{Y}$ is a morphism, and α is a $(k+m)$-cycle on $\mathscr{V}$, then*

$$i_t^!(\alpha_t) = (i^!(\alpha))_t$$

in $A_{k-d}(W_t)$, $\mathscr{W} = f^{-1}(\mathscr{X})$.

(d) *If E is a vector bundle on $\mathscr{Y}$, with restriction E_t on Y_t, and α is a $(k+m)$-cycle on $\mathscr{Y}$, then*

$$c_i(E_t) \cap \alpha_t = (c_i(E) \cap \alpha)_t$$

in $A_{k-i}(Y_t)$.

Proof. Since specialization is a refined Gysin homomorphism, the proposition follows from corresponding compatibilities of refined Gysin homomorphisms, namely Theorem 6.2(a), Theorem 6.2(b), Theorem 6.4, and Proposition 6.3. □

Corollary 10.1. *Assume T is non-singular, $t \in T$ is rational over the ground field, and $\mathscr{Y}$ is smooth over T of relative dimension n. If $\alpha \in A_{k+m}(\mathscr{Y})$, $\beta \in A_{l+m}(\mathscr{Y})$, then*

$$\alpha_t \cdot \beta_t = (\alpha \cdot \beta)_t$$

in $A_{k+l-n}(Y_t)$.

Proof. Let δ (resp. δ_t) be the diagonal imbedding of $\mathscr{Y}$ in $\mathscr{Y} \times \mathscr{Y}$ (resp. Y_t in $Y_t \times Y_t$), and i_t the imbedding of Y_t in $\mathscr{Y}$. Then the two imbeddings $(i_t \times i_t) \circ \delta_t$ and $\delta \circ i_t$ of Y_t in $\mathscr{Y} \times \mathscr{Y}$ are equal, so by functoriality (Theorem 6.5, Example 6.5.2)

$$\delta_t^*(i_t^* \alpha \times i_t^* \beta) = \delta_t^*(i_t \times i_t)^*(\alpha \times \beta) = i_t^* \delta^*(\alpha \times \beta).$$

Since $\mathscr{Y}$ is flat over T, $i_t^*(\gamma) = \gamma_t$ for any cycle γ on $\mathscr{Y}$ (cf. Example 10.1.2), so the displayed equation is equivalent to the required $\alpha_t \cdot \beta_t = (\alpha \cdot \beta)_t$. □

Remark 10.1. These results allow one to deduce identities involving rational equivalence classes for special values of t from the corresponding identities for general values, even if T is not a rational variety. For example, given Y nonsingular, $a, b, c \in A(Y)$, to prove that $c = a \cdot b$, it suffices to find $\mathscr{Y} \to T$ as in Corollary 10.1, with $Y_t = Y$, and α, β, γ on $\mathscr{Y}$ with $\alpha_t = a$, $\beta_t = b$, $\gamma_t = c$, such that $\gamma = \alpha \cdot \beta$. If this is achieved with α meeting β properly, the identity $\gamma = \alpha \cdot \beta$ can be verified generically on T. For a simple application see Example 10.1.9.

Corollary 10.1 is also valid for the product of more than two cyles, as the proof shows.

Example 10.1.1. If $\mathscr{V}$ is a purely $(m+k)$-dimensional closed subscheme of $\mathscr{Y}$, and each irreducible component W_i of V_t has dimension k, then

$$[\mathscr{V}]_t = \sum e_i [W_i]$$

where $e_i = (e_{V_t}\mathscr{V})_{W_i}$ is the multiplicity of $\mathscr{V}$ along V_t at W_i (Example 4.3.4).

Example 10.1.2. Assume that the imbedding i_t of Y_t in $\mathscr{Y}$ is a regular imbedding of codimension m, $m = \dim(T)$. For any $(k+m)$-cycle α on $\mathscr{Y}$,

$$\alpha_t = i_t^*(\alpha)$$

in $A_k(Y_t)$, where i_t^* is the Gysin homomorphism of § 6.2 (Use Theorem 6.2(c)).) This holds, for example, whenever $\mathscr{Y}$ is flat over T (cf. Example A.5.5).

Example 10.1.3. (a) If α is a family of cycles on $Y \times T$, the cycles α_t, as t varies in T, need not be rationally equivalent. If $Y = T$ is a projective non-singular curve of positive genus, and $\alpha = [\Delta]$, where Δ is the diagonal, then α_t is never rationally equivalent to $\alpha_{t'}$ if $t \neq t'$. The cycles α_t are rationally equivalent if T is unirational, however (Example 10.1.7).

(b) If α and β are cycles on $Y \times T$ which are rationally equivalent on $Y \times T$, then α_t is rationally equivalent to β_t on Y for all $t \in T$. (The refined Gysin map $t^!$ preserves rational equivalence.) Doubts about the validity of this fact, or at least the ability to prove it, gave rise to some of the criticism of Severi's methods (cf. Van der Waerden (6)).

Example 10.1.4. Assume $f: \mathscr{X} \to \mathscr{Y}$ is a l.c.i. morphism of relative codimension d, while $f_t: X_t \to Y_t$ is a l.c.i. morphism of relative codimension d'. Let $e = d - d'$, E the excess normal bundle of the diagram

$$\begin{array}{ccc} X_t & \xrightarrow{f_t} & Y_t \\ \downarrow & & \downarrow \\ \mathscr{X} & \xrightarrow[f]{} & \mathscr{Y} \end{array} .$$

Then for all $(k+m)$-cycles α on $\mathscr{Y}$,

$$c_e(E) \cap f_t^!(\alpha_t) = (f^!(\alpha))_t$$

in $A_{k-d}(X_t)$. In particular, if $d = d'$, then $f_t^!(\alpha_t) = (f^!(\alpha))_t$. (See Proposition 6.6.)

Example 10.1.5. Assume T is non-singular, t is rational over the ground field, $p: \mathscr{Y} \to T$ is smooth of relative dimension n, and $f: \mathscr{X} \to \mathscr{Y}$ is a morphism. Let $\alpha \in A_{k+m}\mathscr{X}$, $\beta \in A_{l+m}\mathscr{Y}$. Then

$$\alpha_t \cdot_{f_t} \beta_t = (\gamma_f^*(\alpha \times \beta))_t$$

in $A_{k+l-n}(X_t)$. The product on the left is that of Chap. 8, since $f_t: X_t \to Y_t$, and Y_t is non-singular; on the right $\gamma_f: \mathscr{X} \to \mathscr{X} \times \mathscr{Y}$ is the graph of f, a regular imbedding since $\mathscr{Y}$ is non-singular. (Consider the fibre square

$$\begin{array}{ccc} X_t & \to & \mathscr{X} \times \mathscr{Y} \\ {\scriptstyle f_t}\downarrow & & \downarrow {\scriptstyle f \times 1} \\ Y_t & \to & \mathscr{Y} \times \mathscr{Y} \end{array} .$$

Factor the lower homomorphism into $Y_t \to Y_t \times Y_t \to \mathscr{Y} \times \mathscr{Y}$ and into $Y_t \to \mathscr{Y} \to \mathscr{Y} \times \mathscr{Y}$, and apply Theorem 6.5. In fact, one sees from this that the two classes agree in $A_{k+l-n}((|\alpha| \cap f^{-1}(|\beta|))_t)$.)

Example 10.1.6 (cf. Samuel (3)). Let T be a unirational variety over an algebraically closed field.

(a) Any two points of T can be joined by a chain of rational curves. (By standard reductions, it suffices to prove the following case. If $\pi : T \to \mathbb{A}^m$ is the blow-up of $\mathbb{A}^m$ along a subscheme Z, with exceptional divisor E, and $t \in E$, then there is a morphism $h : \mathbb{A}^1 \to T$ with $h(0) = t$ but $h(\mathbb{A}^1) \not\subset E$. To see this, let C be a non-singular curve, $g : C \to T$ a morphism with $g(P) = t$, $g(C) \not\subset E$. Let $x_1, \ldots, x_m$ be the coordinate on $\mathbb{A}^m$, and let $z_1, \ldots, z_r$ be the generators for the ideal of Z. Choose an integer N larger than $\mathrm{ord}_P(\pi g)^*(z_j))$ for $1 \leqq j \leqq r$. Choose $\bar{h} : \mathbb{A}^1 \to \mathbb{A}^m$ so that the power series expansion of each $\bar{h}^*(x_i)$ at 0 agrees with that of $(\pi g)^*(x_i)$ through order N. Then $\bar{h} = \pi h$, with h as required.)

(b) If T is complete, then $A_0 T = \mathbb{Z}$, generated by the class of any point of T. (Use Theorem 1.4.)

(c) If T is not complete, then $A_0 T = 0$. (If T is open in $\bar{T}$, and $P \in \bar{T} - T$, then $[P]$ generates $A_0 \bar{T}$, but the restriction from $A_0 \bar{T}$ to $A_0 T$ is surjective and maps $[P]$ to 0.)

It had been conjectured by Severi that a complete non-singular surface T with $A_0 T = \mathbb{Z}$ must be rational, but a counter-example is given by Bloch-Kas-Lieberman (1).

Example 10.1.7. Let T be a non-singular variety over an algebraically closed field, such that any two points of T can be joined by a chain of rational curves.

(a) Let Y be a scheme, α a $(k+m)$-cycle on $Y \times T$. Then all the cycle classes $\alpha_t \in A_k(Y \times \{t\}) = A_k Y$ are equal. (Using Theorem 6.5 one may reduce to the case where T is an affine open subset of $\mathbb{A}^1$. In this case the pull-back

$$q^* : A_k Y \to A_{k+1}(Y \times T)$$

is surjective, where $q : Y \times T \to Y$ is the projection (Propositions 1.9 and 1.8). If i_t is the imbedding of Y in $Y \times T$ at t, and $\alpha = q^* \beta$, then $i_t^*(\alpha) = i_t^* q^* \beta = (q i_t)^* \beta = \beta$.)

(b) Let X, Y be schemes, and let

$$f : X \times T \to Y$$

be a morphism, i.e. a family of morphisms $f_t : X \to Y$ parameterized by T. Assume that the morphism

$$F : X \times T \to Y \times T$$

defined by $F(x, t) = (f_t(x), t)$, is proper. Then each f_t is proper, and the induced morphisms

$$(f_t)_* : A_k X \to A_k Y$$

are independent of $t \in T$. (This follows from (a), since, by Proposition 10.1 (a), $(f_t)_*(\alpha) = (F_*(\alpha \times [T]))_t$.)

From (a) is follows in particular that the criterion for rational equivalence in § 1.6 can be enlarged to include families of cycles parametrized by arbitrary unirational varieties (cf. Samuel (5)).

Example 10.1.8. If $\mathscr{Y} \to T$ is given, and α is a positive $(k+m)$-cycle on $\mathscr{Y}$, then each class α_t can be represented by a non-negative cycle. (Choose, in a neighborhood of t, m principal divisors $D_1, \ldots, D_m$ whose intersection is $\{t\}$. Then

$$\alpha_t = D_1 \cdot \ldots \cdot (D_m \cdot \alpha) .$$

By construction, intersecting with a divisor whose normal bundle is trivial preserves non-negative cycles.) See Chap. 12 for generalizations.

Example 10.1.9. Let C be a non-singular projective curve, $C^{(n)}$ its nth symmetric product, whose points are effective divisors of degree n on C. For an effective divisor A on C, set

$$X_A = \{D \in C^{(n)} \mid D \geqq A\} .$$

If $\deg A = a$, X_A is the image of the imbedding of $C^{(n-a)}$ in $C^{(n)}$ which takes E to $A+E$. For any effective divisors A, B on C,

$$(*) \qquad [X_A] \cdot [X_B] = [X_{A+B}]$$

in $A_{n-a-b}(X_A \cap X_B)$, $a = \deg(A)$, $b = \deg(B)$. (If A and B are disjoint, the intersection is transversal. For the general case let $T = C^{(a)}$, with $t \in T$ corresponding to A. Let $\mathscr{X}$ (resp. $\mathscr{X}_B$) be the image of the inclusion of $C^{(n-a)} \times T$ (resp. $C^{(n-a-b)} \times T$) in $C^{(n)} \times T$ which takes (E, D) to $(E+D, D)$ (resp. to $(E+D+B, D)$). Then

$$[\mathscr{X}] \cdot [X_B \times T] = [\mathscr{X}_B]$$

as cycles on $C^{(n)} \times T$, since the intersection is transversal over the open set of T consisting of divisors which are disjoint from B. Since $\mathscr{X}$ specializes to X_A, and $\mathscr{X}_B$ to X_{A+B}, at $t \in T$, (*) follows from the preceding equation and Corollary 10.1.) The corresponding equality in $A^{a+b}(C^{(n)})$ was proved by Mattuck (1), using a linear equivalence between A and $A' - A''$, for divisors A' and A'' which are disjoint from B.

Example 10.1.10. Given $\mathscr{X} \to T$, and a cone $\mathscr{C}$ over $\mathscr{X}$, such that $\mathscr{C}$ is flat over T, then

$$s(\mathscr{C})_t = s(C_t)$$

in $A_*(X_t)$ for $t \in T$. (This follows the definition of Segre classes in § 4.1 and Proposition 10.1 (b), (d), and (a).)

10.2 Conservation of Number

The following proposition is the basic fact on which the principle of continuity depends.

Proposition 10.2. *Let $p: \mathscr{Y} \to T$ be a proper morphism, T an m-dimensional variety as in § 10.1. Let α be an m-cycle on $\mathscr{Y}$. Then the cycle classes $\alpha_t \in A_0(Y_t)$ all have the same degree.*

Note that $\deg(\alpha_t)$ is calculated by regarding Y_t as a scheme over $\varkappa(t)$: if p_t is the induced morphism from Y_t to $\{t\} = \operatorname{Spec}(\varkappa(t))$, then

$$p_{t_*}(\alpha_t) = \deg(\alpha_t) \cdot [\{t\}] .$$

Proof. Let $p_*(\alpha) = N \cdot [T]$, $N \in \mathbb{Z}$. Then by Proposition 10.1 (a),

$$p_{t*}(\alpha_t) = (p_*(\alpha))_t = N \cdot [T]_t = N \cdot [\{t\}] .$$

Therefore $\deg(\alpha_t) = N$, for any $t \in T$. □

We have seen in Proposition 10.1 that our basic intersection operations preserve families of cycle classes. Combined with Proposition 10.2, if a sequence of these operations are applied to families of cycles, resulting in a family of 0-cycles whose support is proper over T, then the degree of these zero cycles will not vary in the family. The following theorem suffices for most applications.

Theorem 10.2. *Let $i: \mathscr{X} \to \mathscr{Y}$ be a regular imbedding of codimension d, $p: \mathscr{Y} \to T$ a flat morphism, such that $p\,i$ is also flat. Let $f: \mathscr{V} \to \mathscr{Y}$ be a morphism, and form the fibre square*

$$\begin{array}{ccc} \mathscr{W} & \to & \mathscr{V} \\ \downarrow & & \downarrow f \\ \mathscr{X} & \underset{i}{\to} & \mathscr{Y} \end{array} .$$

Assume that $\mathscr{W}$ is proper over T. Then for any $(d+m)$-cycle α on $\mathscr{V}$, the degree of the 0-cycle classes

$$X_t \cdot_{Y_t} \alpha_t = i_t^!(\alpha_t)$$

in $A_0(W_t)$ is independent of t. More generally, if P is a polynomial of weight e in the Chern classes of a collection of vector bundles on $\mathscr{W}$, and P_t denotes the same polynomial in the Chern classes of the restrictions of these vector bundles to W_t, and α is any $(d+e+m)$-cycle class on $\mathscr{V}$, then

$$\deg(P_t \cap X_t \cdot_{Y_t} \alpha_t)$$

is independent of $t \in T$.

Proof. By Proposition 10.1 (c) and (d),

$$P_t \cap X_t \cdot_{Y_t} \alpha_t = (P \cap i^!(\alpha))_t .$$

The assertion therefore follows from Proposition 10.2. □

Corollary 10.2.1. *Let Y be a scheme, and let $\mathscr{H}_i \subset Y \times T$ be effective Cartier divisors which are flat over T, $i = 1, \ldots, d$. Let a be a d-cycle on Y. Assume that*

$$\mathscr{H}_1 \cap \ldots \cap \mathscr{H}_d \cap (\operatorname{Supp}(a) \times T)$$

is proper over T. Then

$$\deg((H_1)_t \cdot \ldots \cdot (H_d)_t \cdot a)$$

is independent of t.

Proof. Apply the theorem to the situation

$$\begin{array}{ccc} \cap \mathscr{H}_i \cap \mathscr{V} & \longrightarrow & \mathscr{V} \\ \downarrow & & \downarrow \delta \\ \mathscr{H}_1 \times \ldots \times \mathscr{H}_d & \to & (Y \times T) \times \ldots \times (Y \times T) \end{array}$$

where $\mathscr{V} = \operatorname{Supp}(a) \times T$, δ is the diagonal imbedding, and $\alpha = a \times [T] \in A_{d+m}\mathscr{V}$. □

Corollary 10.2.2. *Let Y be a non-singular n-dimensional variety. Let α_i be (k_i+m)-cycles on $Y \times T$, $1 \leqq i \leqq r$, with $\sum_{i=1}^{r} k_i = (r-1)\,n$, and assume that $\bigcap \operatorname{Supp}(\alpha_i)$ is proper over T. Then*

$$\deg((\alpha_1)_t \cdot \ldots \cdot (\alpha_r)_t)$$

is independent of t.

Proof. Apply the theorem to the situation

$$\begin{array}{ccc} \bigcap \operatorname{Supp}(\alpha_i) & \to & \operatorname{Supp}(\alpha_1) \times \ldots \times \operatorname{Supp}(\alpha_r) \\ \downarrow & & \downarrow \\ (Y \times T) & \underset{\delta}{\to} & (Y \times T) \times \ldots \times (Y \times T) \end{array}$$

with δ the diagonal, $\alpha = \alpha_1 \times \ldots \times \alpha_r$. □

Example 10.2.1. Let $i: \mathscr{X} \to \mathscr{Y}$ be a regular imbedding of codimension d, $\mathscr{V} \subset \mathscr{Y}$ a subscheme with $\mathscr{X}, \mathscr{Y}, \mathscr{V}$ flat over T. Assume that there is a non-empty open subset $T^\circ \subset T$ such that X_t meets V_t properly for all $t \in T^\circ$.

(a) For any $t \in T$, $X_t \cdot_{Y_t} V_t$ is represented by a non-negative cycle. (Since $\mathscr{X}$ meets $\mathscr{V}$ properly, $\mathscr{X} \cdot_{\mathscr{Y}} \mathscr{V}$ is represented by a non-negative cycle. Then use Proposition 10.1 (c) and Example 10.1.8.)

(b) Assume that $\mathscr{X} \bigcap \mathscr{V}$ is proper over T, and $\dim(\mathscr{V}) = d + m$. Let $N = \deg(X_t \cdot V_t)$ for $t \in T^\circ$. Then for any $t \in T$, $X_t \bigcap V_t$ is either positive dimensional, or a finite set of cardinality $\leqq N$. ($\mathscr{X} \cdot \mathscr{V}$ is an effective m-cycle whose support is $\mathscr{X} \bigcap \mathscr{V}$. If $X_t \bigcap V_t$ is finite, $(\mathscr{X} \cdot \mathscr{V})_t$ is an effective cycle of degree N whose support is $X_t \bigcap V_t$.)

In fact, (b) follows from a more general assertion. Assume that the ground field is algebraically closed, that $\dim(\mathscr{V}) = d + m$, and $\mathscr{X} \bigcap \mathscr{V}$ is separated over T. Suppose M is an integer such that $X_t \bigcap V_t$ has at most M points for all $t \in T^\circ$. Then, if $X_t \bigcap V_t$ is finite, it has at most M points. (The essential point for this is that each irreducible component $\mathscr{W}_i$ of $\mathscr{X} \bigcap \mathscr{V}$ has dimension at least m. One may reduce to the case where T is a non-singular curve, and each $\mathscr{W}_i$ is a curve mapping dominantly to T. In this case it is elementary to prove that the number of points in $(\mathscr{W}_i)_t$ is no more than the separable degree of the extension $R(\mathscr{W}_i)/R(T)$. See [EGA]IV.15.5 for generalizations.)

As in Corollary 10.2.2, analogous results hold for families of intersections of pure dimensional schemes on a non-singular variety.

Example 10.2.2. Let $l_1, \ldots, l_4$ be four general lines in $\mathbb{P}^2$, and let P be a point of $\mathbb{P}^2$ not on any l_i. If P is not on either diagonal of the quadrilateral formed by the lines, then there are precisely two non-singular conics tangent to all the l_i and passing through P. If P is on one diagonal there is one such conic, while if P is on both diagonals there are none.

Let $H_i \subset \mathbb{P}^5$ be the hypersurface of conics tangent to l_i, and let H_P be the hyperplane of conics passing through P. For general P, $H_1 \cap \ldots \cap H_4 \cap H_P$ consists of the Veronese of double lines (whose equivalence for the intersection is 14, cf. Example 9.1.8), and two points where the hypersurfaces meet transversally. As P approaches one (or both) diagonals, one (or both) of these points approach the Veronese.

In such circumstances, classical geometers maintained the conservation of number by counting limiting solutions as well; here one (or both) doubled diagonal would be counted as a conic. Since it is not always obvious from the statement of an enumerative problem which degenerate solutions should be counted as "limiting", this contributed to the controversy over the principle.

If the ambient space Y is taken to be the open subspace of $\mathbb{P}^5$ corresponding to non-singular conics, then the intersection of the five hypersurfaces, as P varies, is not proper over $\mathbb{P}^2 - \bigcup_{i=1}^{4} l_i$. For the validity of Theorem 9.2 it is necessary that the "intersection scheme" $\mathscr{W}$ be proper over the parameter space T.

Example 10.2.3 (cf. Study (1)). Over an algebraically closed field, identify the set of 4-tuples of points on $\mathbb{P}^1$ with $\mathbb{P}^4$ as usual: the divisor $\sum_{i=1}^{4} [(s_i : t_i)]$ corresponds to $(x_0 : \ldots x_4)$ if $\sum_{j=0}^{4} x_j S^j T^{4-j} = \prod_{i=1}^{4} (t_i S - s_i T)$. Let T be the open set in $\mathbb{P}^4$ consisting of distinct 4-tuples. The set G of automorphisms of $\mathbb{P}^1$ is identified with $\mathbb{P}^3$, the automorphism with matrix (y_{ij}) corresponding to $(y_{11} : y_{12} : y_{21} : y_{22})$. Let

$$\mathscr{V} = \{(\sigma, D) \in G \times T \mid \sigma(D) = D\}.$$

Then $\mathscr{V}$ is a closed algebraic subset of $G \times T$, and for an open set T° of T, $\mathscr{V}_t$ has 4 distinct points. However, for $t \in T$ corresponding to 4-tuples with cross ratio -1 (resp. a non-trivial cube root of 1), $\mathscr{V}_t$ has 8 (resp. 12) points (cf. Semple and Roth (1) XI.6).

This does not contradict the principle of continuity, since $\mathscr{V}$ cannot arise as a proper intersection; $\mathscr{V}$ is locally defined by 4 equations, and has irreducible components of dimension 3 as well as 4.

Example 10.2.4. Let $Y^2 \subset \mathbb{P}^3$ be the cone $x_3^2 = x_1^2 + x_2^2$, with vertex $P = (1:0:0:0)$. Let V be the line $x_1 = 0$, $x_3 = x_2$. Let $T = \mathbb{A}^1$, with coordinate u, and let

$$\mathscr{X} \subset Y \times T$$

be the family of Cartier divisors $x_2 = u x_0$. The general X_t is a non-singular curve meeting V transversally in one point, but X_0 is the union of two lines, each meeting V at the point P. By Corollary 10.2.2, such a phenomenon cannot occur on a non-singular variety Y. Note that in this example the intersection product $X_0 \cdot V$ of V by the Cartier divisor X_0 is $1 \cdot [P]$.

Example 10.2.5 (cf. Zobel (3)). Let $Y^3 \subset \mathbb{P}^4$ be the cone $x_1 x_4 = x_2 x_3$. Let $\mathscr{V} \subset Y \times \mathbb{A}^1$ be the family of lines defined by the equations

$$x_1 = 0,\ x_3 = 0,\ x_2 = u x_0,$$

with u the coordinate on $\mathbb{A}^1$. Let $V' \subset Y$ be the plane $x_1 = x_2$, $x_3 = x_4$. Then for general $t \in A^1$, V_t and V' are disjoint, while V_0 meets V' in the vertex of the cone. Again, such jumps can occur only at singular points.

Example 10.2.6. Let $T = A^1$, with coordinate u, and let $Y = \mathbb{P}^2$. Let $\mathscr{X} \subset Y \times \mathbb{P}^1$ be the family of lines

$$\mathscr{X} = \{(x : y : z) \times (u) \mid y = u x\}.$$

Let $V \subset Y$ be the line $y = 0$. The classes $X_t \cdot V \in A_0(V)$ are all equal as t varies in T. As happens when a proper intersection degenerates to an improper intersection, the normal cones to $X_t \cap V$ in V, and the distinguished varieties for $X_t \cdot V$, do *not* vary continuously. This phenomenon will be studied in Chap. 11.

Example 10.2.7. In Theorem 10.2, let $\mathscr{W}_1, \ldots, \mathscr{W}_r$ be the connected components of $\mathscr{W}$. Let α be a $(d+m)$-cycle on $\mathscr{V}$, and let

$$X_t \cdot_{Y_t} \alpha_t = \lambda_1(t) + \ldots + \lambda_r(t)$$

with $\lambda_i(t) \in A_0((W_i)_t)$. For any i such that $\mathscr{W}_i$ is proper over T, $\deg(\lambda_i(t))$ is independent of t. (Write $\mathscr{X} \cdot \alpha = \sum \lambda_i$, $\lambda_i \in A_m(\mathscr{W}_i)$; then $\lambda_i(t) = (\lambda_i)_t$.)

Example 10.2.8 ("Compact supports"). For any scheme X, define

$$A_*^c(X) = \varinjlim A_*(Z)$$

where the limit is over all closed subschemes Z of X which are complete. If X is complete, $A_*^c X = A_* X$. For *any* morphism $f: X \to Y$, there are induced homomorphisms

$$f_*: A_*^c X \to A_*^c Y$$

which are functorial. In particular, mapping X to a point, elements of $A_0^c X$ have a well-defined degree.

In addition, there are functorial Gysin maps

$$f^*: A_*^c Y \to A_*^c X$$

for morphisms $f: X \to Y$ which are flat and proper, or l.c.i. and projective. Chern classes of vector bundles on X operate on $A_*^c X$. If Y is non-singular, and $f: X \to Y$ is a morphism, there are cap products $A_*^c X \otimes A^* Y \to A_*^c X$. When X is the complement of the singular locus of a projective variety, such groups have been studied by Collino (2).

Example 10.2.9. Let Y be a scheme, $\mathscr{X}_i \subset Y \times T$ regularly imbedded of codimension d_i, with $\mathscr{X}_i$ flat over T, $1 \leq i \leq r$. Let $f: \mathscr{V} \to Y \times T$ be a morphism, and let $\mathscr{W} = \bigcap f^{-1}(\mathscr{X}_i)$. Let α be a $(k+m)$-cycle on $\mathscr{V}$. Let $e = k - \sum_{i=1}^{r} d_i$. Then there is a $(e+m)$-cycle β on $\mathscr{W}$ such that

$$(X_1)_t \cdot \ldots \cdot (X_r)_t \cdot \alpha_t = \beta_t$$

in $A_e(W_t)$ for all $t \in T$. If P is a polynomial of degree e in Chern classes of vector bundles on $\mathscr{W}$, with restriction P_t to W_t, and $\mathscr{W}$ is proper over T, then

$$\deg(P_t \cap ((X_1)_t \cdot \ldots \cdot (X_r)_t \cdot \alpha_t))$$

is independent of $t \in T$.

10.3 Algebraic Equivalence

In this section the parameter spaces T will be non-singular varieties, and $t \in T$ will denote a point which is rational over the ground field. If X is a scheme, and α is a $(k+m)$-cycle on $X \times T$, $m = \dim T$, then $X \times \{t\} = X$ for $t \in T$, and the class α_t constructed in § 10.1 is a class in $A_k X$.

Definition 10.3. Let X be a scheme. A k-cycle a on X, or a class $a \in A_k X$, is *algebraically equivalent to zero,* written $a \sim_{\text{alg}} 0$, if there is a non-singular variety T, a cycle $\alpha \in A_{k+m}(X \times T)$, $m = \dim T$, and points $t_1, t_2 \in T$ such that

$$a = \alpha_{t_1} - \alpha_{t_2}$$

in $A_k X$. Two k-cycles are algebraically equivalent if their difference is algebraically equivalent to zero (cf. Example 10.3.2). From the following proposition, the cycles algebraically equivalent to zero form a subgroup of $Z_k X$. The group of algebraic equivalence classes will be denoted $B_k X$:

$$B_k X = Z_k X / \sim_{\text{alg}} .$$

Proposition 10.3. *The cycles algebraically equivalent to zero form a subgroup of the group of all cycles on a scheme. This subgroup is preserved by the basic operations:*

(a) *Proper push-forward* (§ 1.4)
(b) *Flat pull-back* (§ 1.7)
(c) *Refined Gysin homomorphisms* (§ 6.2)
(d) *Chern class operations* (§ 3.2).

The groups $B_* X$ therefore satisfy the same formal properties as $A_* X$.

Proof. For the first statement, suppose a and b are two k-cycles on X which are algebraically equivalent to zero. Suppose $a = \alpha_{t_1} - \alpha_{t_2}$ as in the definition, and $b = \beta_{u_1} - \beta_{u_2}$ for a $(k+n)$-cycle β on $X \times U$, $n = \dim U$. Set

$$\gamma = \alpha \times [U] - \beta \times [T],$$

a $(k+m+n)$-cycle on $X \times T \times U$. It suffices to show that

$$a - b = \gamma_{(t_1, u_1)} - \gamma_{(t_2, u_2)}$$

in $A_k X$. This follows immediately from the identity

$$(\alpha \times [U])_{(t,u)} = \alpha_t$$

for any $t \in T$, $u \in U$. This identity in turn follows from the functoriality of refined Gysin maps (Theorem 6.5): if t (resp. u) denotes the morphism from the point Spec(K) to T (resp. U), then (t, u) is the composite $(1_T \times u) \circ t$, so

$$(\alpha \times [U])_{(t,u)} = t^!((1_T \times u)^!(\alpha \times [U]) = t^!(\alpha) = \alpha_t .$$

That the four operations (a)−(d) preserve algebraic equivalence follows from the corresponding parts (a)−(d) of Proposition 10.1. □

Example 10.3.1. If Y is non-singular, the classes in $A^*(Y)$ which are algebraically equivalent to zero form an ideal, so $B^*(Y)$ is a commutative, graded, ring with unit. If $f: X \to Y$ is a morphism, $B_* X$ is a $B^* Y$-module; if X is also non-singular, $f^*: B^* Y \to B^* X$ is a homomorphism of graded rings.

Example 10.3.2. Assume the ground field is algebraically closed. Two k-cycles a, a' on a scheme X are algebraically equivalent in the sense of Definition 10.3 if and only if there is a non-singular variety T, of dimension m, with $(k+m)$-dimensional subvarieties $\mathscr{V}_i$ of $X \times T$, flat over T, $1 \leqq i \leqq r$, and points $t_1, t_2 \in T$, such that

$$a - a' = \sum_{i=1}^{r} [(V_i)_{t_1}] - [(V_i)_{t_2}] .$$

In addition, one may achieve this with T a projective, non-singular, curve. (If a and a' are algebraically equivalent, by connecting points in parameter spaces by chains of curves, one first constructs r non-singular affine curves T_i, subvarieties $\mathscr{W}_i \subset X \times T_i$ projecting dominantly to T_i, and $t_{i1}, t_{i2} \in T$ such that

$$a - a' = \sum_{i=1}^{r} [(W_i)_{t_{i1}}] - [(W_i)_{t_{i2}}] .$$

Here one uses the fact that if $f: T \to T'$ is a morphism of non-singular varieties, $t \in T$, and α is a cycle on $X \times T'$, then, by Theorem 6.5, $\alpha_{f(t)} = (f^! \alpha)_t$. Taking closures of the $\mathscr{W}_i$ in $X \times \bar{T}_i$, where $\bar{T}_i$ is a projective non-singular completion of T_i, one may have the same equation with all T_i projective. Set

$$\begin{aligned} T &= T_1 \times \ldots \times T_r \\ t_j &= t_{1j} \times \ldots \times t_{rj} , \qquad j = 1, 2 \\ \mathscr{V}_i &= \mathrm{pr}_i^{-1}(\mathscr{W}_i) \end{aligned}$$

to achieve the required equation, with $\dim T = r$ (cf. Baldassarri (1)VI.7). To obtain $r = 1$, by induction, it suffices to show that any two points on the product of two curves can be joined by an irreducible curve; for Rosenlicht's simple proof of this last fact, see Weil (5)I: Lemma 5.)

Example 10.3.3. Two positive cycles a, a' on X are algebraically equivalent if and only if there is a positive cycle b on X such that $a+b$ and $a'+b$ are members of an irreducible family (Chow variety) of positive cycles. (Use Example 10.3.2 and proceed as in Example 1.6.2.)

Example 10.3.4. If $Y \subset X$, $U = X - Y$ as in Proposition 1.8, then the sequence

$$B_k Y \to B_k X \to B_k U \to 0$$

is exact. With this and Proposition 10.3 other general calculations for A_* (e.g. § 3.3, § 6.7) extend without change to B_*.

10.4 An Enumerative Problem

A typical problem in enumerative geometry (cf. Schubert (1)§ 1) is to find the number of geometric figures in a given family which satisfy certain conditions. Such a condition would usually require the geometric figure to have a certain relation with a given configuration, which is assumed to be in general position.

If a parameter space for the given family is a non-singular, r-dimensional variety S, those figures satisfying a given condition form a closed subset of S, which will be a hypersurface if the condition is "simple". The problem becomes one of finding the number of points in the intersection of r hypersurfaces $H_1, \ldots, H_r$ in S. This problem of intersection theory is seldom straightforward, however, because one wants to count only non-degenerate solutions to the problem. The parameter space S for non-degenerate solutions will usually not be compact (complete). If the family is compactified to $\bar{S}$, the hypersurfaces $\bar{H}_1, \ldots, \bar{H}_r$ may intersect in the locus $\bar{S} - S$ of degenerate solutions, even with an excess component.

Consider for example the problem of how many (non-singular) plane conics are tangent to five given lines. If the conics are parametrized by $\mathbb{P}^5$, those tangent to a given line form a quadric hypersurface. The five hypersurfaces all contain the Veronese of double lines, whose equivalence for this intersection problem is 31. Of the $2^5 = 32$ solutions predicted by Bézout's theorem, only one corresponds to a non-singular conic (cf. Example 9.1.8). Of course, the fact that there is one such conic can also be seen easily by considering the dual problem.

One may proceed similarly to count the number of conics tangent to five given conics. The conics tangent to a given conic form a hypersurface of degree 6, and the equivalence of the Veronese is 4512, leaving

$$6^5 - 4512 = 3264$$

as the required number (cf. Example 9.1.9).

In general it is difficult to calculate the equivalence of an excess degenerate component directly. For example, if the plane curves of degree d are parametrized by a projective space of dimension $d(d+3)/2$, the locus of curves with multiple components, for large d, is a complicated singular subset.

A classical procedure for such problems is to degenerate the given figures to simpler ones, where the number satisfying the simpler conditions can be computed, and then to appeal to the principle of conservation of number. Of course, for this to be valid, there must be some compactness assumption. One method is to construct a better (non-singular) compactification of S, where the intersections are proper and correspond only to non-degenerate solutions. In the example of conics, the space of "complete conics" – the blow-up of $\mathbb{P}^5$ along the Veronese, is such a compactification.

Using refined intersections and the version of continuity given in § 10.2, it may not be necessary to construct such a compactification. If the degeneration of the given figures is parametrized by a variety T, what is essential is that the intersection of the corresponding varying hypersurfaces form a scheme $\mathscr{W}$ which is proper over T. Or, if a compactification $\bar{S}$ is used, the components of $\mathscr{W}$ contained in $S \times T$ should not meet any components of $\mathscr{W}$ contained in $(\bar{S}-S) \times T$. In these cases, by Corollary 10.2.1, the degree of the intersection cycle of the varying hypersurfaces in S will be independent of $t \in T$. Of course, if non-degenerate solutions (in S) approach degenerate solutions (in $\bar{S}-S$) as t varies, this will no longer be true (cf. Example 10.2.2).

We will illustrate these ideas by sketching a solution to the following problem:

Given an r-dimensional family of plane curves, and r curves in general position in the plane, how many curves in the family are tangent to the r guven curves?

The solution is as follows. Let $D_1, \ldots, D_r$ be the given curves, n_i the degree of D_i, and m_i the class of D_i (i.e., the number of lines passing through a given general point and tangent to D_i at some simple point). If the D_i are in general position, the number of curves in the family tangent to $D_1, \ldots, D_r$ is

$$\prod_{i=1}^{r} (m_i \mu + n_i \nu) .$$

This formula is to be interpreted as follows. Expand the polynomial formally:

$$\prod_{i=1}^{r} (m_i \mu + n_i \nu) = \sum_{k=0}^{r} N_k \mu^k \nu^{r-k} ,$$

so

$$N_k = \sum_{\substack{P \subset \{1, \ldots, r\} \\ \operatorname{card}(P) = k}} \Big(\prod_{i \in P} m_i\Big) \Big(\prod_{j \notin P} n_j\Big) ,$$

and substitute for $\mu^k \nu^{r-k}$ the corresponding *characteristic* of the family, i.e.,

$\mu^k \nu^{r-k} =$ the number of curves in the family passing through k general points and tangent to $r-k$ general lines.

For example if the given family is the family of all conics, then (in characteristic not 2, cf. Example 10.4.3)

$$\mu^5 = \nu^5 = 1 , \qquad \mu^4 \nu = \mu \nu^4 = 2 , \qquad \mu^3 \nu^2 = \mu^2 \nu^3 = 4 .$$

Thus the number of conics tangent to $D_1, \ldots, D_5$ is

$$N_0 + 2 N_1 + 4 N_2 + 4 N_3 + 2 N_4 + N_5$$

with N_k as above. In case the D_i are all non-singular conics (in characteristic not 2), $n_i = m_i = 2$, $N_k = \binom{5}{k} \cdot 2^5$, and the number is

$$2^5 (1 + 2 \cdot 5 + 4 \cdot 10 + 4 \cdot 10 + 2 \cdot 5 + 1) = 3264 .$$

In this problem it shall be assumed that the curves $D_1, \ldots, D_r$ have no multiple components, and that the ground field is algebraically closed of characteristic zero. The proof proceeds in several steps.

Step 1. The set of lines in $\mathbb{P}^2$, denoted $\check{\mathbb{P}}^2$, is identified with a projective plane with homogeneous coordinates a, b, c; the point $(a:b:c)$ corresponds to the line

$$(*) \qquad ax+by+cz=0\,.$$

The *incidence correspondence* $I \subset \mathbb{P}^2 \times \check{\mathbb{P}}^2$,

$$I = \{(P,L) \mid P \in L\}\,,$$

is the non-singular 3-fold defined by the global equation (*). If one defines the bundle E on $\mathbb{P}^2$ by the exact sequence

$$0 \to E \to 1_{\mathbb{P}^2}^{\oplus 3} \xrightarrow{(x\,y\,z)} \mathcal{O}_{\mathbb{P}^2}(1) \to 0\,,$$

then $I = P(E)$. If λ (resp. ζ) is the pull-back of $c_1(\mathcal{O}_{\mathbb{P}^2}(1))$ (resp. $c_1(\mathcal{O}_{\check{\mathbb{P}}^2}(1))$ to I, then

$$\lambda\zeta = \lambda^2 + \zeta^2, \qquad \lambda^3 = \zeta^3 = 0\,,$$

and $A^*(I)$ is a free $\mathbb{Z}$-module with basis

$$1, \lambda, \zeta, \lambda^2, \zeta^2, \lambda^2\zeta = \lambda\zeta^2\,.$$

This follows from the isomorphism (Example 8.3.4)

$$A^*(P(E)) = A^*(\mathbb{P}^2)[\zeta]/\zeta^2 + c_1(E)\,\zeta + c_2(E)\,,$$

and the formula $c(E) = c(\mathcal{O}_{\mathbb{P}^2}(1))^{-1} = 1 - \lambda + \lambda^2$.

For a line M in $\mathbb{P}^2$, set

$$M' = \{(P,L) \in I \mid L = M\}\,,$$
$$M'' = \{(P,L) \in I \mid P \in M\}\,.$$

For a point Q in $\mathbb{P}^2$, set

$$Q' = \{(P,L) \in I \mid P = Q\}\,,$$
$$Q'' = \{(P,L) \in I \mid Q \in L\}\,.$$

Then $\lambda = [M'']$, $\zeta = [Q'']$, $\lambda^2 = [Q']$, $\zeta^2 = [M']$. From this one sees also that $\int \lambda^2\zeta = \int \lambda\zeta^2 = 1$.

Step 2. Let D be a curve in $\mathbb{P}^2$ with no multiple components. Define the (reduced) curve $D' \subset I$ to be the closure of

$$\{(P, L) \in I \mid P \text{ is a simple point of } D, \text{ and } L \text{ is tangent to } D \text{ at } P\}.$$

If n and m are the degree and class of D, then

$$(**) \qquad [D'] = n\,[M'] + m\,[Q'] = n\,\zeta^2 + m\,\lambda^2$$

in $A^2 I$, where M is a line, Q a point, as above. This can be seen by intersecting $[D']$ with the dual basis λ, ζ of $A^1(I)$. For a general line M, D' meets M'' transversally in n points (P_i, L_i), where $P_i \in M \cap D$, and L_i is the tangent to D at P_i; this shows that the coefficient of ζ^2 is n. The dual argument gives the coefficient of λ^2. (The fact that the intersections are transversal follows from the general considerations in Step 3 below.)

The rational equivalence (**) may be constructed explicitly. Choose a general point P_0 and a general line M. Let $Q_1, \ldots, Q_m$ be the points where the tangent lines to D through P_0 meet M.

By gradually deforming D onto M, projecting from P_0, we construct a rational equivalence from $[D']$ to the cycle

$$(**') \qquad n[M'] + \sum_{i=1}^{m} [Q_i'] .$$

We may assume $P_0 = (0:0:1)$ and M is the line $z=0$; let $F(X, Y, Z) = 0$ be the equation for D, and let F_1, F_2, F_3 be the partial derivatives of F. For $t \in \mathbb{A}^1$, $t \neq 0$, let σ_t be the automorphism of $\mathbb{P}^2$ given by

$$\sigma_t(x:y:z) = (x:y:tz) .$$

Then $\sigma_t(D)$ is defined by $F(tX, tY, Z) = 0$. As t varies in $\mathbb{A}^1 - \{0\}$, there is a family of curves $\sigma_t(D)'$ in I. There is a unique extension of this family to a surface $\mathscr{D}'$ in $I \times \mathbb{A}^1$, flat over $\mathbb{A}^1$. The claim is that the fibre D_0' of $\mathscr{D}'$ over $t = 0$ is a curve whose cycle is $n[M'] + \sum [Q_i']$. To verify this, it suffices to show that D_0' is set-theoretically equal to the union of M' and the Q_i', for then the previous calculation (**) determines the coefficients. If $t \neq 0$, a point in $\sigma_t(D)'$ will have the form $(x:y:tz) \times (a:b:c)$ with $ax + by + ctz = 0$, $F(x, y, z) = 0$, and

$$(a:b:c) = (tF_1(tx, ty, tz) : tF_2(tx, ty, tz) : F_3(tx, ty, tz)) ,$$

with not all $F_i(x, y, z) = 0$. The displayed equation is equivalent to the three equations

$$\begin{aligned} aF_2(x, y, z) &= bF_1(x, y, z) , \\ aF_3(x, y, z) &= tcF_1(x, y, z) , \\ bF_3(x, y, z) &= tcF_2(x, y, z) . \end{aligned}$$

One may set $t = 0$ in these equations to find D_0'. For the point $(x:y:0) \times (a:b:c)$ to be in D_0' either $a = b = 0$, i.e., the point is in M'; or there must be a z so that $F(x, y, z) = 0$ and $F_3(x, y, z) = 0$, with $(x:y:z)$ simple on D, i.e., $(x:y:0)$ is one of the Q_i, so the point is in Q_i'. (Singular points of D also give rise to points satisfying the above equations, but since only a finite number of lines through singular points are limits of tangents to nearby simple points, they cannot produce curves in D_0'.)

Step 3. Let $\mathscr{X} \subset \mathbb{P}^2 \times S$ be a family of plane curves, flat over S, with S a non-singular, r-dimensional variety. Assume the general curve X_s of the family has no multiple components. Let S° be any non-empty open subset of S such that X_s is reduced for every $s \in S^\circ$. For example, if the general curve of the family is non-singular, S° may parametrize the non-singular members of the family.

Define the variety $\mathscr{X}(r)$ to be the locally closed subvariety of $I \times \ldots \times I \times S^\circ$ (r copies of I):

$$\mathscr{X}(r) = \{(P_1, L_1) \times \ldots \times (P_r, L_r) \times s \mid \text{each } P_i \text{ is a simple point of } X_s, \text{ and } L_i \text{ is tangent to } X_s \text{ at } P_i\}.$$

Then $\mathscr{X}(r)$ is smooth, of dimension $2r$. Let

$$\varphi : \mathscr{X}(r) \to I \times \ldots \times I \qquad (r \text{ factors})$$

be the projection

Given reduced curves $D_1, \ldots, D_r$ in $\mathbb{P}^2$ form the fibre square

$$(***) \qquad \begin{array}{ccc} W & \to & D_1' \times \ldots \times D_r' \\ \downarrow & & \downarrow \\ \mathscr{X}(r) & \underset{\varphi}{\to} & I \times \ldots \times I \,. \end{array}$$

Let G be the product of r copies of the automorphism group $\mathrm{GL}(\mathbb{P}^2)$ of $\mathbb{P}^2$. Since G acts transitively on $I \times \ldots \times I$, for an open set of $\sigma = (\sigma_1, \ldots, \sigma_r) \in G$, if D_i is replaced by $\sigma_i(D_i)$, the above diagram is differentiably transversal (Appendix B.9.2), and W consists of N (reduced) points. Moreover, if $\bar{\mathscr{X}} \subset \mathbb{P}^2 \times \overline{S^\circ}$ is any compactification of the family, and $\overline{\mathscr{X}(r)}$ is the closure of $\mathscr{X}(r)$ in $I \times \ldots \times I \times \overline{S^\circ}$, and Z is any closed subset of $\overline{\mathscr{X}(r)}$ of dimension less than $2r$ (containing all of $\overline{\mathscr{X}(r)} - \mathscr{X}(r)$), then

$$Z \cap (\bar{\varphi})^{-1}(\sigma_1(D_1') \times \ldots \times \sigma_r(D_r')) = \emptyset \,,$$

for an open set of $\sigma \in G$. In particular, discarding any proper closed subset from S does not affect the number N of solutions.

Step 4. Now construct a degeneration of each D_i to a multiple line as in Step 2, using a different general point and line for each curve. The product of these degenerations is parametrized by $\mathbb{A}^r$. We have a diagram

$$\begin{array}{ccccc} \mathscr{W} & \to & \mathscr{D}_1' \times \ldots \times \mathscr{D}_r' & \to & \mathbb{A}^r \\ \downarrow & & \downarrow & & \\ \overline{\mathscr{X}(r)} & \to & I \times \ldots \times I & & \end{array}$$

with the square a fibre square. Since $\overline{\mathscr{X}(r)}$ is complete, $\mathscr{W}$ is proper over $\mathbb{A}^r$. The fibres of $\mathscr{W}$ over $t_1 = (1, \ldots, 1)$ and $t_0 = (0, \ldots, 0)$ are finite, and disjoint from any given proper closed subset Z of $\overline{\mathscr{X}(r)}$, for generic position of the D_i and the points and lines. Therefore if the above degenerations are restricted to an open neighborhood T of $\{t_0, t_1\}$ in $\mathbb{A}^r$, then $\mathscr{W}$ will be proper over T and disjoint from Z. Now by Corollary 10.2,

$$\deg(\mathscr{X}(r) \cdot_\varphi D_1' \times \ldots \times D_r') = \deg(\mathscr{X}(r) \cdot_\varphi (E_1' \times \ldots \times E_r'))$$

where D_i' (resp. E_i') is the fibre of $\mathscr{D}_i'$ over 1 (resp. 0). The left side is N, the number of tangencies. Writing out the E_i' according to (**′), the right side expands to give the required formula.

One additional point must be verified. This intersection actually gives the number

$$\operatorname{card}\{(P_1, L_1) \times \ldots \times (P_r, L_r) \times s \in \mathscr{X}(r) \mid L_i \text{ is tangent to } D_i \text{ and } X_s \text{ at } P_i\}.$$

To obtain the original solution, one should check that, with the general position assumption, each $s \in S$ will have at most one point in the above set; any bitangents can be put in a proper subvariety Z as above. Similarly, one sees that the N tangents that occur in the solution are all simple tangencies of $\mathscr{X}(s)$ and each D_i.

For other approaches to this problem, we refer to Grayson (3), and Fulton-Kleiman-MacPherson (1), cf. Example 14.7.18.

Example 10.4.1. The class of a non-singular plane curve of degree n (in characteristic zero) is $n(n-1)$ (cf. Example 4.4.5). The number of conics tangent to five non-singular curves of degree n, in general position, is

$$N = n^5((n-1)^5 + 10(n-1)^4 + 40(n-1)^3 + 40(n-1)^2 + 10(n-1) + 1).$$

If $n = 2$, this gives 3264. If $n = 3$, $N = 168{,}399$, while if $n = 11$, $N = 39{,}312{,}710{,}151$.

Example 10.4.2. A *circle* in the plane is a conic passing through the two "ideal" points $(1 : \pm\sqrt{-1} : 0)$. The circles form a $\mathbb{P}^3$, with $a(x^2+y^2) + bxz + cyz + dz^2 = 0$ corresponding to the point $(a : b : c : d)$. The circles tangent to a given line, and the circles tangent to a given circle, form quadric surfaces in $\mathbb{P}^3$. There are 8 circles tangent to three circles (in general position), or to two circles and a line, or to a circle and two lines, but there are only 4 circles tangent to three lines in general position. (In the last case, the double line $z^2 = 0$ corresponds to a point in the intersection of the three quadrics, and the intersection multiplicity of the three quadrics at this point is 4.)

The number of circles tangent to three curves of degrees n_1, n_2, n_3 and classes m_1, m_2, m_3, in general position, is

$$\prod_{i=1}^{3} (m_i \mu + n_i \nu)$$

where $\mu^3 = 1$, $\mu^2\nu = 2$, $\mu\nu^2 = 4$, and $\nu^3 = 4$.

When the three given curves are real, this analysis does not say how many real circles satisfy the conditions. For lines and circles in various positions this number is easy to determine, and the numbers given above can all be achieved by appropriate real configurations. In general counting the number of real solutions to enumerative problems appears to be very difficult.

Example 10.4.3. In characteristic two, the class of a nonsingular conic is 1. The characteristics of the family of all plane conics in characteristic two are

$$\mu^5 = \mu^4\nu = \mu^3\nu^2 = 1, \quad \mu^2\nu^3 = \mu\nu^4 = \nu^5 = 0.$$

The number of conics tangent to five given conics in characteristic two is

$$(\mu + 2\nu)^5 = 1 + 5 \cdot 2 + 10 \cdot 4 = 51$$

(cf. Example 9.1.9, and Vainsencher (3)).

Example 10.4.4 (cf. Example 14.7.18 and Fulton-Kleiman-MacPherson (1)). The analysis of this section extends readily to finding the number of varieties in an r-dimensional family of varieties in $\mathbb{P}^N$ which are tangent to r given varieties in general positon in $\mathbb{P}^n$. For example, given an r-dimensional family of curves in $\mathbb{P}^3$, the number tangent to r given surfaces in general position is

$$\prod_{i=1}^{r} (m_i \nu + n_i \varrho)$$

where n_i is the degree of the ith surface, m_i its first class (i.e. the number of points in a general plane section at which the tangent plane passes through a fixed general point), and the characteristic $\nu^i \varrho^{r-i}$ is the number of curves in the family tangent to i general lines and $r-i$ general planes. For the family of all (plane) conics in $\mathbb{P}^3$, these characteristics were found by Chasles, cf. Schubert (1)§ 20:

$$\nu^8 = 92, \quad \nu^7 \varrho = 116, \quad \nu^6 \varrho^2 = 128, \quad \nu^5 \varrho^3 = 104, \quad \nu^4 \varrho^4 = 64, \quad \nu^3 \varrho^5 = 32,$$
$$\nu^2 \varrho^6 = 16, \quad \nu \varrho^7 = 8, \quad \varrho^8 = 4 .$$

Thus the number of conics tangent to 8 general quadric surfaces is

$$(2\nu + 2\varrho)^8 = 4{,}407{,}296 .$$

Similarly, there are 666,841,088 quadric surfaces in $\mathbb{P}^3$ tangent to 9 given quadrics in general position.

A method for calculating the characteristics for the family of all quadrics of dimension m in $\mathbb{P}^n$ was given by Schubert (4), based on the beautiful geometry of complete quadrics. This was reconsidered by Semple (1) and Tyrell (1), and recently by Demazure, Vainsencher, De Concini and Procesi.

Notes and References

The principle of *conservation of number*, known also as the principle of *special position*, or the *principle of continuity*, has a long and stormy history. From the time of Poncelet, it has been the basic tool of enumerative geometry. Early justifications were based on the fact that the number of solutions of a polynomial equation, when properly counted, remain constant when the coefficients of the polynomial vary continuously.

Controversy has arisen when the principle has been asserted as a general law regarding variations of geometric conditions. For each such formulation, counterexamples were produced (cf. Examples 10.2.3 and 10.2.5), followed by the addition of new hypotheses to rule out such examples, and so forth. When the principle has been founded on intersection theory on appropriate parameter spaces, these difficulties have disappeared. Severi (4) made this point in 1912 to settle the dispute then raging. As we have mentioned, Cayley (2) had advocated an intersection-theoretic approach to enumerative problems in 1868, before most of the controversy arose. Perhaps the desire to keep geometry "pure", unsullied by coordinates or parameter spaces, contributed to the resistance to this idea.

The articles of Zeuthen and Pieri (1), Berzolari (2), Dieudonné (1) and Kleiman (7) are recommended for discussions of the role of the principle of continuity in enumerative geometry, and for additional references.

It should be pointed out that there is not a single principle of continuity for all situations. As intersection theory develops, those constructions which can be proved to vary in families lead to stronger formulations of the principle. For example, Theorem 10.2 gives a version which is valid on singular and non-complete parameter spaces. Severi (11) attempted definitions of intersections on non-complete varieties, or on varieties "modulo" a closed subset. Severi's ideas have been examined by Zobel (2), (4); the ideas developed in § 10.4 resemble those of Zobel.

The theory of algebraic equivalence was developed along with rational equivalence. Modern constructions were given by Weil (5) and Samuel (3) for non-singular projective varieties, using the definition of Example 10.3.2. The use of refined Gysin maps in the present version gives a simpler construction, as well as more general results. The basic reason for this is that, with excess intersections allowed, we can work directly with cycles on the ambient space of a family even if they do not meet all fibres properly.

Results similar to Examples 10.1.6 and 10.1.7 are stated in the seminar of Samuel (5).

Chasles (10) gave the formula for the number of conics in a one-parameter family tangent to a given curve. Halphen, Schubert, and nearly every other enumerative geometer, gave formulas for the number of varieties in a family tangent to given varieties. The general formula of § 10.4 for plane curves appears in Zeuthen (3)§ 165. The proof in § 10.4 represents joint work with MacPherson. For the generalization to higher dimensions see Zobel (4) and Fulton-Kleiman-MacPherson (1).

This tangency problem is only one of an unlimited number of enumerative problems that can be solved using intersection theory. In addition to the classical literature already referred to, the reader will find hundreds of examples in Lemoyne (1). Recently enumerative geometry has played an important role in moduli problems on curves, especially by Arbarello, Cornalba, Griffiths and Harris (1), Harris and Mumford (1), Eisenbud and Harris (1), Harris (1), and Mumford (7).

In 1866 Chasles (5) posed the problem of finding the characteristics $\mu^i \nu^j$ for the family of *all* plane curves of degree n. For $n = 3$ and $n = 4$ answers were given by Maillard and Zeuthen; verifying their answers is the subject of some current research. For $n > 4$, the problem remains open.

For references to the transcendental study of the variation of algebraic cycles, see § 19.3.6.

Chapter 11. Dynamic Intersections

Summary

Let $X \hookrightarrow Y$ be a regular imbedding of codimension d, with normal bundle $N_X Y$; let V be a k-dimensional subvariety of Y, $W = X \cap V$, N the restriction of $N_X Y$ to W, and $C \subset N$ the normal cone to W in V. In Chap. 6 the intersection class $X \cdot V$ in $A_{k-d}(W)$ has been constructed to be $s_N^*[C]$, where $s_N: W \to N$ is the zero-section.

If $X \hookrightarrow Y$ is imbedded in a family $\mathscr{X} \hookrightarrow Y \times T$ of regular imbeddings, with T a non-singular curve, $0 \in T$, $X_0 = X$, and $\mathscr{V} \subset Y \times T$ is a deformation of V, then there is a closed set $\lim\limits_{t \to 0} (X_t \cap V_t)$, contained in W, and a class we denote $\lim\limits_{t \to 0} (X_t \cdot V_t)$ in $A_{k-d}\left(\lim\limits_{t \to 0} X_t \cap V_t\right)$ which refines $X \cdot V$, i.e., maps to $X \cdot V$ in $A_{k-d}(W)$.

The Kodaira-Spencer homomorphism for the deformation $\mathscr{X}$ determines a section $s_{\mathscr{X}}$ of N, and hence a class $s_{\mathscr{X}}^![C]$ in $s_{\mathscr{X}}^{-1}(C)$, which also refines $X \cdot V$. In fact,

$$\lim_{t \to 0} (X_t \cap V_t) \subset s_{\mathscr{X}}^{-1}(C) \subset W$$

and, by these inclusions,

$$\lim_{t \to 0} (X_t \cdot V_t) \to s_{\mathscr{X}}^![C] \to X \cdot V .$$

If X_t meets V_t properly for generic t, then $\lim\limits_{t \to 0} (X_t \cap V_t)$ has dimension $k-d$, so $\lim\limits_{t \to 0} (X_t \cdot V_t)$ is a well-defined cycle representing $X \cdot V$. If $\dim s_{\mathscr{X}}^{-1}(C) = k - d$, this limit cycle must be $s_{\mathscr{X}}^![C]$, in which case the limit cycle is determined by infinitesimal data.

This allows a dynamic interpretation for the distinguished varieties and their equivalences, which can be useful for calculations. For any closed subset Z of X, let $(X \cdot V)^Z$ be the part of $X \cdot V$ supported on Z (§ 6.1). If $N_X Y$ is generated by its sections, there is an open set $\Gamma(Z)$ of sections such that for each $s \in \Gamma(Z)$, $s^![C]$ is a $(k-d)$-cycle, and the part of $s^![C]$ which is supported on Z is precisely $(X \cdot V)^Z$. Thus $(X \cdot V)^Z$ is represented by the part of the limit cycle $\lim\limits_{t \to 0} (X_t \cdot V_t)$ supported on Z, for generic deformations, i.e., deformations whose characteristic section is in $\Gamma(Z)$. Knowing $(X \cdot V)^Z$ for all Z is the same as knowing the equivalences of the distinguished varieties.

For arbitrary cycles on an arbitrary non-singular quasi-projective variety, there is a moving lemma which asserts that the given cycles are rationally equivalent to cycles which meet properly. Intersection cycles constructed by this procedure represent the intersection classes on the ambient variety, but, unlike the cycles arising from geometric deformations, they do not give refinements of the intersection products. The moving lemma has historical importance as the foundation for previous constructions of intersection products, and it can be used to simplify verifications of some intersection formulas.

11.1 Limits of Intersection Classes

In this chapter, the ground field will be assumed to be algebraically closed. We consider families parametrized by a non-singular curve T equipped with a given point $0 \in T$. Let $T^* = T - \{0\}$.

For a scheme $\mathscr{W}$ over T, the fibre of $\mathscr{W}$ over $t \in T$ is denoted W_t. Set

$$\mathscr{W}^* = \mathscr{W} - W_0 = \mathscr{W} \times_T T^* .$$

Let $\mathscr{W}'$ be the closure of $\mathscr{W}^*$ in $\mathscr{W}$. The *limit set* $\lim_{t\to 0} W_t$ is defined to be the fibre of $\mathscr{W}'$ over 0, i.e.,

$$\lim_{t\to 0} W_t = (W')_0 .$$

When $\mathscr{W}$ has irreducible components contained entirely in W_0, $\lim_{t\to 0} W_t$ will be smaller than W_0. If W_t is purely k-dimensional for generic $t \in T^*$, then $\lim_{t\to 0} W_t$ is a purely k-dimensional (or empty) closed subscheme of W_0. Indeed, by discarding a finite number of fibres, one may assume all components of $\mathscr{W}^*$, and hence also $\mathscr{W}'$, map dominantly to T and have dimension $k+1$; the fibres over 0 are Cartier divisors, so of dimension k.

Let $\alpha = \sum n_i[\mathscr{V}_i]$ be a $(k+1)$-cycle on $\mathscr{W}^*$, with $\mathscr{V}_i$ an irreducible subvariety of $\mathscr{W}^*$. We define the *limit cycle* $\lim_{t\to 0} \alpha_t$, a k-cycle on $\lim_{t\to 0} W_t$, as follows:

$$\lim_{t\to 0} \alpha_t = \sum n_i [(\bar{\mathscr{V}}_i)_0]$$

where $\bar{\mathscr{V}}_i$ is the closure of $\mathscr{V}_i$ in $\mathscr{W}'$, and $(\bar{\mathscr{V}}_i)_0$ is the scheme-theoretic fibre of $\bar{\mathscr{V}}_i$ over 0.

Proposition 11.1. (a) *If α and α' are rationally equivalent $(k+1)$-cycles on $\mathscr{W}^*$, then $\lim_{t\to 0} \alpha_t$ and $\lim_{t\to 0} \alpha'_t$ are rationally equivalent k-cycles on $\lim_{t\to 0} W_t$.*

(b) *If $\tilde{\alpha}$ is any cycle on $\mathscr{W}$ which restricts to α on $\mathscr{W}^*$, then $\lim_{t\to 0} \alpha_t$ is rationally equivalent to $\tilde{\alpha}_0$ on W_0.* Here $\tilde{\alpha}_0$ denotes the cycle constructed in § 10.1.

Proof. Let i be the imbedding of 0 in T. There is a diagram

$$\begin{array}{ccccc} A_{k+1}(W_0') & \xrightarrow{i'_*} & A_{k+1}(\mathscr{W}') & \xrightarrow{j'^*} & A_{k+1}(\mathscr{W}^*) \to 0 \\ & & {\scriptstyle i^!}\downarrow & \swarrow{\scriptstyle \sigma} & \\ & & A_k(W_0') & & \end{array}$$

where i' and j' are the inclusion of W_0' and $\mathscr{W}^*$ in $\mathscr{W}'$. The row is the exact sequence of § 1.8. The composite $i^! i'_*$ is zero, since (Theorem 6.2 (a) and Corollary 6.3) $i^! i'_*$ is multiplication by the top Chern class of the pull-back to W_0' of the normal bundle to 0 in T, and this normal bundle is trivial. Therefore there is an induced homomorphism σ as indicated with $\sigma j'^* = i^!$ (see Example 6.3.7 for generalizations). With this notation, $\lim_{t\to 0} \alpha_t$ is $\sigma(\alpha)$. This proves (a), since σ is well-defined on rational equivalence classes. The same argument, using $\mathscr{W}$ in place of $\mathscr{W}'$, proves (b). □

For any class $\alpha \in A_{k+1}(\mathscr{W}^*)$, we will write $\lim_{t\to 0} \alpha_t \in A_k\left(\lim_{t\to 0} W_t\right)$ for the class constructed using the preceding procedure for a cycle which represents α. Note that this limit class is constructed from the class α on the punctured total space $\mathscr{W}^*$, not from the classes α_t on the fibres.

Consider a fibre square

$$\begin{array}{ccc} \mathscr{W} & \xrightarrow{j} & \mathscr{V} \\ {\scriptstyle g}\downarrow & & \downarrow{\scriptstyle f} \\ \mathscr{X} & \xrightarrow[i]{} & \mathscr{Y} \end{array}$$

of schemes over T. Assume that i is a family of regular imbeddings of codimension d, i.e., i and all fibres $i_t\colon X_t \to Y_t$ are regular imbeddings of codimension d. Assume that $\mathscr{V}$ is flat over T of relative dimension k, and f is a closed imbedding. Then for all $t \in T$ we have the intersection class

$$X_t \cdot V_t = X_t \cdot_{Y_t} V_t \in A_{k-d}(W_t) = A_{k-d}(X_t \cap V_t)\,.$$

In particular, $X_0 \cdot V_0 \in A_{k-d}(W_0)$. By Proposition 10.1 (c), $X_t \cdot V_t = (\mathscr{X} \cdot \mathscr{V})_t$, where $\mathscr{X} \cdot \mathscr{V} \in A_{k+1-d}(\mathscr{W})$. We define the *limit intersection class* $\lim_{t\to 0} (X_t \cdot V_t)$ in $A_{k-d}\left(\lim_{t\to 0} W_t\right)$ by setting

$$\lim_{t\to 0} (X_t \cdot V_t) = \lim_{t\to 0} (\mathscr{X}^* \cdot \mathscr{V}^*)_t \in A_{k-d}\left(\lim_{t\to 0} X_t \cap V_t\right).$$

Corollary 11.1. *The inclusion of* $\lim_{t\to 0} (X_t \cap V_t)$ *in* $X_0 \cap V_0$ *maps* $\lim_{t\to 0} (X_t \cdot V_t)$ *to* $X_0 \cdot V_0$.

Proof. Apply (b) of the proposition to $\tilde{\alpha} = \mathscr{X} \cdot \mathscr{V}$. □

If X_t meets V_t properly for generic $t \in T^*$, i.e., $\dim(X_t \cap V_t) = k - d$ for generic $t \in T^*$, then $\lim_{t\to 0} (X_t \cap V_t)$ has pure dimension $k-d$. The class $\lim_{t\to 0} (X_t \cdot V_t)$ is therefore a well-defined $(k-d)$-cycle on $\lim_{t\to 0} X_t \cap V_t$, which we call the *limit intersection cycle*. Since $\mathscr{X}^* \cdot \mathscr{V}^*$ is a positive cycle whose support is $\mathscr{W}^*$, the limit intersection cycle is a non-negative cycle whose support is $\lim_{t\to 0} (X_t \cap V_t)$, and which represents $X_0 \cdot V_0$.

Example 11.1.1 (Fulton-MacPherson (2)). Care must be taken when passing to limits when cycles have negative coefficients, due to possible cancellation in the limit. Let $T = \mathbb{A}^1$, and define families of plane curves $\mathscr{A}, \mathscr{B}$ by

$$\mathscr{A} = V(x(y - tx) + tz^2) \subset \mathbb{P}^2 \times T$$
$$\mathscr{B} = V(y - tx) \subset \mathbb{P}^2 \times T .$$

Let $L = V(x)$, $M = V(y)$. Set $\alpha = [\mathscr{A}] - [\mathscr{B}]$. Then

$$\lim_{t \to 0} \alpha_t = [A_0] - [B_0] = [L] .$$

For all $t \neq 0$,

$$\alpha_t \cdot [M] = [(1:0:1)] + [(-1:0:1)] - [(0:0:1)] .$$

Thus $\lim_{t \to 0} (\alpha_t \cdot [M])$ and $\left(\lim_{t \to 0} \alpha_t\right) \cdot [M]$ are both well-defined zero-cycles, but they are not equal as cycles. If one modifies α_t by adding the constant cycle $[V(x - z)] + [V(x + z)]$, one obtains an example of the same phenomenon where both limit cycles have the same support. Of course, the two cycles are rationally equivalent on M, since α_t is rationally equivalent to α_0.

Example 11.1.2. Let s be a section of a vector bundle E on a scheme X, and let V be a subvariety of E. Form the fibre square

$$\begin{array}{ccc} \mathscr{W} & \rightarrow & V \times \mathbb{A}^1 \\ \downarrow & & \downarrow f \\ X \times \mathbb{A}^1 & \underset{i}{\rightarrow} & E \times \mathbb{A}^1 \end{array}$$

where $i(x, \lambda) = (\lambda \cdot s(x), \lambda)$, and f is the product imbedding. Then $\mathscr{W}$ is the union of $s^{-1}(V) \times \mathbb{A}^1$ and $s_E^{-1}(V) \times 0$, so $W_0 = s_E^{-1}(V)$, while the limit set W_0' is $s^{-1}(V)$. The limit intersection class is $s^![V] \in A_*(s^{-1}(V))$. By Corollary 11.1, $s^![V]$ maps to $s_E^![V]$ in $A_*(s_E^{-1}(V))$. This gives another proof of Corollary 6.5.

11.2 Infinitesimal Intersection Classes

Consider a family of regular imbeddings

$$\mathscr{X} \overset{i}{\hookrightarrow} Y \times T$$

of codimension d, deforming a given imbedding $i_0 : X \hookrightarrow Y$, $X = X_0$. Let $\Lambda = N_0 T$ be the tangent space to T at 0, and let Λ_X be the trivial line bundle on X with fibre Λ; equivalently, $\Lambda_X = N_X \mathscr{X}$. From the fibre diagram

$$\begin{array}{ccccc} X & \xrightarrow{i_0} & Y & \rightarrow & 0 \\ \downarrow & & \downarrow & & \downarrow \\ \mathscr{X} & \xrightarrow{i} & Y \times T & \rightarrow & T \end{array}$$

one sees that $N_X(Y \times T) = N_X Y \oplus N_X \mathscr{X} = N_X Y \oplus \Lambda_X$ (Appendix B.7.4). From the inclusions $X \subset \mathscr{X} \subset Y \times T$ one has an inclusion of normal bundles

$N_X\mathscr{X} \to N_X(Y \times T)$. This takes the form

$$\Lambda_X \xrightarrow{(\varrho, id)} N_X Y \oplus \Lambda_X$$

where $\varrho: \Lambda_X \to N_X Y$ is the *characteristic*, or *Kodaira-Spencer homomorphism.* Fixing a basis $\partial/\partial t$ for Λ, $\varrho(\partial/\partial t)$ is a section of $N_X Y$, which we call the *characteristic section* of the deformation.

Let $\mathscr{V} \subset Y \times T$ be closed subscheme, flat over T of relative dimension k. Let $V = V_0$, $W = X \cap V$, and let N be the restriction of $N_X Y$ to W. Let $s_{\mathscr{X}}$ be the section of N induced by the characteristic section of $N_X Y$. Let $C = C_W V$ be the normal cone to W in V, a closed subcone of N.

We call the class

$$s_{\mathscr{X}}^![C] \in A_{k-d}(s_{\mathscr{X}}^{-1}(C))$$

the *infinitesimal intersection class.*

Theorem 11.2. *With the above notation*

$$\lim_{t \to 0} (X_t \cap V_t) \subset s_{\mathscr{X}}^{-1}(C) \subset X \cap V.$$

These inclusions take the limit intersection class to the infinitesimal intersection class, and the infinitesimal intersection class to the intersection class:

$$\lim_{t \to 0} (X_t \cdot V_t) \to s_{\mathscr{X}}^![C] \to X \cdot V$$

by the induced homomorphisms

$$A_{k-d}\left(\lim_{t \to 0} (X_t \cap V_t)\right) \to A_{k-d}(s_{\mathscr{X}}^{-1}(C)) \to A_{k-d}(X \cap V).$$

Corollary 11.2. *If* $\dim s_{\mathscr{X}}^{-1}(C) = k - d$, *the limit intersection class is a well-defined non-negative* $(k-d)$*-cycle, which depends only on the characteristic homomorphism:*

$$\lim_{t \to 0} (X_t \cdot V_t) = s_{\mathscr{X}}^![C]$$

in $A_{k-d}(s_{\mathscr{X}}^{-1}(C))$. □

Proof of the Theorem. Let $s = \varrho(\partial/\partial t)$ be the characteristic section of $N_X Y$. We deform the given imbedding of X in Y to the imbedding of X in $N_X Y$ given by s, much as in § 5.1. Let $\mathscr{P}$ (resp. $\mathscr{Q}$) be the blow-up of $Y \times T$ along $X \times 0$ (resp. of $\mathscr{V}$ along W), and let $\mathscr{P}^\circ$, (resp. $\mathscr{Q}^\circ$) be the complement of $Bl_X Y$ (resp. $Bl_W V$) in $\mathscr{P}$ (resp. $\mathscr{Q}$). Since $X \subset \mathscr{X} \subset \mathscr{Y}$, $\mathscr{X} \cap Y = X$, and X is a Cartier divisor on $\mathscr{X}$, there is an inclusion $\tilde{i}$ of $\mathscr{X} = Bl_X\mathscr{X}$ in $\mathscr{P}^\circ$, which over $t = 0$, is the inclusion of X in $N_X Y$ given by the section s:

$$\begin{array}{ccccc} X & \xrightarrow{s} & N_X Y & \to & 0 \\ \downarrow & & \downarrow & & \downarrow t_0 \\ \mathscr{X} & \xrightarrow[\tilde{i}]{} & \mathscr{P}^\circ & \to & T. \end{array}$$

Here t_0 denotes the inclusion of 0 in T. In fact, the inclusion of $\mathscr{X}$ in $\mathscr{P}$ induces the inclusion of $X = P(\Lambda_X)$ in $P(N_X(Y \times T)) = P(N_X Y \oplus \Lambda_X)$ determined by $(\varrho, 1)$, where ϱ is the characteristic homomorphism. Choosing a basis $\partial/\partial t$ for Λ identifies the complement of $P(N_X Y)$ in $P(N_X Y \oplus \Lambda_X)$ with $N_X Y$. The imbedding of $\mathscr{V}$ in $Y \times T$ induces an imbedding of $\mathscr{D}$ in $\mathscr{P}$, and hence of $\mathscr{D}^\circ$ in $\mathscr{P}^\circ$. One obtains a fibre diagram

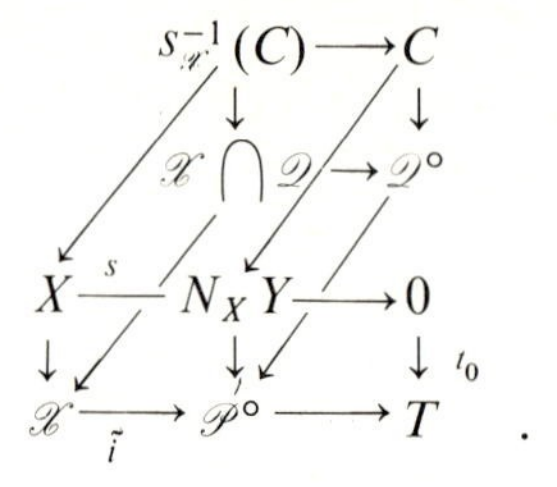

Since $\mathscr{X} \cap \mathscr{D}$ maps properly to $\mathscr{X} \cap \mathscr{V}$, and isomorphically over T^*, $\lim_{t \to 0} (\mathscr{X} \cap \mathscr{D})_t$ maps onto $\lim_{t \to 0} (X_t \cap V_t)$. Since $\lim_{t \to 0} (\mathscr{X} \cap \mathscr{D})_t \subset (\mathscr{X} \cap \mathscr{D})_0$, and $(\mathscr{X} \cap \mathscr{D})_0$ projects (isomorphically) onto $s_{\mathscr{X}}^{-1}(C)$,

$$\lim_{t \to 0} (X_t \cap V_t) \subset s_{\mathscr{X}}^{-1}(C) .$$

For any section $s_{\mathscr{X}} : W \to N$, $s_{\mathscr{X}}^{-1}(C) \subset W$. Likewise, the fact that $s_{\mathscr{X}}^{!}[C]$ maps to $s_N^{!}[C] = (\pi_N^*)^{-1}[C]$ in $A_{k-d}(X \cap V)$ holds for any section of a bundle (Corollary 6.5, or Example 11.1.2); since $X \cdot V = s_N^{!}[C]$, $s_{\mathscr{X}}^{!}[C] = X \cdot V$ in $A_{k-d}(X \cap V)$.

To prove that $\lim_{t \to 0} (X_t \cdot V_t) = s_{\mathscr{X}}^{!}[C]$ in $A_{k-d}(s_{\mathscr{X}}^{-1}(C))$, we apply the commutativity theorem (§ 6.4), giving

$$\tilde{i}^{!} t_0^{!} [\mathscr{D}^\circ] = t_0^{!} \tilde{i}^{!} [\mathscr{D}^\circ]$$

in $A_{k-d}(s_{\mathscr{X}}^{-1}(C))$. Now

$$\tilde{i}^{!} t_0^{!} [\mathscr{D}^\circ] = \tilde{i}^{!}[C] = s_{\mathscr{X}}^{!}[C]$$

by Theorem 6.2 (c). Since $\tilde{i}^{!}[\mathscr{D}^\circ]$ restricts to $\mathscr{X}^* \cdot \mathscr{V}^*$ over T^*,

$$t_0^{!} \tilde{i}^{!} [\mathscr{D}^\circ] = (\tilde{i}^{!}[\mathscr{D}^\circ])_0 = \lim_{t \to 0} (X_t \cdot V_t)$$

in $A_{k-d}(s_{\mathscr{X}}^{-1}(C))$ by Proposition 11.1 (b), which concludes the proof of the theorem. □

11.3 Limits and Distinguished Varieties

Let $X \hookrightarrow Y$ be a regular imbedding of codimension d, $V \subset Y$ a pure k-dimensional subscheme, with $k - d \geqq 0$. Let $W = X \cap V$, $C = C_W V \subset N$, N the restriction of $N_X Y$ to W, $\pi : N \to W$ the projection, $s_N : W \to N$ the zero section. Let $C_1, \ldots, C_r$ be the irreducible components of C, m_i the geometric multiplicity of C_i in C, so $[C] = \sum m_i [C_i]$, and

$$X \cdot V = s_N^*[C] = \sum_{i=1}^{r} m_i s_N^* [C_i]$$

in $A_{k-d}(X \cap V)$. The varieties $Z_i = \pi(C_i)$ are the distinguished varieties for the intersection (§ 6.1). For any closed subset Z of X, set

$$(X \cdot V)^Z = \sum_{Z_i \subset Z} m_i s_N^![C_i] \in A_{k-d}(Z).$$

For any cycle $\alpha = \sum n_i [V_i]$ on X, with V_i irreducible, and $Z \subset X$ closed, set

$$\alpha^Z = \sum_{V_i \subset Z} n_i [V_i].$$

Proposition 11.3. *Assume that Γ is a finite dimensional vector space of sections of $N_X Y$ which generates $N_X Y$. Let Z be a closed subset of X. Then there is a non-empty open subset $\Gamma(Z)$ of Γ such that*

(i) *For all $s \in \Gamma(Z)$, $\dim s^{-1}(C) = k-d$, so $s^![C]$ is a well-defined $(k-d)$-cycle, and*

$$(s^![C])^Z = (X \cdot V)^Z$$

in $A_{k-d}(Z)$.

(ii) *For any deformation $\mathscr{X} \hookrightarrow Y \times T$ of $X \hookrightarrow Y$ whose characteristic section belongs to $\Gamma(Z)$, $\lim_{t \to 0} X_t \cdot V$ is a well-defined $(k-d)$-cycle, and*

$$\left(\lim_{t \to 0} X_t \cdot V\right)^Z = (X \cdot V)^Z$$

in $A_{k-d}(Z)$.

Proof. Apply Serre's Lemma (Appendix B.9.1) to the subvarieties $C_1, \ldots, C_r$ of the bundle N and the closed subsets W and Z of the base. This gives an open set $\Gamma(Z)$ of Γ such that

(a) $\dim s^{-1}(C_i) = k-d$ for all i,

(b) $\dim(s^{-1}(C_i) \cap Z) \leq k-d-1$, if $C_i \not\subset \pi^{-1}(Z)$.

Therefore, if $s \in \Gamma(Z)$, $\dim s^{-1}(C) = k-d$, so $s^![C]$ is a well-defined $(k-d)$-cycle. From (b),

$$(s^![C])^Z = \sum_{C_i \subset \pi^{-1}(Z)} m_i s^![C_i]$$

since $s^{-1}[C_i]$ can have no components in Z if $C_i \not\subset \pi^{-1}(Z)$. Since $s^![C_i]$ represents $s_N^![C_i]$ (Corollary 6.5), $(s^![C])^Z$ is a representative cycle for $(X \cdot V)^Z$, which proves (i). Assertion (ii) follows from (i) and Corollary 11.2. □

Remark 11.3. Assume $N_X Y$ is generated by a subspace Γ of sections, and each section in Γ is characteristic for some deformation. Then the proposition gives a dynamic interpretation for the classes $(X \cdot V)^Z$:

$$(X \cdot V)^Z = \lim_{t \to 0} (X_t \cdot V)^Z$$

is the part of the limit intersection cycle supported on Z, for a *generic* deformation $\mathscr{X} \subset Y \times T$, i.e., a deformation whose characteristic section is in $\Gamma(Z)$.

Knowing the classes $(X \cdot V)^Z$ for all closed subsets Z of X is equivalent to knowing the equivalences of the distinguished varieties Z_i of the intersection of V by X, i.e., the canonical decomposition of $X \cdot V$. This may be described explicitly as follows. For each point $P \in X$, set

$$i(P) = (X \cdot V)^P .$$

Thus $i(P)$ is zero unless P is a distinguished variety, in which case, $i(P)$ is the contribution of P to $X \cdot V$ (cf. Definition 6.1.2). Dynamically, $i(P)$ is the component of the limit cycle $\lim_{t \to 0} X_t \cdot V$ supported on P, for a generic deformation $\mathscr{X}$ of $X \hookrightarrow Y$. For each irreducible curve $C \subset X$, set

$$j(C) = (X \cdot V)^C ,$$
$$i(C) = j(C) - \sum_{P \in C} i(P) .$$

Thus $i(C) = 0$ unless C is a distinguished variety, in which case $i(C)$ is the contribution of C to $X \cdot V$; $j(C)$ is the component of the limit cycle $\lim_{t \to 0} X_t \cdot V$ supported on C, for generic $\mathscr{X}$, while $i(C)$ is the part of $\lim_{t \to 0} X_t \cdot V$ supported on C, but not on any distinguished points of C. Inductively, for any irreducible subvariety Z of X, set

$$j(Z) = (X \cdot V)^Z ,$$
$$i(Z) = j(Z) - \sum_{Z' \subsetneqq Z} i(Z') ,$$

the sum over all proper subvarieties Z' of Z'. If $i(Z) \neq 0$, then Z is a distinguished variety, and $i(Z)$ is the contribution of Z to $X \cdot V$. Conversely, if all the contributions are positive – see § 12.2 for sufficient conditions – the distinguished varieties Z are determined by the nonvanishing of $i(Z)$. Dynamically, $i(Z)$ is the part of the generic limit cycle which is supported on Z, but not on any proper distinguished subvariety of Z.

Suppose in addition that $k = d$, and that a generic section in Γ is characteristic for a deformation $\mathscr{X}$ such that X_t meets V transversally for generic t. Then for each distinguished variety Z, the degree of its contribution $i(Z)$ is the number of points of $X_t \cap V$ which approach Z, but not any proper distinguished subvariety of Z, for a generic deformation $\mathscr{X}$. Equivalently, for any closed subset Z of X,

$$\deg (X \cdot V)^Z \textit{ is the number of points of } X_t \cap V \textit{ which approach } Z, \textit{ for a generic deformation } \mathscr{X}.$$

Example 11.3.1 (Severi (15), Lazarsfeld (1)). Let $H_1, \ldots, H_d$ be hypersurfaces in $\mathbb{P}^n$, defined by forms $F_1, \ldots, F_d$ of degrees $n_1, \ldots, n_d$, and let V be a pure k-dimensional subvariety of $\mathbb{P}^n$. Consider the intersection

$$\begin{array}{ccc} \bigcap_{i=1}^{d} H_i \cap V & \longrightarrow & V \\ \downarrow & & \downarrow \\ H_1 \times \ldots \times H_d & \rightarrow & \mathbb{P}^n \times \ldots \times \mathbb{P}^n . \end{array}$$

Let $X = H_1 \times \ldots \times H_d$, $Y = \mathbb{P}^n \times \ldots \times \mathbb{P}^n$. Then (Appendix B.7.4)

$$N_X Y = \mathscr{O}(n_1) \times \ldots \times \mathscr{O}(n_d) .$$

The space

$$\Gamma = \{(G_1, \ldots, G_d) \mid G_i \text{ is a homogeneous polynomial in } X_0, \ldots, X_n \text{ of degree } n_i\}$$

gives a space of sections of $N_X Y$ satisfying the conditions of Proposition 11.3. The deformation $\mathscr{X} \subset Y \times \mathbb{A}^1$ defined by

$$\mathscr{X} = \{(P_1, \ldots, P_d, t) \in Y \times \mathbb{A}^1 \mid F_i(P_i) + t\,G_i(P_i) = 0 \quad \text{for} \quad i = 1, \ldots, d\}$$

is a deformation whose characteristic section is $(G_1, \ldots, G_d)$. Using such deformations, one has the dynamic interpretation for the distinguished varieties and their intersections given in Remark 11.3. In particular this proves a theorem stated in Lazarsfeld (1)§ 2, that the classes defined dynamically by Severi and Lazarsfeld agree with the refined intersection classes of Fulton-MacPherson (1), (2).

Example 11.3.2 (B. Segre (3), Lazarsfeld (1)). Let H_1, H_2 be plane curves defined by forms

$$F_1 = FA_1, \qquad F_2 = FA_2$$

where A_1 and A_2 have no common factors. Let C be the normal cone to $H_1 \cap H_2$ in $\mathbb{P}^2$. Define Γ as in the preceding example. If $s = (G_1, G_2)$ is a section in Γ such that $A_1 G_2 - A_2 G_1$ and F have no common factors, then $s^![C]$ is a well-defined cycle:

$$(*) \qquad s^![C] = V(A_1 G_2 - A_2 G_1) \cdot V(F) + V(A_1) \cdot V(A_2)\,,$$

where the cycles on the right are intersection cycles of properly intersecting curves. In particular, for any deformations $F_i + t\,G_i + t^2 G_i' + \ldots$ with G_1, G_2 as above, the limit cycle is well-defined and given by (*). Contributions of the distinguished varieties may be deduced from this description (cf. Example 6.1.4). (Consider the deformations $(H_i)_t = F_i + t\,G_i$, $t \in \mathbb{A}^1$. If $\mathscr{W} \subset \mathbb{P}^2 \times \mathbb{A}^1$ is the intersection scheme, the closure $\mathscr{W}'$ of the restriction of $\mathscr{W}$ to $\mathbb{A}^1 - \{0\}$ has equations

$$FA_i + t\,G_i = 0\,, \qquad A_1 G_2 - A_2 G_1 = 0\,.$$

If f, a_i, g_i are local equations for F, A_i, G_i in a local ring $\mathscr{O}$ for $\mathbb{P}^2$ at a point, the exact sequence

$$0 \to \mathscr{O}/(a_1, a_2) \xrightarrow{\cdot f} \mathscr{O}/(f a_1, f a_2) \to \mathscr{O}/(f) \to 0$$

remains exact after tensoring with $\mathscr{O}/(a_1 g_2 - a_2 g_1)$. This implies that $\mathscr{W}'$ is flat over $\mathbb{A}^1$, and that $[W_0']$ is given by formula (*). See Lazarsfeld (1)§ 3 for details.)

Segre's formula (*) is also valid for effective curves on an arbitrary non-singular surface; the equations for the curves should be replaced by sections of appropriate line bundles.

Example 11.3.3 (Lazarsfeld (1)). Consider the plane curves H_1, H_2 defined by the equations $x^2 y = 0$ and $x y^2 = 0$.

(a) The distinguished varieties (for the intersection of the diagonal by $H_1 \times H_2$) are the lines $x = 0$, $y = 0$, and the point $(0:0:1)$. Each contributes a zero-cycle of degree 3 to the intersection. (This may be seen from formula (*)

of the preceding example, for $F = xy$, $A_1 = x$, $A_2 = y$, G_1 and G_2 generic cubic forms.)

(b) For the deformations

$$(H_1)_t = V(x^2y - tz^3), \quad (H_2)_t = V((x - t^2y)(y^2 - t^2x^2))$$

none of the nine points of $(H_1)_t \cap (H_2)_t$ approach the distinguished point $(0:0:1)$ as $t \to 0$. The characteristic section does not satisfy the condition of the preceding example. Contrary to an assertion of Severi (15), the contributions of distinguished varieties may not be described as the *minimum* over all deformations of hypersurfaces. In higher dimensions there are similar examples with irreducible hypersurfaces, see Lazarsfeld (1) p. 283.

(c) If the curves H_1 and H_2 are deformed by the action of the projective linear group acting on $\mathbb{P}^2$, i.e., $(H_i)_t = (\sigma_i)_t(H_i)$, for $(\sigma_i)_t$ generic curves in $GL(\mathbb{P}^2)$ converging to the identity, 7 of the points in $(H_1)_t \cap (H_2)_t$ will approach $(0:0:1)$ as $t \to 0$, while one point approaches each of the lines $x = 0$, $y = 0$. The characteristic sections of such deformations do not generate the normal bundle to $H_1 \times H_2$ in $\mathbb{P}^2 \times \mathbb{P}^2$.

(d) If the intersection class $H_1 \cdot H_2$ is constructed instead from the diagram

$$\begin{array}{ccc} H_1 \cap H_2 & \to & H_1 \times H_2 \\ \downarrow & & \downarrow \\ \Delta_{\mathbb{P}^2} & \to & \mathbb{P}^2 \times \mathbb{P}^2 \end{array}$$

the contribution of $(0:0:1)$ is 7, while each line contributes 1. (In this case deformations by automorphisms of $\mathbb{P}^2$ give enough sections to span the normal bundle $T_{\mathbb{P}^2}$.)

(e) If the intersection class $H_1 \cdot H_2$ is constructed from the diagram

$$\begin{array}{ccc} H_1 \cap H_2 & \to & H_2 \\ \downarrow & & \downarrow \\ H_1 & \to & \mathbb{P}^2, \end{array}$$

by deforming H_1 to $V(x^2y + tG)$ for a generic cubic G, one sees that only the lines $x = 0$, $y = 0$ are distinguished, with $x = 0$ contributing a zero-cycle of degree 3, $y = 0$ a zero-cycle of degree 6.

Example 11.3.4. Let $H_1 = V(FA_1)$, $H_2 = V(FA_2)$ be as in Example 11.3.2. Let P be a point in $V(F) \cap V(A_1)$, but $P \notin V(A_2)$. Consider deformations $(H_1)_t$, $(H_2)_t$ such that $(H_1)_t$ passes through P for t infinitesimal, i.e. $P \in V(G_1)$, where (G_1, G_2) is the characteristic section. For generic (G_1, G_2) satisfying this condition, the limit cycle contains P with multiplicity equal to the multiplicity of F at P. (Use Segre's formula (*).)

For example, of the four points of intersection of the two conics

$$\begin{aligned} xy + t(ax + by)z + t^2(\ldots) &= 0, \\ x(mx + ny + pz) + t(cx + dy + ez)z + t^2(\ldots) &= 0, \end{aligned}$$

only one approaches $(0:0:1)$ as $t \to 0$, if the constants a, b, c, d, e, m, n, p are generic ($p \neq 0$, $e \neq pb$ will do). Severi (12) works out this example, and discussed the subtlety of such limit problems.

Example 11.3.5. The results of § 11.2 and 11.3 extend without essential change to the case where the scheme V maps to Y by an arbitrary morphism f, not necessarily a closed imbedding, with $W=f^{-1}(X)$.

11.4 Moving Lemmas

Let a, b be cycles on a non-singular variety X over an algebraically closed field. If there are families of cycles $\{\alpha_t\}$, $\{\beta_t\}$ parametrized by a non-singular curve T, with $\alpha_0=a$, $\beta_0=b$, then

(i) $$a\cdot b=\lim_{t\to 0}\alpha_t\cdot\beta_t$$

in $A(X)$. Indeed, if α, β are the cycles on $X\times T$ giving the families, then $\alpha_t\cdot\beta_t=(\alpha\cdot\beta)_t$ (Corollary 10.1), and the conclusion follows by Proposition 11.1(b). If α_t meets β_t properly for general t, the right side of (i) gives a cycle which represents $a\cdot b$. If T is a rational curve, then

(ii) $$a\cdot b=\alpha_t\cdot\beta_t$$

for all $t\in T$. For in this case α_t and β_t are rationally equivalent to a and b respectively, and we have seen that the intersection product preserves rational equivalence.

Such dynamic interpretations of the intersection product are particularly valuable when the deformations arise naturally. Typical situations are: (1) divisors may move in their linear systems; (2) subvarieties may move in algebraic families of subvarieties (cf. Example 10.1.9, or § 11.3); (3) if X parametrizes a family of varieties, cycles which parametrize varieties in a certain relation to given geometric figures will move when the geometric figures are deformed (cf. § 10.4); (4) if an algebraic group acts on X, it will move all the cycles on X (cf. Example 11.4.5). In general, however, a positive cycle may not move in any family of positive cycles; for example, a curve on a projective surface with negative self-intersection number cannot be algebraically equivalent to an effective divisor not containing the curve. In its basic form, the moving lemma asserts that, if non-positive cycles are permitted, cycles can be moved to meet other cycles in the expected dimension.

Two cycles, α, β on X are said to meet *properly*, if for each variety V (resp. W) which appears with non-zero coefficient in α (resp. β), V meets W properly, i.e. $\dim(V\cap W)=\dim V+\dim W-\dim X$.

Moving lemma. *If X is non-singular and quasi-projective, and α, β are cycles on X, then there is a cycle α' rationally equivalent to α such that α' meets β properly.*

In particular the intersection product on $A(X)$, for X non-singular and quasi-projective, is uniquely determined by the knowledge of products of properly intersecting cycles. Historically, moving lemmas were used for constructing the intersection product on $A(X)$; for this, a more delicate statement is required, to know that the rational equivalence class of $\alpha' \cdot \beta$ is independent of the choice of α'. Since we do not use the moving lemma either for foundations or applications for intersection theory, we refer to the examples and literature for the proof.

Example 11.4.1. Let X^n be a non-singular quasi-projective variety over an algebraically closed field, $f: X \to U$ a closed imbedding into an open subscheme U of a projective space $\mathbb{P}^m$. Let V, W be irreducible subvarieties of X. For a linear subspace L of $\mathbb{P}^m$ of dimension $m-n-1$, let $C_L(\bar{V})$ be the cone over the closure of $f(V)$, with vertex L, in $\mathbb{P}^m$, and let C_L be the intersection of $C_L(\bar{V})$ with U.

(a) For generic L, C_L meets X properly, and this intersection is generically transversal along V, i.e.,

$$f^*[C_L] = [V] + \gamma_L,$$

where γ_L is a cycle on X which does not contain V.

(b) If $\dim(V \cap W) = \dim V + \dim W - n + e$, with $e > 0$, then, for generic L, if $\gamma_L = \sum n_i [V_i]$,

$$\dim(V_i \cap W) < \dim(V \cap W).$$

The moving lemma follows by induction on the number e, and the fact that cycles on $\mathbb{P}^m$ can be moved (Appendix B.9.2). (a) and (c) are proved by counting constants, much as in Example 8.4.12. For details, as well as the extension to arbitrary ground field, see Roberts (1).

Example 11.4.2. Let X be a variety over an algebraically closed field. We say that two cycles α, β on X meet *transversally* if for each variety V (resp. W) appearing in α (resp. β) with non-zero coefficient, each irreducible component P of $V \cap W$ is simple on X, and V and W meet transversally on an open set of P, i.e., $i(P, V \cdot W; X) = 1$. If X is non-singular and quasi-projective, and α, β are cycles on X, then there is a cycle α' on X, rationally equivalent to α, such that α' meets β transversally. (In the situation of Example 11.4.1, if V and W meet properly, the cycle γ_L meets W transversally, at least after replacing the imbedding of X in $\mathbb{P}^N$ by a suitable Veronese imbedding. See Hoyt (1) for details.)

Example 11.4.3. With the notation of Example 11.4.1, let P be a proper component of the intersection of V and W on X. Then, for generic, L, P is a proper component of the intersection of C_L and W on U, and

$$i(P, V \cdot W; X) = i(P, C_L \cdot W; U).$$

(By the refined projection formula of Example 8.1.7, $f^*[C_L] \cdot_X [W]$ maps to $[C_L] \cdot_U [W]$ in $A_*(C_L \cap W)$.). If $\bar{P}$, $\bar{C}_L$, $\bar{W}$ are the closures of P, C_L, W in $\mathbb{P}^m$, then

$$i(P, V \cdot W; X) = i(\bar{P}, \bar{C}_L \cdot \bar{W}; \mathbb{P}^m).)$$

This method was used by Severi (9) to determine arbitrary intersection multiplicities from the case where the ambient variety is projective space (cf. Example 8.2.6).

Example 11.4.4. *Uniqueness of intersection numbers.* We have defined an intersection number $i(P, V\cdot W; W)$ for a component P of the intersection of subvarieties V, W of a non-singular variety X, when P is proper, and simple on X. If every component of $V \cap W$ is proper the intersection cycle $V \cdot W = \sum i(P, V\cdot W; X)[P]$ is defined. The following properties determine the intersection numbers:

(i) If X° is an open subscheme of X meeting P, and V°, W°, P° are the intersections of X° with V, W, P, then

$$i(P^\circ, V^\circ \cdot W^\circ; X^\circ) = i(P, V\cdot W; X)\,.$$

(ii) (Projection formula). If $f: X \to Y$ is a closed imbedding of non-singular varieties, C is a subvariety of Y meeting X transversally in a variety V, W is a subvariety of X, and P is a proper component of $V \cap W$ on X, then

$$i(P, V\cdot W; X) = i(P, C\cdot W; Y)\,.$$

(iii) (Continuity). If T is a non-singular curve, $\mathscr{V} \subset X \times T$ a subvariety, flat over T, W a subvariety of X such that each V_t meets W properly on X, then $\mathscr{V}$ meets $W \times T$ properly on $X \times T$, and

$$(\mathscr{V} \cdot (W \cdot T))_t = V_t \cdot W$$

for all $t \in T$; here, for a cycle α on $X \times T$, α_t denotes the cycle on X defined in § 10.1.

(iv) (Multiplicity one) If V and W meet transversally in P, then

$$i(P, V\cdot W; X) = 1\,.$$

(To see that (i)–(iv) determine the intersection number, use (i) and (ii) and the argument of Example 11.4.3 to reduce to the case where X is an open subscheme of projective space, and P is the only component of $V \cap W$. Put V in a family $\mathscr{V}$ so that the generic V_t meets W transversally. Then $\mathscr{V}$ meets $W \times T$ transversally, and $V\cdot W = (\mathscr{V} \cdot (W \times T))_0$ is determined by (iii) and (iv).)

Note that by the moving lemma the intersection product characterized by (i)–(iv) uniquely determines the intersection product on A^*X for any non-singular quasi-projective variety X.

Example 11.4.5. Suppose a rational algebraic group G acts on a variety X; e.g., G could be a product of general linear groups $\mathrm{GL}(n)$. For $g \in G$ let $\varphi_g: X \to X$ be multiplication by g. Then, for all $g \in G$,

$$(\varphi_g)_* : A_k X \to A_k X$$

is the identity. (Use Example 10.1.7(b)). If G acts transitively on X, then X is non-singular, and if V, W are subvarieties on X, then $\varphi_g(V)$ meets W properly for generic $g \in G$ (Appendix B.9.2). Therefore the cycles $\varphi_g(V)\cdot(W)$ represent the class $V\cdot W$ in $A_*(X)$.

Example 11.4.6. Let $X \subset \mathbb{A}^3$ be the singular surface defined by $x^2 = yz^2$. Let V be the y-axis. There is no 1-cycle α on X which is rationally equivalent to $[V]$, which meets V properly. (There is a finite morphism $f: \mathbb{A}^2 \to X$ which is an isomorphism off V. If α meets V properly, $\alpha = f_*(\alpha')$ for a 1-cycle α' on $\mathbb{A}^2$. Since $A_1(\mathbb{A}^2) = 0$, $\alpha \sim f_*(0) = 0$, but $A_1(X) = \mathbb{Z}/2\mathbb{Z}$, generated by $[V]$, cf. Example 1.9.5.)

Example 11.4.7. The moving lemma fails on general singular varieties, even if rational coefficients are allowed. Let $X \subset \mathbb{P}^4$ be a cone over a non-singular quadric surface $Q \subset \mathbb{P}^3$, with vertex P. Let $L \subset X$ be the plane spanned by a line in Q and P. Then no non-zero multiple of $[L]$ is rationally equivalent to a cycle which does not pass through P. (The homology class of L in $H_4(X;\mathbb{Q})$ is not dual to any cohomology class in $H^2(X;\mathbb{Q})$, i.e. $cl(L) \neq c \cap cl(X)$ for all $c \in H^2(X;\mathbb{Q})$. Indeed, cycles not meeting P do determine cohomology classes on X (cf. Goresky (1)), and $H^2(X, \mathbb{Z})$ is generated by the cohomolgy class of Q, i.e., by a hyperplane section.) This example was considered by Zobel (3).

If X is a quotient of a non-singular quasi-projective variety by a finite group, one can prove a moving lemma for cycles with rational coefficients (cf. Examples 1.7.6 and 8.3.12). Mumford (7) has extended this to certain varieties – including moduli spaces of stable curves – which are locally such quotients.

Example 11.4.8. Let $f: X \to Y$ be a morphism, with Y non-singular, n-dimensional, and quasi-projective.

(a) If α is a k-cycle on X, β an l-cycle on Y, then there is cycle β' on Y, rationally equivalent to β, such that α meets β' properly, i.e. $\dim(|\alpha| \cap f^{-1}|\beta'|) = k + l - n$. In this case the $(k+l-n)$-cycle $\alpha \cdot_f \beta'$ is defined, and represents the class $\alpha \cdot_f \beta$ in A_*X. (Stratifying the restriction of f to the components of $|\alpha|$, one reduces this to the usual moving lemma, cf. Fulton (2) § 2.3.). We know from § 8.1 that the rational equivalence class of the cycle $\alpha \cdot_f \beta'$ does not change if β' or α is replaced by a rationally equivalent cycle on Y or X. As Demazure pointed out, the proof of this last fact in Fulton (2) § 2.3 Prop. (2) is faulty. Indeed, most foundational treatments of intersection theory based on a moving lemma have failed to take care that all auxilliary constructions preserve properness of intersections.

(b) Suppose X is a variety, f is dominant, and Y° is open in Y, $X^\circ = f^{-1}(Y^\circ)$, such that the restriction $f^\circ: X^\circ \to Y^\circ$ is flat. For any subvariety V of Y whose intersection V° with Y° is non-empty, the class $f^*[V]$ in A_*X, defined in § 8.1, is represented by the cycle $[(f^\circ)^{-1}(V^\circ)]$. (Use proposition 8.1.1 (d) and 8.1.2 (a).) By the moving lemma of (a), this determines the Gysin pull-back $f^*: A_*Y \to A_*X$.

Example 11.4.9 (Severi). Any k-cycle α on $\mathbb{P}^n$ can be written as the proper intersection of $n-k$ divisors $D_1, \ldots, D_{n-k}$:

$$\alpha = D_1 \cdot \ldots \cdot D_{n-k}.$$

These divisors need not be effective, even if α is positive, cf. Example 9.1.2. The proof uses the cone construction of Example 11.4.1. See Samuel (2) II.6.4 for details.

Notes and References

The dynamic nature of intersection multiplicities is evident in all classical discussions of the subject. Severi's definitions of these multiplicities were dynamic, realizing non-transversal or non-proper intersections as limits of transversal intersections. Severi (7) points out that these constructions anticipated their use in topology. Severi used the cone construction of Example 11.4.1 for this dynamic interpretation. Samuel (3) and Chow (1) used this construction to prove moving lemmas, on which they based their construction of the intersection ring ("Chow ring") of a smooth projective variety. Similar proofs of moving lemmas have been given in Chevalley (2), Samuel (5), and Roberts (1). Murre (1) and de Boer (1) used the same cone construction to show how to assign intersection multiplicities to connected components of $V \cap W$, when V and W have complementary dimension on a non-singular projective variety. Severi (15) compares static and dynamic approaches to intersection multiplicities.

The first three sections of this chapter are a presentation of published (Lazarsfeld (1)) and unpublished work of R. Lazarsfeld, who began by analyzing and correcting an idea of Severi (15). Limits of intersections were also studied by B. Segre (3), who gave the formula of Example 11.3.2 for the surprisingly non-trivial case of varying curves on a surface.

Weil (2) Appendix II gave an axiomatic characterization of intersection numbers. His axioms included an associative axiom and a stronger form of the projection formula than the one needed in Example 11.4.4. W.-L. Chow (unpublished) has reworked his intersection theory, based on a strong moving lemma as in Example 11.4.2.

Chapter 12. Positivity

Summary

We have constructed intersection classes by intersecting a cone C in a normal bundle N with the zero-section. If $\sum m_i[C_i]$ is the cycle of C, the intersection class has a corresponding decomposition into $\sum m_i \alpha_i$, $\alpha_i \in A_*(Z_i)$, $Z_i = \mathrm{Supp}(C_i)$. If the bundle N is suitably positive, one can deduce corresponding positivity of the intersection classes, even if the intersections are not proper.

Assume for simplicity that the restriction N_i of N to Z_i is generated by its sections. Then α_i is represented by a non-negative cycle. If N_i is also ample, α_i is represented by a positive cycle. If $N_i \otimes L^\vee$ is generated by its sections, for an ample line bundle L, then using L to compute degrees, the degree of α_i is bounded below by the degree of Z_i.

For intersections on a non-singular variety X, the positivity of its tangent bundle will imply corresponding positivity for all intersection classes on X. For $X = \mathbb{P}^n$, $V_1, \ldots, V_r$ subvarieties, a refined Bézout's theorem follows:

$$V_1 \cdot \ldots \cdot V_r = \sum m_i \alpha_i, \quad \prod_{j=1}^{r} \deg(V_j) = \sum m_i \deg(\alpha_i) \geqq \sum m_i \deg(Z_i),$$

where the Z_i are the distinguished varieties; all irreducible components of $\bigcap_j V_j$ are included among the Z_i.

There are also applications to intersection multiplicities. For example, if $V_1, \ldots, V_r$ meet properly at a non-singular point P of an n-dimensional variety X, and $\tilde{V}_i \subset \tilde{X}$ are the blow-ups at P, then

$$i(P, V_1 \cdot \ldots \cdot V_r; X) = \prod_{j=1}^{r} e_P(V_j) + \deg(\tilde{V}_1 \cdot \ldots \cdot \tilde{V}_r).$$

Here the intersection class $\tilde{V}_1 \cdot \ldots \cdot \tilde{V}_r$ is in $A_0(E)$, $E \cong \mathbb{P}^{n-1}$ the exceptional divisor. The degree of $\tilde{V}_1 \cdot \ldots \cdot \tilde{V}_r$ is always non-negative, and one has lower bounds as in the refined Bézout's theorem, e.g.,

$$\deg(\tilde{V}_1 \cdot \ldots \cdot \tilde{V}_r) \geqq \sum_{i=1}^{s} \deg(W_i),$$

where $W_1, \ldots, W_s$ are the irreducible components of the intersection $\bigcap_j P(C_P V_j)$ of the projective tangent cones. Such positivity is noteworthy since the $\tilde{V}_i$ may have excess intersections, and general intersections on $\tilde{X}$ can be negative. There are similar inequalities for proper intersections of divisors on a possibly singular variety.

Notation. A cycle $\sum n_i[V_i]$ on a scheme X is *non-negative* if each n_i is non-negative, and *positive* if, in addition, at least one n_i is positive. Let $A_k^{\geqq} X$ (resp. $A_k^+ X$) denote the set of classes in $A_k X$ which can be represented by non-negative (resp. positive) cycles. Thus

$$A_k^{\geqq} X = A_k^+ X \bigcup \{0\}.$$

Both sets are closed under addition.

Let L be a line bundle on a complete scheme X. For a k-cycle or cycle class α on X, the *L-degree* of α, denoted $\deg_L(\alpha)$, is defined by

$$\deg_L(\alpha) = \int_X c_1(L)^k \cap \alpha.$$

If V is a subvariety of X, the L-degree of V, $\deg_L(V)$, is defined by

$$\deg_L(V) = \deg_L([V]) = \int_X c_1(L)^{\dim V} \cap [V].$$

12.1 Positive Vector Bundles

A line bundle L on a scheme X is *ample* if there is a positive integer m and a finite morphism f from X to a projective space $\mathbb{P}^N$ such that $L^{\otimes m} = f^* \mathcal{O}_{\mathbb{P}^N}(1)$. In fact one may then find such m, f with f a closed imbedding, although we do not use this fact.

Lemma 12.1. *If L is an ample line bundle on X, and α is a non-negative (resp. positive) k-cycle on X, then*

$$\deg_L(\alpha) \geqq 0 \qquad (\textit{resp.}\ \deg_L(\alpha) > 0).$$

Proof. With $f\colon X \to \mathbb{P}^N$, $L^{\otimes m} = f^* \mathcal{O}(1)$ as above,

$$m^k \deg_L(\alpha) = \int_X c_1(f^* \mathcal{O}(1)) \cap \alpha = \int_{\mathbb{P}^N} c_1(\mathcal{O}(1))^k \cap f_*(\alpha).$$

by Proposition 2.5 (c). Since push-forward by finite morphisms preserves positivity of cycles, we are reduced to the familiar fact that every subvariety V of projective space $\mathbb{P}^N$ has positive degree. This can be verified by induction on the dimension k of V, as follows. If $k > 0$, choose a hyperplane H which meets V but does not contain V. Then the intersection cycle

$$H \cdot V = \sum i(W; H \cdot V; \mathbb{P}^N) \cdot [W]$$

is a positive cycle which represents $c_1(\mathcal{O}(1)) \cap [V]$, so

$$\deg V = \deg(H \cdot V) > 0$$

by induction. □

In particular, if X carries an ample line bundle, then $0 \notin A_k^+ X$. If X is not projective, however, it may happen that $0 \in A_k^+ X$ (cf. Example 12.1.1).

A vector bundle E is *ample* if the line bundle $\mathcal{O}_{E^\vee}(1)$ on $P(E^\vee)$ is an ample line bundle. In case E has a finite dimensional vector space V of sections which generate E, this is equivalent to saying that the induced morphism from $P(E^\vee)$ to the projective space $P(V^\vee)$ is a finite morphism. Any quotient bundle of an ample bundle is ample, and any direct sum of ample bundles is ample. If E is a vector bundle, and L an ample line bundle such that $E \otimes L^\vee$ is generated by its sections, then E is ample.

Theorem 12.1. *Let E be a vector bundle of rank r on a scheme X, $\pi: E \to X$ the projection, $s_E: X \to E$ the zero-section. Let V be a k-dimensional subvariety of E, $k \geqq r$.*

(a) *If E is generated by its sections, then*

$$s_E^*[V] \in A_{k-r}^{\geqq}(X).$$

(b) *If L is an ample line bundle on X such that $E \otimes L^\vee$ is generated by its sections, then*

$$\deg_L(s_E^*[V]) \geqq \sum_{i=1}^{q} \deg_L(W_i) > 0,$$

where $W_1, \ldots, W_q$ are the distinct irreducible components of $s_E^{-1}(V)$.

(c) *If E is ample and generated by its sections, then*

$$s_E^*[V] \in A_{k-d}^{+}(X).$$

(d) *If E is ample, and L is any ample line bundle on X, then*

$$\deg_L(s_E^*[V]) > 0.$$

Proof. (a) By assumption there is a surjection $\varrho: F \to E$ from a trivial bundle F onto E. Let $s = s_E$, $t = s_F$ denote the zero sections. The morphism ϱ is smooth of relative dimension $n - r$, $n = \operatorname{rank}(F)$, so $\varrho^*[V] = [\varrho^{-1}(V)]$. Since $s = \varrho \circ t$,

$$s^*[V] = t^* \varrho^*[V] = t^*[\varrho^{-1}(V)]$$

by Proposition 6.5(b). Thus we are reduced to proving the assertion for F; i.e., we may assume E is trivial.

The proof proceeds by induction on the rank r of E. Write $E = G \oplus 1$, where G is trivial of rank $r - 1$, let j be the inclusion of G in $E: j(g) = (g, 0)$; and let u be the zero section of G. Then $s = j \circ u$, so (Theorem 6.5)

$$s^*[V] = u^*(j^*[V]),$$

so it suffices to show that $j^*[V] \in A_{k-1}^{\geqq} G$. The class $j^*[V]$ is constructed by intersecting V by the effective divisor G on E (Definition 2.3). If $V \not\subset G$, then $j^*[V]$ is represented by the non-negative intersection cycle $G \cdot V$ on $G \cap V$. If $V \subset G$, then $j^*[V] = 0$ since the normal bundle to G in E is trivial.

(b) Mapping a trivial bundle F onto $E \otimes L^\vee$ determines a surjection from $F \otimes L$ onto E. Arguing as in (a), one is reduced to the case where $E = F \otimes L$, F trivial of rank r. The proof proceeds by induction on r.

Assume $r=1$. In this case s_E imbeds X as a Cartier divisor on $E=L$. If $V \subset s_E(X)$, then $s_E^{-1}(V)=V$, and

$$s_E^*[V]=c_1(L) \frown [V].$$

Therefore $\deg_L(s_E^*[V])=\int c_1(L)^{k-1}\cdot c_1(L) \frown [V]=\deg_L(V)$, which is positive by Lemma 12.1. If V meets the zero section properly, then

$$s_E^*[V]=\sum_{i=1}^{q} m_i [W_i]$$

where m_i is the (positive) intersection multiplicity. Therefore

$$\deg_L s_E^*[V]=\sum m_i \deg_L(W_i) \geqq \sum \deg_L(W_i)\ .$$

To complete the case $r=1$, we verify that V must meet the zero section. If not, let $\bar{V}$ be the closure of V in $P(L\oplus 1)$, using the canonical imbedding of L in $P(L\oplus 1)$, and let $f\colon \bar{V}\to X$ be the projection. Since V does not meet the zero section, $\bar{V}$ is contained in the complement of $s_E(X)$ in $P(L\oplus 1)$, which may be identified with $L^\vee$. Therefore f is an affine as well as projective morphism, so f is finite (Appendix B.2.4); in particular, f^*L is an ample line bundle. However, the pull-back of $L^\vee$ to $L^\vee$ is trivial, so f^*L is trivial. A trivial line bundle on a positive dimensional variety cannot be ample, however, as follows from Lemma 12.1.

For $r>1$, we proceed by induction on r. Choose a splitting $F=F'\oplus 1$, with F' trivial of rank $r-1$, so that if $G=F'\otimes L$, and j is the induced imbedding of G in E, then $j(G)$ meets V. (For the existence of such a splitting the ampleness of L and positive dimensionality of V are irrelevant. One may assume V is a point, and restrict L to an open subset of X on which L is trivial, in which case the assertion is obvious.) If $j(G)$ meets V properly, and $V_1,\dots,V_p$ are the irreducible components of $j^{-1}(V)$, with intersection multiplicities $m_1,\dots,m_p$ and u is the zero section of G, then

$$s_E^*[V]=\sum_{j=1}^{p} m_j\cdot u^*[V_j],$$

so

$$\deg_L s_E^*[V]\geqq \sum_{j=1}^{p} \deg_L u^*[V_j].$$

Since $s_E^{-1}(V)=\bigcup_{j=1}^{p} u^{-1}(V_j)$. each W_i appears among the irreducible components of some $u^{-1}(V_j)$. By the inductive assumption for G,

$$\sum_{j=1}^{p} \deg_L u^*[V_j] \geqq \sum_{i=1}^{t} \deg_L(W_i)>0,$$

which concludes the proof in this case. Otherwise V is contained in $j(G)$, so

$$j^*[V]=c_1(\eta^*L)\frown [V]$$

where η is the projection from G to X, since η^*L is the normal bundle to G in E. Therefore

$$s_E^*[V]=u^*(c_1(\eta^*L)\frown [V])=c_1(L)\frown u^*[V],$$

by Proposition 6.3. Therefore $\deg_L s_E^*[V] = \deg_L u^*[V]$. Since $u^{-1}(V) = s_E^{-1}(V)$, the inductive assumption applies to give the desired inequalities.

(c) By (a), it sufficies to show $s_E^*[V] \neq 0$. It suffices to prove this after base extension from the ground field to its algebraic closure, so we may assume the ground field is algebraically closed (cf. Example 6.2.9). For any section s of E such that $s(X)$ meets C,

$$s_E^*[V] = s^*[V] = s_E^*[C]$$

where C is the normal cone to V along $V \cap s(X)$ (Example 6.2.3 and Corollary 6.5). Thus we may assume C is an irreducible cone. The proof proceeds by induction on the dimension of X, being trivial if $\dim X = 0$. We may assume C projects onto X. We assume also $C \neq E$, since the claim is trivial when $C = E$.

Suppose there is a section s of E such that $s(X)$ meets C, but $s(X) \not\subset C$. Let $X' = s^{-1}(C)$. Then

$$s_E^*[C] = s^*[C] = s_E^*[C']$$

where C' is the normal cone to C along X'. Since C' has smaller support, one is done by induction.

Suppose there is no such section, i.e. any section either misses C or is contained in C. Let Γ be the space of sections, ϱ the canonical surjection from the trivial bundle $X \times \Gamma$ onto E, H the kernel of ϱ. By assumption, $\varrho^{-1}(C)$ is a "constant" cone, i.e. $\varrho^{-1}(C) = X \times D$ where D is a cone in Γ. For each $x \in X$, the fibre $H(x)$ of H at x is contained in D, and satisfies $H(x) + D \subset D$. But for any cone D in a vector space Γ, $L = \{v \in \Gamma \mid v + D \subset D\}$ is a linear subspace of Γ which is contained in D. Therefore H is contained in a proper trivial subbundle $X \times L$ of $X \times \Gamma$. Therefore E surjects onto the trivial bundle $X \times (\Gamma/L)$. From the definition of ampleness, however, an ample vector bundle on a positive dimensional variety cannot have a trivial quotient bundle.

(d) If V is a subcone of E, this is proved in Fulton-Lazarsfeld (3) Corollary 2.2. The proof, in addition to intersection theory, uses constructions of Bloch-Gieseker (1), and relies on the Hard Lefschetz theorem; it will not be repeated here. If V meets the zero section of E, then

$$s_E^*[V] = s_E^*[C],$$

where C is the normal cone to $s_E^{-1}(V)$ in V, so the previous case applies.

To show that V must meet the zero section, we deform to the zero section and apply the previous case. Let $T = \mathbb{A}^1$, $T^* = \mathbb{A}^1 - \{0\}$. Let $\mathscr{V}^* \subset E \times T^*$ be the image of $V \times T^*$ by the closed imbedding $(v, \lambda) \to (\lambda v, \lambda)$. Let $\mathscr{V}$ be the closure of $\mathscr{V}^*$ in $E \times T$. Then V is the fibre V_1 of $\mathscr{V}$ over 1, and the fibre V_0 of $\mathscr{V}$ over 0 meets the zero section at least in the projection of V on X. Then $[V_1] = [V_0]$ in $A_* E$ (say by Example 10.1.7), so $s_E^*[V] = s_E^*[V_0]$, and it suffices to apply the preceding case to an irreducible component of V_0 which meets the zero section. $\square$

Example 12.1.1. (a) Let $X = \mathbb{A}^1 \times \mathbb{P}^1$. Then $[\{0\} \times \mathbb{P}^1] \sim 0$, so $A_1^{\geqq} X = A_1^+ X$.

(b) Hironaka has constructed a complete, non-singular variety X over $\mathbb{C}$ with a positive 1-cycle rationally equivalent to 0, so $A_1^{\geqq} X = A_1^+ X$ (cf. Hartshorne (5) p. 443). Such a variety cannot be projective, of course.

Example 12.1.2. Let i imbed X as a Cartier divisor on Y, and let V be a k-dimensional subvariety of Y, $k \geqq 1$. There are three possibilities for the intersection class $X \cdot V = i^*[V]$ in $A_{k-1}(X)$:

(i) X meets V properly. Then $i^*[V]$ is a positive cycle supported on $X \cap V$.
(ii) X doesn't meet V. Then $i^*[V] = 0$.
(iii) X contains V. Then $i^*[V] = c_1(L) \cap [V]$ where $L = i^*\mathcal{O}_Y(X)$ is the normal bundle.

These cases can all arise when $Y = L$ is a line bundle on X, and i is the zero section.

Example 12.1.3. (a) Let E be a vector bundle of rank r on a scheme X, which is generated by its sections, and let $\pi: E \to X$ be the projection. Then

(i) $A_k^{\geqq} E = \pi^*(A_{k-r}^{\geqq} X)$.
(ii) If E is ample, then $A_k^+(E) = \pi^*(A_{k-r}^+ X)$. (These are equivalent to (a) and (c) of the theorem.)
(b) If E is an ample vector bundle of rank r on X, and $V \subset E$ is a subvariety of dimension $\geqq r$, then $s^{-1}(V) \neq \emptyset$ for any section s of E. (Apply (d) and Corollary 6.5.) In particular, $0 \notin A_k^+ E$ if $k \geqq r$.

Example 12.1.4. The conclusion (c) of Theorem 12.1 may not hold if E is not generated by its sections. Let D be a divisor of positive degree on a projective non-singular curve X, which is not linearly equivalent to an effective divisor, let $E = L = \mathcal{O}_X(D)$, and let $V = X$ imbedded by the zero section. Then E is ample, and $E \otimes L^\vee$ is generated by its sections, but $s_E^![V] = [D]$ is not in $A_0^+ X$.

In the situation of Theorem 12.1 (c), some positive multiple of $s_E^*[V]$ is in $A_{k-r}^+(X)$. (Argue as in the proof of (b).) We do not know if this is true in the situation of (d).

Example 12.1.5. Following Sommese (1), a line bundle L on a complete scheme X is called *n-ample* if there is a positive integer m and a morphism $f: X \to \mathbb{P}^N$ such that $L^{\otimes m} = f^*\mathcal{O}_{\mathbb{P}^N}(1)$, and all fibres of f have dimension $\leqq n$. A vector bundle E is n-ample if $\mathcal{O}_{E^\vee}(1)$ on $P(E^\vee)$ is n-ample. Note that 0-ample is the same as ample.

If E is n-ample and generated by its sections, and V is a subvariety of E, with $\dim(V) \geqq \operatorname{rank}(E) + n$, then $s_E^*[V]$ is in $A_*^+(X)$. (Argue as in (c).)

If L is an n-ample line bundle, and $E \otimes L^\vee$ is generated by its sections, with V as above, some positive multiple of $s_E^*[V]$ is in $A_*^+(X)$ (cf. Example 12.1.4).

Example 12.1.6. *Existence of degeneracy loci* (Fulton-Lazarsfeld (3)). Let E, F be vector bundles of ranks e, f on a variety X. Assume

$$\operatorname{Hom}(E, F) = E^\vee \otimes F$$

is a ample vector bundle. Let Σ be a subvariety of $\operatorname{Hom}(E, F)$ of codimension d. If $\dim X \geqq d$, then for any homomorphism $\sigma: E \to F$ of vector bundles,

$$\{x \in X \mid \sigma(x) \in \Sigma\}$$

is a *non-empty* closed set, all of whose components have dimension at least $\dim(X) - d$. For example, if $\sum_k$ is the variety of maps of rank $\leqq k$, then $\sum_k$ has codimension $(e-k)(f-k)$ (Lemma A.7.2), so

$$D_k(\sigma) = \{x \in X \mid \operatorname{rank} \sigma(x) \leqq k\}$$

is non-empty whenever $\dim X \geqq (e-k)(f-k)$. (Apply Theorem 12.1 (d) to the section of $E^\vee \otimes F$ determined by σ.) This last result first appeared in Fulton-Lazarsfeld (2).

Similarly suppose E is a vector bundle of rank r, L a line bundle, such that $S^2 E \otimes L$ (resp. $\Lambda^2 E \otimes L$) is ample, and

$$\sigma: E^\vee \to E \otimes L$$

is a symmetric (resp. skew-symmetric) vector bundle homomorphism (cf. Example 14.4.11). If $\dim X \geqq \binom{r-k+1}{2}$ (resp. $\binom{r-k}{2}$, with k even), then $D_k(\sigma)$ is non-empty. See Example 14.4.13 for refinements.

Example 12.1.7. *Polynomials in Chern classes of ample vector bundles* (Fulton-Lazarsfeld (3)). Let E be a vector bundle of rank e on X. Let λ be a partition of an integer n in integers $\leqq e$:

$$e \geqq \lambda_1 \geqq \lambda_2 \geqq \ldots \geqq \lambda_n \geqq 0,$$

$\sum \lambda_i = n$. Define the Schur polynomial $\Delta_\lambda(E)$ by

$$\Delta_\lambda(E) = \det(c_{\lambda_i + j - i}(E))_{1 \leqq i, j \leqq n},$$

an isobaric polynomial of weight n in the Chern classes of E. Examples include $c_n(E)$, for $\lambda = (n, 0, \ldots, 0)$, and $(-1)^n s_n(E)$, for $\lambda = (1, \ldots, 1)$, cf. § 14.5.

Let α be a positive k-cycle on X, with $k \geqq n$.

(a) if E is generated by its sections, then $\Delta_\lambda(E) \cap \alpha \in A_{k-n}^{\geqq}(X)$.
(b) If E is ample and generated by its sections, then $\Delta_\lambda(E) \cap \alpha \in A_{k-n}^{+}(X)$.
(c) If E is ample, and L is any ample line bundle on X, then

$$\deg_L(\Delta_\lambda(E) \cap \alpha) > 0.$$

(One may assume X is a variety, and $\alpha = [X]$. If V is the trivial bundle of rank $n+e$ on X, there is a subvariety Ω of the bundle $H = \operatorname{Hom}(V, E)$, with codim $(\Omega, H) = n$, such that $\Delta_\lambda(E) \cap [X] = s_H^*[\Omega]$. This is proved in Example 14.3.2. Since H satisfies the same positivity assumptions as E, the conclusions follow from Theorem 12.1 (a), (c), and (d).)

In fact positive linear combinations of the polynomials Δ_λ are the only polynomials which have positive degree for all ample bundles on all varieties. The first few Δ_λ are:

$$c_1;\ c_2,\ c_1^2 - c_2;\ c_3,\ c_1 c_2 - c_3,\ c_1^2 - 2c_1 c_2 + c_3;$$

note that $c_1^2 - 2c_2$ is *not* such a positive combination.

Example 12.1.8. Let E be a vector bundle which is generated by its sections on an n-dimensional variety X over an algebraically closed field. Let p be a positive integer. Assume that $0 \notin A^+_{n-p}(X)$, which holds for all p if X is projective. Then the following are equivalent:

(i) $c_p(E) \cap [X] = 0$.
(ii) there is a trivial subbundle F of E such that $\operatorname{rank}(E/F) < p$.
(iii) $c_j(E) \cap \alpha = 0$ for all $j \geqq p$, all $\alpha \in A_* X$.

(Realize $c_p(E) \cap [X]$ as a dependency locus of generic sections of E, cf. Example 14.4.3.)

The implication (i) $\Rightarrow$ (iii) fails for affine varieties (cf. Kumar-Murthy (1)).

Example 12.1.9. Let E be an ample vector bundle of rank r on a scheme X over an algebraically closed field K. If V is a subvariety of E with $\dim V > r$, then $\Gamma(V, \mathcal{O}_V) = K$. (Let $Z = V \cap s_E(X)$. If $f\colon V \to \mathbb{A}^1$ is a non-constant morphism, then $f(Z)$ is a finite set. For general $\lambda \in \mathbb{A}^1 - f(Z)$, $f^{-1}(\lambda)$ contradicts Example 12.1.3 (b).)

Example 12.1.10. The ampleness of E on all of X is not needed for the truth of (c). Let E be a vector bundle, on a projective variety X over an algebraically closed field, which is generated by its sections. Following Lazarsfeld, define the *disamplitude locus* $\operatorname{Damp}(E)$ to be the set of points $x \in X$ such that for some irreducible curve Z through x, $E|_Z$ has a trivial quotient line bundle. Equivalently, if V is the space of sections of E, $\operatorname{Damp}(E)$ is the projection in X of the locus in $P(E^\vee)$ where the canonical map from $P(E^\vee)$ to $P(V^\vee)$ is not finite, which shows that $\operatorname{Damp}(E)$ is a closed subset of X. Call E *generically ample* if $\operatorname{Damp}(E) \neq X$.

If X is an irreducible subcone of E, with $\dim(C) \geqq \operatorname{rank}(E)$, and the support of C is not contained in $\operatorname{Damp}(X)$, then $s_E^*[C] \in A_*^+(X)$. (The proof is similar to the proof of Theorem 12.1 (c).) It follows that Schur polynomials in Chern classes of generically ample vector bundles are positive.

If $X \subset G_1(\mathbb{P}^4)$ is the Fano surface of lines on a cubic 3-fold (cf. Example 14.7.13), one can show that the universal quotient bundle, and the dual of the universal subbundle on $G_1(\mathbb{P}^4)$ restrict to generically ample, but not ample, vector bundles on X.

Example 12.1.11. Let E be a vector bundle of rank r on a scheme X over an infinite field K, which is generated by sections $s_1, \ldots, s_N$. For $t = (t_1, \ldots, t_N) \in K^N$, let s_t be the section $\sum t_i s_i$ of E. Let V be a purely m-dimensional subscheme of E. Then for generic $t \in K^N$, $\dim s_t^{-1}(V) = m - r$ (or $s_t^{-1}(V)$ is empty), and

$$s_t^![V] = [s_t^{-1}(V)].$$

(If $\varrho\colon X \times \mathbb{A}^N \to E$ is the surjection determined by $s_1, \ldots, s_N$, then $s_t = \varrho \circ i_t$, where $i_t(x) = (x, t)$. By Proposition 6.5 (b), $s_t^![V] = i_t^![W]$, with $W = \varrho^{-1}(V)$. Let $T_1, \ldots, T_N$ be the coordinate functions on $\mathbb{A}^N$. Given any closed subscheme W of $X \times \mathbb{A}^N$, one shows by induction on N that for generic $t \in K^N$, the functions $T_1 - t_1, \ldots, T_N - t_N$ form a regular sequence on $\mathcal{O}_W$. When $N = 1$ and X is affine, this says that $T_1 - t_1$ should not belong to any of the associated primes of the ideal of W; since any proper ideal of $A[T_1]$ contains $T_1 - t_1$ for at most one $t_1 \in K$, at most a finite number of t_1 need be discarded.)

This argument also shows that a generic section s_t of E is a regular section.

12.2 Positive Intersections

Consider our standard intersection set-up: a fibre square

$$\begin{array}{ccc} W & \xrightarrow{j} & V \\ {\scriptstyle g}\downarrow & & \downarrow{\scriptstyle f} \\ X & \xrightarrow[i]{} & Y \end{array}$$

with i a regular imbedding of codimension d, N the pull-back of $N_X Y$ to W, V a purely k-dimensional scheme, C the normal cone to W in V, $[C] = \sum m_i [C_i]$ its cycle on N,

$$X \cdot V = \sum m_i \alpha_i$$

the canonical decomposition of the intersection cycle, with

$$\alpha_i = s_N^! [C_i] \in A_{k-d}(Z_i),$$

$Z_i = s_N^{-1}(C_i)$; the Z_i are the distinguished varieties (not necessarily distinct).

Theorem 12.2. *Fix a distinguished variety Z_i, and let N_i denote the restriction of N to Z_i.*

(a) *If N_i is generated by its sections, then α_i is represented by a non-negative cycle on Z_i.*

(b) *If $N_i \otimes L^\vee$ is generated by its sections for some ample line bundle L on Z_i, then*

$$\deg_L(\alpha_i) \geqq \deg_L(Z_i) > 0.$$

(c) *If N_i is ample and generated by its sections, then α_i is represented by a positive cycle on Z_i.*

(d) *If N_i is ample, and L is any ample line bundle on Z_i, then*

$$\deg_L(\alpha_i) > 0.$$

Proof. By the construction of the α_i, these are immediate consequences of Theorem 12.1. □

In particular, if N itself satisfies any of these three positivity assertions, all the terms in the canonical decomposition have the corresponding positivity.

Let X be a non-singular, n-dimensional variety, and let $V_1, \ldots, V_r$ be pure-dimensional subvarieties of X, with

$$m = \sum_{j=1}^{r} \dim(V_i) - (r-1)\, n \geqq 0.$$

Construct the intersection product $V_1 \cdot \ldots \cdot V_r$ from the diagram

$$\begin{array}{ccc} \bigcap V_j & \to & V_1 \times \ldots \times V_r \\ \downarrow & & \downarrow \\ X & \xrightarrow[\delta]{} & X \times \ldots \times X \end{array}$$

with canonical decomposition $V_1 \cdot \ldots \cdot V_r = \sum m_i \alpha_i$, $\alpha_i \in A_m(Z_i)$, Z_i distinguished. Let T_i denote the restriction of the tangent bundle T_X to Z_i.

Corollary 12.2. *Fix a distinguished variety Z_i.*

(a) *If T_i is generated by its sections, then $\alpha_i \in A_m^{\geqq}(Z_i)$.*

(b) *If L is an ample bundle on Z_i such that $T_i \otimes L^\vee$ is generated by its sections, then*

$$\deg_L(\alpha_i) \geqq \deg_L(Z_i) > 0.$$

(c) *If T_i is ample and generated by its sections, then $\alpha_i \in A_m^+(Z_i)$.*

(d) *If T_i is ample, and L is any ample line bundle on Z_i, then*

$$\deg_L(\alpha_i) > 0.$$

Proof. This follows from the theorem, and the fact that the normal bundle to δ is the sum of $(r-1)$ copies of T_X (Appendix B.7.4). □

It should emphasized that, in (b) $\deg_L(\alpha_i)$ denotes the degree of α_i as a cycle of dimension m, while $\deg_L(Z_i)$ is the degree of Z_i as a variety, which may have dimension larger than m. Note also that each irreducible component of $\bigcap V_j$ appears as a distinguished variety, regardless of its dimension.

If T_X (or its restriction to $\bigcap V_j$) is generated by its sections, the intersection class can be represented by a non-negative cycle. If conditions (b), (c), or (d) hold for T_X (or its restriction to $\bigcap V_j$), corresponding positivity holds for each term in the canonical decomposition.

Example 12.2.1. (a) The class of non-singular varieties X for which T_X is generated by its sections includes all projective spaces, Grassmannians, flag manifolds (complete or partial), and abelian varieties. If an algebraic group G acts transitively on a variety X, in characteristic zero, then T_X is generated by its sections; the same holds in positive characteristic provided one assumes that, for $x \in X$, the morphism $f_x: G \to X$, $f_x(g) = g \cdot x$, is smooth. (The derivative of f_x at e maps T_eG onto T_xX. In characterstic zero, f_x is always smooth.)

(b) Let X be a smooth hypersurface of degree d in $\mathbb{P}^{n+1}$, over an algebraically closed field of characteristic not dividing d. If $n \geqq 2$ (resp. $n = 1$), then T_X is generated by its sections if and only if $d \leqq 2$ (resp. $d \leqq 3$). (If F is a form defining X, define the bundle E to be the kernel of the vector bundle homomorphism

$$\mathcal{O}_X(1)^{\oplus(n+2)} \xrightarrow{(\partial F/\partial X_0, \ldots, \partial F/\partial X_{n+1})} \mathcal{O}_X(d).$$

There results an exact sequence $0 \to \mathcal{O}_X \to E \to T_X \to 0$, from which sections of T_X may be computed.)

(c) The property of having tangent bundle generated by its sections is inherited by arbitrary Cartesian products and arbitrary open subschemes of varieties with this property.

(d) If $X = \mathbb{P}^n$, $L = \mathcal{O}(1)$, then T_X is ample, generated by its sections, and $T_X \otimes L^\vee$ is generated by its sections (Appendix B.5.8). By a theorem of Mori (1), $\mathbb{P}^n$ is the only variety with an ample tangent bundle.

For $X \neq \mathbb{P}^n$, the restriction of T_X to certain subvarieties may be ample, in which case positivity can be deduced for corresponding contributions to intersection products.

(e) If $X^n \subset \mathbb{P}^m$ is non-singular, and $T_X \otimes \mathcal{O}(-1)$ is generated by its sections, then X is a linear subspace, or a plane conic. (One may induct on n, taking generic hyperplane sections, or appeal to Mori's theorem.) In particular, T_X is not a direct summand of $T_{\mathbb{P}^m}|_X$ unless X is linear.

Example 12.2.2. Let G be the Grassmannian of n-planes in $\mathbb{P}^N$. To give a morphism $f: X \to G$ is equivalent to giving a subbundle S of rank $n+1$ of the trivial bundle $X \times \mathbb{A}^{N+1}$. If Q denotes the quotient bundle, then (cf. Appendix B.5.8) $f^* T_G = \operatorname{Hom}(S, Q) = S^\vee \otimes Q$. The bundle $f^* T_G$ is generated by its sections. It is ample if and only if for each flag

$$0 \subset L \subset H \subset \mathbb{A}^{N+1}$$

with $\dim L = 1$, $\dim H = N$, the set $\{x \in X \mid L \subset S(x) \subset H\}$ is finite. (If $V = \mathbb{A}^{N+1}$, $P(f^* T_G^\vee) = P(\operatorname{Hom}(Q, S))$ has a canonical map to $P(\operatorname{Hom}(V, V))$; the above set is the largest possible fibre of this map.)

Example 12.2.3. If the imbeddings $X \hookrightarrow Y$, $V \to Y$ can be deformed to $\mathscr{X} \to \mathscr{Y}$, $\mathscr{V} \to \mathscr{Y}$ over a parameter space T as in Chap. 11, such that for generic t, X_t meets V_t properly, then $X \cdot V = \lim_{t \to 0} (X_t \cdot V_t)$ can be represented by a non-negative cycle. For example, this gives another proof that the intersection product preserves non-negativity on a homogeneous space.

Example 12.2.4 (cf. Fulton-Lazarsfeld (4)). Let $i: X \hookrightarrow Y$ be a regular imbedding of codimension d such that the normal bundle $N_X Y$ is ample.

(a) If α is a k-cycle on Y with $k \geqq d$ and $i^*(\alpha) = 0$, then α is not algebraically equivalent to any positive k-cycle whose support meets X. (Use Theorem 12.2(d) and Proposition 10.3(c).)

(b) If an algebraic group acts transitively on Y, then any subvariety of Y of dimension at least d must meet X. Over $\mathbb{C}$, this was proved by Lübke (1).

Example 12.2.5. Assume an algebraic group acts transitively on the variety Y. Let X be a variety, $f: X \to Y$ a morphism such that $f^* T_Y$ is an ample vector bundle. Then for any subvariety V of Y with $\dim X + \dim V \geqq \dim Y$,

$$f^*[V] \neq 0.$$

In particular, $f(X)$ must meet V. (Apply the previous example to the graph imbedding $\gamma_f: X \to X \times Y$ and the subvariety $X \times V$ of $X \times Y$.)

More generally, if f^*T_Y is n-ample (cf. Example 12.1.5) and

$$\dim V \geqq \dim Y - \dim X + n,$$

then $f^*[V] \neq 0$.

Example 12.2.6. Consider a residual intersection situation, with notation as in Theorem 9.2. If the restriction of $N \otimes \mathscr{O}(-D)$ to R is generated by its sections, then the residual intersection class $\mathbb{R}$ is represented by a non-negative cycle on R. Conclusions as in Theorem 2.12(b), (c), (d) may likewise be drawn if $N \otimes \mathscr{O}(-D)$ is suitably ample. (Use Example 9.2.2.)

Example 12.2.7. Let $H_1, \ldots, H_d$ be effective Cartier divisors on a scheme X, and let V be a purely k-dimensional subscheme of X. Let $W = H_1 \cap \ldots \cap H_d \cap V$, and form the intersection class $H_1 \cdot \ldots \cdot H_d \cdot V$ in $A_{k-d}W$ using the diagram

$$\begin{array}{ccc} W & \longrightarrow & V \\ \downarrow & & \downarrow \delta \\ H_1 \times \ldots \times H_d & \rightarrow & X \times \ldots \times X. \end{array}$$

Let $H_1 \cdot \ldots \cdot H_d \cdot V = \sum m_i \alpha_i$ be the canonical decomposition, $\alpha_i \in A_{k-d}(Z_i)$, Z_i distinguished.

(a) If the restriction of each line bundle $\mathscr{O}_X(H_j)$ to Z_i is generated by its sections (resp. and ample) then α_i is in $A^{\geqq}_{k-d}(Z_i)$ (resp. $A^{+}_{k-d}(Z_i)$). If each such restriction is ample, then $\deg_L(\alpha_i) > 0$ for any ample L on Z_i.

(b) Suppose W contains a divisor D of the form $D = H \cap V$, for H an effective divisor on X. Let R be the residual scheme to D in W with respect to V, and let $\mathbb{R} \in A_{k-d}(R)$ be the residual intersection class. If the restriction of each $\mathscr{O}_X(H_j - H)$ to R is generated by its sections (resp. and ample), then $\mathbb{R} \in A^{\geqq}_{*}(R)$ (resp. $A^{+}_{*}(R)$).

If $X = \mathbb{P}^n$, and $\deg H_j \geqq \deg H$ for all j, then $\mathbb{R} \in A^{\geqq}_{k-d}(R)$. If $\deg H_j > \deg H$ for all j, then $\mathbb{R} \in A^{+}_{k-d}(R)$, and

$$\deg(\mathbb{R}) \geqq \sum_{i=1}^{q} \deg(R_i),$$

where $R_1, \ldots, R_q$ are the irreducible components of R.

Example 12.2.8. Given $V_1, \ldots, V_r \subset X$ as in Corollary 12.2, assume T_X is generated by its sections. Let L be a line bundle on X. Then there are irreducible m-dimensional subvarieties $W_1, \ldots, W_s$ in $\bigcap_{j=1}^{r} V_j$, and non-negative integers $n_1, \ldots, n_s$, so that

$$\sum_{i=1}^{s} n_i \deg_L(W_i) = \prod_{j=1}^{r} \deg_L(V_j).$$

(Let $\sum n_i[W_i]$ be a non-negative representative for $V_1 \cdot \ldots \cdot V_r$.) For $X = \mathbb{P}^n$, $L = \mathscr{O}(1)$, this implies that $\bigcap V_j$ must contain m-dimensional varieties of small degree.

Example 12.2.9. In the situation at the beginning of § 12.2, let Z be an irreducible component of W, and let

$$\alpha(Z) = \sum_{Z_i = Z} m_i \alpha_i$$

be the contribution of Z to $X \cdot V$. Assume that $N_Z \otimes L^\vee$ is generated by its sections, for an ample line bundle L on Z. Then

$$\text{(i)} \qquad \deg_L \alpha(Z) \geqq (e_W V)_Z \cdot \deg_L(Z),$$

where $(e_W V)_Z$ is the multiplicity of V along W at Z (Example 4.3.4). If $\mathcal{O}_{Z,V}$ is Cohen-Macaulay, then

$$\text{(ii)} \qquad \deg_L \alpha(Z) \geqq l(\mathcal{O}_{Z,W}) \cdot \deg_L(Z).$$

(Let $\beta = \sum m_i [P(C_i)]$, the sum over those components C_i whose support is Z. We may assume $\dim(Z) = k - r$, $0 < r \leqq d$. If Q is the universal quotient bundle on $P(N_Z)$, and p the projection to Z, then

$$\alpha(Z) = p_*(c_{d-1}(Q) \cap \beta).$$

By the formula for Chern classes of a tensor product (Example 3.2.2) and the fact that $Q \otimes p^* L^\vee$ is generated by its sections,

$$\begin{aligned} \deg_L \alpha(Z) &= \sum_{i=0}^{d-1} \int c_1(p^* L)^{k-d+i} c_{d-1-i}(Q \otimes p^* L^\vee) \cap \beta \\ &\geqq \int c_1(p^* L)^{k-r} c_{r-1}(Q \otimes p^* L^\vee) \cap \beta \\ &= \int c_1(p^* L)^{k-r} c_{r-1}(Q) \cap \beta \\ &= (e_W V)_Z \cdot \deg_L(Z), \end{aligned}$$

cf. Fulton-Lazarsfeld (4). Inequality (ii) follows from (i) and Example 4.3.5 (c).

Example 12.2.10. Let $X \hookrightarrow Y$ be a regular imbedding of projective varieties, such that $N_X Y$ is ample. Let $f: Y \to S$ be a surjective morphism onto a variety S. If $f(X) \neq S$, then $\dim(f(X)) = \dim(X)$. (Following f by a projection, one may replace S by $\mathbb{P}^m$. Consider linear $L \subset \mathbb{P}^m$ with $\dim(L) + \dim(f(X)) = m - 1$. If $\dim(f(X)) < \dim(X)$, then $\dim(f^{-1}(L)) \geqq \dim(Y) - \dim(X)$ for all L. But some $f^{-1}(L)$ are disjoint from X, and others meet X, which leads to a contradiction of Theorem 12.2 (d).)

Example 12.2.11. There is a strengthening of (c) and (d) of Theorem 12.1 and 12.2:

(i) Let L be an ample line bundle on X, E a vector bundle of rank r on X such that $S^m E \otimes L^\vee$ is generated by its sections for some $m > 0$. Let C be an irreducible cone in E, with $Z = \text{Supp}(C)$. Let $k = \dim(C)$, $l = \dim(Z)$, and assume that $k \geqq r$. Then

$$\deg_L(s_E^*[C]) \geqq m^{k-r-l} \cdot e(C) \cdot \deg_L(Z),$$

where $e(C)$ is the multiplicity of C along its zero section.

(ii) In the situation of Theorem 12.2, if $S^m N_i \otimes L^\vee$ is generated by its sections, with L ample on Z_i, then

$$\deg_L(\alpha_i) \geqq m^{-\dim(Z_i)} \cdot \deg_L(Z_i).$$

One therefore has the analogous sharpening of Corollary 12.2. Note that the condition that $S^m E \otimes L^\vee$ be generated by its sections for some $m > 0$ and some ample L is equivalent to the ampleness of E, so this result implies (d). For a sketch of the proof, see Fulton-Lazarsfeld (4); when $m = 1$, the argument of Example 12.2.9 suffices.

12.3 Refined Bézout Theorem

On $\mathbb{P}^n$ there is an exact sequence (cf. Appendix B.5.8)

$$0 \to \mathcal{O} \to \mathcal{O}(1)^{\oplus(n+1)} \to T_{\mathbb{P}^n} \to 0.$$

Thus $T_{\mathbb{P}^n}$ is ample, generated by its sections, and $T_{\mathbb{P}^n} \otimes \mathcal{O}(-1)$ is generated by its sections, so all the results of the preceding section are valid for arbitrary intersections on $\mathbb{P}^n$.

Theorem 12.3. *Let $V_1, \ldots, V_r$ be pure dimensional subschemes of $\mathbb{P}^n$, with*

$$m = \sum_{j=1}^{r} \dim(V_j) - (r-1)n \geqq 0.$$

Let $Z_1, \ldots, Z_t$ be the distinguished varieties of the intersection,

$$V_1 \cdot \ldots \cdot V_r = \sum_{i=1}^{t} m_i \alpha_i$$

the canonical decomposition, $\alpha_i \in A_m(Z_i)$. Then α_i is represented by a positive cycle on Z_i, and

$$\deg(\alpha_i) \geqq \deg(Z_i).$$

In particular,

$$\prod_{j=1}^{r} \deg(V_i) = \sum_{i=1}^{t} m_i \deg(\alpha_i) \geqq \sum_{i=1}^{t} m_i \deg(Z_i) > 0. \quad \square$$

Example 12.3.1. Let $W_1, \ldots, W_s$ be the irreducible components of $\bigcap V_j$. Since each W_i is distinguished,

$$\prod_{j=1}^{r} \deg(V_j) \geqq \sum_{i=1}^{s} m_i \deg(W_i) \geqq \sum_{i=1}^{r} \deg(W_i),$$

as we saw in Example 8.4.6. (To recover this inequality also in case $m < 0$, imbed $\mathbb{P}^n$ linearly in $\mathbb{P}^N$, $N = n - m$, and intersect the cones over the V_j with vertex L^{N-n-1} disjoint from $\mathbb{P}^n$.)

Example 12.3.2. (a) Suppose certain of the distinguished varieties $Z_1, \ldots, Z_q$ of the intersection of $V_1, \ldots, V_r$ on $\mathbb{P}^n$ are known, together with the corresponding m_i and $\deg(\alpha_i)$. If

$$\sum_{i=1}^{q} m_i \deg(\alpha_i) = \prod_{j=1}^{r} \deg(V_j),$$

then $Z_1, \ldots, Z_q$ are the only distinguished varieties. In particular, $\bigcap_{j=1}^{r} V_j = Z_1 \bigcup \ldots \bigcup Z_q$.

(b) Suppose $H_1, \ldots, H_n$ are hypersurfaces in $\mathbb{P}^n$, $d_i = \deg(H_i)$, and a smooth curve Z of degree d and genus g is s scheme-theoretic connected component of $\bigcap_{j=1}^{n} H_j$. Suppose that

$$\left(\sum_{j=1}^{n} d_j - (n+1)\right) d + 2 - 2g = \prod_{j=1}^{n} d_j .$$

Then $Z = \bigcap_{j=1}^{n} H_j$. (Use Example 9.1.1 (ii).)

(c) The canonical imbedding of a non-hyperelliptic curve of genus 5 is a curve of degree 8 in $\mathbb{P}^4$. If the curve is not trigonal, an argument using Riemann-Roch shows there are three quadrics whose intersection contains the curve as a component. It follows that the curve is the complete intersection of the three quadrics (cf. Griffiths-Harris (1) p. 535).

Example 12.3.3. Consider an intersection of r hypersurfaces $H_1, \ldots, H_r$ in $\mathbb{P}^n$, $d_i = \deg H_i$, with a purely k-dimensional subscheme V of $\mathbb{P}^n$. Let $Z_1, \ldots, Z_t$ be the distinguished varieties for the intersection product $H_1 \cdot \ldots \cdot H_r \cdot V$ constructed as in Example 12.2.7. Let $H_1 \cdot \ldots \cdot H_r \cdot V = \sum_{i=1}^{t} m_i \alpha_i$ be the canonical decomposition. If $\dim Z_i = k - r + e_i$, and $d = \min(d_1, \ldots, d_r)$, then

$$\deg(\alpha_i) \geqq d^{e_i} \deg(Z_i).$$

(Apply Theorem 12.2. (b) with $L = \mathcal{O}(d)$, noting that $\deg_L(\beta) = d^m \deg \beta$ for an m-cycle β.)

Example 12.3.4. Suppose a non-singular curve Z is an irreducible component of the intersection of three surfaces H_1, H_2, H_3 in $\mathbb{P}^3$. Let $d_i = \deg H_i$, $d = \deg Z$, $g = \text{genus}(Z)$. Even if Z is a transversal intersection of two of the surfaces at generic points of Z, the postulated contribution of Z to the intersection $H_1 \cdot H_2 \cdot H_3$, i.e.,

$$\left(\sum d_i - 4\right) d + 2 - 2g = \int \prod_{j=1}^{3} (1 + H_j) \cap s(Z, \mathbb{P}^3)$$

may be *greater* than the actual equivalence of Z to the intersection.

For example, let $H_1 = V(x_2 x_3)$, $H_2 = V(x_1 x_3)$, $H_3 = V(x_1 x_2)$. The postulated contribution of each of the three lines in the intersection is 4, but the total intersection number is only 8. In fact, the actual contribution of each line is 2; the point of intersection of the lines is also distinguished, and contributes

2 to the total intersection. (By the previous example, the contribution of each line must be at least 2. By symmetry they must all be the same. By positivity of all contributions, they cannot be more than 2.)

Example 12.3.5. The bound $\deg(\alpha_i) \geqq \deg(Z_i)$ is seldom sharp when Z_i has excess dimension, i.e., $\dim(Z_i) > m$ (cf. Example 12.3.3). It *is* possible for all inrreducible components $W_1, \ldots, W_s$ of V_j to have dimension $m + e$, $e > 0$, and

$$\prod_{j=1}^{r} \deg V_j = \sum_{i=1}^{s} \deg(W_i).$$

(Thus all Z_i are irreducible components, $m_i = 1$, and $\deg \alpha_i = \deg(Z_i)$.) This happens if there is a linear space $\mathbb{P}^{n-e}$ containing all the varieties V_i, and the V_i meet generically transversally as subvarieties of $\mathbb{P}^{n-e}$. However, this is essentially the only way this can happen. If no pair of the varieties $V_1, \ldots, V_r$ are contained in a hyperplane and $W_1, \ldots, W_s$ are the irreducible components of $\bigcap_{j=1}^{r} V_j$, Lazarsfeld has shown that

$$(*) \qquad \sum_{i=1}^{s} \deg(W_i) \leqq \prod_{j=1}^{r} \deg(V_j) - e$$

where $e = \max(\dim W_i) - \sum_{j=1}^{r} \dim(V_i) + (r-1)n$. An extreme case, when $V = V_1$ is a variety spanning $\mathbb{P}^n$, and V_2 is a point on V, is the well-known inequality

$$\deg(V) \geqq \operatorname{codim}(V, \mathbb{P}^n) + 1$$

(cf. Example 8.4.6). For $V_1 = V$ a curve, V_2 a line, (*) says that no secant line can meet V in more than $\deg(V) - n + 2$ points.

Example 12.3.6. Assume X, Y, Z are non-singular varieties of $\mathbb{P}^n$, with $Z = X \bigcap Y$ (scheme-theoretically), and

$$\dim Z = \dim X + \dim Y - n + e.$$

Let E be the excess normal bundle for the intersection

$$\begin{array}{ccc} Z & \to & Y \\ \downarrow & & \downarrow \\ X & \to & \mathbb{P}^n. \end{array}$$

For each $z \in Z$, let Λ_z be the span of the (projective) tangent spaces of X and Y at z, so Λ_z is a $(n-e)$-plane in $\mathbb{P}^n$. For $0 \leqq i \leqq e$, and a general $(e-i)$-plane L in $\mathbb{P}^n$, set

$$S_i = \{z \in Z \mid \Lambda_z \text{ meets } L\}.$$

Then S_i has a natural scheme structure, and for general L, $\operatorname{codim}(S_i, Z) = i$, and

$$\text{(i)} \qquad c_i(E \otimes \mathcal{O}(-1)) \cap [Z] = [S_i].$$

(Let $\mathbb{P}^n = P(V)$; L corresponds to a subspace $\tilde{L}$ of V. Consider the diagram

$$\begin{array}{ccc} & & \tilde{L}\otimes\mathcal{O}(1) \hookrightarrow V\otimes\mathcal{O}(1) \\ & & \downarrow \\ 0 \to T_Z \to T_X|_Z \oplus T_Y|_Z & \to & T_{\mathbb{P}^n}|_Z \to E \to 0. \end{array}$$

Then S_i is the locus where the composite $\tilde{L}\otimes\mathcal{O}(1)\to E$ has rank $\leqq e-i$; tensoring by $\mathcal{O}(-1)$, this is the locus where $e-i+1$ sections of $E(-1)$ become dependent (cf. Example 14.4.3).)

(ii)
$$X\cdot Y = \sum_{i=0}^{e} c_1(\mathcal{O}(1))^{e-i}\cap[S_i].$$

$(c_e(E) = c_e(\mathcal{O}(1)\otimes E(-1)) = \sum c_1(\mathcal{O}(1))^{e-i}c_i(E(-1)).)$

(iii)
$$\deg(X\cdot Y) = \deg(Z) + \sum_{i=1}^{e}\deg([S_i]).$$

In particular, $\deg(Z) = \deg(X\cdot Y)$ if and only if Λ_z is a constant $(n-e)$-plane. By Example 12.3.5, this can happen only when X and Y are contained in an $(n-e)$-plane.

In case $X=Y=Z$, the S_i are the polar varieties, and $E = N_X\mathbb{P}^n$; (i) gives the relation between the polar classes $[S_i]$ and the Chern classes of X (cf. Example 14.4.15), while (iii) specializes to the equation

$$(\deg(X))^2 = \sum_{i=0}^{e}\deg[S_i]$$

if $2\dim(X)\geqq n$, $e = n-\dim(X)$.

Example 12.3.7. In the situation of Theorem 12.3, let Z be an irreducible component of $\bigcap V_j$, and let $\alpha(Z) = \sum m_i\alpha_i$, the sum over those α_i such that $Z_i = Z$. Then

(i)
$$\deg\alpha(Z)\geqq e(Z)\cdot\deg(Z),$$

where $e(Z)$ denotes the multiplicity of $V_1\times\ldots\times V_r$ along $\bigcap V_j$ at Z (Example 4.3.4). If $V_1,\ldots,V_r$ are Cohen-Macaulay schemes, then

(ii)
$$\deg\alpha(Z)\geqq l(Z)\cdot\deg(Z),$$

where $l(Z)$ is the length of the local ring of $\bigcap V_j$ along Z. In particular, if $V_1,\ldots,V_r$ are Cohen-Macaulay, then

(iii)
$$\prod_{j=1}^{r}\deg(V_j)\geqq\sum l(Z)\cdot\deg(Z),$$

the sum over the irreducible components Z of $\bigcap V_j$. (Apply Example 12.2.9.) Note that (iii) may fail if one of the V_j is not Cohen-Macaulay, even when the intersection is proper (Example 7.1.5).

12.4 Intersection Multiplicities

For pure-dimensional subschemes $V_1, \ldots, V_r$ of a non-singular variety X, with $\sum_{i=1}^{r} \dim(V_i) = (r-1)n$, the intersection class $V_1 \cdot \ldots \cdot V_r$ in $A_0(\bigcap V_i)$ is constructed from the diagram

$$\begin{array}{ccc} \bigcap_{i=1}^{r} V_i & \rightarrow & V_1 \times \ldots \times V_r \\ \downarrow & & \downarrow \\ X & \xrightarrow[\delta]{} & X \times \ldots \times X . \end{array}$$

Assume P is an isolated point of $\bigcap_{i=1}^{r} V_i$, which is rational over the ground field. The intersection number $i(P, V_1 \cdot \ldots \cdot V_r; X)$ is the coefficient of $[P]$ in the class $V_1 \cdot \ldots \cdot V_r$. Shrinking X if necessary, we assume the V_i meet only at P.

Let $\pi: \tilde{X} \rightarrow X$ be the blow-up of X at P, E the exceptional divisor, $\eta: E \rightarrow P$ the induced map; $E = P(T_P X)$ is a projective space $\mathbb{P}^{n-1}$ over the field $K = \varkappa(P)$ with $\mathscr{O}(1) = \mathscr{O}(-E)|_E$. Degrees of subvarieties and cycles on E are calculated as usual on $\mathbb{P}^{n-1}$.

Let $\tilde{V}_i \subset \tilde{X}$ be the blow-up of V_i at P. The exceptional divisor is

$$\tilde{V}_i \cap E = P(C_P V_i),$$

where $C_P V_i$ is the tangent cone to V_i at P. Let $e_P V_i$ denote the multiplicity of V_i at P (§ 4.3). Note that $\bigcap_{i=1}^{r} \tilde{V}_i = \bigcap_{i=1}^{r} \tilde{V}_i \cap E = \bigcap_{i=1}^{r} P(C_P V_i)$.

Theorem 12.4. (a) *With the preceding assumptions,*

$$V_1 \cdot \ldots \cdot V_r = \sum_{i=1}^{r} e_P(V_i) \cdot [P] + \eta_*(\tilde{V}_1 \cdot \ldots \cdot \tilde{V}_r)$$

in $A_0(P) = \mathbb{Z}$.

(b) *There are varieties* $W_1, \ldots, W_s$, *whose union is* $\bigcap_{i=1}^{r} P(C_P V_i)$, *and positive 0-cycles* β_j *on* W_j, *and positive integers* m_j, *satisfying:*

(1) $\tilde{V}_1 \cdot \ldots \cdot \tilde{V}_r = \sum_{j=1}^{s} m_j \beta_j$ *in* $A_0(\bigcap \tilde{V}_i)$.

(2) *For any* W_j *which is an isolated point of* $\bigcap P(C_P V_i)$, $\beta_j = [W_j]$ *and* $m_j = i(W_j, \tilde{V}_1 \cdot \ldots \cdot \tilde{V}_r; \tilde{X})$.

(3) *For all* $j = 1, \ldots, s$, $\deg \beta_j \geqq \deg(W_j) > 0$.

Corollary 12.4. *With notation as in the theorem,*

$$\begin{aligned} i(P, V_1 \cdot \ldots \cdot V_r; X) &= \sum_{i=1}^{r} e_P(V_i) + \sum_{j=1}^{s} m_j \deg(\beta_j) \\ &\geqq \sum_{i=1}^{r} e_P(V_i) + \sum_{j=1}^{s} m_j \deg(W_j). \end{aligned}$$

The term $\sum m_j \deg(W_j)$ *is at least as large as the sum of the degrees of the irreducible component of* $\bigcap_{i=1}^{r} P(C_P V_i)$. *In particular,*

$$i(P, V_1 \cdot \ldots \cdot V_r; X) \geqq e_P(V_1) \cdot \ldots \cdot e_P(V_r)$$

with equality if and only if $\bigcap_{i=1}^{r} P(C_P V_i) = \emptyset$. □

Proof of the theorem. (Another proof is sketched in Example 12.4.6.) The proof of (a) uses the deformation to the normal bundle (§ 5). Let M (resp. M_i) be the blow-up of $X \times \mathbb{P}^1$ (resp. $V_i \times \mathbb{P}^1$) along $P \times \infty$, denote the exceptional divisor of this blow-up by X' (resp. V_i'), so $X' = P(T_P X \oplus 1)$, $V_i' = P(C_P V_i \oplus 1)$. We have a diagram of closed imbeddings:

$$\begin{array}{ccccc} & & X' & \supset & V_i' \\ & & {\scriptstyle j}\downarrow & & \downarrow \\ V_i \subset X & \xrightarrow[i]{} & M & \supset & M_i \\ & & {\scriptstyle k}\uparrow & & \uparrow \\ & & \tilde{X} & \supset & \tilde{V}_i \end{array}$$

with i the imbedding at 0, j, k the canonical imbeddings of X' and $\tilde{X}$ in M; i, j, and k are inclusions of non-singular Cartier divisors on M. We have equalities of cycles

$$i^* [M_i] = [V_i], \qquad j^* [M_i] = [V_i'], \qquad k^* [M_i] = [\tilde{V}_i].$$

For $q = 1, \ldots, r$, let i_q be the inclusion of $V_q \bigcap \ldots \bigcap V_r$ in $M_q \bigcap \ldots \bigcap M_r$ induced by i. Using the refined projection formula (Proposition 8.1.1 (c)),

$$\begin{aligned} i_{1*}([V_1] \cdot \ldots \cdot [V_r]) &= i_{1*}(i^* [M_1] \cdot [V_2] \cdot \ldots \cdot [V_r]) \\ &= [M_1] \cdot i_{2*}([V_2] \cdot \ldots \cdot [V_r]) = \ldots = [M_1] \cdot \ldots \cdot [M_{r-1}] \cdot i_{r*}[V_r] \end{aligned}$$

in $A_0(\bigcap M_i)$. Similar, if j_q (resp. k_q) is the inclusion of $V_q' \bigcap \ldots \bigcap V_r'$ (resp. $\tilde{V}_q \bigcap \ldots \bigcap \tilde{V}_r$) in $M_q \bigcap \ldots \bigcap M_r$,

$$\begin{aligned} j_{1*}([V_1'] \cdot \ldots \cdot [V_r']) &= [M_1] \cdot \ldots \cdot [M_{r-1}] \cdot j_{r*}[V_r'], \\ k_{1*}([\tilde{V}_1] \cdot \ldots \cdot [\tilde{V}_r]) &= [M_1] \cdot \ldots \cdot [M_{r-1}] \cdot k_{r*}[\tilde{V}_r]. \end{aligned}$$

A basic identity from the deformation is (cf. Example 5.1.1)

$$i_{r*}[V_r] = j_{r*}[V_r'] + k_{r*}[\tilde{V}_r]$$

in $A_*(M_r)$. Substituting this in the preceding three equations yields

$$i_{1*}([V_1] \cdot \ldots \cdot [V_r]) = j_{1*}([V_1'] \cdot \ldots \cdot [V_r']) + k_{1*}([\tilde{V}_1] \cdot \ldots \cdot [\tilde{V}_r])$$

in $A_0(\bigcap M_i)$. The projections $M \to X \times \mathbb{P}^1 \to X$ induce a proper morphism p from $\bigcap M_i$ to $\bigcap V_i$. The desired formula (a) follows by applying p_* to the last equation. Indeed, $[V_i'] = [P(C_P V_i \oplus 1)]$ has degree $e_P(V_i)$ on $X' = P(T_P X \oplus 1) = \mathbb{P}^n$, by the definition of multiplicity. By Bézout's theorem, $[V_1'] \cdot \ldots \cdot [V_r']$ is represented by a 0-cycle of degree $\sum_{i=1}^{r} e_P(V_i)$ on X'.

We turn now to (b). Unfortunately, the normal bundle to the diagonal imbedding $\tilde{\delta}$ of $\tilde{X}$ in $\tilde{X} \times \ldots \times \tilde{X}$ is not ample (see Example 12.4.7). We find another intersection which yields the same class, but which has an ample normal bundle. For simplicity of notation, assume $r = 2$; if $r > 2$ the same proof is valid, replacing each two-fold Cartesian product by the corresponding r-fold product, e.g. $\tilde{V}_1 \times \tilde{V}_2$ by $\tilde{V}_1 \times \ldots \times \tilde{V}_r$. Consider the diagram

$$\begin{array}{ccccc} & \overset{\mu}{\nearrow} & Y & \leftarrow & W \\ & & \downarrow \lambda & & \downarrow \\ \tilde{X} & \underset{\tilde{\delta}}{\rightarrow} & \tilde{X} \times \tilde{X} & \leftarrow & \tilde{V}_1 \times \tilde{V}_2 \\ \pi \downarrow & & \downarrow \pi \times \pi & & \\ X & \underset{\delta}{\rightarrow} & X \times X. & & \end{array}$$

Here $\lambda : Y \to \tilde{X} \times \tilde{X}$ is the blow-up of $\tilde{X} \times \tilde{X}$ along $E \times E$, and W is the blow-up of $\tilde{V}_1 \times \tilde{V}_2$ along $P(C_P V_1) \times P(C_P V_2)$. Since $\tilde{\delta}^{-1}(E \times E) = E$ is locally principal, $\tilde{\delta}$ factors into $\lambda \circ \mu$, as shown. Shrink X if necessary so X is affine and its tangent bundle, which is the normal bundle N_δ to δ, is trivial.

Since $\tilde{V}_1 \times \tilde{V}_2$ meets $E \times E$ in $P(C_P V_1) \times P(C_P V_1)$, which is regularly imbedded in $\tilde{V}_1 \times \tilde{V}_2$, with the same codimension as the imbedding of $E \times E$ in $\tilde{X} \times \tilde{X}$, $W = \lambda^{-1}(\tilde{V}_1 \times \tilde{V}_2)$; by a simple case of the blow-up formula (cf. Corollary 6.7.2)

(i) $$\lambda^! [\tilde{V}_1 \times \tilde{V}_2] = [W]$$

in $A_*(W)$. Since $\mu^! \lambda^! = \tilde{\delta}^!$ (Theorem 6.5),

(ii) $$\tilde{V}_1 \cdot \tilde{V}_2 = \tilde{\delta}^! [\tilde{V}_1 \times \tilde{V}_2] = \mu^! [W].$$

We need also a formula for the normal bundle N_μ to μ:

(iii) $$N_\mu = \pi^* N_\delta \otimes \mathcal{O}(-E),$$

which will be discussed after completing the proof of (b). Since N_δ is trivial, and $\mu^{-1}(W) \subset E = \mathbb{P}^{n-1}$, and $\mathcal{O}(-E)$ restricts to $\mathcal{O}(1)$ on E, Theorem 12.2(b) applies to the intersection diagram

$$\begin{array}{ccc} \bigcap P(C_P(V_i)) & \rightarrow & W \\ \downarrow & & \downarrow \\ \tilde{X} & \underset{\mu}{\rightarrow} & Y, \end{array}$$

producing a canonical decomposition $\mu^! [W] = \tilde{X} \cdot W = \sum_{j=1}^{s} m_j \beta_j$ and distinguished varieties W_j satisfying (1) and (2) of (b). Although we cannot assert that this is the canonical decomposition obtained by intersecting $\tilde{V}_1 \times \tilde{V}_2$ by the diagonal of $\tilde{X} \times \tilde{X}$, (b) (3) follows from the fact that the two classes are equal in $A_0(\bigcap \tilde{V}_i)$.

It remains to verify (iii). Let F be the exceptional divisor in Y for the map λ. Let $\alpha : Y \to X \times X$ be the composite of λ and $\pi \times \pi$. We claim that μ imbeds $\tilde{X}$

as the residual scheme to F in $\alpha^{-1}(\delta(X))$, i.e., their ideal sheaves are related by

(iv) $$\mathscr{I}(\alpha^{-1}(\delta(X)) = \mathscr{I}(F) \cdot \mathscr{I}(\mu(\tilde{X})).$$

Equation (iii) follows from (iv) by the construction of normal bundles (cf. Example 9.2.2). Identity (iv) may be verified by a straightforward calculation in local coordinates. More conceptually, it may be seen as follows. Let Y_1 be the blow-up of $X \times X$ along $P = P \times P$, with exceptional divisor F_1. There is a commutative diagram of birational morphisms:

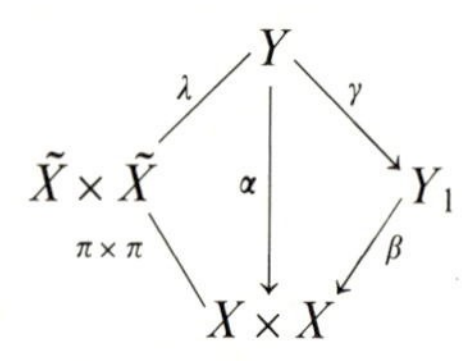

with γ an isomorphism in a neighborhood of $\mu(\tilde{X})$. Identity (iv) therefore pulls back via γ from the analogous identity on Y_1:

$$\mathscr{I}(\beta^{-1}(\delta(X))) = \mathscr{I}(F_1) \cdot \mathscr{I}(\gamma\mu(\tilde{X})).$$

This last formula simply expresses the fact that $\gamma\mu(\tilde{X})$ is the proper transform of X in Y_1 (Appendix B.6.10). □

Remark 12.4. For the intersection of divisors, on a possibly singular variety, a similar theorem is proved more simply in Example 12.4.8.

Example 12.4.1. A particular case of the corollary is the criterion for multiplicity one (Proposition 8.2(c)):

$$i(P, V_1 \cdot \ldots \cdot V_r; X) \geqq 1,$$

with equality if and only if all the V_i are non-singular and meet transversally at P.

Example 12.4.2. If V_1, V_2 are curves on a surface X, the curves $\tilde{V}_1$, $\tilde{V}_2$ must meet properly in $\tilde{X}$, and one may repeat the process. One recovers the formula of M. Noether (2):

$$i(P, V_1 \cdot V_2; X) = \sum e_Q(V_1)\, e_Q(V_2)$$

the sum over all infinitely near points Q of X.

Example 12.4.3. The varieties $\tilde{V}_1, \ldots, \tilde{V}_r$ may not intersect properly, even though $V_1, \ldots, V_r$ do. If $X = \mathbb{A}^4$, $V_1 = V(y^2 - x^3, w)$, $V_2 = V(w^2 - z^3, y)$, then V_1 and V_2 meet properly at the origin P, but $\tilde{V}_1 \cap \tilde{V}_2 = P(C_P V_1) \cap P(C_P V_2)$ is a line l in $P(T_P X) = \mathbb{P}^3$. In fact, $e_P(V_i) = 2$, $i(P, V \cdot W; \mathbb{A}^4) = 9$, and $\tilde{V}_1 \cdot \tilde{V}_2$ is represented by a 0-cycle of degree 5 on l.

Example 12.4.4. The equation (a) of the theorem (and its proof) can be generalized as follows:

Let $V_1, \ldots, V_r$ be pure-dimensional subschemes of a non-singular X^n, with

$$m = \sum_{i=1}^{r} \dim(V_i) - (r-1)n \geqq 0.$$

Let P be a non-singular subvariety of X, $\pi: \tilde{X} \to X$ the blow-up of X along P. Let $W = \bigcap_{i=1}^{r} V_i$, $W' = \pi^{-1}(W)$, and $\eta: W' \to W$ the map induced by π. Let $X' = P(N_P C \oplus 1)$, $X'_W = X' \bigcap W'$, and ϱ the projection from X'_W to W. Let $\tilde{V}_i \subset \tilde{X}$ be the blow-up of V_i along $V_i \bigcap P$, and let $V'_i \subset X'$ be $P(C_{V_i \cap P} V_i \oplus 1)$. Then

$$V_1 \cdot \ldots \cdot V_r = \varrho_* (V'_1 \cdot \ldots \cdot V'_r) + \eta_* (\tilde{V}_1 \cdot \ldots \cdot \tilde{V}_r)$$

in $A_m(W)$. If W_α is an irreducible component of W of dimension m, equating coefficients of $[W_\alpha]$ in this equation gives a formula for the intersection number $i(W_\alpha; V_1 \cdot \ldots \cdot V_r; X)$; if $W_\alpha = P$ has dimension m, then

$$i(P, V_1 \cdot \ldots \cdot V_r; X) = \prod e_P(V_i) + q,$$

where q is the coefficient of $[P]$ in $\eta_* (\tilde{V}_1 \cdot \ldots \cdot \tilde{V}_r)$.

Example 12.4.5 (cf. Samuel (2) II § 6.2b). Let V be a subvariety of Y, with Y affine or projective space over an algebraically closed field, and let W be a subvariety of V. Then

(a) $$e_W(V) = \min i(W, X \cdot V; Y),$$

the minimum taken over all subvarieties X of Y such that W is a proper component of $X \bigcap V$. (Find X an intersection of hypersurfaces so that $e_W(X) = 1$, and $P(C_W V) \bigcap P(C_W X)$ has no irreducible component mapping onto W.) If $W = P$ is a point, one need only consider linear spaces X through P.

(b) $e_W(V) = 1$ if and only if W is not contained in the singular locus of V.

Example 12.4.6. Let $i: X \hookrightarrow Y$ be a regular imbedding of schemes of codimension d, $V \subset Y$ a subvariety of dimension k, and assume $W = X \bigcap V$ has only one irreducible component P, of dimension $k-d$. Assume the imbedding of P in X is a regular imbedding. Let $\tilde{X}$, $\tilde{Y}$, $\tilde{V}$ denote the blow-ups of X, Y, V along P. Let D be the exceptional divisor in V, $\eta: D \to P$ the projection. Then

$$X \cdot_Y V = e_P(V) \cdot [P] + \eta_* (\tilde{X} \cdot_{\tilde{Y}} \tilde{V})$$

in $A_{k-d}(P) = \mathbb{Z}$. (Apply Theorem 9.2 to the residual intersection diagram

$$\begin{array}{ccccc} & & R & & \\ & & \downarrow & & \\ D & \to & f^{-1}(X) & \to & \tilde{V} \\ & & \downarrow & & \downarrow \\ & & X & \to & Y, \end{array}$$

giving $X \cdot_Y V = \{c(N) \cap s(D, \tilde{V})\}_{k-d} + \mathbb{R}$. The first term gives $e_P(V) \cdot [P]$. By Example 9.2.2, or § 17.6, $\mathbb{R} = \tilde{X} \cdot_{\tilde{Y}} \tilde{V}$.)

This general formula implies Theorem 12.4(a). (Take $Y = X \times \ldots \times X$, $V = V_1 \times \ldots \times V_r$. Use Example 4.3.10 to calculate the first term, and argue as in the end of the proof of Theorem 12.4(b) to interpret the second.)

Example 12.4.7. The normal bundle to the diagonal imbedding of $\tilde{X}$ in $\tilde{X} \times \tilde{X}$, i.e., the tangent bundle $T_{\tilde{X}}$, does not restrict to an ample bundle on E. There is an exact sequence

$$0 \to T_E \to T_{\tilde{X}}|_E \to N_E \tilde{X} \to 0$$

and $N_E \tilde{X} = \mathcal{O}(-1)$ on $E = \mathbb{P}^{n-1}$. If C is the normal cone to $\tilde{V}_1 \times \tilde{V}_2$ along $\tilde{V}_1 \bigcap \tilde{V}_2$, then C is a closed subscheme of $T_{\tilde{X}}|_E$. If one knew that the irreducible components C_i of C met T_E properly, one could deduce the positivity of the corresponding contributions from the ampleness of T_E; this would strengthen part (b) of the theorem. On the other hand, if $C_i \subset T_E$, its intersection with the zero section in $T_{\tilde{X}}|_E$ would be strictly negative.

Example 12.4.8. *Intersection multiplicities for divisors.* For effective divisors $D_1, \ldots, D_d$ on an d-dimensional variety X, the intersection class $D_1 \cdot \ldots \cdot D_d$ in $A_0(\bigcap D_i)$ is the class constructed from the diagram

$$\begin{array}{ccc} \bigcap D_i & \to & X \\ \downarrow & & \downarrow \delta \\ D_1 \times \ldots \times D_d & \to & X \times \ldots \times X \end{array}$$

by the procedure of § 6.1. Alternatively, this class may be constructed inductively:

$$D_1 \cdot \ldots \cdot D_d = D_1 \cdot (D_2 \cdot \ldots \cdot D_d),$$

using the simple description of § 2.3 (cf. Example 6.5.1).

Assume P is an isolated point of the intersection of $D_1, \ldots, D_d$ on X. The intersection multiplicity $i(P, D_1 \cdot \ldots \cdot D_d; X)$ is the coefficient of $[P]$ in the class $D_1 \cdot \ldots \cdot D_d$. Shrinking X if necessary, we assume P is the only point in D_i. Let $\pi: \tilde{X} \to X$ be the blow-up of X at P, E the exceptional divisor, $\eta: E \to P$ the map induced by π. Regard E as an algebraic scheme over the residue field $\varkappa(P)$ of P. The bundle $\mathcal{O}(1) = \mathcal{O}(-E)|_E$ is an ample line bundle on E, which we use to compute degrees of subvarieties and cycles on E.

Theorem. Let $\pi^* D_i = n_i E + R_i$ *for some positive integers* n_i, *and some effective divisors* R_i *on* $\tilde{X}$. *Then*

(a) $$D_1 \cdot \ldots \cdot D_d = n_1 \cdot \ldots \cdot n_d \cdot e_P(X) \cdot [P] + \eta_* (R_1 \cdot \ldots \cdot R_d).$$

Let $R_1 \cdot \ldots \cdot R_d = \sum_{i=1}^{t} m_i \alpha_i$ *be the canonical decomposition for the intersection of the* R_i *on* $\tilde{X}$, $\alpha_i \in A_0(Z_i)$, $Z_1, \ldots, Z_t$ *the distinguished varieties. Then, for* $i = 1, \ldots, t$,

(b) $$\deg(\alpha_i) \geqq \deg(Z_i).$$

Corollary. *With notation as in the theorem,*

$$i(P, D_1 \cdot \ldots \cdot D_d; X) = n_1 \cdot \ldots \cdot n_d \cdot e_P(X) + \sum_{i=1}^{t} m_i \deg(\alpha_i)$$

$$\geqq n_1 \cdot \ldots \cdot n_d \cdot e_P(X) + \sum_{i=1}^{t} m_i \deg(Z_i).$$

In particular,

$$i(P, D_1 \cdot \ldots \cdot D_n; X) \geqq n_1 \cdot \ldots \cdot n_d \cdot e_P(X)$$

with equality if and only if $\bigcap R_i = \emptyset$.

(For the proof of (a) write $R_i = \pi^* D_i - n_i E$, expand $R_1 \cdot \ldots \cdot R_d$, and use the projection formula to calculate $\eta_*(R_1 \cdot \ldots \cdot R_d)$, noting that $\eta_*(E^d) = (-1)^{d-1} e_P(X) \cdot [P]$ by § 4.3; (b) follows from Theorem 12.2(b) and the ampleness of the restriction of $\mathcal{O}(R_i)$ to E. We refer to Fulton-Lazarsfeld (4) for details.)

The theorem also generalizes to the proper intersection of d divisors on a variety of dimension $> d$, much as in Example 12.4.4.

The last inequality of Corollary 12.4 may be deduced from the preceding Corollary, by writing the diagonal as an intersection of divisors, in a neighborhood of P.

Example 12.4.9. In the situation of the preceding example, let A be the local ring of X at P, m its maximal ideal, and let f_i be a local equation for D_i at P. A natural choice for the integers n_i are the largest integers such that $f_i \in m^{n_i}$. In this case the inequality

$$(*) \qquad i(P, D_1 \cdot \ldots \cdot D_d; X) \geqq n_1 \cdot \ldots \cdot n_d \cdot e_P(X)$$

may also be deduced algebraically from the formula of Lech (1), stated in the Notes and References to Chap. 7. Let $\mathrm{Gr}(A) = \oplus\, m^n/m^{n+1}$, and let $\bar{f}_i \in m^{n_i}/m^{n_i+1}$ be the residue class of f_i. Then equality holds in (*) precisely when $V(\bar{f}_1, \ldots, \bar{f}_d) = \emptyset$ in $\mathrm{Proj}(\mathrm{Gr}(A))$. If $\mathrm{Gr}(A)$ is Cohen-Macaulay – e.g. if A is regular – the latter is equivalent to $\bar{f}_1, \ldots, \bar{f}_d$ being a regular sequence. Note that the left side of (*) is the multiplicity $e_A(f_1, \ldots, f_d)$ (Example 7.1.2).

When P is singular on X, however, larger n_i may sometimes satisfy the conditions of the theorem. On the other hand, the n_i need not be chosen maximally, in which case the R_i may contain the exceptional divisor E. Note that for general divisors containing E, the degree of the intersection class may well be negative.

Example 12.4.10. Let X, V be non-singular subvarieties of a non-singular variety Y, meeting properly at a point P. Suppose that $T_P V \subset T_P X$. Then

$$i(P, X \cdot V; Y) \geqq 2^d,$$

$d = \dim(V)$. (Write X as an intersection of d divisors near P. These divisors meet V in divisors D_i which are singular at P. Then

$$i(P, X \cdot V; Y) = i(P, D_1 \cdot \ldots \cdot D_d; V) \geqq \prod_{i=1}^{d} e_P(D_i).)$$

Notes and References

The results of this chapter represent joint work with R. Lazarsfeld, cf. Fulton-Lazarsfeld (3), (4). The refined Bézout theorem of § 12.3 developed from our work with R. MacPherson. In addition, Examples 12.1.9, 12.1.10, 12.2.2, 12.2.10, 12.3.5 and 12.3.6 were proposed by Lazarsfeld, and the proof of Theorem 12.1 (c) is his. The inequality (iii) of Example 12.3.7 was proved by Patil and Vogel (1) by other methods.

A special case of the corollaries in § 12.4 and Example 12.4.8 appears frequently in the literature. Let $H_1, \ldots, H_n$ be hypersurfaces in $\mathbb{A}^n$ meeting properly at $P = 0$, and let m_i be the multiplicity of H_i at P. Then

$$i(P, H_1 \cdot \ldots \cdot H_n; A^n) \geqq m_1 \cdot \ldots \cdot m_n,$$

with equality if and only if the leading forms of the H_i have no common non-trivial solutions. Indeed, nearly every proposed definition of intersection multiplicity has been tested by this inequality. For $n = 3$, see Berzolari (1), and for the general case: Zariski (2), Perron (1), B. Segre (7), Kirby (1) and Northcott (1), among others. For varieties of larger codimension, the inequality was considered by Severi (9), Samuel (2) II.6.2, and Griffiths and Harris (1) p. 393. Using Serre's definition of intersection multiplicity, Tennison (1) showed that equality holds if the projective tangent cones do not intersect (cf. Example 20.4.3). Teissier (1), (2) has studied the inequality for divisors on a singular variety.

For the intersection of two plane curves, M. Noether (2) gave the formula for the difference between the intersection number and the product of the multiplicities as the sum of the intersection numbers in the first infinitesimal neighborhood (cf. Example 12.4.2). In higher dimensions, the proper transforms can have excess intersection; in this case we have not found classical precedents for the inequalities of § 12.4.

The definition of ample vector bundle used in § 12.1 is that of Hartshorne (1), cf. Griffiths (1). Special cases of the positivity theorem of Example 12.1.7 were proved by Kleiman, Bloch and Gieseker, Griffiths, Usui and Tango, among others; see Griffiths (1), (3), and Fulton-Lazarsfeld (3) for references.

Chapter 13. Rationality

Summary

Refined intersection products can be used to prove the existence of rational solutions of algebraic equations, either in the given ground field K, or in extensions of restricted degrees.

Suppose $V_1, \ldots, V_r$ are subvarieties of a complete nonsingular variety X, with $\sum_{j=1}^{r} \operatorname{codim}(V_i, X) = \dim(X)$. By our construction, the intersection cycle $V_1 \cdot \ldots \cdot V_r$ is represented by a 0-cycle on $\bigcap_{j=1}^{r} V_j$. Therefore there are points $P_1, \ldots, P_t$ in $\bigcap_{j=1}^{r} V_j$, and integers $n_1, \ldots, n_t$ such that

$$(*) \qquad \sum_{i=1}^{t} n_i [K(P_i) : K] = \deg([V_1] \cdot \ldots \cdot [V_r]) .$$

For example, if $K = \mathbb{R}$, and the right side is odd, $\bigcap_{j=1}^{r} V_j$ must contain real points. If some part of the intersection class is known, similar conclusions are valid for the rest of it. Each isolated point P of $\bigcap_{j=1}^{r} V_j$ appears in (*), with coefficient the intersection multiplicity of the V_j at P. With suitable positivity assumptions on the tangent bundle of X, the coefficients n_i can all be taken to be non-negative, even when the intersections are improper.

Notation. If W is a complete scheme, i.e. W is proper over the ground field K, and $\alpha = \sum n_P[P]$ is a 0-cycle on W, the degree of α is the sum

$$\deg(\alpha) = \sum n_P [K(P) : K]$$

where $K(P)$ is the residue field of $\mathcal{O}_{P,W}$ at P, and $[K(P) : K]$ is the degree of the field extension. Rationally equivalent 0-cycles have the same degree (§ 1.4).

For example, if $K = \mathbb{R}$ is the field or real numbers (or an arbitrary real closed field), then $K(P) = \mathbb{R}$ if P is a real point of W, $K(P) = \mathbb{C}$ if P is complex, and

$$\deg(\alpha) = \sum_{P \text{real}} n_P + 2 \sum_{P \text{complex}} n_P .$$

In particular, if $\deg(\alpha)$ is odd, W must contain real points.

Let $i: X \to Y$ be a regular imbedding of codimension d, V a pure d-dimensional subscheme of Y, $W = X \bigcap V$. Let N be the restriction of $N_X Y$ to W, $C = C_W V$ the normal cone to W in V, $[C] = \sum m_i [C_i]$ its cycle on N. Recall that a subvariety of W is distinguished if it is the support of an irreducible component C_i of C. For each distinguished variety Z set

$$\alpha(Z) = \sum_{\mathrm{supp}(C_i)=Z} m_i s_N^! [C_i]$$

in $A_0(Z)$; here $s_N: W \to N$ is the zero section. Thus

$$X \cdot V = \sum_Z \alpha(Z)$$

is the canonical decomposition of $X \cdot V$. If W is complete,

$$\deg(X \cdot V) = \sum_Z \deg(\alpha(Z)).$$

Thus one may deduce the existence of points on the distinguished varieties from the knowledge of $\deg(X \cdot V)$. The results of Chaps. 7 and 12 can be used to restrict the possibilities for the $\alpha(Z)$.

We apply this to the intersection of r pure-dimensional subschemes $V_1, \ldots, V_r$ of a smooth variety X over K, forming the intersection product $V_1 \cdot \ldots \cdot V_r$ by intersecting the Cartesian product of the V_j by the diagonal:

$$\begin{array}{ccc} \bigcap V_j & \hookrightarrow & V_1 \times \ldots \times V_r \\ \downarrow & & \downarrow \\ X & \underset{\delta}{\hookrightarrow} & X \times \ldots \times V \end{array}.$$

Proposition 13. *Assume* $\sum_{j=1}^{r} \dim V_j = (r-1)\, n$, $n = \dim X$, *and* $\bigcap V_j$ *is complete. Let*

$$V_1 \cdot \ldots \cdot V_r = \sum \alpha(Z)$$

be the canonical decomposition of the intersection product, summed over the distinguished varieties $Z \subset W$, $\alpha(Z) \in A_0(Z)$.

(a) *If* Z *is a point which is isolated in* $\bigcap V_j$, *then*

$$\alpha(Z) = i(Z, V_1 \cdot \ldots \cdot V_r; X)\,[Z].$$

(b) *If the restriction of the tangent bundle* T_X *to* Z *is generated by its section, then* α_Z *is represented by a non-negative cycle.*

Proof. Note that isolated points in $\bigcap V_i$ are proper components of the intersection. (a) was proved more generally in § 7.1. (b) is a special case of Corollary 12.2. □

The following corollary suffices for many applications.

Corollary 13.1. *Suppose* $P_1, \ldots, P_s$ *are distinct points which are isolated in* $\bigcap V_i$, *and*

$$m_k = i(P_k, V_1 \cdot \ldots \cdot V_r; X)$$

is the intersection multiplicity at P_k.

(a) *Among the other components of* $\bigcap V_j$, *there are points* $Q_1, \ldots, Q_t$ *and integers* $n_1, \ldots, n_t$ *so that*

$$(*) \qquad \sum_{i=1}^{t} n_i[K(Q_i):K] = \deg(V_1 \cdot \ldots \cdot V_r) - \sum_{k=1}^{s} m_k[K(P_k):K].$$

(b) *If* T_X *is generated by its sections, one may find such* Q_i, n_i, *with* $n_i > 0$. □

The field K is called a *p-field*, for a prime number p, if every finite extension of K has degree over K which is a power of p. For example, the field of real numbers, or any real closed field, is a 2-field.

Corollary 13.2. *Suppose* $P_1, \ldots, P_s$ *are distinct K-rational points which are isolated points of* $\bigcap V_j$, *and K is a p-field. If*

$$\sum_{k=1}^{s} i(P_k, V_1 \cdot \ldots \cdot V_r; X) \not\equiv \deg(V_1 \cdot \ldots \cdot V_r) \pmod{p}$$

then $\bigcap V_j$ *contains additional K-rational points.*

In particular, $\bigcap V_j$ must contain K-rational points whenever $\deg(V_1 \cdot \ldots \cdot V_r)$ is not divisible by p. Corollary 13.2 follows from Corollary 13.1 (a) since, when (*) is taken mod p, all points which are not K-rational are ignored. □.

Remark 13.1. Similar conclusions may be drawn in other contexts where classes are constructed on loci of interest, for example, on residual schemes (cf. Example 13.11), double point loci (§ 9.3), or degeneracy loci (§ 14).

Remark 13.2. A reader more comfortable with geometry over algebraically closed fields may work directly with the variety $X_{\bar{K}}$ over an algebraic closure $\bar{K}$ of K. The point then is that if subvarieties $\bar{V}_i$ of $X_{\bar{K}}$ are defined over K, the intersection classes and the contributions $\alpha(Z)$ of distinguished varieties are represented by zero-cycles which are rational over K. (A 0-cycle $\sum n_P[P]$ on $X_{\bar{K}}$ is rational over K if each conjugate of P over K occurs with the same coefficient as P; in addition, if K is not a perfect field, each coefficient must be divisible by the corresponding degree of inseparability.)

Example 13.1. Let K be a p-field.

(a) If W is a subscheme of $\mathbb{P}^n$ which contains a k-cycle α whose degree is not divisible by p, then W has a K-rational point. (If $k = 0$ this is obvious. If $k > 0$, $c_1(\mathcal{O}(1))^k \cap \alpha$ is represented by a zero-cycle with the same degree.)

(b) If $V_1, \ldots, V_r$ are pure-dimensional subschemes of $\mathbb{P}^n$, with $\sum \operatorname{codim}(V_j) \leq n$, and $\prod \deg(V_j)$ is not divisible by p, then $\bigcap V_j$ has a K-rational point. (Apply (a) to a representative for $V_1 \cdot \ldots \cdot V_r$.)

(c) (Pfister (1)). If $H_1, \ldots, H_r$ are homogeneous polynomials in $K[X_0, \ldots, X_r]$ of degrees prime to p, and $r \leq n$, then there is a non-zero solution $x = (x_0, \ldots, x_n)$ to the system of equations

$$H_1(x) = \ldots = H_r(x) = 0$$

with all $x_i \in K$.

In particular, (cf. Behrend (1), Lang (1)) if $K = \mathbb{R}$, or K is any real closed field, r form of odd degrees in more than r variables must have common non-zero real solutions.

Example 13.2. Assume K is a p-field.

(a) If W is a subscheme of $\mathbb{P}^m \times \mathbb{P}^n$ which contains a cycle α with some bidegree not divisible by p, then W has a K-rational point. (If $\alpha \sim \sum n_{ij} s^i t^j$, with the notation of Example 8.4.2, then $s^{m-a} t^{n-b} \alpha$ is represented by a zero cycle on W of degree n_{ab}.)

(b) If $V_1, \ldots, V_r$ are pure-dimensional subschemes of $\mathbb{P}^m \times \mathbb{P}^n$, and some bidegree of $[V_1] \cdot \ldots \cdot [V_r]$ is not divisible by p, then $\bigcap V_j$ has a K-rational point.

The analogous assertions are valid on arbitrary multiprojective spaces.

Example 13.3 (Behrend (1)). Let $K = \mathbb{R}$, or a real closed field. Let $H_1, \ldots, H_r$ be bihomogeneous polynomials in $K[X_1, \ldots, X_m, Y_1, \ldots, Y_n]$ of odd bidegrees. If the binomial coefficient $\binom{r}{k}$ is odd for any k with $r - n < k < m$, then there are real solutions (x, y) to the system

$$H_1(x, y) = \ldots = H_r(x, y) = 0$$

with $x \neq 0$, $y \neq 0$. In particular, if $m = n = r$, and r is not a power of 2, then there are always non-trivial solutions. (With coefficients mod 2, the intersection class of the corresponding hypersurfaces in $\mathbb{P}^{m-1} \times \mathbb{P}^{n-1}$ is congruent to

$$(s+t)^r = \sum \binom{r}{k} s^k t^{r-k},$$

the sum over $k < m$, $r - k < n$. If $m = n = r$, and $(s+t)^r \equiv s^r + t^r \bmod p$, then r is a power of p.)

Example 13.4. Let K be a p-field. Assume there is a bilinear mapping

$$\varphi: U \times V \to W$$

of finite dimensional vector spaces U, V, W over K, with no zero-divisors, i.e., $\varphi(x, y) = 0$ implies $x = 0$ or $y = 0$. If m, n, r are the dimensions of U, V, W, then p divides $\binom{r}{k}$ for all $r - n < k < m$. In particular, if the three dimensions are equal, they must be a power of p.

It follows that, if K is a real closed field, a bilinear form $K^n \times K^n \to K^n$ must have zero divisors unless n is a power of 2. (Cf. Behrend (1); for $K = \mathbb{R}$, earlier topological proofs had been given by Stiefel and Whitney.) If $K = \mathbb{R}$, in fact, n must be 1, 2, 4, or 8; the only proof known at present uses topology, cf. Milnor (1). (Taking bases, φ is given by r polynomials of bidegree (1, 1). Argue as in Example 13.3.)

Example 13.5. An arbitrary polynomial f has a unique expression $f = \sum f^{(m)}$, with $f^{(m)}$ homogeneous of degree m. The non-zero term of minimal degree is called the *initial form* of f, its degree the *initial degree;* the non-zero term of maximal degree ($= \deg(f)$) may be called the *final form* of f.

Let $f_1, \ldots, f_n$ be polynomials in $K[X_1, \ldots, X_n]$, not all forms. Assume that the n initial forms of these polynomials have no common non-zero solutions in an algebraic closure $\bar{K}$ of K, and likewise for the n final forms. Then the system of equations

$$f_1(x) = \ldots = f_n(x) = 0$$

has non-zero solutions $x_\alpha = (x_{\alpha 1}, \ldots, x_{\alpha n}) \in \bar{K}^n$, with

$$\sum n_\alpha [K(x_\alpha) : K] = \prod_{i=1}^{n} d_i - \prod_{i=1}^{n} c_i,$$

where $d_i = \deg(f_i)$, c_i is the initial degree of f_i, and the n_α are positive integers. (For a polynomial $f = \sum f^{(m)}$ of degree d, let $F = \sum X_0^{d-m} f^{(m)}$ be the homogenized form of f, a form of degree d in $K[X_0, \ldots, X_n]$. The n hypersurfaces defined by $F_i = 0$ have no common points on the hyperplane $X_0 = 0$. The intersection multiplicity of the hypersurfaces at $(1:0:\ldots:0)$ is $\prod_{i=1}^{n} c_i$ by Theorem 12.5. Apply Corollary 13.1.)

For example, if K is real, and $\prod d_i - \prod c_i$ is odd, there must be zero-real solutions.

Example 13.6. It is possible for n real polynomials in n variables to have a finite number of real solutions, whose number is larger than the product of the degrees. Let $n = 3$, $m \geqq 3$, and

$$f_1 = \sum_{i=1}^{m} (X_1 - i)^2 + \sum_{i=1}^{m} (X_2 - i)^2, \quad f_2 = f_3 = X_3.$$

Then there are m^2 real solutions, while $\prod \deg(f_i) = 2m$. Or one may set $f_2 = X_1 X_3$, $f_3 = X_2 X_3$, to obtain a similar example with f_1, f_2, f_3 algebraically independent. In any such example, some of the given real solutions must lie on a positive-dimensional variety of complex solutions.

Are there such examples with all f_i of the same degree? There are upper bounds on the sum of the Betti numbers of such varieties (cf. Milnor (2)); the related question for the sum of the Betti numbers is apparently still open.

Example 13.7. Let X be a complete n-dimensional variety, $D_1, \ldots, D_n$ Cartier divisors on X. Then there are points $P_1, \ldots, P_t$ in the intersection of the supports of the divisors, and integers $n_1, \ldots, n_t$, such that

$$\sum n_i [K(P_i) : K] = \int_X D_1 \cdot \ldots \cdot D_n.$$

Example 13.8. Let $H_1, \ldots, H_r$ be real hypersurfaces in $\mathbb{P}^r$, $\deg H_i = n_i$. Suppose a non-singular curve Z of degree d is a scheme-theoretic component of $\bigcap H_i$. Then $\bigcap H_i$ must contain real points outside Z, provided

$$\prod_{i=1}^{r} n_i \not\equiv \left(\sum_{i=1}^{r} n_i - r - 1 \right) d \pmod 2.$$

(Use Example 9.1.1.)

Example 13.9. *Algebraic Borsuk-Ulam* (Arason-Pfister (1), cf. Knebusch (2))

(a) Let $g_1, \ldots, g_n \in R[X_1, \ldots, X_{n+1}]$ be odd polynomials, i.e. $g_i(-x) = -g_i(x)$, with R a real closed field. Let

$$S^n = \{x = (x_1, \ldots, x_{n+1}) \mid x_1^2 + \ldots + x_{n+1}^2 = 1\}.$$

Then there is a point $x \in S^n$ with $g_i(x) = 0$, $i = 1, \ldots, n$. (For an odd polynomial q, write $q = \sum q^{(j)}$, with $q^{(j)}$ a homogeneous polynomial of odd degree j. If $\deg(q) = d$, define a form $\bar{q}$ by

$$\bar{q}(X_1, \ldots, X_{n+1}) = \sum (X_1^2 + \ldots + X_{n+1}^2)^{(d-j)/2} \cdot q^{(j)}.$$

By Example 13.1 (c), the forms $\bar{q}_1, \ldots, \bar{q}_n$ must have a non-trivial common solution $(a_1, \ldots, a_{n+1})$. Set $x_i = a_i/a$, where $a^2 = \sum a_i^2$.)

(b) For any $f_1, \ldots, f_n \in R[X_1, \ldots, X_{n+1}]$, there are points $x \in S^n$ with $f_i(x) = f_i(-x)$ for $i = 1, \ldots, n$. (Apply (a) to the odd parts of the f_i.)

(c) If $R = \mathbb{R}$, the topological Borsuk-Ulam theorem follows. (Approximate given continuous functions on S^n by polynomials, and apply (b).)

Example 13.10. Let X be a non-singular variety over $\mathbb{R}$ such that $X_{\mathbb{C}} \cong \mathbb{P}_{\mathbb{C}}^{2n}$. Then X must have real points. (Since

$$\int_X c_1(T_X)^{2n} = (2n+1)^n$$

is odd, X has odd 0-cycles.) Therefore X is isomorphic to $\mathbb{P}^{2n}$ (cf. Serre (6) p. 168).

Example 13.11. Consider a residual intersection situation as in Definition 9.2.2, assuming also $k = d$ and W is complete. Then there are points P_i in the residual set R, and integers n_i so that

$$\sum n_i [K(P_i) : K] = \deg(X \cdot V) - \int_Z c(N) \cap s(Z, V).$$

Example 13.12. Let L be a finite extension of K. For a scheme X over K, the base extension

$$X_L = X \otimes_K L = L \times_{\operatorname{Spec}(K)} \operatorname{Spec}(L)$$

is a scheme over L. There is a canonical morphism $\pi: X_L \to X$ (of K-schemes) which is finite and flat of degree $[L:K]$. The composite

$$A_*X \xrightarrow{\pi^*} A_*(X_L) \xrightarrow{\pi_*} A_*X$$

is multiplication by $[L:K]$ (cf. Example 1.7.4).

Example 13.13 (Colliot-Thélène and Ischebeck (1)). Let X be a complete $\mathbb{R}$-scheme, $X_{\mathbb{C}}$ its complexification. Assume that the space $X(\mathbb{R})$ of real points of X has s connected components.

(a) There is an exact sequence

$$A_0(X_{\mathbb{C}}) \xrightarrow{\pi_*} A_0(X) \xrightarrow{\beta} (\mathbb{Z}/2\mathbb{Z})^s \to 0$$

where β takes a 0-cycle $\sum n_P [P]$ to a vector whose components are the sums, mod 2, of the coefficients of the points P on the corresponding components of $X(\mathbb{R})$. (It suffices to verify this for X a smooth curve over $\mathbb{R}$, in which case the exactness is a theorem of Witt, cf. Knebusch (1) p. 70.)

(b) There are 0-cycles of degree $s-2$ which are not rationally equivalent to any positive 0-cycle on X. (Choose P_i in the ith component of $X(\mathbb{R})$. Then, by (a),

$$P_1 + \ldots + P_{s-1} - P_s$$

is such a zero cycle.)

(c) If $s \geqq 1$, $\pi_*(\tilde{A}_0(X_{\mathbb{C}})) = 2\tilde{A}_0(X)$ is the largest divisible subgroup of $\tilde{A}_0(X)$. (Use Examples 13.12 and 1.6.6.)

(d) If any two points of $X_{\mathbb{C}}$ can be joined by a chain of rational curves (e.g. if $X_{\mathbb{C}}$ is a unirational variety), then

$$\tilde{A}_0(X) = (\mathbb{Z}/2\mathbb{Z})^{s-1}.$$

Two points of $X(\mathbb{R})$ are rationally equivalent if and only if they belong to the same connected component of $X(R)$. (Use Example 10.1.6.)

Using results of M. Knebusch and H. Delfs, Colliot-Thélène and Ischebeck (1) prove analogous results for arbitrary real closed fields.

Notes and References

When all intersections are isolated (proper), results like those in this chapter are routine, once one has an intersection product that preserves rationality over a given ground field, as in Weil (2). When sufficient moving is available, one may deduce the existence of solutions in general from the proper case. Several cases of this appear in the literature, when the ambient space is a product of projective spaces:

(i) If $K = \mathbb{R}$, one may use compactness arguments (cf. Shafarevich (1) IV § 2.2).

(ii) If K is real closed, Behrend (1) used Hensel's lemma and specialization arguments. See also Lang (1).

(iii) If K is a p-field, Pfister (1) used valuations and specialization.

Even when such moving is available the present intersection theory gives a simple, direct approach. For more general intersections, the results here are new.

The applications to p-fields were suggested by J.-L. Colliot-Thélène. S. Kleiman urged working directly over the ground field, rather than descending from an algebraic closure as in Remark 13.2. Applications to real forms (cf. Example 13.10) were suggested by J. Harris and A. Landman. Example 13.6 came from discussions with R. Lazarsfeld and D. Eisenbud. Generalizations of Example 13.9 were given by Terjanian (1).

Chapter 14. Degeneracy Loci and Grassmannians

Summary

Let $\sigma: E \to F$ be a homomorphism of vector bundles of ranks e and f on a variety X, and let $k \leqq \min(e, f)$. The degeneracy locus

$$D_k(\sigma) = \{x \in X \mid \operatorname{rank}(\sigma(x)) \leqq k\}$$

has codimension at most $(e-k)(f-k)$ in X, if it is not empty. We construct a class

$$\mathbb{D}_k(\sigma) \in A_m(D_k(\sigma)),$$

$m = \dim(X) - (e-k)(f-k)$, whose image in $A_m(X)$ is given by the Thom-Porteous formula:

$$\mathbb{D}_k(\sigma) = \Delta_{f-k}^{(e-k)}(c(F-E)) \cap [X].$$

Here $\Delta_q^{(p)}(c)$ denotes the determinant of the p by p matrix $(c_{q+j-i})_{1 \leqq i,j \leqq p}$. If $\dim(D_k(\sigma)) = m$, and X is non-singular, or, more generally, if a suitable depth condition is satisfied, then $\mathbb{D}_k(\sigma)$ is the m-cycle determined by the natural scheme structure on $D_k(\sigma)$. In general the formation of $\mathbb{D}_k(\sigma)$ commutes with other intersection operations. These properties determine $\mathbb{D}_k(\sigma)$ in case $\dim(D_k(\sigma)) > m$.

If $A_1 \subset \ldots \subset A_d \subset E$ is a flag of sub-bundles of E, the *determinantal locus* is

$$\Omega(\underline{A}; \sigma) = \{x \in X \mid \dim(\operatorname{Ker}(\sigma(x)) \bigcap A_i(x)) \geqq i \quad \text{for} \quad 1 \leqq i \leqq d\}.$$

Similarly, there are classes $\mathbf{\Omega}(\underline{A}; \sigma)$ in $A_*(\Omega(\underline{A}; \sigma))$, whose images in $A_*(X)$ are given by certain determinants in Chern classes. If $c(E/A_i) = 1$, the formula is

$$\mathbf{\Omega}(\underline{A}; \sigma) = \Delta_\lambda(c(F-E)) \cap [X],$$

where $\lambda = (\lambda_1, \ldots, \lambda_d)$, $\lambda_i = f - \operatorname{rank}(A_i) + i$, and Δ_λ is the *Schur polynomial*

$$\Delta_\lambda(c) = \det \begin{pmatrix} c_{\lambda_1} & c_{\lambda_1+1} & \cdots & c_{\lambda_1+d-1} \\ c_{\lambda_2-1} & c_{\lambda_2} & \cdots & c_{\lambda_2+d-2} \\ c_{\lambda_d-d+1} & & & c_{\lambda_d} \end{pmatrix}.$$

A special case of degeneracy locus is the zero set of a section of a vector bundle. In this case the degeneracy class localizes the top Chern class of the bundle (§ 14.1). The construction of general degeneracy loci is reduced to the case of sections of bundles on Grassmannians; proving the formulas requires some Gysin computations (§ 14.2).

Formal identities among Schur polynomials determine formulas for intersecting determinantal classes. When applied to Grassmann bundles, these formulas yield generalizations of classical formulas of Schubert calculus: the basis theorem, duality, Pieri's formula, and Giambelli's formula.

Notation. The fibre of a vector bundle E over a scheme X at a point $x \in X$ is denoted $E(x)$; it is a vector space over $\varkappa(x)$. If $\sigma: E \to F$ is a vector bundle homomorphism, $\Lambda^k\sigma$ denotes the induced homomorphism on k^{th} exterior powers.

If $\sigma: E \to F$ is a homomorphism of vector bundles on a scheme X, the *zero scheme* of σ will be denoted $Z(\sigma)$. On an affine open set U where E and F are trivial, σ is defined by a matrix of elements in the coordinate ring of U, which generate the ideal of $Z(\sigma)$ on U. In particular, if s is a section of a bundle E, i.e., a homomorphism from the trivial line bundle to E, its zero scheme is denoted $Z(s)$ (cf. Appendix B.3.2).

More generally, for a non-negative integer k, we have the kth *degeneracy locus*

$$D_k(\sigma) = \{x \in X \mid \operatorname{rank}(\sigma(x)) \leqq k\} = Z(\Lambda^{k+1}(\sigma)) .$$

The second description determines the scheme structure on $D_k(\sigma)$: locally its ideal is generated by $(k+1)$-minors of a matrix for σ.

Let $\underline{A}$ be a flag of subbundles of E:

$$0 \subsetneqq A_1 \subsetneqq \ldots \subsetneqq A_d \subset E .$$

Given $\sigma: E \to F$, set

$$\begin{aligned} \Omega(\underline{A}; \sigma) &= \{x \in X \mid \dim(\operatorname{Ker}(\sigma(x)) \cap A_i(x)) \geqq i, \;\; 1 \leqq i \leqq d\} \\ &= \bigcap_{i=1}^{d} Z(\Lambda^{a_i - i+1}(\sigma_i)) . \end{aligned}$$

Here $\sigma_i: A_i \to F$ is the restriction of σ to A_i, and a_i is the rank of A_i. The second description determines the scheme structure.

For vector bundles E, F on X, set

$$c(F-E) = c(F)/c(E) = 1 + (c_1(F) - c_1(E)) + \ldots$$

and let $c_i(F-E)$ be the term of degree i in this expansion.

If $\lambda_1, \ldots, \lambda_d$ are integers, and $c^{(1)}, \ldots, c^{(d)}$ are formal sums:

$$c^{(i)} = \sum_j c_j^{(i)}$$

with $c_j^{(i)}$ of degree j, $j \in \mathbb{Z}$, set

$$\Delta_{\lambda_1, \ldots, \lambda_d}(c^{(1)}, \ldots, c^{(d)}) = \det \begin{pmatrix} c_{\lambda_1}^{(1)} & c_{\lambda_1+1}^{(1)} & \cdots & c_{\lambda_1+d-1}^{(1)} \\ & c_{\lambda_2}^{(2)} & & \\ & & \ddots & \\ c_{\lambda_d-d+1}^{(d)} & & & c_{\lambda_d}^{(d)} \end{pmatrix} .$$

If $c^{(1)} = \ldots = c^{(d)} = c$, we denote this $\Delta_{\lambda_1, \ldots, \lambda_d}(c)$, or $\Delta_\lambda(c)$, i.e. $\Delta_\lambda(c) = |c_{\lambda_i+j-i}|$. If, in addition, $\lambda_1 = \ldots = \lambda_d = e$, this is $\Delta_e^{(d)}(c)$.

14.1 Localized Top Chern Class

Let E be a vector bundle of rank e on a purely n-dimensional scheme X, and let $s: X \to E$ be a section, with zero scheme $Z(s)$. We construct a class

$$\mathbb{Z}(s) \in A_{n-e}(Z(s)),$$

called the *localized top Chern class* of E with respect to s, whose image in $A_{n-e}(X)$ is $c_e(E) \cap [X]$. In case s is a regular section, i.e. the e functions locally defining s form a regular sequence, $\mathbb{Z}(s)$ will be the cycle $[Z(s)]$ determined by the natural scheme structure on $Z(s)$.

Consider the fibre square

$$\begin{array}{ccc} Z(s) & \rightarrow & X \\ {\scriptstyle i}\downarrow & & \downarrow{\scriptstyle s} \\ X & \underset{s_E}{\rightarrow} & E \end{array}$$

where s_E is the zero section, i the inclusion. Since s_E is a regular imbedding of codimension e, the construction of § 6.1 gives a refined intersection product $s_E^!([X])$ in $A_{n-e}(Z(s))$. Define

$$\mathbb{Z}(s) = s_E^!([X]).$$

Proposition 14.1. (a) $i_*(\mathbb{Z}(s)) = c_e(E) \cap [X]$.

(b) *Each irreducible component of* $Z(s)$ *has codimension at most* e *in* X. *If* $\operatorname{codim}(Z(s), X) = e$, *then* $\mathbb{Z}(s)$ *is a positive cycle whose support is* $Z(s)$.

(c) *If* s *is a regular section, then*

$$\mathbb{Z}(s) = [Z(s)].$$

(d) *Let* $f: X' \to X$ *be a morphism,* $s' = f^* s$ *the induced section of* $f^* E$, g *the induced morphism from* $Z(s')$ *to* $Z(s)$.

(i) *If* f *is flat, then* $g^* \mathbb{Z}(s) = \mathbb{Z}(s')$.

(ii) *If* f *is a local complete intersection morphism, then* $f^! \mathbb{Z}(s) = \mathbb{Z}(s')$.

(ii) *If* f *is proper, and* X' *and* X *are varieties, then* $g_* \mathbb{Z}(s') = \deg(X'/X) \cdot \mathbb{Z}(s)$.

Proof. (a) follows from the self-intersection formula:

$$i_* s_E^![X] = s_E^* s_*[X] = s^* s_*[X] = c_e(E) \cap [X]$$

by Theorem 6.2(a), Corollary 6.5, and Corollary 6.3, respectively. (b) and (c) are special cases of Proposition 7.1(a) and (b). In the situation of (d) there is a fibre diagram

$$\begin{array}{ccc} Z(s') & \xrightarrow{i'} & X' \\ {\scriptstyle i'}\downarrow & & \downarrow{\scriptstyle s'} \\ X' & \underset{s_{f^*E}}{\longrightarrow} & f^* E \\ {\scriptstyle f}\downarrow & & \downarrow \\ X & \underset{s_E}{\longrightarrow} & E, \end{array}$$

from which it follows that $\mathbb{Z}(s') = s_E^![X']$ (Theorem 6.2(c)). (d) therefore follows from the corresponding properties of intersection products: Theorem 6.2(b), Theorem 6.4 and Theorem 6.6, and Theorem 6.2(a) respectively. □

Example 14.1.1. If X is Cohen-Macaulay, and $\dim Z(s) = n-e$, then s is regular, and $\mathbb{Z}(s) = [Z(s)]$ (Proposition 7.1).

Example 14.1.2. $\mathbb{Z}(s)$ can also be defined to be $s^![X]$, this time using the zero section to imbed X in E (Theorem 6.4).

Example 14.1.3. Let $E = E_1 \oplus E_2$, $s = s_1 \oplus s_2$, s_i a section of E_i, $e_i = \operatorname{rank} E_i$. Then the image of $\mathbb{Z}(s)$ in $A_{n-e}(Z(s_2))$ is

$$c_{e_1}(E_1) \cap \mathbb{Z}(s_2)\,.$$

(See Example 17.4.8 for a generalization.)

Example 14.1.4. *Residual formula for top Chern classes.* Let s be a section of a rank e vector bundle E on a purely n-dimensional scheme X. Assume that $Z(s)$ contains D, an effective Cartier divisor on X. Then there is a section s' of $E \otimes \mathscr{O}(-D)$ such that the canonical homomorphism from $E \otimes \mathscr{O}(-D)$ to E takes s' to s. (Locally the functions defining s are divisible by an equation for D, and the quotients define s'). In addition, $Z(s')$ is the residual scheme to D in $Z(s)$, and

$$\mathbb{Z}(s) = \mathbb{Z}(s') + \sum_{i=1}^{e} (-1)^{i-1} c_{e-i}(E) \cap D^{i-1} \cdot [D]$$

in $A_{n-e}(Z(s))$. In particular, if s' is a regular section, then $\mathbb{Z}(s') = [Z(s')]$, which gives an explicit formula for $\mathbb{Z}(s)$. (Apply Theorem 9.2 in the situation

$$\begin{array}{ccccc} & & Z(s') & & \\ & & \downarrow & & \\ D & \to & Z(s) & \to & X \\ & & \downarrow & & \downarrow s \\ & & X & \underset{s_E}{\to} & E\,. \end{array}$$

That $\mathbb{Z}(s')$ is the residual class follows from Example 9.2.2.)

Example 14.1.5. *Euler characteristics and Milnor numbers.* Let X be a non-singular, n-dimensional variety over an algebraically closed field K, $f: X \to C$ a proper morphism onto a non-singular curve. The induced mapping on tangent bundles

$$df: T_X \to f^* T_C$$

determines a section s_f of $f^* T_C \otimes T_X^\vee$. Assume f is smooth except over a finite set $\{t_1, \ldots, t_m\}$ in C. The localized top Chern class $\mathbb{Z}(s_f)$ is a 0-cycle class on the singularity locus of f.

(a) Assume C is complete. Then

$$\deg \mathbb{Z}(s_f) = (-1)^n (\chi(X) - \chi(C)\, \chi(X_t))\,,$$

where X_t is a general fibre of f, and χ denotes the Euler characteristic (degree of top Chern class of tangent bundle). (One has

$$\deg \mathbb{Z}(s_f) = \int_X c_n(f^* T_C \otimes T_X^\vee) = \int_X (-1)^n c_n(T_X) + \int_X (-1)^{n-1} (c_1(f^* T_C)\, c_{n-1}(T_X)\,.$$

Since $\int\limits_X f^*[t] \cdot c_{n-1}(T_X) = \int\limits_{X_t} c_{n-1}(T_X|_{X_t}) = \int\limits_{X_t} c_{n-1}(T_{X_t})$ for general $t \in C$, the formula follows.

(b) In the complex case ($K = \mathbb{C}$), this formula may be rewritten:

$$\deg \mathbb{Z}(s_f) = (-1)^n \sum_{i=1}^{m} (\chi(X_{t_i}) - \chi(X_t)),$$

i.e. $\mathbb{Z}(s_f)$ measures the change in Euler characteristic contributed by the singularities of f. (Choose small disks D_i surrounding the t_i, and apply Mayer-Vietoris to the covering of X by $f^{-1}(D_1), \ldots, f^{-1}(D_m)$, and $f^{-1}\left(C - \bigcup\limits_{i=1}^{m} \bar{D}_i\right)$; cf. Beauville (2) p. 95.)

(c) In the complex case, the part of $\mathbb{Z}(s_f)$ supported on X_{t_i} has degree $(-1)^n(\chi(X_{t_i}) - \chi(X_t))$. More generally, if C is replaced by a neighborhood of t_i so that T_C is trivial, $\mathbb{Z}(s_f)$ is a localized top Chern class for the cotangent bundle of X, whose degree is $(-1)^n(\chi(X_{t_i}) - \chi(X_t))$. (One may prove this by using outward pointing vector fields, as in MacPherson (1).) Note that if C is not complete, the global Chern classes may vanish, while the local Chern classes still carry useful information.

(d) If $x \in X$ is an isolated critical point of f, the equivalence of $\mathbb{Z}(s_f)$ in $A_0(\{x\})$ is $\mu_x(f) \cdot [x]$,

$$\mu_x(f) = \dim_K(\mathscr{O}_{x,X}/(\partial f/\partial z_1, \ldots, \partial f/\partial z_n))$$

where the partial derivatives are defined in terms of local coordinates $z_1, \ldots, z_n$ at x, and a local coordinate for Y at $f(x)$. The number $\mu_x(f)$ is the *Milnor number*, cf. Milnor (3). For example, $\mu_x(f) = 1$ precisely when the Hessian $(\partial^2 f/\partial z_i \partial z_j)$ is non-singular at x. For a general discussion of Milnor numbers, see Orlik (1).

(e) (cf. Iversen (2)). More generally, if the singularity locus of f contains an effective divisor D, let s_f' be the section of $f^* T_C \otimes T_X^\vee \otimes \mathscr{O}(-D)$ induced by s_f (Example 14.1.4). Then

$$\mathbb{Z}(s) = \mathbb{Z}(s_f') + (-1)^{n-1} \sum_{i=1}^{n} D^i \cdot c_{n-i}(T_X).$$

(This follows from Example 14.1.4 and the fact that $f^*[t] \cdot D = 0$ for $t \in C$.) Therefore

$$\deg \mathbb{Z}(s_f') = (-1)^n \left(\chi(X) - \chi(C) \cdot \chi(X_t) + \sum_{i=1}^{n} \int\limits_X D^i \cdot c_{n-i}(T_X)\right).$$

If x is an isolated point of $Z(s_f')$, Iversen calls x a *moderate* critical point. In this case the equivalence of $\mathbb{Z}(s_f')$ at x is

$$\dim_K \mathscr{O}_{x,X}/(g_1, \ldots, g_n)$$

with $g_i = (\partial f/\partial z_i)/h$, $\partial f/\partial z_i$ as in (d), and h the greatest common divisor of the partial derivatives. Iversen (2) gives explicit formulas for these numbers when X is a surface, in characteristic zero.

14.2 Gysin Formulas

There are many formulas for pushing forward classes from Grassmann or flag bundles. In this section we prove two such formulas, for use later in the chapter.

Let $\underline{A}$ be a flag of vector bundles on X, i.e, a chain

$$0 \subsetneqq A_1 \subsetneqq A_2 \subsetneqq \ldots \subsetneqq A_d = A$$

of subbundles. Let $\varphi: Fl(\underline{A}) \to X$ be the associated flag bundle, whose fibre over $x \in X$ consists of flags of subspaces of $A(x)$:

$$D_1(x) \subset \ldots \subset D_d(x)$$

with $\dim D_i(x) = i$ and $D_i(x) \subset A_i(x)$. On $Fl(\underline{A})$ there is a universal flag $D_1 \subset \ldots \subset D_d$, with rank $D_i = i$ and $D_i \subset \varphi^* A_i$. An inductive construction of $Fl(\underline{A})$ will be given in the proof of the following proposition. Let $a_i = \operatorname{rank}(A_i)$.

Proposition 14.2.1. *Let $M_1, \ldots, M_d$ be vector bundles on X. Let $i_1, \ldots, i_d$ be integers. Then for all $\alpha \in A_* X$,*

$$\begin{aligned} &\varphi_*(c_{i_1}(\varphi^* M_1 - D_1) \cdot \ldots \cdot c_{i_d}(\varphi^* M_d - D_d) \cap \varphi^* \alpha) \\ &= c_{j_1}(M_1 - A_1) \cdot \ldots \cdot c_{j_d}(M_d - A_d) \cap \alpha\,, \end{aligned}$$

with $j_p = i_p - a_p + p$, $1 \leqq p \leqq d$.

Proof. The inductive construction of $Fl(\underline{A})$ is as follows. If $d=1$, $Fl(\underline{A}) = P(A_1)$, with $D_1 = \mathcal{O}_{A_1}(-1)$ the universal line bundle. Assume

$$F' = Fl(A_1 \subset \ldots \subset A_{d-1})$$

has been constructed, with projection $\varphi': F' \to X$, and universal flag $D_1' \subset \ldots \subset D_{d-1}'$. Then $F = Fl(\underline{A})$ may be constructed as a projective bundle over F':

$$F = P(\varphi'^* A_d / D_{d-1}') \xrightarrow{\varrho} F'\,.$$

Let $\varphi = \varphi' \circ \varrho$, $D_i = \varrho^* D_i'$ for $i < d$, and determine D_d so D_d/D_{d-1} is the universal sub-bundle of $\varphi^* A_d / D_{d-1}$. We claim that

$$\text{(i)} \qquad \varrho_*(c_j(\varphi^* M_d - D_d) \cap \varrho^* \beta) = c_{j-a_d+d}(\varphi'^* M_d - \varphi'^* A_d) \cap \beta$$

for $\beta \in A_* F'$. From the exact sequence defining D_d, and the Whitney sum formula, one has

$$c(\varphi^* M_d - D_d) = c(\varphi^* M_d - \varphi^* A_d) \cdot c(Q)\,,$$

where Q is the universal quotient bundle of rank $a_d - d$ on the projective bundle F over F'. By the projection formula,

$$\varrho_*(c_j(\varphi^* M_d - D_d) \cap \varrho^* \beta) = \sum_{i=0}^{j} c_{j-i}(\varphi'^* M_d - \varphi'^* A_d) \cdot \varrho_*(c_i(Q) \cap \varrho^* \beta)\,.$$

Since $\varrho_*(c_i Q \cap \varrho^* \beta)$ is 0 if $i \neq a_d - d$, and it equals β if $i = a_d - d$ (Example 3.3.3), (i) follows. The general formula then follows from (i) by induction on d. $\square$

Corollary 14.2. *If $\lambda_1, \ldots, \lambda_d$ are integers, and $\alpha \in A_* X$, then*

$$\varphi_* (\Delta_{\lambda_1, \ldots, \lambda_d}(c(\varphi^* M_1 - D_1), \ldots, c(\varphi^* M_d - D_d)) \cap \varphi^* \alpha)$$
$$= \Delta_{\mu_1, \ldots, \mu_d}(c(M_1 - A_1), \ldots, c(M_d - A_d)) \cap \alpha$$

where $\mu_i = \lambda_i - a_i + i$.

Proof. Apply the proposition to each monomial in the expansion for the determinants. □

Let E be a vector bundle of rank n on X, $d \leqq n$, $G_d(E)$ the Grassmann bundle of d-planes in E, π the projection formula from $G_d(E)$ to X. Let S be the universal subbundle of $\pi^* E$, of rank d. Let $k = n - d$.

Proposition 14.2.2. *Let F be a vector bundle of rank f on X. Then for all $\alpha \in A_* X$,*

$$\pi_* (c_{df}(S^\vee \otimes \pi^* F) \cap \pi^* \alpha) = \Delta_{f-k}^{(d)}(c(F - E)) \cap \alpha .$$

Proof. (See Example 14.2.1 for another proof.) By the splitting principle (Remark 3.2.2) it suffices to prove this when E has a flag $\underline{A}$ of subbundles:

$$A_1 \subset \ldots \subset A_d = E$$

with $\operatorname{rank} A_i = k + i$. As usual we may assume that $\alpha = [X]$, X a variety. There is a canonical morphism

$$\mu : Fl(\underline{A}) \to G_d(E)$$

with $\mu^* S = D_d$. Over the open set consisting of d-dimensional spaces L in fibres $E(x)$ such that $L \bigcap (A_i(x))$ has dimension i for all i, μ is an isomorphism. The composite $\varphi = \pi \circ \mu$ is the projection to X. Since μ is birational,

$$c_{df}(S^\vee \otimes \pi^* F) \cap [G_d(E)] = \mu_* (c_{df}(D_d^\vee \otimes \varphi^* F) \cap [Fl(\underline{A})] .$$

Therefore

$$\pi_* (c_{df}(S^\vee \otimes \pi^* F) \cap \pi^* \alpha) = \varphi_* (c_{df}(D_d^\vee \otimes \varphi^* F) \cap \varphi^* \alpha) .$$

If $x_i = c_1(D_i/D_{i-1})$, and $y_1, \ldots, y_f$ are Chern roots for $\varphi^* F$, then

$$c_{df}(D_d^\vee \otimes \varphi^* F) = \prod_{i=1}^{d} \prod_{j=1}^{f} (y_j - x_i) .$$

It follows from Lemma A.9.1 (ii) that

$$c_{df}(D_d^\vee \otimes \varphi^* F) = \Delta_f^{(d)}(c(\varphi^* F - D_1), \ldots, c(\varphi^* F - D_d)) .$$

By Corollary 14.2, since $a_p - p = k$ for all p,

$$\varphi_* (\Delta_f^{(d)}(c(\varphi^* F - D_1), \ldots, c(\varphi^* F - D_d)) \cap \varphi^* \alpha)$$
$$= \Delta_{f-k}^{(d)}(c(F - A_1), \ldots, c(F - A_d)) \cap \alpha .$$

By Lemma A.9.1 (i),

$$\Delta_{f-k}^{(d)}(c(F - A_1), \ldots, c(F - A_d)) = \Delta_{f-k}^{(d)}(c(F - E), \ldots, c(F - E)) ,$$

which concludes the proof. □

Example 14.2.1. (a) If Q is the universal quotient bundle on $G_d(E)$, and $\alpha \in A_*(X)$, then

$$\pi_*(c_{i_1}(Q) \cdot \ldots \cdot c_{i_d}(Q) \cap \pi^* \alpha) = \begin{cases} \alpha & \text{if } i_1 = \ldots = i_d = k \\ 0 & \text{otherwise.} \end{cases}$$

(If $i > k$, $c_i Q = 0$. If $\sum i_j < dk$, the left side vanishes for dimensional reasons. To prove that $\pi_*(c_k(Q)^d \cap \pi^* \alpha) = \alpha$, one may assume E is trivial, X a point, and one may induct on d, cf. Example 3.3.3.)

(b) From (a), one has another proof of Proposition 14.2.2. Indeed, since

$$c_{df}(S^\vee \otimes \pi^* F) = \Delta_f^{(d)}(c(\pi^* F - S)) = \Delta_f^{(d)}(c(\pi^* F - \pi^* E) \cdot c(Q)),$$

one may apply (a) to all the terms appearing in the expansion of this determinant.

Example 14.2.2. Józefiak, Lascoux and Pragacz (1) prove the following generalization of Proposition 14.2.2. For any integers $\lambda_1, \ldots, \lambda_d$,

$$\pi_*(\Delta_{\lambda_1, \ldots, \lambda_d}(c(\pi^* F - S)) \cap \pi^* \alpha) = \Delta_{\mu_1, \ldots, \mu_d}(c(F - E)) \cap \alpha,$$

where $\mu_i = \lambda_i - n + d$.

14.3 Determinantal Formula

Let $\sigma: E \to F$ be a homomorphism of vector bundles of ranks e, f on a purely n-dimensional scheme X. Let $\underline{A}$ be a flag of subbundles of E:

$$0 \subsetneqq A_1 \subsetneqq \ldots \subsetneqq A_d \subset E$$

and let $Fl(\underline{A})$ be the associated flag bundle (§ 14.2), with universal flag $D_1 \subset \ldots \subset D_d$, $\operatorname{rank} D_i = i$, and projection φ from $Fl(\underline{A})$ to X. The induced homomorphism

$$D_d \hookrightarrow \varphi^* E \xrightarrow{\varphi^* \sigma} \varphi^* F$$

determines a section s_σ of $D_d^\vee \otimes \varphi^* F$. Then φ maps the zero set $Z(s_\sigma)$ onto the determinantal locus $\Omega(\underline{A}; \sigma)$. Let

$$\eta: Z(s_\sigma) \to \Omega(\underline{A}; \sigma)$$

be the morphism (of underlying reduced schemes) induced by φ. The construction of § 14.1 gives a localized top Chern class

$$\mathbb{Z}(s_\sigma) \in A_m(Z(s_\sigma)),$$

$m = \dim Fl(\underline{A}) - df$. Define the *determinantal class* $\boldsymbol{\Omega}(\underline{A}; \sigma) \in A_m(\Omega(\underline{A}; \sigma))$ by the formula

$$\boldsymbol{\Omega}(\underline{A}; \sigma) = \eta_*(\mathbb{Z}(s_\sigma)).$$

Theorem 14.3. *Let* $a_i = \operatorname{rank}(A_i)$, $\lambda_i = f - a_i + i$, $h = \sum_{i=1}^{d} \lambda_i$, *so that* $m = n - h$. *Assume that* $f - a_d + d \geqq 0$. *Then*

(a) *The image of* $\mathbf{\Omega}(\underline{A};\sigma)$ *in* $A_{n-h}(X)$ *is*

$$\Delta_{\lambda_1,\ldots,\lambda_d}(c(F-A_1),\ldots,c(F-A_d))\cap[X].$$

(b) *Each irreducible component of* $\Omega(\underline{A};\sigma)$ *has codimension at most* h *in* X. *If* $\operatorname{codim}(\Omega(\underline{A};\sigma);X)=h$, *then* $\mathbf{\Omega}(\underline{A};\sigma)$ *is a positive cycle whose support is* $\Omega(\underline{A};\sigma)$.

(c) *If* $\operatorname{codim}(\Omega(\underline{A};\sigma),X)=h$, *and* X *is Cohen-Macaulay, then* $\Omega(\underline{A};\sigma)$ *is Cohen-Macaulay, and*

$$\mathbf{\Omega}(\underline{A};\sigma)=[\Omega(\underline{A};\sigma)].$$

(d) *The formation of* $\mathbf{\Omega}(\underline{A};\sigma)$ *commutes with Gysin maps and proper push forward.*

Proof. Assertion (d) means that if $f:X'\to X$ is a morphism, E', F', A'_i, σ' the pull-backs to X', g the induced morphism from $\Omega(\underline{A}';\sigma')$ to $\Omega(\underline{A};\sigma)$, then formulas (i), (ii), (iii) of Proposition 14.1 (d) are valid, where $\mathbb{Z}(\sigma)$ (resp. $\mathbb{Z}(s'))$ is replaced by $\mathbf{\Omega}(\underline{A};\sigma)$ (resp. $\mathbf{\Omega}(\underline{A}';\sigma')$). To prove it, form the fibre square

$$\begin{array}{ccc} Fl(\underline{A}') & \xrightarrow{\tilde{f}} & Fl(\underline{A}) \\ \varphi'\downarrow & & \downarrow\varphi \\ X' & \xrightarrow[f]{} & X \end{array}.$$

Then $\tilde{f}^*(s_\sigma)=s_{\sigma'}$, so (d) follows from Proposition 14.1 (d) and the commutativity of pushing forward with Gysin maps (Proposition 1.7, Theorem 6.2 (a)) and the functoriality of push-forward (§ 1.4).

By Proposition 14.1 (a),

$$\mathbb{Z}(s_\sigma)=c_{df}(D_d^\vee\otimes F)\cap[Fl(\underline{A})]$$

in $A_m(Fl(\underline{A}))$. By Lemma A.9.1 (ii),

$$c_{df}(D_d^\vee\otimes F)=\Delta_{f,\ldots,f}(c(\varphi^*F-D_1),\ldots,c(\varphi^*F-D_d)).$$

By Corollary 14.2,

$$\begin{aligned}\varphi_*(\Delta_{f,\ldots,f}&(c(\varphi^*F-D_1),\ldots,c(\varphi^*F-D_d))\cap[Fl(\underline{A})]\\ &=\Delta_{\lambda_1,\ldots,\lambda_d}(c(F-A_1),\ldots,c(F-A_d))\cap[X],\end{aligned}$$

which proves (a).

We next prove (b) and (c) in a "universal local case", namely: $X=\mathbb{A}^{ef}$ is affine space, with coordinate functions (x_{ij}), $1\leqq i\leqq f$, $1\leqq j\leqq e$; E and F are trivial, E with basis $v_1,\ldots,v_e$; A_i is the trivial subbundle of E with basis $v_1,\ldots,v_{a_i}$; σ is given by the matrix (x_{ij}). In this case $\Omega=\Omega(\underline{A};\sigma)$ is defined by the vanishing of all (a_p-p+1)-minors of the matrix consisting of the first a_p columns of (x_{ij}), for $p=1,\ldots,d$. This determinantal scheme is an irreducible variety of codimension h in $\mathbb{A}^{ef}$ and the morphism $\eta:Z(s_\sigma)\to\Omega$ is a birational morphism. Since $Fl(\underline{A})$ is non-singular, and $\operatorname{codim}(Z(s_\sigma),Fl(\underline{A}))=df$, s_σ is a regular section of $D_d^\vee\otimes\varphi^*F$ (Lemma A.7.1). In fact, Eagon and Hochster have proved that Ω is a Cohen-Macaulay variety (cf. Lemma A.7.2).

By Proposition 14.1 (c),

$$\mathbb{\Omega}(\underline{A};\sigma)=\eta_*\mathbb{Z}(s_\sigma)=\eta_*[Z(s_\sigma)]=[\Omega].$$

So (b) and (c) are proved in this case.

To prove (b) and (c) in general, it suffices to prove them locally on X, since the construction of Ω is compatible with restriction to open subsets (by (d)). Thus we may assume there is a morphism $f: X \to \mathbb{A}^{ef}$, such that E, F, A_i and σ are pull-backs from the universal case just considered. Since $\mathbb{A}^{ef}$ is non-singular, f is a l.c.i. morphism. Let $\tilde{\Omega}$ be the universal determinantal subscheme of $\mathbb{A}^{ef}$, and let $\gamma_f: X \to X \times \mathbb{A}^{ef}$ be the graph of f. By (d), and the case just considered,

$$\mathbb{\Omega}(\underline{A};\sigma)=f^![\tilde{\Omega}]=\gamma_f^![X\times\tilde{\Omega}].$$

Since γ_f is a regular imbedding, (b) and (c) follow from Lemma 7.1 and Proposition 7.1. □

Remark 14.3. An alternative construction of the class $\mathbb{\Omega}(\underline{A};\sigma)$ may be given as follows. Let $H = \operatorname{Hom}(E, F)$, p the projection from H to X. On H there is a canonical (tautological) homomorphism u from p^*E to p^*F. Then $\Omega(p^*\underline{A}; u)$ has pure codimension h in H; indeed, H is locally a product of X by the universal case considered in the proof of the theorem. A given homomorphism $\sigma: E \to F$ determines a section $t_\sigma: X \to H$ such that $t_\sigma^* u = \sigma$. Then

$$(*)\qquad \mathbb{\Omega}(\underline{A};\sigma)=t_\sigma^![\Omega(p^*\underline{A};u)].$$

Indeed, (c) implies that $\mathbb{\Omega}(p^*\underline{A};u)=[\Omega(p^*\underline{A};u)]$, and then (*) follows from (d). This argument also proves the following corollary.

Corollary 14.3. *The classes* $\mathbb{\Omega}(\underline{A};\sigma)$ *are uniquely determined by properties* (c) *and* (d) *of the theorem.* □

Example 14.3.1. The general criterion for $\mathbb{\Omega}(\underline{A};\sigma)$ to be the cycle determined by $\Omega(\underline{A};\sigma)$ depends on the notion of *depth*. For a closed subscheme Y of a scheme X, $\operatorname{depth}(Y,X)$ is defined to be the largest integer d such that, for all $x \in Y$, the ideal of Y in the local ring of X at x contains a regular sequence of length d.

Let $\Omega = \Omega(\underline{A};\sigma)$, $\mathbb{\Omega}=\mathbb{\Omega}(\underline{A};\sigma)$. If $\operatorname{codim}(\Omega,X)=h$, and $[\Omega]=\sum m_i[\Omega_i]$ is the cycle of Ω, then $\mathbb{\Omega}=\sum e_i[\Omega_i]$, with $1 \leqq e_i \leqq m_i$ for all i. In general,

$$\operatorname{depth}(\Omega,X)\leqq h,$$

with equality if and only if $\operatorname{codim}(\Omega,X)=h$ and $\mathbb{\Omega}=[\Omega]$. For E a trivial line bundle, and $d=a_1=1$, one recovers Proposition 14.1 (c). For X Cohen-Macaulay, this contains Theorem 14.3 (c) and a converse to it. (The assertions are local, so one may assume X is affine and one is in the situation at the end of the proof of Theorem 14.3:

$$\begin{array}{ccccc}\Omega & \to & X\times\tilde{\Omega} & \to & \tilde{\Omega}\\ \downarrow & & \downarrow & & \downarrow\\ X & \underset{\gamma_f}{\to} & X\times\mathbb{A}^{ef} & \to & \mathbb{A}^{ef}.\end{array}$$

There is a projective resolution $0 \to F_h \to \ldots \to F_1 \to I \to 0$ over the coordinate ring of $\mathbb{A}^{ef}$, where I is the ideal of the Schubert variety $\tilde{\Omega}$. The regular sequence defining X in $X \times \mathbb{A}^{ef}$ remains a regular sequence on $X \times \tilde{\Omega}$ if and only if the pull-back complex $f^*(F.)$ remains a resolution of the ideal of Ω in the coordinate ring of X; each condition is equivalent to the vanishing of higher Tor of the coordinate rings of X and $\tilde{\Omega}$ over the coordinate ring of $\mathbb{A}^{ef}$. These conditions occur when depth$(\Omega, X) = h$ (cf. Kempf-Laksov (1) Corollary 8). The conclusions then follow from Proposition 7.1.)

Example 14.3.2. Let E be a vector bundle of rank r on an n-dimensional variety X. Let $s_1, \ldots, s_N$ be sections of E, $N \geqq 2r$. For partitions $\lambda : \lambda_1 \geqq \ldots \geqq \lambda_r \geqq 0$, set

$$\Omega_\lambda = \{x \in X \mid \dim \operatorname{Span}(s_1(x), \ldots, s_{r+i-\lambda_i}(x)) \leqq r - \lambda_i \text{ for } i = 1, \ldots, r\}.$$

(a) Codim$(\Omega_\lambda, X) \leqq h$, where $h = \sum \lambda_i$, and there is a class $\mathbf{\Omega}_\lambda$ in $A_{n-h}(\Omega_\lambda)$ such that

$$\mathbf{\Omega}_\lambda = \Delta_\lambda(c(E)) \cap [X]$$

in $A_{n-h}(X)$. If $\dim(\Omega_\lambda) = n - h$, and X is Cohen-Macaulay, or depth$(\Omega_\lambda, X) = h$, then $\mathbf{\Omega}_\lambda = [\Omega_\lambda]$, where the scheme structure on Ω_λ is defined by vanishing of appropriate determinants. (Let F be the trivial bundle with basis $e_1, \ldots, e_N$, A_i the subbundle of F spanned by $e_1, \ldots, e_{a_i}$, $a_i = r + i - \lambda_i$. Apply Theorem 14.3 to the mapping from F to E given by $(s_1, \ldots, s_N)$.)

(b) If $\lambda = (p, 0, \ldots, 0)$, then

$$\Omega_\lambda = \{x \in X \mid \dim \operatorname{Span}((s_1(x), \ldots, s_{r+1-p}(x)) \leqq r - p\},$$

and $\mathbf{\Omega}_\lambda$ represents $c_p(E) \cap [X]$.

(c) If $\lambda = (1, \ldots, 1, 0, \ldots, 0)$, with p 1's, then

$$\Omega_\lambda = \{x \in X \mid \dim \operatorname{Span}(s_1(x), \ldots, s_{r+p-1}(x)) \leqq r - 1\},$$

and $\mathbf{\Omega}_\lambda$ represents $(-1)^p s_p(E) \cap [X]$.

(Use Lemma A.9.2 to calculate $\Delta_{1, \ldots, 1}$.)

(d) If E is generated by its sections, and the ground field is infinite, then for generic sections $s_1, \ldots, s_N$ of E, codim$(\Omega_\lambda, X) = h$, and

$$\Delta_\lambda(c(E)) \cap [X] = \mathbf{\Omega}_\lambda = [\Omega_\lambda].$$

(With F as in (a), $H = \operatorname{Hom}(F, E)$, let V be the corresponding degeneracy locus for the canonical homomorphism on H. In light of Remark 14.3, it suffices to apply Example 12.1.11 to the subscheme V of the bundle H.)

Example 14.3.3. *Projective characters.* Let X be a non-singular d-dimensional subvariety of $\mathbb{P}^n$, and let

$$\underline{A} : A_0 \subsetneqq \ldots \subsetneqq A_d \subset \mathbb{P}^n$$

be a flag of subspaces of $\mathbb{P}^n$, with $a_i = \dim A_i$. Set

$$X(\underline{A}) = \{x \in X \mid \dim(T_x(X) \cap A_i) \geqq i,\ 0 \leqq i \leqq d\}.$$

Here T_xX is the tangent d-plane to X in $\mathbb{P}^n$. Let $\lambda_i = n - d - a_i + i$, $|\lambda| = \sum_{i=0}^{d} \lambda_i$. For general $\underline{A}$, $X(\underline{A})$ is a subscheme of X of codimension $|\lambda|$, and

(1) $$[X(\underline{A})] = \Delta_\lambda(c^{(0)}, \ldots, c^{(d)})$$

where $c^{(i)} = (1+h)^{n-a_i}/c(T_X)$, $h = c_1(\mathscr{O}_X(1))$. (Let $\mathbb{P}^n = P(V)$, $A_i = P(\tilde{A}_i)$, and consider the diagram

$$\begin{array}{l} \tilde{A}_0 \otimes \mathscr{O}_X(1) \subset \ldots \subset \tilde{A}_d \otimes \mathscr{O}_X(1) \subset V \otimes \mathscr{O}_X(1) \\ \qquad\qquad\qquad\qquad\qquad\qquad\qquad \downarrow \quad \searrow^{\sigma} \\ \qquad\qquad\qquad 0 \to T_X \to T_{\mathbb{P}^n}|_X \to N_X\mathbb{P}^n \to 0 . \end{array}$$

Then $X(\underline{A}) = \Omega(\underline{\tilde{A}} \otimes \mathscr{O}_X(1); \sigma)$.)

The degree of $X(\underline{A})$ is denoted $X(a_0, \ldots, a_d)$, or simply $(a_0, \ldots, a_d)$, and is called a *projective character* of $X \subset \mathbb{P}^n$ (cf. Severi (2)). From (1), we have

(2) $$X(a_0, \ldots, a_d) = \int_X h^{d-|\lambda|} \Delta_\lambda(c^{(0)}, \ldots, c^{(d)}) ,$$

which gives the extrinsic invariants $X(a_0, \ldots, a_d)$ in terms of the intrinsic invariants $c_i(T_X)$ and the hyperplane class h.

Let L be a generic linear subspace of $\mathbb{P}^n$ of dimension m, $Y = X \cap L$. Then the characters of $X \subset \mathbb{P}^n$ determine the characters of $Y \subset L$. The character $Y(a_0, \ldots, a_{d-n+m})$ for $Y \subset L$ is the same as the character $X(a_0, \ldots, a_d)$ for $X \subset \mathbb{P}^n$, with $a_i = n - d + i$ for $i > d - n + m$. (This follows from the geometric definition, or from (2).) Similarly, if X is a projection from a variety Y in a larger projective space, the characters of X are the same as certain characters of Y (cf. Severi (2)).

The ith *class* of X, $0 \leqq i \leqq d$, denoted ϱ_i, is defined by

$$\varrho_i = X(n-d-1, n-d, \ldots, n-d+i-2, n-d+i, \ldots, n) .$$

The ith *rank* of X, for $0 \leqq i \leqq \min(d, n-d)$, is

$$\omega_i = X(n-d-i, n-d+1, \ldots, n) .$$

Thus $\varrho_0 = \omega_0 = \deg(X)$, and

$$\varrho_i = \deg\{x \in X \mid \dim T_xX \cap A \geqq i-1\}$$

for a given general $(n-d+i-2)$-plane A, while

$$\omega_i = \deg\{x \in X \mid T_xX \text{ meets } B\}$$

for a given general $(n-d-i)$-plane B. Thus ϱ_i corresponds to the partition $\lambda = (1, \ldots, 1, 0, \ldots, 0)$ with i 1's, and ω_i to its conjugate $(i, 0, \ldots, 0)$. From (2) and Lemmas A.9.1 and A.9.2 (cf. Examples 14.4.15 and 12.3.6)

$$\varrho_i = \sum_{j=0}^{i} (-1)^j \binom{d-j+1}{i-j} \deg(c_j(T_X)) ,$$

$$\omega_i = \sum_{j=0}^{i} \binom{d+i}{i-j} \deg(s_j(T_X)) .$$

14.4 Thom-Porteous Formula

Let $\sigma: E \to F$ be a homomorphism of vector bundles over a purely n-dimensional scheme X; let $e = \operatorname{rank} E$, $f = \operatorname{rank} F$, and let $k \leqq \min(e,f)$. We construct the kth *degeneracy class*

$$\mathbb{D}_k(\sigma) \in A_m(D_k(\sigma)),$$

with $m = n - (e-k)(f-k)$, $D_k(\sigma)$ the kth degeneracy locus. Let $d = e - k$, and let $G_d(E)$ be the Grassmannian of d-planes in E; let π be the projection from $G_d(E)$ to X, and let S be the universal subbundle of $\pi^* E$. The composite

$$S \hookrightarrow \pi^* E \xrightarrow{\pi^*\sigma} \pi^* F$$

determines a section, denoted s_σ, of $S^\vee \otimes \pi^* F$. The zero set $Z(s_\sigma)$ maps onto $D_k(\sigma)$. Let η be the induced morphism from $Z(s_\sigma)$ to $D_k(\sigma)$. The localized top Chern class $\mathbb{Z}(s_\sigma)$ is in $A_m(Z(s_\sigma))$, since $m = \dim(G) - df$. Set

$$\mathbb{D}_k(\sigma) = \eta_*(\mathbb{Z}(s_\sigma))$$

in $A_m(D_k(s_\sigma))$.

Theorem 14.4. (a) *The image of* $\mathbb{D}_k(\sigma)$ *in* $A_m(X)$ *is*

$$\Delta_{f-k}^{(e-k)}(c(F-E) \cap [X].$$

(b) *Each irreducible component of* $D_k(\sigma)$ *has codimension at most* $(e-k)(f-k)$, *in* X. *If* $\operatorname{codim}(D_k(\sigma), X) = (e-k)(f-k)$, *then* $\mathbb{D}_k(\sigma)$ *is a positive cycle whose support is* $D_k(\sigma)$.

(c) *If* $\operatorname{codim}(D_k(\sigma), X) = (e-k)(f-k)$, *and* X *is Cohen-Macaulay, then* $D_k(\sigma)$ *is also Cohen-Macaulay and*

$$\mathbb{D}_k(\sigma) = [D_k(\sigma)].$$

(d) *The formation of* $\mathbb{D}_k(\sigma)$ *commutes with Gysin maps and proper push-forward.*

Proof. The interpretation and proof of (d) are the same as in Theorem 14.3. Since

$$\mathbb{Z}(s_\sigma) = c_{df}(S^\vee \otimes \pi^* F) \cap [G_d(E)],$$

(a) follows from Proposition 14.2.2 (cf. Example 14.2.1).

To prove (b) and (c), we assume first that E contains a flag of the form $A_1 \subset \ldots \subset A_d = E$, with rank $A_i = k + i$. In this case

$$D_k(\sigma) = \Omega(\underline{A}; \sigma)$$

as schemes. We claim that

$$\mathbb{D}_k(\sigma) = \mathbf{\Omega}(\underline{A}; \sigma)$$

as well. To see this, let $Fl(\underline{A})$ be the flag bundle,

$$\mu: Fl(\underline{A}) \to G_d(E)$$

the canonical birational morphism constructed in the proof of Proposition 14.2.2, with $\mu^*(S) = D_d$, $\varphi = \pi \circ \mu$, and $\mu^*(s_\sigma)$ the section of $D_d^\vee \otimes \varphi^* F$ used in the construction of $\Omega(\underline{A}; \sigma)$. Then

$$\Omega(\underline{A}; \sigma) = \pi_* \mu_* (\mathbb{Z}(\mu^*(s_\sigma))) = \pi_* (\mathbb{Z}(s_\sigma)) = \mathbb{D}_k(\sigma)$$

by Proposition 14.1 (d) (iii). Now (b) and (c) follow from the corresponding parts of Theorem 14.3.

In the general case, by the splitting principle there is a morphism $f: X' \to X$, which is a succession of projective bundles, so that $f^* E$ contains a flag of the type just considered. Since $f^!(\mathbb{D}_k(\sigma)) = \mathbb{D}_k(f^* \sigma)$ by (d), the assertions for $f^* \sigma$ imply the same for σ. □

As in Remark 14.3, $\mathbb{D}_k(\sigma)$ may also be defined as $t_\sigma^! [D_k(u)]$, t_σ the section of $\operatorname{Hom}(E, F)$ determined by σ, and u the tautological homomorphism on $\operatorname{Hom}(E, F)$. Similarly one has the following corollary.

Corollary 14.4. *The classes* $\mathbb{D}_k(\sigma)$ *are uniquely determined by properties* (c) *and* (d) *of the theorem.* □

Later we shall see that the classes $\mathbb{D}_k(\sigma)$ actually live in certain bivariant groups, which better expresses their nature as "cohomology" classes (cf. Example 17.4.2). This will also provide a sharper formulation of the assertion (d) that their formation commutes with other intersection operations.

Example 14.4.1. If $e \leqq f$, and $k = e - 1$,

$$D_k(\sigma) = \{x \in X \mid \sigma(x) \text{ is not injective}\}.$$

Then $D_k(\sigma)$ has codimension at most $f - e + 1$, and

$$\mathbb{D}_k(\sigma) = c_{f-e+1}(F - E) \cap [X]$$

in $A_{n-f+e-1}(X)$.

Example 14.4.2. *Localized Chern classes.* Let E be a vector bundle of rank r on an n-dimensional variety X, let $p \leqq r$, and let $s_1, \ldots, s_{r-p+1}$ be sections of E. Set

$$D(\underline{s}) = \{x \in X \mid s_1(x), \ldots, s_{r-p+1}(x) \text{ are dependent}\}.$$

Then $D(\underline{s})$ has a natural scheme structure, all components have codimension at most p in X, and there is a class $\mathbb{D}(\underline{s})$ in $A_{n-p}(D(\underline{s}))$ such that

$$\mathbb{D}(\underline{s}) = c_p(E) \cap [X].$$

If $\operatorname{codim}(D(\underline{s}), X) = p$, $\mathbb{D}(\underline{s})$ is a positive cycle supported on $D(\underline{s})$. If the sections are nowhere dependent, then $\mathbb{D}(\underline{s}) = 0$, so $c_p(E) = 0$. if $\operatorname{codim}(D(\underline{s}), X) = p$, and X is Cohen-Macaulay or $\operatorname{depth}(D(\underline{s}), X) = p$, then $\mathbb{D}(\underline{s}) = [D(\underline{s})]$. The formation of $\mathbb{D}(\underline{s})$ commutes with Gysin maps and proper push-forward. (The sections determine a morphism σ from a trivial bundle of rank $r - p + 1$ to E, with $D(\underline{s}) = D_{r-p}(\sigma)$.)

Example 14.4.3. *Geometric construction of Chern classes.* Let E be a vector bundle of rank r on a quasi-projective variety X over an algebraically closed field.

(i) For a suitable line bundle L, there are sections $s_1, \ldots, s_{r+1}$ of $E \otimes L$ such that for any $p \leqq r$,

$$D_p = \{x \in X \mid s_1(x), \ldots, s_{r-p+1}(x) \text{ are dependent}\}$$

has pure dimension p in X, or is empty, and

$$[D_p] = c_p(E \otimes L) \cap [X].$$

(Choose L so $E \otimes L$ is generated by its sections, and argue as in Example 14.3.2 (d).)

(ii) The classes $c_p(E)$ are determined from $c_i(E \otimes L)$ and $c_1(L)$ by the formula (Example 3.2.2)

$$c_p(E) = \sum_{i=0}^{p} (-1)^{p-i} \binom{r-i}{p-i} c_1(L)^{p-i} c_i(E \otimes L).$$

Of course, if this is taken as the definition of Chern classes, work equivalent to that in Chap. 3 must be done to prove that they are well-defined.

Example 14.4.4. If $\operatorname{codim}(D_k(\sigma), X) = (e-k)(f-k)$, and $[D_k(\sigma)] = \sum m_i [D_i]$ is the cycle of $D_k(\sigma)$, then $\mathbb{D}_k(\sigma) = \sum e_i [D_i]$, with $1 \leqq e_i \leqq m_i$. In general,

$$\operatorname{depth}(D_k(\sigma), X) \leqq (e-k)(f-k)$$

with equality if and only if $\operatorname{codim}(D_k(\sigma), X) = (e-k)(f-k)$ and $\mathbb{D}_k(\sigma) = [D_k(\sigma)]$. (The notation and proof are the same as in Example 14.3.1.)

Example 14.4.5. *Special divisors* (cf. Kempf (1), Kleiman-Laksov (2), (3), Arbarello-Cornalba-Griffiths-Harris (1)). Let J be the Jacobian of a curve C, and let $w_i = [W_{g-i}]$, with notation as in Example 4.3.3. Let $W_d^r \subset J$ be the locus of special divisor classes:

$$W_d^r = \{L \in J \mid \dim H^0(L \otimes \mathcal{O}(dP_0)) \geqq r+1\}.$$

Let $\mathscr{P}$ be the Poincaré bundle on $J \times C$, normalized to be trivial on $J \times P_0$. Set

$$E_m = p_*(\mathscr{P} \otimes q^*(\mathcal{O}(m P_0))), \qquad F_t = p_*(\mathscr{P} \otimes q^*(\mathcal{O}_{tP_0})),$$

where $p: J \times C \to J$, $q: J \times C \to C$ are projections. *Set* $m = t + d$. From the exact sequence

$$0 \to \mathcal{O}(dP_0) \to \mathcal{O}(m P_0) \to \mathcal{O}_{tP_0} \to 0$$

on C, applying $p_*(\mathscr{P} \otimes q^*(-))$, one deduces an exact sequence

$$0 \to E_d \to E_m \xrightarrow{\sigma} F_t.$$

For $t \geqq 2g-1$, E_m and F_t are locally free of ranks $m+1-g$ and t respectively, and

$$W_d^r = D_k(\sigma), \qquad k = m - r - g.$$

Applying the Porteous formula to σ, there is a class

$$\mathbb{W}_d^r \in A_\varrho(W_d^r), \qquad \varrho = g - (r+1)(g-d+r),$$

whose image in $A_\varrho(J)$ is $\Delta_{g-d+r}^{(r+1)}(c(F_t-E_m))\cap[J]$. Since F_t is a successive extension of trivial bundles, $c(F_t)=1$. By Example 4.3.3, $s_i(E_m)=w_i$. Therefore

$$\mathbb{W}_d^r = \det \begin{vmatrix} w_{g-d+r} & \cdots & w_{g-d+2r} \\ \vdots & & \vdots \\ w_{g-d} & \cdots & w_{g-d+r} \end{vmatrix}$$

in $A_\varrho(J)$. If $\dim W_d^r=\varrho$, then $\mathbb{W}_d^r=[W_d^r]$. Both W_d^r and $\mathbb{W}_d^r$ are independent of choice of $t\geqq 2g-1$.

Poincaré's formula asserts that $i!\,w_i$ is numerically (or homologically) equivalent to θ^i, where $\theta=w_1$. It follows (cf. Example A.9.3) that $\mathbb{W}_d^r$ is numerically equivalent to

$$\frac{0!\cdot 1!\cdot\ldots\cdot r!}{(g-d+r)!\cdot\ldots\cdot(g-d+2r)!}\cdot\theta^{g-\varrho}.$$

In particular, $\mathbb{W}_d^r\neq 0$, and so $W_d^r\neq 0$, whenever $\varrho\geqq 0$.

The non-emptiness of W_d^r also follows from the fact that $E^\vee\otimes F_t$ is an ample vector bundle (cf. Example 12.1.6, and Fulton-Lazarsfeld (2)).

If C is a curve of general moduli, Griffiths and Harris (2) have shown that the W_d^r are reduced, of dimension ϱ. Arguing by specialization, as in Example 12.2.3 it follows that for any C, $\mathbb{W}_d^r$ can be represented by a positive cycle, if $\varrho\geqq 0$.

If C is smooth over a given ground field, and P_0 is a rational point, the discussion of Chap. 13 can be applied to these classes. If $\varrho\geqq 0$, the locus W_d^r always supports a zero-cycle of degree $N(g,r,d)$, where

$$N(g,r,d)=\frac{1!\cdot 2!\cdot\ldots\cdot r!\cdot g!}{(g-d+r)!\cdot\ldots\cdot(g-d+2r)!}.$$

(In fact, $\theta^\varrho\cdot\mathbb{W}_d^r$ is represented by such a cycle.) For example, if the ground field is $\mathbb{R}$, $\varrho\geqq 0$, and $N(g,r,d)$ is odd, then W_d^r must contain real points.

Example 14.4.6 (Beauville (3)). Continuing the notation of the previous example, assume also that $\varrho=0$. Then

$$\deg \mathbb{W}_d^r = N(g,r,d).$$

If $S: A_0(J)\to J$ is the homomorphism of Example 1.4.4, Beauville has proved that if $p_1,\ldots,p_n$ are positive integers adding to g, then

$$S(w_{p_1}\cdot\ldots\cdot w_{p_n})=\frac{(g-2)!}{p_1!\cdot\ldots\cdot p_n!}\left(\frac{1}{2}\sum_{i=1}^n p_i(g-p_i)\right)\cdot(K-(2g-2)P_0)$$

where K is the canonical divisor. Therefore

$$S(\mathbb{W}_d^r)=\frac{d}{2g-2}N(g,r,d)\cdot(K-(2g-2)P_0)$$

in J.

If W_d^r is not infinite, W_d^r contains at most $N(g,r,d)$ points. In fact $[W_d^r]$ is a zero cycle of degree $N(g,r,d)$. If C has general moduli, W_d^r consists of $N(g,r,d)$ distinct points.

Example 14.4.7. *Excess Porteous formula.* Let $\sigma: E \to F$, and let k be an integer such that $D_{k-1}(\sigma) = \emptyset$. Then on $D = D_k(\sigma)$, there is an exact sequence

$$0 \to K \to E_D \to F_D \to C \to 0$$

with K, C vector bundles of ranks $e-k, f-k$. Then

(a) $$\mathbb{D}_k(\sigma) = \{c(K^\vee \otimes C) \cap s(D_k(\sigma), X)\}_m$$

in $A_m(D_k(\sigma))$, $m = n - (e-k)(f-k)$. If D is a local complete intersection in X, of codimension d with normal bundle N, then

(b) $$\mathbb{D}_k(\sigma) = c_p(K^\vee \otimes C - N) \cap [D_k(\sigma)]$$

where $p = (e-k)(f-k) - d$. (Let $H = \operatorname{Hom}(E, F)$, u, t_σ as in the discussion preceding Corollary 14.4, and let $\tilde{H} \subset H$ be the open subscheme consisting of maps of rank at least k, and set $\tilde{D} = D_k(u) \bigcap \tilde{H}$. On $\tilde{D}$ there is an exact sequence

$$0 \to \tilde{K} \to E_{\tilde{D}} \to F_{\tilde{D}} \to \tilde{C} \to 0$$

and $\tilde{D}$ is regularly imbedded in $\tilde{H}$ with normal bundle $\tilde{K}^\vee \otimes \tilde{C}$ (cf. Ronga (1)). Now t_σ is assumed to map X into $\tilde{H}$. The class $\mathbb{D}_k(\sigma)$ may be constructed from the fibre square

$$\begin{array}{ccc} D_k(\sigma) & \to & X \\ \downarrow & & \downarrow {\scriptstyle t_\sigma} \\ \tilde{D} & \to & \tilde{H} \end{array}$$

by Theorem 6.4. Since $\tilde{K}$ and $\tilde{C}$ restrict to K and C on D, the conclusions follow from the basic formulas for excess intersections (§ 6.1, 6.3).) See Harris-Tu (2) for a study of Chern numbers of K and C.

Example 14.4.8. Let $f: X^n \to Y^m$ be a morphism of non-singular varieties. Let

$$S_k(f) = \{x \in X \mid \operatorname{rank} df(x) \leqq k\},$$

i.e., $S_k(f) = D_k(df)$, with $df: T_X \to f^* T_Y$ the associated map of tangent bundles. If $S_k(f) \neq \emptyset$, then $\dim S_k(f) \geqq N$, $N = n - (n-k)(m-k)$. The scheme $S_k(f)$ carries a cycle class $\mathbb{S}_k(f)$ with

$$\mathbb{S}_k(f) = \Delta_{m-k}^{(n-k)}(c(f^* T_Y)/c(T_X)) \cap [X]$$

in $A_N(X)$. If $\dim S_k(f) = N$, then $\mathbb{S}_k(f) = [S_k(f)]$.

In case $m \geqq n$, $k = n-1$, one recovers the formula for the ramification class (Example 9.3.12).

If $m = 1$, $k = 0$. $\mathbb{S}_k(f)$ is the class constructed in Example 14.1.5.

With appropriate transversality assumptions, one may use the Porteous formula to study higher Thom-Boardman singularities (cf. Lascoux (6), Ronga (1).)

When $Y = \mathbb{P}^m$, f is given by a linear system without basepoint. For more general linear systems, see Porteous (3) and Piene (3).

Example 14.4.9. If $\sigma: E \to F$ is a vector bundle homomorphism, one has the dual homomorphism $\sigma^\vee: F^\vee \to E^\vee$, with $D_k(\sigma^\vee) = D_k(\sigma)$. Then

(i) $$\mathbb{D}_k(\sigma^\vee) = \mathbb{D}_k(\sigma)$$

in $A_m(D_k(\sigma))$. ($\mathbb{D}_k(\sigma^\vee)$ satisfies (c) and (d), so equals $\mathbb{D}_k(\sigma)$ by Corollary 14.4.) This corresponds to the formal identity (Lemma 14.5.1)

(ii) $$\Delta_{e-k}^{(f-k)}(c(E^\vee - F^\vee)) = \Delta_{f-k}^{(e-k)}(c(F-E)).$$

Replacing $c(E^\vee - F^\vee)$ by $c(E-F)$ in the left side of (ii) changes the sign by $(-1)^{(e-k)(f-k)}$, giving an alternative for Theorem 14.4 (a):

(iii) $$\mathbb{D}_k(\sigma) = (-1)^{(e-k)(f-k)} \Delta_{e-k}^{(f-k)} c(E-F) \cap [X].$$

Example 14.4.10. Another construction of $\mathbb{D}_k(\sigma)$ may be given as follows. Let $G_k(F)$ be the Grassmann bundle of k-planes in F, π' the projection to X, Q the universal quotient of $\pi'^* F$. The composite

$$\pi'^* E \xrightarrow{\pi'^* \sigma} \pi'^* F \to Q$$

determines a section s'_σ of $\pi'^* E^\vee \otimes Q$. Then $Z(s'_\sigma)$ maps onto $D_k(\sigma)$, by a restriction η' of π', and

$$\mathbb{D}_k(\sigma) = \eta'_*(\mathbb{Z}(s'_\sigma)).$$

(Denote the right side by $\mathbb{D}'_k(\sigma)$. Under the canonical duality isomorphism of $G_k F$ with $G_{f-k}(F^\vee)$ s'_σ corresponds to the section $s_{\sigma^\vee}$ constructed from the dual homomorphism $\sigma^\vee$. Therefore $\mathbb{D}'_k(\sigma) = \mathbb{D}_k(\sigma^\vee)$, and the assertion follows from Example 14.4.9.)

Example 14.4.11. *Symmetric and skew-symmetric degeneracy loci* (cf. Harris-Tu (1), Józefiak-Lascoux-Pragacz (1)). Let E be a vector bundle of rank e on a purely n-dimensional scheme X. A bundle map $\sigma: E^\vee \to E$ is *symmetric* if $\sigma^\vee = \sigma$. Such σ correspond to sections s_σ of $\mathrm{Sym}^2 E$. Let $k \leqq e$, $d = e - k$. There is a subcone $D_k^s(E)$ of $\mathrm{Sym}^2 E$, of codimension $\binom{d+1}{2}$, locally defined by the vanishing of all $(k+1)$-minors of the corresponding symmetric e by e matrices.

Given $\sigma: E^\vee \to E$ symmetric, let $D_k^s(\sigma)$ be the locus where σ has rank at most k, and define $\mathbb{D}_k^s(\sigma)$ by

$$\mathbb{D}_k(\sigma) = s_\sigma^![D_k^s(E)]$$

in $A_m(D_k^s(\sigma))$, $m = n - \binom{d+1}{2}$. The analogues of Theorem 14.4 and Example 14.4.4 are valid for these classes, with (a) replaced by the formula

$$\mathbb{D}_k^s(\sigma) = 2^d \cdot \Delta_{d,d-1,\ldots,1}(c(E)) \cap [X].$$

Similarly, if $\sigma: E^\vee \to E$ is skew-symmetric, i.e. $\sigma^\vee = -\sigma$, and $D_k^{ss}(\sigma)$ is the locus where the rank of σ is at most k, with k even, one constructs a class

$\mathbb{D}_k^{ss}(\sigma)$ in $A_m(D_k^{ss}(\sigma))$, with $m = n - \binom{d}{2}$, and

$$\mathbb{D}_k^{ss}(\sigma) = \Delta_{d-1,\ldots,1}(c(E)) \cap [X].$$

(If S is the universal subbundle on $G_k E$, $\mathrm{Sym}^2(S)$ is a subbundle of $\mathrm{Sym}^2(\pi^* E)$, and

$$\mathbb{D}_k^s(\sigma) = \pi_*(c_{\mathrm{top}}(\mathrm{Sym}^2(\pi^* E)/\mathrm{Sym}^2(S)) \cap [G_k E]).$$

Similarly for $\mathbb{D}_k^{ss}$, with Λ^2 replacing Sym^2. For the calculation of these Gysin formulas we refer to the cited references; see also Example 14.5.1. A proof more along the lines of §14.4 has been given by J. Damon (unpublished).)

There are useful extensions of these formulas to symmetric and skew-symmetric maps from $E^\vee$ to $E \otimes L$, for L a line bundle, for which we refer the reader to Harris-Tu (1).

Example 14.4.12. For any positive integers a, b, the polynomial $\Delta_b^{(a)}(c)$ is characterized as the unique polynomial of weight ab such that for all vector bundles A, B of ranks a, b,

$$c_{ab}(A^\vee \otimes B) = \Delta_b^{(a)}(c(B-A)).$$

If σ is a vector bundle map from A to B on X, $\mathbb{D}_0(\sigma)$ is represented by $\Delta_b^{(a)}(c(B-A)) \cap [X]$. Porteous' formula says that the same polynomial works for higher degeneracy loci: if E, F have ranks $k+a$, $k+b$, and $\sigma: E \to F$, then

$$\mathbb{D}_k(\sigma) = \Delta_b^{(a)}(c(F-E)) \cap [X].$$

Similarly, for any positive integer a, $2^a \Delta_{a,\ldots,1}(c)$ is the unique polynomial of weight $\binom{a}{2}$ such that for all vector bundles A of rank a

$$c_{\binom{a}{2}}\mathrm{Sym}^2(A) = 2^a \Delta_{a,\ldots,1}(c(A))$$

(see Example 14.5.1). If σ is a symmetric map from $A^\vee$ to A, this says that $\mathbb{D}_0(\sigma)$ is represented by $2^a \Delta_{a,\ldots,1}(c(A))$. The formula of the preceding example states that for any bundle of rank $k+a$, and $\sigma: E^\vee \to E$ symmetric,

$$\mathbb{D}_k^s(\sigma) = 2^a \Delta_{a,\ldots,1}(c(E)) \cap [X].$$

The analogous assertion is also true for alternating maps. Is there a proof of these three formulas which uses only these characterizations of the polynomials involved?

Example 14.4.13. *Positivity of degeneracy classes.* Positivity assumptions on the bundle $E^\vee \otimes F = \mathrm{Hom}(E, F)$ force corresponding positivity on the classes $D_k(\sigma)$. Let $m = n - (e-k)(f-k) \geqq 0$.

(a) If $E^\vee \otimes F$ is generated by its sections, then $\mathbb{D}_k(\sigma)$ is represented by a non-negative m-cycle on $D_k(\sigma)$.

(b) If $E^\vee \otimes F$ is ample, and L is an ample line bundle, then

$$\int_X c_1(L)^m \cap \mathbb{D}_k(\sigma) > 0.$$

In particular, if $E^\vee \otimes F$ is ample, and $m \geqq 0$, then $D_k(\sigma) \neq \emptyset$ for any σ (Example 12.1.5). Corresponding assertions are also valid for symmetric and skew-symmetric bundle maps. (Realize $\mathbb{D}_k(\sigma)$ as the intersection of a section of $\mathrm{Hom}(E, F)$ with the universal degeneracy locus in $\mathrm{Hom}(E, F)$, and apply Theorem 12.1.)

Example 14.4.14 (Giambelli (2), (3), Harris-Tu (1)). Let $m \leqq n$. The variety of m by n matrices, modulo scalars, is a projective space $\mathbb{P}^{mn-1}$. Let $V_k(m, n) \subset \mathbb{P}^{mn-1}$ be the subvariety of matrices of rank at most k. Then

$$\deg(V_k(m, n)) = \prod_{i=0}^{m-k-1} \frac{(n+i)!}{(m-1-i)!\,(n-k+i)!} \cdot 1! \cdot \ldots \cdot (m-k-1)!\,.$$

(Apply Porteous to $\sigma: 1^{\oplus m} \to \mathscr{O}(1)^{\oplus n}$ on $\mathbb{P}^{mn-1}$. To calculate the resulting determinant use Example A.9.4.) There are similar formulas for varieties of symmetric and skew-symmetric matrices. For references to the classical literature see Baker (2) p. 111.

Example 14.4.15. *Classes, ranks, and polar loci.* (a) Let X be a non-singular d-dimensional subvariety of $\mathbb{P}^n$, and let L be a linear subspace of $\mathbb{P}^n$ of dimension $n-d+m$. For $p \geqq 0$, and $-m \leqq p \leqq n-d$, set

$$X(L, p) = \{x \in X \mid \dim(T_x X \cap L) \geqq m+p\}\,.$$

For general L, $X(L, p)$ is a subscheme of X of codimension $p(m+p+1)$, and

$$[X(L, p)] = \Delta_p^{(m+p+1)}((1+h)^{d-m}/c(T_X)) = \Delta_{m+p+1}^{(p)}(c(T_X^\vee)/(1-h)^{d-m})\,.$$

(As in Example 14.3.3, if $\mathbb{P}^m = P(V)$, $L = P(\tilde{L})$, there is a canonical morphism from $\tilde{L} \otimes \mathscr{O}(1)$ to $N_X\mathbb{P}^n$, and $X(L, p)$ is the locus where this morphism has rank at most $n-d-p$.)

(b) The cases where $p=1$ give formulas for the *polar classes* of X. Fix i, $0 \leqq i \leqq d$, and let L be a subspace of dimension $n-d+i-2$. Set

$$X(L) = \{x \in X \mid \dim T_x X \cap L \geqq i-1\}\,.$$

For general L, $\mathrm{codim}(X(L), X) = i$, and $X(L)$ is called an ith *polar locus.* For general L,

$$[X(L)] = \sum_{j=0}^{i} (-1)^j \binom{d-j+1}{i-j} h^{i-j} c_j(T_X)\,.$$

The ith *class* ϱ_i of X is defined to be the degree of $X(L)$, for general L. Thus the preceding formula refines the formula for ϱ_i given in Example 14.3.3. For other approaches and generalizations to singular varieties see Example 4.4.5, Todd (8), Pohl (1), Porteous (3), Piene (3), and Kleiman (8).

(c) The cases where $m+p+1=1$ determine the ranks of X. Fix $0 \leqq i \leqq \min(d, n-d)$, let $\dim L = n-d-i$, and set

$$X'(L) = \{x \in X \mid T_x X \text{ meets } L\}\,.$$

Then for general L, $\mathrm{codim}(X'(L), X) = i$, and

$$[X'(L)] = \sum_{j=0}^{i} \binom{d+i}{i-j} h^{i-j} s_j(T_X)\,.$$

The ith *rank* ω_i of X is defined to be the degree of $X'(L)$, for general L. This refines the formula for ω_i given in Example 14.3.3. Note that if $p_L: X \to \mathbb{P}^{d+i-1}$ is induced by projection from L, $X'(L)$ is the ramification locus $S_{d-1}(p_L)$ considered in Example 14.4.8.

For example, if X is a curve in $\mathbb{P}^n$, the (first) rank of X is the number of tangents to X meeting a general $(n-2)$-plane. If $N = \deg(X)$, $g = \text{genus}(X)$, the rank is $2N + 2g - 2$.

Example 14.4.16. Let E be a vector bundle of rank e on an n-dimensional variety X, with a subbundle A of rank a. Let $G_d(E)$ be the Grassmannian of d-planes in E, and let $\Omega_k(A)$ be the subscheme of $G_d(E)$:

$$\Omega_k(A) = \{L \subset E(x) \mid x \in X, \ \dim L \cap A(x) \geqq k\}.$$

Let $p = e - d - a + k$. Then $\Omega_k(A)$ has pure codimension pk in $G_d(E)$, and $\Omega_k(A)$ has a natural scheme structure so that

$$[\Omega_k(A)] = \Delta_p^{(k)}(c(Q - \pi^* A)) \cap [G_d(E)],$$

where Q is the universal quotient bundle on G_dE, and π the projection from $G_d(E)$ to X. (If σ is the canonical map from $\pi^* A$ to Q, then $\Omega_k(A)$ is $D_{a-k}(\sigma)$. The scheme determines the cycle, as in Example 14.3.1.)

Example 14.4.17. *Linear systems on curves* (cf. Schwarzenberger (2), Arbarello-Cornalba-Griffiths-Harris (1)). (a) Let C be a non-singular projective curve, L a line bundle on C. An $(s+1)$-dimensional subspace V of $\Gamma(C, L)$ determines an s-dimensional linear system $|V|$ on C. Set

$$V_d^r = \{D \in C^{(d)} \mid \dim V \cap \Gamma(C, L(-D)) \geqq s + 1 - d + r\},$$

i.e., V_d^r consists of those effective divisors of degree d which impose at most $d - r$ conditions on V. One may realize V_d^r as a degeneracy locus as follows. Let $\mathscr{D} \subset C \times C^{(d)}$ be the universal divisor: $\mathscr{D} = \{(P, D) \mid P \in D\}$. Let p and q be the projections from $\mathscr{D}$ to $C^{(d)}$ and C. Set

$$E_L = p_* q^* L.$$

Then E_L is a bundle of rank d on $C^{(d)}$, whose fibre at D is $\Gamma(C, L/L(-D))$. There is a canonical homomorphism of vector bundles

$$\sigma: V \to E_L$$

on $C^{(d)}$, and $V_d^r = D_{d-r}(\sigma)$. From Porteous's formula (Theorem 14.4) there is a class $\mathbb{V}_d^r$ in $A_m(V_d^r)$, with $m = d - r(s + 1 - d + r)$, so that

$$\mathbb{V}_d^r = \Delta_{s+1-d+r}^{(r)} c(E_L)$$

in $A(C^{(d)})$.

(b) The Chern class of E_L, or the Chern polynomial $c_t(E_L)$ may be calculated as follows. For $P \in C$, let $C^{(d-1)}(P)$ be the image of the imbedding of $C^{(d-1)}$ in $C^{(d)}$ by $D \to D + P$, and let $x_P \in A^1(C^{(d)})$ be the class of $C^{(d-1)}(P)$. For any line bundle M on C set

$$c_M(t) = \prod (1 + x_P t)^{n_P} \in A^*(C^{(d)})[t],$$

where $\sum n_P[P]$ is a divisor whose line bundle is M. Then

(i) $c_t(E_L) = c_{K-L}(t) \cdot c_t(\Omega^1_{C^{(d)}}) = c_{K-L}(t) \cdot (1-x\,t)^{d-g+1}/u_d^* w(-t/(1-x\,t))$,

where K is the canonical line bundle on C, $x = x_{P_0}$, and $w(\tau) = \sum w_i \tau^i$, $w_i \in A^i(J)$ as in Example 4.3.3. Since all x_P are algebraically equivalent to x,

(ii) $$c_t(E_L) \equiv (1-x\,t)^{d+g-n-1}/u_d^* w(-t/(1-x\,t))$$

modulo algebraic equivalence, where $n = \deg(L)$. Modulo homological or numerical equivalence, Poincaré's formula gives

(iii) $$c(E_t) \equiv (1-x\,t)^{d+g} \cdot e^{t\theta/(1-xt)},$$

where $\theta = u_d^*(w_1)$. (If $L = K$, then $E_L = \Omega^1_{C^{(d)}}$, and (i) follows from Example 4.3.3(b). For P in C, there is an exact sequence

$$0 \to E_{L(-P)} \to E_L \to \mathcal{O}_{C^{(d-1)}(P)} \to 0,$$

so $c_t(E_L) = c_t(E_{L(-P)}) \cdot (1 - x_P t)$. The validity of (i) for L or for $L(-P)$ is therefore equivalent, so the verification for one bundle K suffices for all.) Formula (iii) is proved using the Grothendieck-Riemann-Roch theorem in Arbarello-Cornalba-Griffiths-Harris (1), where several enumerative applications may be found.

14.5 Schur Polynomials

Let $c_1, c_2, c_3, \ldots$ and $s_1, s_2, s_3, \ldots$ be sequences of commuting variables, related by the identity

$$(1 + c_1 t + c_2 t^2 + \ldots) \cdot (1 - s_1 t + s_2 t^2 - s_3 t^3 \ldots) = 1.$$

For example, if $c_i = c_i(E)$ are the Chern classes of a vector bundle E, then $s_i = s_i(E^\vee)$ are the Segre classes of $E^\vee$. If $c_i = c_i(E - F)$, for vector bundles E, F, then

$$s_i = (-1)^i c_i(F - E) = c_i(F^\vee - E^\vee).$$

A *partition* is a finite sequence $\lambda = (\lambda_1, \ldots, \lambda_d)$ of non-negative integers, arranged in non-increasing order. Set $|\lambda| = \lambda_1 + \ldots + \lambda_d$, the integer being partitioned. To λ corresponds a *Young diagram*, consisting of λ_i boxes in the ith row from the top, lined up on the left. Thus

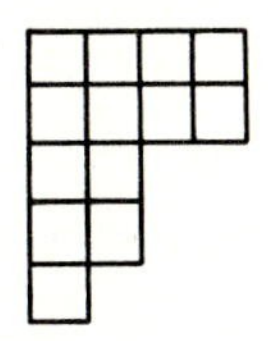

is the Young diagram for (4,4,2,2,1). The Young diagram for the *conjugate* partition is obtained by interchanging rows and columns in the Young diagram. Thus (5,4,2,2) is the conjugate to (4,4,2,2,1). Denote the conjugate of λ by $\tilde{\lambda}$.

For a partition $\lambda = (\lambda_1, \ldots, \lambda_d)$, set

$$\Delta_\lambda(c) = \Delta_{\lambda_1, \ldots, \lambda_d}(c) = \det(c_{\lambda_i + j - i})$$

the corresponding *Schur polynomial* (or S-function); $\Delta_\lambda(c)$ is isobaric of weight $|\lambda|$ in the c_i. Note that adding an arbitrary string of zeros to λ does not change $\Delta_\lambda(c)$.

Lemma 14.5.1. *Let λ and μ be conjugate partitions. Then*

$$\Delta_\lambda(c) = \Delta_\mu(s) .$$

This is proved in Lemma A.9.2. We note several consequences. Corresponding to $\lambda = (1, \ldots, 1)$, $\mu = (d)$, one has

(1)
$$s_d = \det \begin{pmatrix} c_1 & c_2 & \ldots & c_d \\ 1 & c_1 & \ldots & c_{d-1} \\ 0 & 1 & \ldots & \\ 0 & \ldots & 1 & c_1 \end{pmatrix} .$$

If λ partitions de into d equal pieces of size e, the lemma gives

(2)
$$\Delta_e^{(d)}(c) = \Delta_d^{(e)}(s) = (-1)^{de} \Delta_d^{(e)}(c^{-1}) .$$

(For the second equality see the last step in the proof of Lemma A.9.2.)

If r is an integer such that $s_i = 0$ for all $i > r$, and $\lambda_{r+1} > 0$, then

(3)
$$\Delta_\lambda(c) = 0 .$$

(Indeed, if μ is the conjugate to λ, then $\mu_1 > r$, so the first row of the matrix whose determinant is $\Delta_\mu(s)$ is zero.)

Lemma 14.5.2. *Let $\lambda = (\lambda_1, \ldots, \lambda_d)$ be a partition, m a non-negative integer. Then*

$$\Delta_\lambda(c) \cdot c_m = \sum \Delta_\mu(c) ,$$

the sum over all partitions $\mu = (\mu_1, \ldots, \mu_{d+1})$ such that

$$\mu_1 \geqq \lambda_1 \geqq \mu_2 \geqq \lambda_2 \geqq \ldots \geqq \mu_d \geqq \lambda_d \geqq \mu_{d+1} \geqq 0$$

and $|\mu| = |\lambda| + m$.

The Young diagrams for the partitions μ appearing in the above formula are those which one obtains by adding m boxes to the Young diagram for λ, to the right of the given rows, such that the new diagram still has rows of non-increasing length, and such that no two of the new boxes appear in the same column. For the proof see Lemma A.9.4.

The rule for multiplying general S-functions is somewhat more complicated. A *Young tableau* is a Young diagram together with a collection of symbols in the boxes. A *simple m-expansion* of a tableau is one obtained from a given tableau by adding m new boxes according to the prescription of the preceding paragraph, putting the same new symbol in each. A $(\mu_1, \ldots, \mu_e)$-expansion of a tableau is the result of e successive simple expansions, first adding μ_1 boxes with the symbol α_1, then μ_2 boxes with the symbol α_2, etc. Given a $(\mu_1, \ldots, \mu_e)$ expansion, write down the added symbols in order, proceeding from right to left in the first row, then right to left in the second, and so on, obtaining a monomial

$$\alpha_{i_1}\alpha_{i_2} \ldots \alpha_{i_t}$$

where $t = \mu_1 + \ldots + \mu_e$. The expansion is called *strict* if for every $1 \leqq j \leqq t$, and $1 \leqq k < e$, the number of α_k's occuring among the first j terms of this monomial is no less than the number of α_{k+1}'s occuring in these first j terms.

Lemma 14.5.3. *Let $\lambda = (\lambda_1, \ldots, \lambda_d)$, $\mu = (\mu_1, \ldots, \mu_e)$ be partitions. Then*

$$\Delta_\lambda(c) \cdot \Delta_\mu(c) = \sum_\varrho N_{\lambda,\mu,\varrho} \Delta_\varrho(c)$$

where the sum is over all partitions ϱ with $|\varrho| = |\lambda| + |\mu|$, and $N_{\lambda,\mu,\varrho}$ is the number of ways a Young tableau on the Young diagram of ϱ arises by a strict $(\mu_1, \ldots, \mu_e)$-expansion of the diagram for λ.

This rule for computing the coefficients is known as the Littlewood-Richardson rule. See Appendix A.9 for references.

Example 14.5.1 (Lascoux (7); cf. Macdonald (3) p. 31). Let E be a vector bundle of rank n. Let $N = \binom{n}{2}$. Then, with the notation of Example A.9.1,

(a) $$c(S^2 E) = 2^{-N} \cdot \sum_{\mu \subset \varepsilon} d_{\varepsilon\mu} 2^{|\mu|} \Delta_{\tilde{\mu}}(c(E))$$

(b) $$c(\Lambda^2 E) = 2^{-N} \cdot \sum_{\mu \subset \delta} d_{\delta\mu} 2^{|\mu|} \Delta_{\tilde{\mu}}(c(E)) .$$

In particular, one has formulas of Giambelli:

(a′) $$c_{\binom{n+1}{2}}(S^2 E) = 2^n \Delta_{n, n-1, \ldots, 1}(c(E))$$

(b′) $$c_{\binom{n}{2}}(\Lambda^2 E) = \Delta_{n-1, \ldots, 0}(c(E)) .$$

(For (a) let $x_1, \ldots, x_n$ be Chern roots. Then

$$c(S^2 E) = \prod_{i \leqq j} (1 + x_i + x_j) = \prod_{i=1}^{n} (1 + 2x_i) \cdot 2^{-N} \prod_{i<j} (1 + 2x_i + 1 + 2x_j)$$

$$= 2^{-N} s_\varepsilon(1 + 2x_1, \ldots, 1 + 2x_n) = 2^{-N} \sum_{\mu \subset \varepsilon} d_{\varepsilon\mu} 2^{|\mu|} s_\mu(x_1, \ldots, x_n) ,$$

the last two steps using Example A.9.1 (b) and (a). The proof of (b) is similar.)

Example 14.5.2 (Lascoux (7); cf. Macdonald (3) p. 37). Let E, F be vector bundles of ranks n and m. Then

$$c(E \otimes F) = \sum d_{\lambda\mu} \Delta_{\tilde{\mu}}(c(E)) \Delta_{\lambda'}(c(F))$$

the sum over partitions $\mu \subset \lambda$ with $m \geqq \lambda_1 \geqq \ldots \geqq \lambda_n \geqq 0$; here $\lambda' = (m - \lambda_n, m - \lambda_{n-1}, \ldots, m - \lambda_1)$, and coefficients $d_{\lambda\mu}$ are as in Example

A.9.1. (The formula follows from Example A.9.2(c).) For $m=1$, the formula specializes to that of Example 3.2.2.

Example 14.5.3. Let λ, μ be partitions of length d, and assume $s_i=0$ for all $i>d$. Then

$$\Delta_\lambda \cdot \Delta_\mu = \left| c_{\lambda_i+\mu_{d+1-j}-i+j} \right|_{1 \leqq i,j \leqq d}.$$

This formula dates from Jacobi, and was rediscovered by Porteous, cf. Lascoux (3), Macdonald (3) p. 30.

Example 14.5.4. *Skew Schur functions* (cf. Lascoux (5), Stanley (1), Macdonald (3) p. 39). If λ, μ are partitions with $\mu \subset \lambda$, define

$$\Delta_{\lambda/\mu} = \left| c_{\lambda_i-\mu_j-i+j} \right|.$$

Then

$$\Delta_{\lambda/\mu} = \sum N_{\nu,\mu,\lambda} \Delta_\nu ,$$

where $N_{\nu,\mu,\lambda}$ are the integers appearing in Lemma 14.5.3, and the sum over all ν with $|\nu|+|\mu|=|\lambda|$.

14.6 Grassmann Bundles

Let E be a vector bundle of rank n on a scheme X, and let d be a positive integer less than n. Let $G=G_d(E)$ be the Grassmann bundle of d-planes in E, with projection π from G to X. There is a universal exact sequence

$$0 \to S \to \pi^* E \to Q \to 0$$

on G, with S the universal rank d subbundle, and Q the universal rank $(n-d)$ quotient bundle. Set

$$c_i = c_i(Q - \pi^* E).$$

The corresponding s_i for the discussion of § 14.5 are

$$s_i = c_i(S^\vee) = (-1)^i c_i(S),$$

as follows from the Whitney sum formula.

In this section a partition λ will be a sequence $(\lambda_1, \ldots, \lambda_d)$ of d terms, $\lambda_1 \geqq \ldots \geqq \lambda_d \geqq 0$. Note by consequence (3) of Lemma 14.4.1 that $\Delta_{\lambda_1,\ldots,\lambda_q}(c)=0$ if $q>d$ and $\lambda_q>0$, so no non-zero Schur polynomials are lost by this restriction. Given λ, set

$$\Delta_\lambda = \Delta_\lambda(c) = \Delta_{\lambda_1,\ldots,\lambda_d}(c(Q-\pi^* E)).$$

Proposition 14.6.1 (Pieri's formula). *For any partition λ, and any $m \geqq 0$,*

$$\Delta_\lambda \cdot c_m = \sum_\mu \Delta_\mu$$

the sum over all partitions μ with

$$\mu_1 \geqq \lambda_1 \geqq \mu_2 \geqq \lambda_2 \geqq \ldots \geqq \mu_d \geqq \lambda_d$$

and $|\mu| = |\lambda| + m$.

This follows from Lemma 14.5.2 and the vanishing noted preceding the proposition. Likewise, the following proposition follows from Lemma 14.5.3.

Proposition 14.6.2 (Product formula). *For any partition λ, μ,*

$$\Delta_\lambda \cdot \Delta_\mu = \sum_\varrho N_{\lambda,\mu,\varrho} \Delta_\varrho$$

the sum over all partitions ϱ with $|\varrho| = |\lambda| + |\mu|$, and $N_{\lambda,\mu,\varrho}$ given by the Littlewood-Richardson rule. □

Proposition 14.6.3 (Duality theorem). *Let λ and μ be partitions with $|\lambda| + |\mu| \leqq d(n-d)$, and let $\alpha \in A_*(X)$. Then*

$$\pi_*(\Delta_\lambda \cdot \Delta_\mu \cap \pi^* \alpha) = \begin{cases} \alpha & \text{if } \lambda_i + \mu_{d-i+1} = n-d \text{ for } 1 \leqq i \leqq d \\ 0 & \text{otherwise}. \end{cases}$$

Proof. By compatibility of π_* and π^* with proper inclusions, we may assume $\alpha = [X]$, X a variety; therefore $\pi^*[X] = [G]$. If $|\lambda| + |\mu| < d(n-d)$ the conclusion is clear for dimensional reasons. If $|\lambda| + |\mu| = d(n-d)$, then

$$\pi_*(\Delta_\lambda \cdot \Delta_\mu \cap [G]) = m[X]$$

for some integer m. This equation remains valid if X is replaced by any open subvariety, so we may assume E is trivial. In this case we claim that

$$\Delta_\lambda \cdot \Delta_\mu = \begin{cases} \Delta^{(d)}_{n-d} & \text{if } \lambda_i + \mu_{d-i+1} = n-d \text{ for } 1 \leqq i \leqq d \\ 0 & \text{otherwise}. \end{cases}$$

Here $\Delta^{(d)}_{n-d}$ corresponds to the partition of $d(n-d)$ into d equal parts. This follows readily from the Littlewood-Richardson rule (or see Example 14.6.1). Given λ, the only strict $(\mu_1, \ldots, \mu_d)$-extension of λ to a tableau on a d by $n-d$ rectangle is obtained by adding, at each stage, a new symbol to every column which has not been filled in previous stages.

To conclude the proof, we show that $\pi_*(\Delta^{(d)}_{n-d} \cap [G]) = [X]$. Choose a flag $\underline{A}$ of trivial subbundles of E, with rank $A_i = i$, and let σ be the projection from $\pi^* E$ to Q. By the determinantal formula,

$$[\Omega(\pi^* \underline{A}; \sigma)] = \Delta^{(d)}_{n-d} \cap [G].$$

Here $\Omega(\pi^* \underline{A}; \sigma) = \{L \in G_d(E) \mid \dim L \cap A_i \geqq i\} = G_d(A_d)$. Therefore π maps $\Omega(\pi^* \underline{A}; \sigma)$ isomorphically onto X, so

$$\pi_*[\Omega(\pi^* \underline{A}; \sigma)] = [X],$$

as required. □

Proposition 14.6.4 (Giambelli's formula). *Let $\underline{A}$ be a flag of subbundles of E:*

$$0 \subsetneqq A_1 \subsetneqq A_2 \subsetneqq \ldots \subsetneqq A_d \subset E.$$

Let $\Omega(\underline{A}) = \Omega(\pi^* \underline{A}; \sigma)$, where σ is the projection from $\pi^* E$ onto Q, i.e.,

$$\Omega(\underline{A}) = \{L \in G_d(E) \mid \dim L \cap A_i \geqq i, 1 \leqq i \leqq d\}.$$

Let $\lambda_i = n-d+i-\operatorname{rank}(A_i)$. *Assume* X *is pure-dimensional. Then* $\Omega(\underline{A})$ *has pure codimension* $|\lambda|$ *in* $G_d(E)$. *If* $c_i(E-A_j)=0$ *for all* $i>0$ *and all* $j=1,\ldots,d$, *then*

$$[\Omega(\underline{A})] = \Delta_\lambda \cap [G_d(E)] .$$

Proof. The determinantal formula (Theorem 14.3) gives

$$\boldsymbol{\Omega}(\pi^*\underline{A};\sigma) = \Delta_{\lambda_1,\ldots,\lambda_d}(c(Q-\pi^*A_1),\ldots,c(Q-\pi^*A_d)) \cap [G_d(E)] .$$

The assumption on Chern classes implies that

$$c(Q-\pi^*A_i) = c(Q-\pi^*E)\cdot c(\pi^*E-\pi^*A_i) = c(Q-\pi^*E) .$$

To conclude, it must be verified that $\boldsymbol{\Omega}(\pi^*\underline{A};\sigma) = [\Omega(A)]$. This verification is local, where the bundles are trivial and $G_d(E)$ is a product $X\times G_d(\mathbb{A}^n)$. On the non-singular variety $G_d(\mathbb{A}^n)$ the equality is known by Theorem 14.3(c), and it pulls back by flat projection to $G_d(E)$. □

Proposition 14.6.5 (Basis theorem). *For each* $k\geqq 0$, *there is a canonical isomorphism*

$$A_k(G_d(E)) \cong \bigoplus_\lambda A_{k-d(n-d)+|\lambda|}(X) ,$$

the sum over all partitions $\lambda=(\lambda_1,\ldots,\lambda_d)$ *with*

$$n-d \geqq \lambda_1 \geqq \ldots \geqq \lambda_d \geqq 0 .$$

Each element in $A_k(G_d(E))$ *has a unique expression in the form*

$$\sum_\lambda \Delta_\lambda \cap \pi^*(\alpha_\lambda) ,$$

$\alpha_\lambda \in A_{k-d(n-d)+|\lambda|}(X)$, λ *as above.*

Proof. (See also Example 14.6.2.) Let $\varphi\colon \bigoplus_\lambda A_{k-d(n-d)+|\lambda|}(X) \to A_kX$ be the homomorphism taking $\oplus\,\alpha_\lambda$ to $\sum \Delta_\lambda \cap \pi^*\alpha_\lambda$. If $\oplus\,\alpha_\lambda$ is a non-zero element in the kernel of φ, choose $\bar{\lambda}$ with $|\bar{\lambda}|$ maximal such that $\alpha_{\bar{\lambda}} \neq 0$. Define μ to be the partition with $\mu_i + \bar{\lambda}_{d-i+1} = n-d$, $1\leqq i\leqq d$. By duality (Proposition 14.6.3),

$$0 = \pi_*(\Delta_\mu \cdot \sum \Delta_\lambda \cap \pi^*\alpha_\lambda) = \alpha_{\bar{\lambda}} ,$$

a contradiction.

To show that φ is surjective, as in the proof of Theorem 3.3(b) – which is the case $d=1$ of this proposition – it suffices to verify this when X is irreducible and E is trivial. Choose a basis $e_1,\ldots,e_n$ for E, which identifies $G_d(E)$ with $X\times G$, $G=G_d(\mathbb{A}^n)$. For each sequence $\underline{a}$, $0<a_1<\ldots<a_d\leqq n$, let A_i be the space spanned by $e_1,\ldots,e_{a_i}$, and

$$\Omega(\underline{a}) = \{L\in G \mid \dim L \textstyle\bigcap A_i \geqq i,\ 1\leqq i\leqq d\} .$$

Note that $\Omega(\underline{b}) \subset \Omega(\underline{a})$ if $\underline{b} \leqq \underline{a}$ (i.e., $b_i \leqq a_i$ for all i). We have seen that $\Omega(\underline{a})$ is a subvariety of G, and that

$$\Delta_\lambda \cap [G_d(E)] = [X\times\Omega(\underline{a})] ,$$

with $\lambda_i = n - d + i - a_i$. To prove φ surjective, it therefore suffices to show that for each $\underline{a}$ the homomorphism

$$\sum_{\underline{b} \leq \underline{a}} A_{k-|\underline{b}|}(X) \to A_k(X \times \Omega(\underline{a}))$$

taking $\oplus \alpha_{\underline{b}}$ to $\sum \alpha_{\underline{b}} \times [\Omega(\underline{b})]$, is surjective; here $|\underline{b}|$ denotes $\sum (b_i - i)$.

Let V be the span of $e_{a_1}, e_{a_2}, \ldots, e_{a_d}$, W the span of the other basic vectors. Let

$$G_V = \{L \in G \mid L \text{ projects isomorphically to } V\}.$$

Then G_V is an affine open subvariety of G, canonically isomorphic to $\operatorname{Hom}(V, W)$. Any L in G_V has a unique basis projecting to the basis $e_{a_1}, \ldots, e_{a_d}$ of V; L determines a d by n matrix

$$\begin{array}{c} \\ \\ \\ \end{array}
\begin{pmatrix}
* & \cdots & * & \overset{a_1}{1} & * & \cdots & * & \overset{a_2}{0} & * & \cdots & * & \overset{a_d}{0} & * & \cdots & * \\
\cdot & & & 0 & & & & 1 & & & & \cdot & & & \cdot \\
\cdot & & & & & & & 0 & & & & \cdot & & & \cdot \\
\cdot & & & \cdot & & & & \cdot & & & & \cdot & & & \cdot \\
\cdot & & & \cdot & & & & \cdot & & & & \cdot & & & \cdot \\
\cdot & & & \cdot & & & & \cdot & & & & & & & \cdot \\
\cdot & & & \cdot & & & & \cdot & & & & 0 & & & \cdot \\
* & \cdots & * & 0 & * & \cdots & * & 0 & * & \cdots & * & 1 & * & \cdots & *
\end{pmatrix}$$

whose i^{th} row gives the coefficients of the ith basic vector of L in terms of $e_1, \ldots, e_n$. Conversely, any such matrix determines an L in G_V. Set

$$\Omega(\underline{a})^\circ = \Omega(\underline{a}) \bigcap G_V.$$

This is a linear subspace of $G_V = \operatorname{Hom}(V, W)$ of dimension $\sum (a_i - i)$; in the above matrix description, it is given by the vanishing of all entries of all rows to the right of the indicated 1's. There is a disjoint union

$$\Omega(\underline{a}) - \Omega(\underline{a})^\circ = \coprod_{j=1}^{d} \Omega(a_1, \ldots, a_j - 1, \ldots, a_d)$$

where any sequence which is not a strictly increasing set of positive integers is discarded. One concludes by induction on $|\underline{a}|$ and the exact sequence (Proposition 1.8)

$$A_k(X \times (\Omega(\underline{a}) - \Omega(\underline{a})^\circ)) \to A_k(X \times \Omega(\underline{a})) \to A_k(X \times \Omega(\underline{a})^\circ) \to 0. \quad \square$$

Example 14.6.1. The duality theorem may also be given a simple geometric proof. One may assume E trivial, with basis $e_1, \ldots, e_n$. Let A_i be spanned by the first a_i basis elements, and let B_i be spanned by the last b_i elements. Let $\lambda_i = n - d + i - a_i$, $\mu_i = n - d + i - b_i$. By Giambelli's formula,

$$\Delta_\lambda \cap [G] = [\Omega(\underline{a})], \qquad \Delta_\mu \cap [G] = [\Omega(\underline{b})].$$

Then $\Omega(\underline{a}) \bigcap \Omega(\underline{b}) = \emptyset$ unless $a_i + b_{d-i+1} = n + 1$ for all i, in which case $\Omega(\underline{a}) \bigcap \Omega(\underline{b})$ is one (reduced) point. One concludes by Proposition 8.2.1.

Example 14.6.2. The basis theorem may also be proved by induction on d and n. If $E = F \oplus 1$, with F trivial, one has an inclusion i of $G_{d-1}(F)$ in

$G_d(E)$; the complement U is an affine bundle over $G_d(F)$, equipped with a section s. One has a diagram

$$\begin{array}{ccccc} A_k(G_{d-1}F) & \xrightarrow{i_*} A_k(G_dE) & \xrightarrow{j_*} & A_k(U) \to 0 \\ & & & p^*\uparrow \downarrow s^* \\ & & & A_{k-d}(G_dF) \; . \end{array}$$

Computing how the classes Δ_λ transport via the indicated maps, and using the formula $i_* i^*(\alpha) = c_{n-d}(Q) \cap \alpha$, the knowledge of generators for $A_k(G_{d-1}F)$ and $A_{k-d}(G_dF)$ determines generators for $A_k(G_d(E))$. (For another proof based on this geometric construction, see Laksov (2).)

Example 14.6.3. If $\Omega(\underline{a})^\circ$ is the open affine subvariety of the Schubert variety $\Omega(\underline{a})$ constructed in the proof of the basis theorem, then

$$\Omega(\underline{a})^\circ \subset \{L \in G \mid \dim(L \cap A_i) = i\} .$$

This inclusion may be strict; the right side need not be an affine variety.

Example 14.6.4. The classes $\Delta_\lambda(c(Q))$ may be used in place of $\Delta_\lambda(c(Q - \pi^* E))$ in the basis theorem. It can happen that $\Delta_\lambda(c(Q)) \neq 0$ for λ a partition with more than d non-zero terms, so Pieri's formula for these classes must be stated in the modified form of Lemma 14.5.2. If $c(E) = 1$, there is no distinction, and $\Delta_\lambda = 0$ if $\lambda_1 > n - d$ or $\lambda_{d+1} > 0$.

Example 14.6.5. There is a canonical duality isomorphism

$$\varphi: G_{n-d}(E^\vee) \to G_d(E)$$

such that $\varphi^* S^\vee$ is the universal quotient bundle on $G_{n-d}(E^\vee)$, and $\varphi^* Q^\vee$ the universal subbundle. Then

$$\varphi^* \Delta_\lambda(c(Q - \pi^* E)) = \Delta_{\tilde{\lambda}} c(\varphi^* S^\vee) ,$$

where $\tilde{\lambda}$ is the conjugate partition to λ. In particular, if $c(E) = 1$, then

$$\varphi^* \Delta_\lambda = \Delta_{\tilde{\lambda}} .$$

Example 14.6.6 (cf. Grothendieck (1), Jouanolou (1)). If X is non-singular, E a vector bundle of rank n on X, then $A^*(G_dE)$ is the algebra over A^*X generated by elements $a_1, \ldots, a_d, b_1, \ldots, b_{n-d}$, modulo the relations

$$\sum_{i=0}^{k} a_i b_{k-i} = c_k(E)$$

for $k = 1, \ldots, n$. (Take $a_i = c_i(S)$, $b_j = c_j(Q)$, with S and Q the universal bundles.)

More generally, if $F = F(d_1, \ldots, d_k)$ is a flag bundle of flags in E, with universal flag $D_1 \subset \ldots \subset D_k = E_F$, $\operatorname{rank}(D_i) = d_i$, then $A^*(F)$ is the $A^*(X)$-algebra generated by the Chern classes of the quotient bundles D_i/D_{i-1}, with relations determined by the identity

$$\prod_{i=1}^{k} c(D_i/D_{i-1}) = c(E_F) .$$

14.7 Schubert Calculus

The Grassmann variety of d-planes in $\mathbb{P}^n$, denoted $G_d(\mathbb{P}^n)$, is identified with $G_{d+1}(\mathbb{A}^{n+1})$. The universal quotient bundle Q on $G = G_d(\mathbb{P}^n)$ has rank $n-d$. The classes

$$\sigma_m = c_m(Q) \in A^m G, \qquad 1 \leqq m \leqq n-d,$$

are called the *special Schubert* classes. For each $\lambda = (\lambda_0, \ldots, \lambda_d)$, with $n-d \geqq \lambda_0 \geqq \ldots \geqq \lambda_d \geqq 0$, define the *Schubert class* $\{\lambda\}$ or $\{\lambda_0, \ldots, \lambda_d\}$ by the formula

$$\{\lambda_0, \ldots, \lambda_d\} = \Delta_\lambda(\sigma) = \det(\sigma_{\lambda_i+j-i})_{0 \leqq i,j \leqq d}.$$

Thus $\sigma_m = \{m, 0, \ldots, 0\}$. By the results of the previous section, the Schubert classes form a free $\mathbb{Z}$-basis of $A^* G$. In particular, $\operatorname{Pic}(G) = A^1 G = \mathbb{Z}$, generated by σ_1. The multiplication is determined by *Pieri's formula:*

$$\{\lambda\} \cdot \sigma_m = \sum \{\mu\}$$

the sum over μ with $n-d \geqq \mu_0 \geqq \lambda_0 \geqq \ldots \geqq \mu_d \geqq \lambda_d$, $|\mu| = |\lambda| + m$. More generally,

$$\{\lambda\} \cdot \{\mu\} = \sum N_{\lambda,\mu,\varrho} \{\varrho\}$$

where the $N_{\lambda,\mu,\varrho}$ are determined by the Littlewood-Richardson rule. If $|\lambda| + |\mu| = (d+1)(n-d)$, one has the *duality theorem:*

$$\{\lambda\} \cdot \{\mu\} = \begin{cases} \{n-d, \ldots, n-d\} & \text{if } \lambda_i + \mu_{d-i} = n-d \text{ for } 0 \leqq i \leqq d \\ 0 & \text{otherwise}. \end{cases}$$

If $A_0 \subsetneqq A_1 \subsetneqq \ldots \subsetneqq A_d \subset \mathbb{P}^n$ is a flag of subspaces, with $a_i = \dim A_i$, let

$$\Omega(A_0, \ldots, A_d) = \{L \in G_d(\mathbb{P}^n) \mid \dim L \cap A_i \geqq i, 0 \leqq i \leqq d\}.$$

Then $\Omega(A_0, \ldots, A_d)$ is a subvariety of G of dimension $\sum_{i=0}^{d} (a_i - i)$, i.e., $\sum_{i=0}^{d} a_i - \frac{1}{2} d(d+1)$, called a *Schubert variety*. Its class in $A_*(G)$ depends only on $a_0, \ldots, a_d$, and it is denoted by $(a_0, \ldots, a_d)$; for $0 \leqq a_0 < a_1 < \ldots < a_d \leqq n$,

$$(a_0, \ldots, a_d) = [\Omega(A_0, \ldots, A_d)].$$

Giambelli's formula reads:

$$\{\lambda_0, \ldots, \lambda_d\} \cap [G] = (a_0, \ldots, a_d)$$

with $\lambda_i = n - d + i - a_i$. The notations $\{\lambda_0, \ldots, \lambda_d\}$ and $(a_0, \ldots, a_d)$ originate with Schubert; both remain in use, together with many others.

These results are all special cases of the formulas of the preceding section; the shift in indexing from $(\lambda_1, \ldots, \lambda_{d+1})$ to $(\lambda_0, \ldots, \lambda_d)$ compensates for the loss of dimension in passing from affine to projective spaces.

The formulas of this section can also be proved by direct geometric argument, as in Hodge-Pedoe (1). The approach followed here, on the other hand, shows how the formulas are consequences of the basic determinantal formula, and of general polynomial identities, together with the vanishing of certain higher Chern classes and inverse Chern classes that occurs on Grassmannians.

Example 14.7.1. Let $G = G_d(\mathbb{P}^n)$, $\alpha \in A_k(G)$. For each Schubert class $(b_0, \ldots, b_d)$ of codimension k, let

$$\alpha_{b_0, \ldots, b_d} = \int_G \alpha \cdot (b_0, \ldots, b_d) .$$

Then the expression of α in terms of the Schubert classes of dimension k is

$$\alpha = \sum \alpha_{n-a_d, \ldots, n-a_0} (a_0, \ldots, a_d) .$$

(This is a formal consequence of duality.) If β is a class of complementary dimension, then

$$\int_G \alpha \cdot \beta = \sum \alpha_{n-a_d, \ldots, n-a_0} \beta_{a_0, \ldots, a_d} ,$$

the sum over all Schubert classes of dimension k.

Example 14.7.2. For $G_1(\mathbb{P}^3)$, Schubert used a special notation:

$$\begin{aligned}
1 &= (2, 3) = \{0, 0\} = [G_1(\mathbb{P}^3)] \\
g &= (1, 3) = \{1, 0\} = \sigma_1 \\
g_p &= (0, 3) = \{2, 0\} = \sigma_2 \\
g_e &= (1, 2) = \{1, 1\} \\
g_s &= (0, 2) = \{2, 1\} \\
G &= (0, 1) = \text{class of a point in } G_1(\mathbb{P}^3) .
\end{aligned}$$

The products are: $g^2 = g_p + g_e$, $g \cdot g_p = g \cdot g_e = g_s$, $g_p^2 = g_e^2 = G$, $g_p \cdot g_e = 0$, $g \cdot g_s = G$.

It follows that $g^4 = 2G$: there are two lines in $\mathbb{P}^3$ which meet 4 given lines in general position.

Example 14.7.3. The special Schubert classes are

$$\sigma_m = (n-d-m, n-d+1, n-d+2, \ldots, n) .$$

If A is a subspace of $\mathbb{P}^n$ of codimension $d+m$, σ_m is represented by the special Schubert variety

$$\{L \in G_d(\mathbb{P}^n) \mid L \cap A \neq \emptyset\} .$$

Example 14.7.4. If $(a_0, \ldots, a_d)$ and $(b_0, \ldots, b_d)$ are Schubert cycles of complementary dimensions, then

$$\int_G (a_0, \ldots, a_d) \cdot (b_0, \ldots, b_d) = \begin{cases} 1 & \text{if } a_i + b_{d-i} = n, \ 0 \leqq i \leqq d \\ 0 & \text{otherwise} . \end{cases}$$

Example 14.7.5 (cf. Hodge-Pedoe (1)XIV.2). Let $\varphi : G_{n-d-1}(\mathbb{P}^n) \to G_d(\mathbb{P}^n)$ be the duality isomorphism of Example 14.6.5. Then

$$\varphi^* ((a_0, \ldots, a_d)) = (b_0, \ldots, b_{n-d-1}) ,$$

where the integers $b_0, \ldots, b_{n-d-1}$ form the complement of $n-a_d, \ldots, n-a_0$ in the set of integers from 0 to n. (This follows formally from Giambelli's formula and Example 14.6.5, using the combinatorial fact which begins the proof of Lemma A.9.2.)

Example 14.7.6. (a) Let C be a reduced curve of degree d in $\mathbb{P}^3$. Let

$$V_C = \{l \in G_1(\mathbb{P}^3) \mid l \text{ meets } C\}.$$

Then V_C is a hypersurface in $G = G_1 \mathbb{P}^3$, and

$$[V_C] = d \cdot g$$

with $g = \sigma_1$ as in Example 14.7.2.

(Consider the incidence correspondence $I \subset \mathbb{P}^3 \times G$, consisting of pairs (P, l), $P \in l$, with projection $p_1 : I \to \mathbb{P}^3$, $p_2 : I \to G$. Then

$$V_C = p_2 p_1^{-1}(C).$$

Since p_2 maps $p_1^{-1}(C)$ birationally onto V_C,

$$[V_C] = p_{2*} p_1^* [C] = p_{2*} p_1^* (d \cdot [l_0]) = d \cdot g.$$

where l_0 is a fixed line in $\mathbb{P}^3$. Or one may simply use Example 14.7.1.)

(b) If $C_1, \ldots, C_4$ are fixed curves in $\mathbb{P}^3$, in general position with respect to the action of the projective linear group, then there are

$$2 \prod_{i=1}^{4} \deg(C_i)$$

lines meeting all four curves. (Use (a) and $\int_G g^4 = 2$.)

Example 14.7.7. (a) Let C be an irreducible non-planar curve in $\mathbb{P}^3$. Let W_C be the surface in $G = G_1(\mathbb{P}^3)$ which is the closure of the set of lines which meet C in two or more points. Then, with the notation of Example 14.7.2,

$$[W_C] = d \cdot g_p + \frac{n(n-1)}{2} \cdot g_e,$$

where d is the apparent number of double points of C (the number of chords to C through a general point), and n is the degree of C (so there are $n(n-1)/2$ chords to C in a general plane).

(b) Given curves C, C_1, C_2 in general position, with d, n as above for C, and $n_i = \deg C_i$, there are

$$n_1 n_2 (d + n(n-1)/2)$$

chords to C which meet C_1 and C_2.

(c) If C' is another such curve, with d' apparent double points, and degree n', in general position with respect to C, the number of common chords to C and C' is

$$dd' + n n' (n-1)(n'-1)/4.$$

Example 14.7.8. The rank of a curve C in $\mathbb{P}^3$ is the number r of tangents to C which meet a given general line (cf. Example 14.4.15). The number of tangents to C meeting a given curve C' of degree n', in general position, is $r n'$.

For higher dimensional applications of Schubert calculus to tangency problems, see Fulton-Kleiman-MacPherson (1).

Example 14.7.9. Let λ, μ be partitions, with $\mu \subset \lambda$. Set (see Example 14.5.4)

$$\Delta_{\lambda/\mu} = |\sigma_{\lambda_i-\mu_j-i+j}|_{0 \leqq i,j \leqq d} = |\sigma_{b_j-a_i}|,$$

where $a_i = n-d+i-\lambda_i$, $b_j = n-d+j-\mu_j$, $a_i \leqq b_i$. Then $\Delta_{\lambda/\mu}$ is dual to the class

$$(n-a_d, \ldots, n-a_0) \cdot (b_0, \ldots, b_d).$$

(If $|\nu| = |\lambda| - |\mu|$, by Example 14.5.4, $\Delta_{\lambda/\mu} \cdot \Delta'_\nu = N_{\nu,\mu,\lambda} = \Delta_\nu \cdot \Delta_\mu \cdot \Delta'_\lambda$, where Δ'_ν, Δ'_λ are the classes dual to Δ_ν, Δ_λ.)

Example 14.7.10. A general formula for multiplying Schubert classes is given by Example 14.5.3:

$$\{\lambda_0, \ldots, \lambda_d\} \cdot \{\mu_0, \ldots, \mu_d\} = |\sigma_{\lambda_i+\mu_{d-j}-i+j}|_{0 \leqq i,j \leqq d}.$$

Example 14.7.11 (Schubert (1), cf. Hodge-Pedoe (1)XIV.7). Let $G = G_d(\mathbb{P}^n)$, $N = (d+1)(n-d)$. For any Schubert variety $(a_0, \ldots, a_d)$, of dimension k, let

$$\deg(a_0, \ldots, a_d) = \int_G \sigma_1^k \cdot (a_0, \ldots, a_d).$$

(The Plücker imbedding of G is determined by the line bundle $\Lambda^{n-d}(Q)$, so $\sigma_1 = c_1(Q)$ is the class of a hyperplane section.)

(i)
$$\deg(a_0, \ldots, a_d) = \sum_{i=0}^{d} \deg(a_0, \ldots, a_i - 1, \ldots, a_d)$$

where $\deg(b_0, \ldots, b_d)$ is defined to be zero if the conditions $0 \leqq b_0 < b_1 < \ldots < b_d \leqq n$ are not all satisfied. (Use Pieri's formula.)

(ii)
$$\deg(a_0, \ldots, a_d) = \frac{k!}{a_0! \ldots a_d!} \prod_{i<j} (a_j - a_i)$$

where $k = \sum_{i=0}^{d} a_i - d(d+1)/2$. (Use (i) and induction on k.)

(iii)
$$\deg(G) = \int_G \sigma_1^N = \frac{1!\,2! \ldots d!\,N!}{(n-d)!\,(n-d+1)! \ldots n!}.$$

(Set $a_i = n-d+i$.) This number is the number of d-planes meeting N given $(n-d-1)$-planes in general position in $\mathbb{P}^n$. For $d=1$, this number is $\frac{1}{n-1}\binom{2n-2}{n}$.

More generally, Schubert (3) uses Pieri's formula and induction on k to prove

(iv)
$$(a_0, \ldots, a_d) \cdot \sigma_1^k = \sum (k!)\, |1/(a_i - b_j)!|\, (b_0, \ldots, b_d),$$

where the sum is over all $0 \leqq b_0 < \ldots < b_d \leqq n$ with $\sum b_i = \sum a_i - k$; in the determinant, $1/(a_i - b_j)!$ is taken to be 0 if $a_i < b_j$. Equivalently,

(v)
$$\deg((a_0, \ldots, a_d) \cdot (b_0, \ldots, b_d)) = k!\, |1/(a_i + b_{d-j} - n)!|,$$

where $k = \sum (a_i - i) + \sum (b_j - j) - (d+1)(n-d)$.

Example 14.7.12. Parameter spaces for many enumerative problems can be formed by a finite number of projective, Grassmann, and flag bundle constructions. Let $\mathbb{P}^n = P(V)$, $\dim V = n+1$, $G = G_d\mathbb{P}^n = G_{d+1}(V)$, with universal subbundle S.

(a) The variety of hypersurfaces of degree m in d-planes in $\mathbb{P}^m$ may be identified with the projective bundle $P(\mathrm{Sym}^m(S^\vee))$ over G.

(b) If $M = P(W)$ is a subspace of $P(V)$, $W \subset V$, the incidence variety

$$I_M = \{(P, L) \in M \times G \mid P \in L\}$$

is a Grassmann bundle over M. If A is the universal rank 1 subbundle of W on M, I_M is $G_d(V_M/A)$. If $\dim M = n - d$, the projection from I_M to G is birational.

(c) The space $X = P(\mathrm{Sym}^m(S^\vee)) \times_G I_M$ represents triples (H, L, P), H a hypersurface in L as in (a), P a point of $L \cap M$. The subvariety D of triples with $P \in H$ is given by the vanishing of the composite

$$A_X^{\otimes m} = \mathrm{Sym}^m(A_X) \hookrightarrow \mathrm{Sym}^m(S_X) \to \mathscr{O}(1)_X$$

with $\mathscr{O}(1)$ the universal line bundle on $P(\mathrm{Sym}^m S^\vee)$. The image of D in $P(\mathrm{Sym}^m(S^\vee))$ represents hypersurfaces $H \subset L$ which meet M.

(d) For $n = 3$, $d = 2$, $m = 2$, one has the 8-dimensional variety $P(\mathrm{Sym}^2 S^\vee)$ of conics in $\mathbb{P}^3$. With X as in (c), one may calculate that

$$\int_X c_1((A_X^\vee)^{\otimes 2} \otimes \mathscr{O}(1)_X)^8 = 92$$

which corresponds to the fact that there are 92 conics meeting 8 lines in general position (cf. Example 3.2.22).

Example 14.7.13. *Fano schemes* (cf. Altman-Kleiman (2)). Let $\mathbb{P}^n = P(E)$, E a vector space of dimension $n+1$. Let $G = G_d(\mathbb{P}^n) = G_{d+1}(E)$, S the universal subbundle of E_G.

A hypersurface $X \subset \mathbb{P}^n$ of degree m is given by a section of $\mathrm{Sym}^m(E^\vee)$ on $\mathbb{P}^n$. Since $S^\vee$ is a quotient of $E_G^\vee$, this determines a section of $\mathrm{Sym}^m(S^\vee)$ on G, whose zero scheme is the *Fano scheme* F of d-planes in X. If $\mathrm{codim}(F, G) = \binom{m+d}{m}$, then the class of $[F]$ in A^*G is

$$[F] = c_{\binom{m+d}{m}}(\mathrm{Sym}^m(S^\vee))\,.$$

In case $m = 3$, $d = 1$, $[F] = c_4(\mathrm{Sym}^3(S^\vee)) = 18\, s_1^2 s_2 + 9\, s_2^2$, where $s_i = c_i(S^\vee)$. The degree of F is

$$\int [F] \cdot \sigma_1^{2n-6} = 18 \deg(G_1(\mathbb{P}^{n-1})) + 9 \deg(G_1(\mathbb{P}^{n-2}))$$

$$= \frac{27\,(2n-6)!}{(n-3)!\,(n-1)!} \cdot (3n-7)\,.$$

For $n = 3$, this gives 27, the number of lines on a cubic surface. For $n = 4$, it gives 45, the degree of the Fano surface of lines on a cubic three-fold.

Example 14.7.14 (cf. Hodge-Pedoe (1)XIV.7, Altman-Kleiman (2)).

(a) The lines in $\mathbb{P}^n$ which lie in a given non-singular quadric hypersurface form a subvariety F (a Fano variety) of codimension 3 of $G = G_1(\mathbb{P}^n)$, with

$$[F] = 4\{2, 1\} = 4(n-3, n-1).$$

In fact $[F] \cdot (0, 4) = 0$ since no line in the quadric passes through a general point, and $\int_G [F] \cdot (1, 3) = 4$ since four lines in a general section by a 3-plane meet a general line in the 3-plane. Or one may construct F as the zero scheme of a section of $\mathrm{Sym}^2(S^\vee)$, as in the previous example, so (Example 14.5.1)

$$[F] = c_3(\mathrm{Sym}^2(S^\vee)) = 2^2 \Delta_{2,1}(c(S^\vee)) = 4\Delta_{2,1}(c(Q)).$$

(b) There are 16 lines common to two general quadrics in $\mathbb{P}^4$. $\left(\int_G (1, 3)^2 = 1.\right)$

Example 14.7.15. If $2d \leqq n-1$, the d-planes in $\mathbb{P}^n$ which lie in a given non-singular quadric hypersurface form a subscheme F of codimension $(d+1)(d+2)/2$ in $G = G_d(\mathbb{P}^n)$, whose class in A^*G is

$$[F] = 2^{d+1}\{d+1, d, \ldots, 1\} = 2^{d+l}(n-2d-1, n-2d+1, \ldots, n-1).$$

(As in the previous example, this is the top Chern class of $\mathrm{Sym}^2(S^\vee)$.) If $n = 2d+2$, two general quadrics in $\mathbb{P}^{2d+2}$ have 4^{d+1} d-planes in common.

Example 14.7.16. *Flag manifolds.* Fix n, and fix points $e_0, \ldots, e_n$ in $\mathbb{P}^n$ which span $\mathbb{P}^n$. For any integer m between 0 and n, let $[m]$ denote the m-plane spanned by $e_0, e_1, \ldots, e_m$. For any set $\underline{a}$ of integers between 0 and n, number the elements of $\underline{a}$ in increasing order:

$$0 \leqq a_0 < a_1 < \ldots < a_d \leqq n, \quad d = \#\underline{a} - 1.$$

Let $[\underline{a}]$ denote the corresponding Schubert variety:

$$[\underline{a}] = \{L \in G_d(\mathbb{P}^n) \mid \dim L \cap [a_i] \geqq i, 0 \leqq i \leqq d\}.$$

Denote by $\underline{a}^*$ the dual sequence:

$$\underline{a}^* = \{n-a_d, n-a_{d-1}, \ldots, n-a_0\},$$

so that $[\underline{a}^*]$ is the dual Schubert variety to $[\underline{a}]$.

If $\underline{a}$ and $\underline{b}$ are sequences, write $\underline{b} < \underline{a}$ if $\underline{b}$ is a proper subset of $\underline{a}$.

For integers $0 \leqq d_1 < d_2 < \ldots < d_r < n$, let $F = F(d_1, \ldots, d_r; n)$ denote the flag manifold whose points are flags of linear subspaces

$$L_1 \subset L_2 \subset \ldots \subset L_r \subset \mathbb{P}^n$$

with $\dim L_i = d_i$. Consider nests of r sets:

$$\underline{a}^1 < \underline{a}^2 < \ldots < \underline{a}^r, \quad \#\underline{a}^i = d_i + 1.$$

For each such nest, let

$$[\underline{a}^1; \ldots; \underline{a}^r] = \{(L_1, \ldots, L_r) \in F \mid L_i \in [\underline{a}^i], 1 \leqq i \leqq r\}.$$

Then $[\underline{a}^1; \ldots; \underline{a}^r]$ is an irreducible subvariety of F, with

$$\dim [\underline{a}^1; \ldots; \underline{a}^r] = \sum_{i=1}^{r} \sum_{j=0}^{d_i}{}' (a_j^i - j) .$$

In this sum any term $(a_j^i - j)$ is omitted if a_j^i appears in the previous $\underline{a}^{i-1}$.

Let $(\underline{a}^1; \ldots; \underline{a}^r)$ denote the class of $[\underline{a}^1; \ldots; \underline{a}^r]$ in $A_* F$. This class is independent of the choice of basis $e_0, \ldots, e_n$.

(i) (Basis) The classes $(\underline{a}^1; \ldots; \underline{a}^r)$ form a free basis for $A_* F$.

(ii) (Duality) If $\dim [\underline{a}^1; \ldots; \underline{a}^r] + \dim [\underline{b}^1; \ldots; \underline{b}^r] = \dim (F)$, then

$$\int (\underline{a}^1; \ldots; \underline{a}^r) \cdot (\underline{b}^1; \ldots; \underline{b}^r) = \begin{cases} 1 & \text{if } \underline{b}^i = \underline{a}^{i*} \text{ for } 1 \leqq i \leqq r \\ 0 & \text{otherwise} . \end{cases}$$

Thus if $\alpha \in A_k F$,

$$\alpha = \sum (\alpha \cdot [\underline{a}^{1*}; \ldots; \underline{a}^{r*}]) [\underline{a}^1; \ldots; \underline{a}^r]$$

the sum over all $[\underline{a}^1; \ldots; \underline{a}^r]$ of dimension k. (Realizing the flag manifold as a succesion of Grassmann bundles, one verifies that $A_* F$ is a free abelian group with the same number of generators as the number of nests $\underline{a}^1 < \ldots < \underline{a}^r$, $\# \underline{a}^i = d_i + 1$. It therefore suffices to prove (ii). In case

$$a_j^i + b_{d_i - j}^i = n \text{ for all } 1 \leqq i \leqq r, \ 0 \leqq j \leqq d_i ,$$

or $a_j^i + b_{d_i - j}^i < n$ for one such i, j, the conclusion follows by using complementary flags, as in Example 14.6.1. In any other case, $\dim [\underline{a}^1, \ldots, \underline{a}^r] + \dim [\underline{b}^1, \ldots, \underline{b}^r] > \dim F$; equivalently, if $a_j^i \leqq c_j^i$ for all i, j, with some inequality strict, then $\dim [\underline{a}^1, \ldots, \underline{a}^r] < \dim [\underline{c}^1, \ldots, \underline{c}^r]$.)

Example 14.7.17. *Incidence varieties* (Martinelli (2)). Fix $d \leqq n$, and let

$$I = \{(P, L) \in \mathbb{P}^n \times G_d(\mathbb{P}^n) \mid P \in L\} ,$$

a smooth variety of dimension $d(n-d) + n$. For $0 \leqq a_0 < \ldots < a_d \leqq n$, and $0 \leqq k \leqq d$, let

$$[a_0, \ldots, \overset{*}{a}_k, \ldots, a_d] = \{(P, L) \in I \mid \dim L \cap [a_i] \geqq i, \ 0 \leqq i \leqq d, \text{ and } P \in [a_k]\}$$

with $[a_p]$ as in the previous example. This is a variety of dimension $\sum (a_i - i) + k$. The classes $(a_0, \ldots, \overset{*}{a}_k, \ldots, a_d)$ of these varieties form a basis of $A_* I$. The class

$$(n - a_d, \ldots, n \overset{*}{-} a_k, \ldots, n - a_0)$$

is dual to $(a_0, \ldots, \overset{*}{a}_k, \ldots, a_d)$. To know the coefficients of a class in $A_k I$, it suffices to compute the intersection number with all dual classes. This basis is often more convenient than that obtained by realizing I as a Grassmann bundle over $\mathbb{P}^n$ (Example 14.7.12 (b)) and applying Proposition 14.6.5. The appealing notation was used by Martinelli. (The assertions are special cases of the preceding example.)

Example 14.7.18. Fix d, n and I as in the preceding example. Let V be a subvariety of $\mathbb{P}^n$ of codimension $e \leqq d + 1$. Let $V_\circ$ be the non-singular locus of V, and let $V' \subset I$ be the closure of

$$\{(P, L) \in I \mid P \in V_\circ, \dim (T_P V \cap L) \geqq d - e + 1\} .$$

Then V' is a subvariety of I of codimension $d+1$. Let $\mu(V)$ be the class of $[V']$ in $A^{d+1}I$. Let $\mu_k = \mu(M)$, where M is a linear subspace of codimension $d+1-k$. Then

$$\mu(V) = m_{d-e+1} \cdot \mu_0 + m_{d-e} \cdot \mu_1 + \ldots + m_0 \mu_{d-e+1}$$

where m_i is the ith class of V (Example 14.4.15).

From this one may deduce the formula for the number of varieties in an r-dimensional family tangent to r given varieties in general position, in terms of the corresponding numbers for r given linear varieties (cf. § 10.4). (For the proof, intersect both sides with the basis for $A_{d+1}(I)$ described in the preceding example. For details see Fulton-Kleiman-MacPherson (1).)

Notes and References

In topology, for suitably generic maps of vector bundles on an oriented manifold, Thom showed that the degeneracy loci must be Poncaré dual to some universal polynomials in the characteristic classes of the bundles. These universal polynomials were determined by Porteous (2).

In algebraic geometry, on a non-singular projective variety X, and when the degeneracy loci have the expected dimensions, the formulas for their classes in $A_*(X)$ were proved by Kempf and Laksov (1). The extension to arbitrary singular varieties, with possibly excess degeneracy loci, as given in § 14.3 and § 14.4, is new here. In fact, once one has the machinery of Chaps. $1-7$ to handle excess intersections on possibly singular varieties, the proof of Kempf and Laksov goes over without difficulty. The particular case of the localized top Chern class was discussed in Fulton-MacPherson (1).

The realization that Schubert calculus is essentially the same as the algebra of Schur polynomials (resulting from representation theory of the symmetric group) has occurred more than once. Giambelli (2) was apparently the first to express general determinantal loci by such polynomials. The polynomials themselves go back to Jacobi. The connection was explicitly pointed out by Lesieur (1), after Ehresmann (1) had worked out the cell structure and cohomology ring of Grassmannians, cf. Horrocks (1). Recently the importance of Schur functions has been illuminated by Lascoux (1)−(7). The seminars in Strasbourg (1976) and Torún (1980) (cf. Stanley (1) and Dieudonné (2)) are good sources for this. For combinatorial facts about Schur functions, the book of Macdonald (3) is also recommended. It should be pointed out that, although much of this algebra goes back to Schubert, Pieri, Giambelli, and other classical algebraic geometers, the general rule for intersecting Schubert varieties was not given before the Littlewood-Richardson rule.

For symmetric and skew-symmetric degeneracy loci (cf. Example 14.4.11), early formulas were also given by Giambelli (3). Barth and Tjurin gave special cases, with applications, for vector bundle maps. General formulas have been given by Józefiak, Lascoux, and Pragacz (1), and by Harris and Tu (1), to which we refer for additional applications and references.

An important generalization of Schubert calculus is from Grassmannians and flag varieties to homogeneous varieties of the form G/P, for P a parabolic subgroup of a semi-simple linear algebraic group G. For this we refer to papers of Berstein-Gel'fand-Gel'fand (1), Demazure (1), Lakshmibai-Musili-Seshadri (1), Marlin (1), and Hiller (1), (2), (3). The basis for flag manifolds described in Example 14.7.16 was given by Ehresmann (1); it agrees with that arising from the description of a flag manifold as $\mathrm{SL}(n+1)/P$.

An interesting calculation of the cohomology ring of Grassmannians via zeros of vector fields is given by Carrell and Lieberman (1).

The article by Kleiman (3) contains general constructions and functorial properties of Grassmann bundles, as well as applications to smoothing cycles. Formulas for Gysin push-forward homomorphisms for Grassmann and flag bundles are given by Damon (1), Ilori (1), Józefiak-Lascoux-Pragacz (1), and Harris-Tu (1).

The residual formula for Chern classes in Example 14.1.4 is new. The excess Porteous formula of Example 14.4.7 is the result of conversations with G. Ellingsrud, J. Harris, and R. Lazarsfeld. Examples 14.4.5 and 14.4.17 include strengthening and simplification of known facts for linear systems on curves.

Even to summarize the applications of Schubert calculus to enumerative geometry could double the length of this book; it is hoped that the examples give some hint of the possibilities. From the classical period there are the books and papers of Schubert, Zeuthen, Pieri, Giambelli, and Severi. The books of Semple and Roth (1), Hodge and Pedoe (1), and Baker (1), (2) contain hundreds of these applications. For a modern survey the article of Kleiman (8) is recommended. Other classical as well as new applications may be found in Griffiths and Harris (1) and Arbarello, Cornalba, Griffiths, and Harris (1).

For a sampling of recent enumerative applications of the Porteous formula and related ideas, some references are: Harris and Mumford (1), Laksov (4), (5), Eisenbud and Harris (1).

Chapter 15. Riemann-Roch for Non-singular Varieties

Summary

The Grothendieck-Riemann-Roch theorem (GRR) states that for a proper morphism $f: X \to Y$ of non-singular varieties,

$$\operatorname{ch}(f_* \alpha) \cdot \operatorname{td}(T_Y) = f_* (\operatorname{ch}(\alpha) \cdot \operatorname{td}(T_X))$$

for all α in the Grothendieck group of vector bundles, or of coherent sheaves, on X. When Y is a point, one recovers Hirzebruch's formula (HRR) for the Euler characteristic of a vector bundle E on X:

$$\sum (-1)^i \dim H^i(X, E) = \int_X \operatorname{ch}(E) \cdot \operatorname{td}(T_X).$$

The aim of this chapter is to show how the geometry of the deformation to the normal cone leads to a simple proof of GRR when f is a closed imbedding. The same proof gives the corresponding theorem without denominators, which in turn yields a simple proof of the formula for blowing up Chern classes.

The reader of this chapter is assumed to have some familiarity with the cohomology of coherent sheaves, although the necessary facts are reviewed in the first section. In addition, the proof of GRR when f is a projection is only sketched briefly. The first nine sections of the article of Borel and Serre (1) are recommended for a detailed discussion of these points.

Although the theorem is stated here for arbitrary non-singular varieties, the proof in this chapter makes an additional assumption of projectivity. The general case will be considered, together with singular varieties, in Chap. 18.

15.1 Preliminaries

For any scheme X, $K^\circ X$ denotes the *Grothendieck group of vector bundles* (locally free sheaves) on X. Each vector bundle E determines an element, denoted $[E]$, in $K^\circ X$. $K^\circ X$ is the free abelian group on the set of isomorphism classes of vector bundles, modulo the relations

$$[E] = [E'] + [E'']$$

whenever E' is a subbundle of a vector bundle E, with quotient bundle $E'' = E/E'$. The tensor product makes $K^\circ X$ a ring: $[E] \cdot [F] = E \otimes F]$. For any morphism $f: Y \to X$ there is an induced homomorphism

$$f^*: K^\circ X \to K^\circ Y,$$

taking $[E]$ to $[f^*E]$, where f^*E is the pull-back bundle; this makes K° a contravariant functor from schemes to commutative rings.

The *Grothendieck group of coherent sheaves* on X, denoted by $K_\circ X$, is defined to be the free abelian group on the isomorphism classes $[\mathscr{F}]$ of coherent sheaves on X, modulo the relations

$$[\mathscr{F}] = [\mathscr{F}'] + [\mathscr{F}'']$$

for each exact sequence

$$0 \to \mathscr{F}' \to \mathscr{F} \to \mathscr{F}'' \to 0$$

of coherent sheaves on X. Tensor product makes $K_\circ X$ a $K^\circ X$-module:

$$K^\circ X \otimes K_\circ X \to K_\circ X,$$

$[E] \cdot [\mathscr{F}] = [E \otimes_{\mathscr{O}_X} \mathscr{F}]$.

For any *proper* morphism $f: X \to Y$, there is a homomorphism

$$f_*: K_\circ X \to K_\circ Y$$

which takes $[\mathscr{F}]$ to $\sum_{i \geqq 0} (-1)^i [R^i f_* \mathscr{F}]$. Here $R^i f_* \mathscr{F}$ is Grothendieck's higher direct image sheaf, the sheaf associated to the presheaf

$$U \to H^i(f^{-1}(U), \mathscr{F})$$

on Y. It is a basic fact that the $R^i f_* \mathscr{F}$ are coherent when $\mathscr{F}$ is coherent and f is proper ([EGA]III.3.2.1). The fact that this push-forward f_* is well-defined on $K_\circ X$ results from the long exact cohomology sequence for the $R^i f_*$. The fact that $K_\circ$ is a covariant functor for proper morphisms uses the spectral sequence for composite morphisms.

The push-forward and pull-back are related by the usual projection formula:

$$f_*(f^*\alpha \cdot \beta) = \alpha \cdot f_*\beta$$

for $f: X \to Y$ proper, $\alpha \in K^\circ Y$, $\beta \in K_\circ X$. This follows from the formula $R^i f_*(f^*E \otimes \mathscr{F}) = E \otimes R^i f_* \mathscr{F}$ for E locally free on Y, $\mathscr{F}$ coherent on X.

On any X there is a canonical "duality" homomorphism:

$$K^\circ X \to K_\circ X$$

which takes a vector bundle to its sheaf of sections. When X is *non-singular*, this duality map is an isomorphism. The reason for this is that a coherent sheaf $\mathscr{F}$ on a non-singular X has a finite resolution by locally free sheaves, i.e., there is an exact sequence

$$0 \to E_n \to E_{n-1} \to \ldots \to E_1 \to E_0 \to \mathscr{F} \to 0$$

with $E_0, \ldots, E_n$ locally free. The inverse homomorphism from $K_\circ X$ to $K^\circ X$ takes $[\mathscr{F}]$ to $\sum_{i=0}^{n} (-1)^i [E_i]$, for such a resolution. (See Appendix B.8.3 for details.)

From now on we consider schemes X which are smooth over a given ground field K. For such X we identify $K^\circ X$ and $K_\circ X$, and write simply $K(X)$. For $X = \operatorname{Spec}(K)$, $K(X) = \mathbb{Z}$, and we make this identification. We write $A(X)$ for the ring of cycles modulo rational equivalence on X, and $A(X)_{\mathbb{Q}}$ for $A(X) \otimes_{\mathbb{Z}} \mathbb{Q}$.

There is a homomorphism, called the *Chern character* (cf. Examples 3.2.3, 15.1.2)

$$\mathrm{ch} : K(X) \to A(X)_{\mathbb{Q}}$$

determined by the following properties:

(i) ch is a homomorphism of rings;

(ii) if $f : Y \to X$, $\mathrm{ch} \circ f^* = f^* \circ \mathrm{ch}$;

(iii) if L is a line bundle on X,

$$\mathrm{ch}[L] = \exp(c_1(L)) = \sum_{i \geq 0} (1/i!) c_1(L)^i .$$

For a vector bundle E we write either $\mathrm{ch}\, E$ or $\mathrm{ch}[E]$.

The Chern character does not commute with proper push forward. A fundamental insight of Grothendieck was to phrase the Riemann-Roch problem as the problem of comparing $\mathrm{ch} \circ f_*$ with $f_* \circ \mathrm{ch}$, for a proper morphism $f : X \to Y$. When Y is a point, and E is a vector bundle on X, $\mathrm{ch}\, f_*[E]$ is

$$\chi(X, E) = \sum_{i \geq 0} (-1)^i \dim_K H^i(X, E)$$

in $A(\operatorname{Spec}(K))_{\mathbb{Q}} = \mathbb{Q}$.

Model for closed imbeddings. To motivate the general formula, we work out a special case of a closed imbedding $f : X \to Y$ in which all the terms may be calculated explicitly. In this model, X is arbitrary, and Y is $P(N \oplus 1)$, where N is an arbitrary vector bundle of rank d on X. The imbedding f is the zero section imbedding of X in N, followed by the canonical open imbedding of N in $P(N \oplus 1)$:

$$f : X \to N \subset P(N \oplus 1) = Y .$$

Let p be the bundle projection from Y to X, and let Q be the universal quotient bundle, of rank d, on Y. Let s be the section of Q determined by the projection of the trivial factor in $p^*(N \oplus 1)$ to Q. This section s vanishes precisely on X, i.e., $Z(s) = X$. In particular, for any $\alpha \in A(Y)$,

$$f_*(f^* \alpha) = \alpha \cdot f_*[X] = c_d(Q) \cdot \alpha \tag{1}$$

(Proposition 14.1 (a)). In addition, the Koszul complex determined by s:

$$0 \to \Lambda^d Q^\vee \to \ldots \Lambda^2 Q^\vee \to Q^\vee \xrightarrow{s^\vee} \mathscr{O}_Y \to f_* \mathscr{O}_X \to 0$$

is a resolution of the sheaf $f_* \mathscr{O}_X$ (Lemma A.7.1). For any vector bundle E on X, we therefore have an explicit resolution of $f_* E$:

$$0 \to \Lambda^d Q^\vee \otimes p^* E \to \ldots \to Q^\vee \otimes p^* E \to p^* E \to f_* E \to 0 .$$

Therefore

$$\text{(2)} \qquad \operatorname{ch} f_*[E] = \sum_{p=0}^{d} (-1)^p \operatorname{ch}(\Lambda^p Q^\vee) \cdot \operatorname{ch}(p^* E) .$$

To write this as the image of a class on X, we want, by (1), to divide the right side by $c_d(Q)$. In fact, the *Todd class* $\operatorname{td}(Q)$ is determined by the identity

$$\text{(3)} \qquad \sum_{p=0}^{d} (-1)^p \operatorname{ch}(\Lambda^p Q^\vee) = c_d(Q) \cdot \operatorname{td}(Q)^{-1} .$$

(see Examples 3.2.4 and 3.2.5).

Combining (1), (2), and (3),

$$\operatorname{ch} f_* E = c_d(Q) \operatorname{td}(Q)^{-1} \cdot \operatorname{ch}(p^* E) = f_*(f^* \operatorname{td}(Q)^{-1} \cdot f^* \operatorname{ch}(p^* E)) .$$

Since $f^* Q = N$, and $f^* p^* E = E$, this can be rewritten

$$\text{(4)} \qquad \operatorname{ch} f_* E = f_*(\operatorname{td}(N)^{-1} \cdot \operatorname{ch}(E)) .$$

The Todd class, like the total Chern class, takes sums to products. Therefore

$$\operatorname{td}(N)^{-1} = f^* \operatorname{td}(T_Y)^{-1} \cdot \operatorname{td}(T_X)$$

(cf. Appendix B.7.2). The right side of (4) is therefore

$$f_*(f^* \operatorname{td}(T_Y)^{-1} \cdot \operatorname{td}(T_X) \cdot \operatorname{ch} E) = \operatorname{td}(T_Y)^{-1} \cdot f_*(\operatorname{td}(T_X) \cdot \operatorname{ch} E) ,$$

and (4) may be rewritten:

$$\text{(5)} \qquad \operatorname{ch}(f_* E) \cdot \operatorname{td}(T_Y) = f_*(\operatorname{ch}(E) \cdot \operatorname{td}(T_X)) .$$

The Grothendieck-Riemann-Roch theorem we shall prove is the assertion that (5) holds for an arbitrary proper morphism $f: X \to Y$.

Example 15.1.1. (a) The group $K_\circ(\mathbb{P}^m)$ is generated by the classes $[\mathcal{O}(n)]$, $0 \leqq n \leqq m$. If p maps $\mathbb{P}^m$ to a point, then

$$p_*[\mathcal{O}(n)] = \chi(\mathbb{P}^m, \mathcal{O}(n)) = \binom{n+m}{m} .$$

(b) For any X, Y there is an exterior product

$$K_\circ X \otimes K_\circ Y \xrightarrow{\times} K_\circ(X \times Y)$$

with $[\mathcal{F}] \times [\mathcal{G}] = \mathrm{pr}_1^*(\mathcal{F}) \otimes \mathrm{pr}_2^*(\mathcal{G})$, pr_i the projections. If $Y = \mathbb{P}^m$, this product

$$K_\circ X \otimes K_\circ \mathbb{P}^m \to K_\circ(X \times \mathbb{P}^m)$$

is surjective.

(The proofs are quite similar to the proofs of corresponding facts for rational equivalence given in § 3.3. In fact, if E is a vector bundle on X, the pull-back $K_\circ X \to K_\circ E$ is an isomorphism, and $K_\circ P(E)$ is a direct sum of e copies of $K_\circ X$. An elegant proof of this is given by Quillen (2) § 8.)

Example 15.1.2. (a) Let c_i be the ith elementary symmetric function in variables $x_1, \ldots, x_e$. Let $p_k = x_1^k + \ldots + x_e^k$. Then (cf. Macdonald (3) p. 20)

$$p_k = \det \begin{pmatrix} c_1 & 1 & 0 & \ldots & 0 \\ 2c_2 & c_1 & 1 & & \vdots \\ \vdots & \vdots & & & 1 \\ k\,c_k & (k-1)\,c_{k-1} & & \ldots & c_1 \end{pmatrix}.$$

Thus $p_0 = e$, $p_1 = c_1$, $p_2 = c_1^2 - 2c_2$, $p_3 = c_1^3 - 3c_1c_2 + 3c_3, \ldots$.

(b) If E is a vector bundle of rank e on X, $c_i = c_i(E)$, and p_k is defined as in (a), then (cf. Example 3.2.3)

$$\operatorname{ch}(E) = \sum_{k=0}^{\infty} (1/k!)\, p_k = e + c_1 + \tfrac{1}{2}(c_1^2 - 2c_2) + \ldots.$$

(c) If $c_i = 0$ for $0 < i < d$, then $p_d = (-1)^{d-1}\, d\, c_d$. If E is a vector bundle of rank e, and $c_i(E) = 0$ for $0 < i < d$, then

$$\operatorname{ch}(E) = e + ((-1)^{d-1}/(d-1)!)\, c_d(E) + \ldots.$$

(d) If X is any connected scheme, the function which assigns to each vector bundle its rank is additive on exact sequences, and defines a homomorphism

$$\operatorname{rk}: K^\circ X \to \mathbb{Z}.$$

With this notion of rank, and the determinantal definition of p_k in (a), the formulas of (b) and (c) extend to arbitrary elements of $K^\circ X$. Note that any $\xi \in K^\circ X$ has Chern classes $c_i(\xi)$ by Example 3.2.7.

Example 15.1.3. For any partition $\lambda: \lambda_1 \geqq \ldots \geqq \lambda_e \geqq 0$ of e, and x_i, c_i as in Example 15.1.2, define $p_\lambda(c) = p_\lambda(c_1, \ldots, c_e)$ to be the sum of all the distinct monomials in $x_1, \ldots, x_e$ which are obtained from $x_1^{\lambda_1} \cdot \ldots \cdot x_e^{\lambda_e}$ by permutation of the variables $x_1, \ldots, x_e$. For example, if $\lambda = (k, 0, \ldots, 0)$, p_λ is the polynomial p_k of the preceding example.

If E is a vector bundle, let $p_\lambda(c(E))$ be the polynomial obtained by substituting $c_i(E)$ for c_i. If

$$0 \to E' \to E \to E'' \to 0$$

is an exact sequence of vector bundles, then a result of Thom (cf. Milnor-Stasheff (1) § 16.2) states:

$$p_\lambda(c(E)) = \sum_{\alpha\beta=\lambda} p_\alpha(c(E'))\, p_\beta(c(E'')),$$

where the sum is over all pairs of partitions α, β whose juxtaposition $\alpha\beta$ is λ. This generalizes the formula $\operatorname{ch}(E) = \operatorname{ch}(E') + \operatorname{ch}(E'')$.

Example 15.1.4 (cf. Hirzebruch (1) Lemma 1.7.1, or Borel-Serre (1) Prop. 10). On $\mathbb{P}^m$, let $x = c_1(\mathcal{O}_{\mathbb{P}^m}(1))$. Then for all $n \geqq 0$,

$$\int_{\mathbb{P}^m} e^{nx}\, x^{m+1}/(1 - e^{-x})^{m+1} = \binom{n+m}{n}.$$

(The integrand is a power series in x, for which one wants the coefficient of x^m. Divide the integrand by x^{m+1}, and compute the residue by changing variables: $y = 1 - e^{-x}$.)

Example 15.1.5. On any scheme X, the *topological filtration* on $K_\circ X$ is defined by letting $F_k K_\circ X$ be the subgroup generated by coherent sheaves whose support has dimension at most k. Equivalently, $F_k K_\circ X$ is generated by the classes $[\mathcal{O}_V]$, as V ranges over closed subvarieties of X of dimension at most k. If $f: X \to Y$ is a proper morphism, then

$$f_*(F_k K_\circ X) \subset F_k K_\circ Y.$$

The associated graded groups $\mathrm{Gr}_k K_\circ X = F_k K_\circ X / F_{k-1} K_\circ X$ are covariant for proper morphisms.

If $\mathscr{F}$ is a coherent sheaf whose support has dimension at most k, $\mathscr{F}$ determines a k-cycle $Z_k(\mathscr{F})$ on X:

$$Z_k(\mathscr{F}) = \sum_{\dim V = k} m_V(\mathscr{F})\,[V]$$

where $m_V(\mathscr{F})$ is the length of the stalk of $\mathscr{F}$ at the generic point of V, as a module over the local ring $\mathcal{O}_{V,X}$ of X along V.

There is a unique, surjective homomorphism

$$\varphi: Z_k X \to \mathrm{Gr}_k K_\circ X$$

which takes $[V]$ to $[\mathcal{O}_V]$, and $Z_k(\mathscr{F})$ to $[\mathscr{F}]$ for any coherent sheaf $\mathscr{F}$ whose support has dimension at most k. This homomorphism is covariant for proper morphisms, and passes to rational equivalence, determining

$$\varphi: A_k X \to \mathrm{Gr}_k K_\circ X$$

which is surjective, and commutes with push-forward by proper morphisms. (For the covariance, reduce to the case where $f: X \to Y$ is a proper morphism of varieties, and $\mathscr{F} = \mathcal{O}_X$; replacing Y by an open subvariety, one may assume f is finite, and $\mathcal{O}_X$ is free over $\mathcal{O}_Y$, in which case $R^i f_* \mathcal{O}_X = 0$ for $i > 0$ and $f_*[\mathcal{O}_X] = \deg(X/Y)\,[\mathcal{O}_Y]$. To show that φ passes to rational equivalence, use Example 1.6.4: if $f: X \to \mathbb{P}^1$ is dominant, X a variety, $D_0 = f^{-1}(0)$, $D_\infty = f^{-1}(\infty)$, then

$$0 \to f^*\mathcal{O}(-1) \to \mathcal{O}_X \to \mathcal{O}_{D_t} \to 0$$

for $t = 0, \infty$; therefore $[\mathcal{O}_{D_0}] = \mathcal{O}_{D_\infty}]$ in $K_\circ X$. Since $[D] = Z_k(\mathcal{O}_D)$ for D an effective Cartier divisor on a $(k+1)$-dimensional variety, $\varphi[D_0] = \varphi[D_\infty]$, as required.)

Example 15.1.6. If X is an arbitrary algebraic scheme, and $\alpha \in K_\circ X$, there is a quasi-projective scheme X', a projective morphism $f: X' \to X$, and a class $\alpha' \in K_\circ X'$ such that $f_* \alpha' = \alpha$. If the ground field has characteristic zero, one may even take X' to be smooth (but not necessarily connected); for this one uses Hironaka's resolution of singularities. See Lemma 18.3 or Fulton (4) for a proof.

Example 15.1.7 (see [SGA6] or Baum-Fulton-MacPherson (2) App. 2).

(a) If X is a scheme, X_{red} the associated reduced scheme, then the inclusion of X_{red} in X induces an isomorphism

$$K_\circ(X_{\text{red}}) \cong K_\circ(X).$$

(b) If X is a closed subscheme of a scheme Y, then $K_\circ(X)$ is canonically isomorphic to the Grothendieck group of coherent sheaves on Y whose support is contained in X.

(c) If X is a closed subscheme of a non-singular scheme Y, then $K_\circ(X)$ is canonically isomorphic to the Grothendieck group of bounded complexes of locally free sheaves on Y which are exact off X, modulo the subgroup generated by complexes which are exact on all of Y.

Example 15.1.8 (cf. [SGA 6]III). There is a class of morphisms $f\colon X \to Y$, of possibly singular schemes, called *perfect* morphisms, for which there are functorial Gysin homomorphisms:

$$f_*\colon K^\circ X \to K^\circ Y \quad \text{and} \quad f^*\colon K_\circ Y \to K_\circ X.$$

For the first, f is assumed to be proper. For simplicity, we describe these concepts for schemes which are quasi-projective over a fixed non-singular base scheme.

If f is a closed imbedding, f is perfect if and only if $f_*\mathcal{O}_X$ may be resolved by a finite complex $E.$ of locally free sheaves on Y. It follows that every locally free sheaf F on X has a resolution $G.$ on Y, and

$$f_*[F] = \sum (-1)^i [G_i].$$

For any coherent sheaf $\mathcal{F}$ on Y,

$$f^*[\mathcal{F}] = \sum (-1)^i [\operatorname{Tor}_i^Y(\mathcal{O}_X, \mathcal{F})],$$

where $\operatorname{Tor}_i^Y(\mathcal{O}_X, \mathcal{F})$ is the ith homology sheaf of the complex $E. \otimes_{\mathcal{O}_Y} \mathcal{F}$.

A general morphism $f\colon X \to Y$ of imbeddable schemes is perfect if, when f is factored into a closed imbedding $i\colon X \to P$ followed a smooth morphism $p\colon P \to Y$, i is perfect in the above sense. If f is proper, one may take $P = P(E)$, E a vector bundle on Y, p the projection. One defines $f_* = p_* \circ i_*$, $f^* = i^* \circ p^*$. Here $p^*[\mathcal{F}] = [p^*\mathcal{F}]$ for a coherent sheaf $\mathcal{F}$ on Y. The homomorphism

$$p_*\colon K^\circ(P(E)) \to K^\circ Y$$

is uniquely determined by the projection formula $p_*(p^*\alpha \cdot \beta) = \alpha \cdot p_*\beta$ for $\alpha \in K^\circ Y$, $\beta \in K^\circ(P(E))$, and the identity $p_*[\mathcal{O}_E(n)] = [\operatorname{Sym}^n E^\vee]$.

15.2 Grothendieck-Riemann-Roch Theorem

Theorem 15.2. *Let $f\colon X \to Y$ be a proper morphism of non-singular varieties. Then for all $\alpha \in K(X)$,*

$$\operatorname{ch}(f_*\alpha) \cdot \operatorname{td}(T_Y) = f_*(\operatorname{ch}(\alpha) \cdot \operatorname{td}(T_X))$$

in $A(Y)_{\mathbb{Q}}$.

Proof. We shall assume f factors into a composite

$$X \xrightarrow{g} Y \times \mathbb{P}^m \xrightarrow{p} Y$$

where g is a closed imbedding, and p is the projection. Such a factorization exists whenever X is quasi-projective, for if $i: X \to \mathbb{P}^m$ is a locally closed imbedding, then $g = (f, i)$ is a closed imbedding of X in $Y \times \mathbb{P}^m$. For the general case, see Example 15.2.9 and Corollary 18.3.1 (c).

The theorem may be interpreted to say that the homomorphism

$$\tau_X : K(X) \to A(X)_{\mathbb{Q}}$$

defined by

$$\tau_X(\alpha) = \mathrm{ch}(\alpha) \cdot \mathrm{td}(T_X)$$

commutes with proper push-forward: $f_* \circ \tau_X = \tau_Y \circ f_*$. It follows that, if the theorem is valid for g and for p, then it is valid for $f = p \circ g$.

For the projection, consider more generally the projection $f: Y \times Z \to Y$, for Z non-singular. There is a commutative diagram

$$\begin{array}{ccc} K(Y) \otimes K(Z) & \xrightarrow{\tau_Y \otimes \tau_Z} & A(Y)_{\mathbb{Q}} \otimes A(Z)_{\mathbb{Q}} \\ \times\downarrow & & \times\downarrow \\ K(Y \times Z) & \xrightarrow{\tau_{Y \times Z}} & A(Y \times Z)_{\mathbb{Q}} . \end{array}$$

The point here is that $\mathrm{td}(T_{X \times Y}) = \mathrm{td}(T_X) \times \mathrm{td}(T_Y)$, as follows from the multiplicative property of the Todd class. Now if $Z = \mathbb{P}^m$, the left vertical map is surjective, and $K(\mathbb{P}^m)$ is generated by $[\mathcal{O}(n)]$, $n \geqq 0$ (Example 15.1.1). One is therefore reduced to verifying the theorem for the mapping p from $\mathbb{P}^m$ to a point, and $\alpha = [\mathcal{O}(n)]$, i.e., to verifying the formula

$$\int_{\mathbb{P}^m} \mathrm{ch}(\mathcal{O}(n)) \cdot \mathrm{td}(T_{\mathbb{P}^m}) = \chi(\mathbb{P}^m, \mathcal{O}(n)) .$$

Since $\mathrm{td}(T_{\mathbb{P}^m}) = (x/1 - e^{-x})^{m+1}$, where $x = c_1(\mathcal{O}_{\mathbb{P}^m}(1))$ (cf. Appendix B.5.8) this is the content of Examples 15.1.1 (a) and 15.1.4.

Riemann-Roch for closed imbeddings. We turn now to the case where $f: X \to Y$ is a closed imbedding. Let N be the normal bundle to X in Y. We shall use the deformation to the normal bundle to deform the imbedding f into the imbedding $\bar{f}: X \to P(N \oplus 1)$ discussed in the "model" at the end of the previous section. From § 5.1, we construct a commutative diagram

$$\begin{array}{ccccc} X & \xrightarrow{\bar{f}} & P(N \oplus 1) + \tilde{Y} = M_\infty & \to & \{\infty\} \\ {}^{i_\infty}\downarrow & & \searrow^{k} \quad \downarrow^{l} \quad \swarrow_{j_\infty} & & \downarrow \\ X \times \mathbb{P}^1 & \xrightarrow{F} & M & \longrightarrow & \mathbb{P}^1 \\ {}_{i_0}\uparrow & & \uparrow j_0 & & \uparrow \\ X & \xrightarrow{f} & Y = M_0 & \to & \{0\} \end{array}$$

where M is the blow-up of $Y \times \mathbb{P}^1$ along $X \times \{\infty\}$. We may assume $\alpha = [E]$, with E a vector bundle on X. Let $\tilde{E} = p^* E$, where p is the projection from $X \times \mathbb{P}^1$ to X. Choose a resolution $G.$ of $F_*(\tilde{E})$ on M:

$$0 \to G_n \to G_{n-1} \to \ldots \to G_0 \to F_*(\tilde{E}) \to 0 .$$

Since $X \times \mathbb{P}^1$ and M are flat over $\mathbb{P}^1$, it follows from Lemma A.4.2 that the restrictions of this exact sequence to the fibres M_0 and M_∞ remain exact. Therefore $j_0^* G.$ is a resolution of $j_0^* F_*(\tilde{E})$, and $j_\infty^* G.$ is a resolution of $j_\infty^* F_*(\tilde{E})$. Since $j_0^* F_* \tilde{E} = f_* i_0^* \tilde{E} = f_*(E)$,

(i) $j_0^* G.$ resolves $f_*(E)$ on $Y = M_0$.

Similarly, $j_\infty^* G.$ resolves $\bar{f}_*(E)$ on M_∞. But $\bar{f}(X)$ is disjoint from $\tilde{Y}$. Therefore

(ii) $k^* G.$ resolves $\bar{f}_*(E)$ on $P(N \oplus 1)$, and
(iii) $l^* G.$ is acyclic.

For a complex $F.$ of vector bundles, we write $\operatorname{ch}(F.)$ for $\sum (-1)^i \operatorname{ch}(F_i)$. We compute the image of $\operatorname{ch}(f_* E)$ in $A(M)_{\mathbb{Q}}$:

$$\begin{aligned} j_{0*}(\operatorname{ch}(f_* E)) &= j_{0*}(\operatorname{ch}(j_0^* G.)) && \text{by (i)} \\ &= \operatorname{ch}(G.) \cdot j_{0*}[Y] && \text{(projection formula)} \\ &= \operatorname{ch}(G.) \cdot (k_*[P(N \oplus 1)] + l_*[\tilde{Y}]) \end{aligned}$$

by the basic fact that $[M_\circ] - [M_\infty] = [\operatorname{div}(\varrho)] = 0$ in $A(M)_{\mathbb{Q}}$ (cf. Example 5.1.1). Continuing, with the projection formula again:

$$\begin{aligned} &= k_*(\operatorname{ch}(k^* G.)) + l_*(\operatorname{ch}(l^* G.)) \\ &= k_*(\operatorname{ch}(\bar{f}_* E)) + 0 \qquad \text{by (ii) and (iii).} \end{aligned}$$

For $\bar{f}$, $\operatorname{ch}(\bar{f}_* E)$ was calculated in the model of the previous section (formula (5)). Therefore

(iv) $j_{0*}(\operatorname{ch}(f_* E)) = k_*(\bar{f}_*(\operatorname{td}(N)^{-1} \cdot \operatorname{ch}(E))$ in $A(M)_{\mathbb{Q}}$.

Let $q: M \to Y$ be the composite of the blow-down map from M to $Y \times \mathbb{P}^1$, followed by the projection to Y. By construction of M, $q \circ j_0 = \operatorname{id}_Y$ and $q \circ k \circ \bar{f} = f$. Applying q_* to (iv), we deduce

$$\operatorname{ch}(f_* E) = f_*(\operatorname{td}(N)^{-1} \cdot \operatorname{ch}(E)) .$$

As in the transition from formula (4) to (5) in the previous section, this is equivalent to the theorem. □

Corollary 15.2.1 (Hirzebruch-Riemann-Roch). *Let E be a vector bundle on a non-singular, complete variety X. Then*

$$\chi(X, E) = \int_X \operatorname{ch}(E) \cdot \operatorname{td}(T_X) .$$

Proof. This is the content of the theorem for the mapping of X to a point. □

Corollary 15.2.2. *If X is a non-singular, complete, n-dimensional variety, then*

$$\chi(X, \mathcal{O}_X) = \int_X \operatorname{td}_n(T_X) . \quad \square$$

Example 15.2.1. Let X be a non-singular complete curve of genus g, i.e., $2 - 2g = \int_X c_1(T_X)$. Then

$$\chi(X, \mathcal{O}_X) = \frac{1}{2} \int_X c_1(T_X) .$$

Since $\dim H^0(X, \mathcal{O}_X) = 1$, this is equivalent to $\dim H^1(X, \mathcal{O}_X) = g$.

Let E be a vector bundle of rank e on X. Then by HRR

$$\chi(X, E) = \int_X c_1(E) + e(1-g)\,.$$

In particular, if $L = \mathcal{O}(D)$ is a line bundle,

$$\chi(X, \mathcal{O}(D)) = \deg(D) + 1 - g\,.$$

If E, F are bundles of ranks e, f on X, then (cf. Weil (1))

$$\chi(X, \operatorname{Hom}(E, F)) = e\int_X c_1(F) - f\int_X c_1(E) + (ef)(1-g)\,.$$

Example 15.2.2. Let X be a non-singular complete surface. Then

$$\operatorname{td}(T_X) = 1 + \frac{1}{2}c_1 + \frac{1}{12}(c_1^2 + c_2)$$

with $c_i = c_i(T_X)$. Therefore

$$\chi(X, \mathcal{O}_X) = \frac{1}{12}\int_X c_1^2 + c_2\,.$$

If $K = -c_1(T_X)$ is a canonical divisor, $\chi = \int_X c_2(T_X)$ is the topological Euler characteristic, $p_g = \dim H^2(X, \mathcal{O}_X)$, and $q = \dim H^1(X, \mathcal{O}_X)$, this reads

$$(*) \qquad \chi(X, \mathcal{O}_X) = 1 - q + p_g = \frac{1}{12}((K \cdot K) + \chi)\,.$$

If E is a vector bundle of rank e on X, with $d_i = c_i(E)$, then

$$\chi(X, E) = \frac{1}{2}\int_X (d_1^2 - 2d_2 + c_1 d_1) + e\,\chi(X, \mathcal{O}_X)\,.$$

In particular, if D is a divisor on X,

$$\chi(X, \mathcal{O}(D)) = \tfrac{1}{2}((D \cdot D) - (K \cdot D)) + \chi(X, \mathcal{O}_X)\,.$$

If D is an effective divisor on X,

$$\chi(X, \mathcal{O}_D) = -\tfrac{1}{2}((D \cdot D) + (K \cdot D))\,.$$

(Use the exact sequence $0 \to \mathcal{O}(-D) \to \mathcal{O}_X \to \mathcal{O}_D \to 0$.) If D is an irreducible curve on X, and $p_a(D) = \dim H^1(D, \mathcal{O}_D)$ is its arithmetic genus, then

$$p_a(D) = \tfrac{1}{2}((D \cdot D) + (K \cdot D)) + 1\,.$$

For example, a plane curve of degree n has arithmetic genus $\frac{1}{2}(n-1)(n-2)$. A curve of bidegree (m, n) in $\mathbb{P}^1 \times \mathbb{P}^1$ has arithmetic genus $(m-1)(n-1)$.

Example 15.2.3. Let $f\colon X^1 \to Y^2$ be as in Example 9.3.2, with $f(X) = \bar{X}$. Then

$$\deg \mathbb{D}(f) = 2(p_a\bar{X} - p_a X)\,.$$

(By the double point formula, $\deg \mathbb{D}(f) = \bar{X} \cdot \bar{X} + K_Y \cdot \bar{X} - \deg K_X$. Then use Examples 15.2.1 and 15.2.2.) Comparing with Example 9.3.2, this is the well-known formula for the degree of the conductor (cf. Serre (5) IV).

Example 15.2.4. *Index theorem.* Let X be a non-singular projective surface, H an ample divisor on X. Let D be a divisor on X, such that $\int D \cdot H = 0$. Then

$\int D \cdot D \leqq 0$, and the following are equivalent: (i) $\int D \cdot D = 0$. (ii) $\int D \cdot E = 0$ for all divisors E on X (iii) a non-zero multiple of D is algebraically equivalent to zero.

Here we write $\int D \cdot E$ for the intersection number $(D \cdot E)_X$. One may replace H by a multiple if necessary to assume H is represented by a smooth hyperplane section of X. The proof is in several steps:

1) There is a $c = c(X, H)$ so that $H^2(X, \mathscr{O}(D)) = 0$ for all D with $\int D \cdot H \geqq c$. This follows from the cohomology sequence of the exact sequence

$$0 \to \mathscr{O}(D + (k-1)H) \to \mathscr{O}(D + kH) \to \mathscr{O}(D + kH)|_H \to 0$$

using Serre's vanishing theorem, and the vanishing of $H^1(L)$ for line bundles L of degree larger than $2g - 2$ on a curve of genus g (cf. Example 15.2.7).

2) If $\int D \cdot D > 0$, then $\int D \cdot H > 0$ if and only if $H^0(\mathscr{O}(nD)) \neq 0$ for some $n > 0$. For if $nD \sim E$ with E effective, then $E \cdot H > 0$. Conversely, by Riemann-Roch, if $\int D \cdot D > 0$ then $\chi(\mathscr{O}(nD)) \to \infty$ as $n \to \infty$.

3) If $\int D \cdot D > 0$, and $\int D \cdot H > 0$ for one ample divisor H, then $\int D \cdot H > 0$ for all ample divisors H.

4) If $\int D \cdot D > 0$ and $\int D \cdot H = 0$, then $\int (nD - H)^2 > 0$, $\int (nD - H) \cdot H < 0$, and

$$\int (nD - H)(D + mH) = n \int D \cdot D - m \int H \cdot H > 0$$

for $n \gg m \gg 0$, contracting 3). This proves the first assertion.

5) If $\int D \cdot D = \int D \cdot H = 0$ but $\int D \cdot E \neq 0$, replace E by $aE + bH$ so that $\int E \cdot H = 0$. By 4),

$$\int (mD + E) \cdot (mD + E) = 2m \int D \cdot E + \int E \cdot E \leqq 0$$

for all integers m, which is absurd.

6) To see (ii) $\Rightarrow$ (iii), consider $D_{m,n} = mD + nH$. By Riemann-Roch, for $n \geqq n_0$,

$$\dim H^0(X, \mathscr{O}(D_{m,n})) + \dim H^2(X, \mathscr{O}(D_{m,n})) > 0\,.$$

By 1), for $n > n_0$, there are effective divisors E_m linearly equivalent to $D_{m,n}$. All E_m have the same Hilbert polynomial, and since such curves are parametrized by an algebraic scheme (cf. Mumford (2)), there are E_{m_1} and E_{m_2} in the same algebraic family.)

A generalization appears in Example 19.3.1. For the singular case, see Kleiman's Appendix to Exposé XIII of [SGA 6]. The theorem was discovered and proved by Hodge using De Rahm theory, as in Griffiths-Harris (1). B. Segre (1) gave a proof more like the one sketched here, cf. Mumford (2).

Example 15.2.5 (a) If $\dim(X) = 3$,

$$\operatorname{td}(T_X) = 1 + \frac{1}{2} c_1 + \frac{1}{12}(c_1^2 + c_2) + \frac{1}{24} c_1 c_2,$$

with $c_i = c_i(T_X)$. If E is a vector bundle of rank e on X, with Chern classes $d_i = c_i(E)$, then

$$\chi(X, E) = \int_X \frac{1}{6}(d_1^3 - 3d_1 d_2 + 3d_3) + \frac{1}{4} c_1 (d_1^2 - 2d_2) + \frac{1}{12}(c_1^2 + c_2) d_1 + \frac{e}{24} c_1 c_2.$$

If $E = \mathcal{O}(D)$, this simplifies to

$$\chi(X, \mathcal{O}(D)) = \int_X \frac{1}{6} D^3 + \frac{1}{4} c_1 \cdot D^2 + \frac{1}{12}(c_1^2 + c_2) \cdot D + \frac{1}{24} c_1 c_2 .$$

(b) If $X = \mathbb{P}^3$, and E is a vector bundle on X, let $c_i(E) = n_i h^i$, h a hyperplane class, $n_i \in \mathbb{Z}$. Then

$$n_1^3 - 3 n_1 n_2 + 3 n_3 + 11 n_1$$

is divisible by 6. For example, if E has rank 2, $n_1 n_2$ must be even.

Example 15.2.6. Let X be an abelian variety of dimension n, E a vector bundle on X. Then

$$\chi(X, E) = \int_X \mathrm{ch}_n(E) .$$

If L is a line bundle, then

$$\chi(X, L) = \frac{1}{n!} \int_X c_1(L)^n .$$

In particular, for any divisor D on X, the self-intersection number $\int_X D^n$ is divisible by $n!$.

Example 15.2.7. The Hirzebruch-Riemann-Roch theorem is most useful when one has some control on some of the cohomology groups. Some basic facts in this direction are:

(a) For $X \subset \mathbb{P}^N$, a theorem of Serre (1) implies that $H^i(X, E(n)) = 0$ for $n \gg 0$. Therefore

$$\dim H^0(X, E(n)) = \int_X \mathrm{ch}(E(n)) \cdot \mathrm{td}(T_X)$$

for n sufficiently large.

(b) Serre duality gives isomorphisms

$$H^i(X, E) \cong H^{n-i}(X, K \otimes E^\vee) .$$

Here $n = \dim(X)$, and $K = \Omega_X^n = T_X^\vee$ is the canonical line bundle (cf. Serre (2), Grothendieck (4), Hartshorne (2)).

(c) In characteristic zero, the Kodaira vanishing theorem states that if L is an ample line bundle on X, then

$$H^i(X, K \otimes L) = 0 \quad \text{for} \quad i > 0 .$$

More generally, Le Potier (cf. Verdier (2)) has shown that, for E an ample vector bundle on X,

$$H^i(X, \Omega_X^j \otimes E) = 0$$

for $i + j \geqq \dim(X) + \mathrm{rank}(E)$, $\Omega_X^j = \Lambda^j T_X^\vee$.

Example 15.2.8. Let $f\colon X \to Y$ be a smooth proper morphism of non-singular varieties, and let $T_f = \mathrm{Ker}(T_X \to f^* T_Y)$ be the relative tangent bundle. Then for all $\alpha \in K(X)$,

$$\mathrm{ch}(f_* \alpha) = f_*(\mathrm{ch}(\alpha) \cdot \mathrm{td}(T_f)) .$$

(Since $\mathrm{td}(T_f) = \mathrm{td}(T_X) \cdot f^* \mathrm{td}(T_Y)^{-1}$, this is equivalent to GRR for f, as in the equivalence of formulas (4) and (5) of § 15.1.) For example, if $X = P(E)$, for E

a vector bundle on Y, and f is the projection, then (cf. Appendix B.5.8)

$$\operatorname{ch}(f_*\alpha) = f_*(\operatorname{ch}(\alpha)\operatorname{td}(f^*E \otimes \mathcal{O}(1))).$$

Similarly, if $X \to Y$ is a family of abelian varieties, $T_f = f^*F$ for a vector bundle F on Y, and

$$\operatorname{ch}(f_*\alpha) = f_*(\operatorname{ch}(\alpha)) \cdot \operatorname{td}(F).$$

Example 15.2.9. In characteristic zero, the proof of Theorem 15.2 can be completed, in case X is not quasi-projective, as follows. Given $\alpha \in K(X)$, there is a non-singular, quasi-projective X', a proper morphism $g: X' \to X$, and an element $\alpha' \in K(X')$ with $g_*(\alpha') = \alpha$ (Example 15.1.6). The theorem as proved applies to g and to $f \circ g$. Therefore

$$g_*(\operatorname{ch}(\alpha') \cdot \operatorname{td}(T_{X'})) = \operatorname{ch}(\alpha) \cdot \operatorname{td}(T_X),$$

and

$$(fg)_*(\operatorname{ch}(\alpha') \cdot \operatorname{td}(T_{X'})) = \operatorname{ch}((fg)_*(\alpha')) \cdot \operatorname{td}(T_Y).$$

Since $(fg)_* = f_* g_*$, the theorem follows.

For a proof without resolution of singularities, valid in arbitrary characteristic, see Corollary 18.3.1.

Example 15.2.10. The arithmetic genus satisfies a fundamental "modular" law:

If X, Y, and Z are non-singular divisors on a non-singular projective variety M, such that X is linearly equivalent to $Y + Z$, and Y meets Z transversally in a disjoint union of varieties $V_1, \ldots, V_r$, then

$$\chi(X, \mathcal{O}_X) = \chi(Y, \mathcal{O}_Y) + \chi(Z, \mathcal{O}_Z) - \sum_{i=1}^{r} \chi(V_i, \mathcal{O}_{V_i}).$$

In fact, there is only one way to assign a number to every non-singular projective variety, taking the value 1 on a point, and satisfying this modular law. This is proved in Washnitzer (2) and Fulton (7).

If X is a non-singular subvariety of a non-singular variety Y, deformation to the normal bundle gives a linear equivalence between Y and $P(N \otimes 1) + \tilde{Y}$ with $P(N \oplus 1)$ and $\tilde{Y}$ meeting transversally in $P(N)$. Since $\chi(P(E), \mathcal{O}_{P(E)}) = \chi(X, \mathcal{O}_X)$ for any vector bundle E on X, it follows that

$$\chi(\tilde{Y}, \mathcal{O}_{\tilde{Y}}) = \chi(Y, \mathcal{O}_Y).$$

Example 15.2.11. *Signatures.* The signature (index) $\sigma(X)$ of a compact oriented manifold is defined to be zero if its (real) dimension is not a multiple of 4, and the index of the quadratic form given by cup product on $H^{2k}(X; \mathbb{R})$, if $\dim(X) = 4k$. If X is a complex projective variety, Hodge theory gives the formula

$$\sigma(X) = \sum_{p,q} (-1)^q \dim H^q(X, \Omega_X^p)$$

(cf. Hirzebruch (1) 15.8.2, or Griffiths and Harris (1) p. 126.).

The geometry of deformation to the normal bundle can be used to prove a formula for the signature of the blow-up $\tilde{Y}$ of a complex manifold Y along a submanifold X. Namely,

$$\sigma(\tilde{Y}) = \begin{cases} \sigma(Y) & \text{if } \operatorname{codim}(X, Y) \text{ is odd} \\ \sigma(Y) - \sigma(X) & \text{if } \operatorname{codim}(X, Y) \text{ is even}. \end{cases}$$

(With the notation of the previous example,

$$\sigma(Y) = \sigma(\tilde{Y}) + \sigma(P(N \oplus 1)).$$

This follows from Hirzebruch (1)11.3.1, since Y does not meet $P(N)$ in the deformation space M, and, for any vector bundle E on X, $\sigma(P(E))$ is zero if $\operatorname{rank}(E)$ is odd, and is $\sigma(X)$ if $\operatorname{rank}(E)$ is even.) A similar method may be used for Hirzebruch's general T_y-genus.

Example 15.2.12. *Todd classes.* If X is a non-singular variety set

$$\operatorname{Td}(X) = \operatorname{td}(T_X) \in A(X)_{\mathbb{Q}}.$$

Then

(i) $\int_X \operatorname{Td}(X) = \chi(X, \mathcal{O}_X)$,

(ii) $\operatorname{Td}(X \times Y) = \operatorname{Td}(X) \times \operatorname{Td}(Y)$.

In particular, $\chi(X \times Y, \mathcal{O}_{X \times Y}) = \chi(X, \mathcal{O}_X)\, \chi(Y, \mathcal{O}_Y)$.

(iii) If $f: X \to Y$ is a closed imbedding of codimension d, and X is the intersection of d Cartier divisors $D_1, \ldots, D_d$ on Y, then

$$\operatorname{Td}(X) = f^*(\operatorname{Td}(Y) \cdot \prod_{i=1}^{d} (1 - \exp(-D_i)/D_i),$$

so

$$f_*(\operatorname{Td}(X)) = \operatorname{Td}(Y) \cdot \prod_{i=1}^{d} (1 - \exp(-D_i)).$$

Example 15.2.13. *Chern numbers.* Each isobaric polynomial $P = P(c_1, \ldots, c_n)$ of weight n determines a Chern number $P(X)$ for any n-dimensional non-singular complete variety X:

$$P(X) = \int_X P(c_1(T_X), \ldots, c_n(T_X)).$$

For example, the nth Todd polynomial td_n gives the arithmetic genus

$$\operatorname{td}_n(X) = \chi(X, \mathcal{O}_X).$$

(a) If p_λ are the polynomials defined in Example 15.1.3, then (cf. Milnor-Stasheff (1)§ 16.4) for non-singular X, Y

$$p_\lambda(X \times Y) = \sum_{\alpha\beta = \lambda} p_\alpha(X)\, p_\beta(Y),$$

a result of Thom. In particular, if $\lambda = (k, 0, \ldots, 0)$, and neither X nor Y is a point, then $p_\lambda(X \times Y) = 0$.

(b) An isobaric polynomial P of weight n is determined by its values on X, where X varies over all n-dimensional Cartesian products of projective spaces. (The p_λ form a basis for the polynomials of weight n, and

$$p_\lambda(\mathbb{P}^{\mu_1} \times \ldots \times \mathbb{P}^{\mu_m}) = \begin{cases} 0 & \text{if } \lambda \text{ is not a refinement of } \mu \\ \prod (\lambda_i + 1) & \text{if } \lambda = \mu. \end{cases}$$

(c) (cf. Hirzebruch (1)0.3) The nth Todd polynomial is the only Chern number whose value is 1 on all products of projective spaces. (Use (b) and Example 15.2.10.) Hence multiples of td_n are the only Chern numbers which are invariant under blow-ups along non-singular centers.

Example 15.2.14. In characteristic zero, at least, the arithmetic genus is a birational invariant. If X is a non-singular complex projective variety,

$$\dim H^i(X, \mathscr{O}_X) = h^{0i} = h^{i0} = \dim H^0(X, \Omega_X^i),$$

So $\chi(X, \mathscr{O}_X) = \sum (-1)^i \dim H^0(X, \Omega_X^i)$. The individual numbers h^{i0} are birational invariants (cf. Hirzebruch (1)).

Example 15.2.15. Let $f: X \to Y$ be a regular imbedding of possibly singular schemes, with normal bundle N. Assume that Y admits a closed imbedding into a non-singular variety, so that any coherent sheaf on Y is quotient of a locally free sheaf (Appendix B.8.1). There is a homomorphism

$$f_*: K^\circ X \to K^\circ Y$$

which takes $[\mathscr{F}]$ to $\sum (-1)^i [E_i]$, if E resolves $f_* \mathscr{F}$. Then for all $\alpha \in K^\circ X$

$$\mathrm{ch}(f_* \alpha) \cap [Y] = f_*(\mathrm{td}(N)^{-1} \cdot \mathrm{ch}(\alpha) \cap [X])$$

in $A_* Y_{\mathbb{Q}}$. (The proof is the same as that given for the non-singular case; see § 18.3 for generalizations.)

Example 15.2.16. Give $A_* X$ its natural filtration:

$$F_k A_* X = \sum_{i \leq k} A_i X,$$

so the associated graded group $F_k A_* / F_{k-1} A_*$ is A_k.

(a) If X is non-singular, and V is a k-dimensional subvariety of X, then

$$\mathrm{ch}([\mathscr{O}_V]) = f_*([V] + \alpha)$$

where f is the inclusion of V in X, and $\alpha \in F_{k-1} A_* V$. (Let S be a proper closed subscheme of V, so that $V - S$ is regularly imbedded in $X - S$. Since

$$A_* S \to A_* X \to A_*(X - S) \to 0$$

is exact, and the image of $A_* S$ is in $f_*(F_{k-1} A_* V)$, it suffices to prove the assertion when S is empty. In this case

$$\mathrm{ch}([\mathscr{O}_V]) = f_*(\mathrm{td}(N)^{-1} \cap [V]),$$

where N is the normal bundle to V in X.)

(b) It follows from (a) that the homomorphism $\varphi: A_* X \to \mathrm{Gr}_* K_\circ X$ of Example 15.1.5 becomes an isomorphism after tensoring with $\mathbb{Q}$, and that the Chern character determines an isomorphism

$$\mathrm{ch}: K(X)_{\mathbb{Q}} \xrightarrow{\sim} A(X)_{\mathbb{Q}}$$

of $\mathbb{Q}$-algebras. (By (a) ch maps $F_k K_\circ X$ into $F_k A_*(X)_\mathbb{Q}$. This induces a homomorphism

$$\operatorname{Gr} K_\circ X \to \operatorname{Gr} A_*(X)_\mathbb{Q} = A_*(X)_\mathbb{Q}$$

such that the composite with φ is the natural inclusion of $A_* X$ in $A_* X_\mathbb{Q}$. Since φ is surjective, both maps are isomorphisms after tensoring with $\mathbb{Q}$. Since, after tensoring with $\mathbb{Q}$, ch determines an isomorphism on associated graded groups, the same must hold on the original groups. For a more conceptual proof see Corollary 18.3.2.)

Example 15.2.17. Let E be a vector bundle on a scheme X, $p: P(E) \to X$ the associated vector bundle, $\mathcal{O}(1)$ the universal quotient bundle of $p^* E^\vee$. Then for all $\alpha \in A_* X$, and all $n \geqq 0$,

$$\operatorname{ch}(\operatorname{Sym}^n(E^\vee)) \cap \alpha = p_*(\operatorname{ch}(\mathcal{O}(n)) \cdot \operatorname{td}(p^* E \otimes \mathcal{O}(1)) \cap p^* \alpha)\,.$$

(The equality in each degree amounts to a formal identity in the Chern classes of a vector bundle. One may deduce it without calculation as follows: (1) it suffices to prove it when $\alpha = [X]$, X a smooth projective variety; indeed, one may use any Grassmannian where monomials of given degree in Chern classes of the universal quotient bundle are independent. (2) Since $\operatorname{Sym}^n E^\vee = p_* \mathcal{O}(n)$, and $R^i p_* \mathcal{O}(n) = 0$ for $i > 0$, the desired formula follows from Riemann-Roch (Example 15.2.8). For later use we note that only the projective case of Riemann-Roch is used in this argument.)

Example 15.2.18. Let $G = G_1(\mathbb{P}^n)$, I the incidence variety of $(P, L) \in \mathbb{P}^n \times G$ with $P \in L$ (cf. Example 14.7.12); $I = P(S)$ is a $\mathbb{P}^1$ bundle over G, with projection $p: I \to G$, and line bundle $\mathcal{O}(1)$. Let $X = I \times_G I$, p_1, p_2 the two projections from X to I. For any $d \geqq 1$, set

$$E_d = p_{2*}(p_1^*(\mathcal{O}(d)) \otimes \mathcal{O}_X / \mathcal{I}(\Delta)^{2n-1})$$

where $\mathcal{I}(\Delta)$ is the ideal sheaf of the diagonal imbedding of I in X. Then E_d is a vector bundle of rank $2n-1$ on I, and

$$c(E_d) = \prod_{i=0}^{2n-2} (1 + (d-2i)\, c_1(\mathcal{O}(1)) + i\, p^*(\sigma_1))\,,$$

with σ_1 as in § 14.7. (Apply GRR to $p_2: X \to I$, and calculate $\operatorname{ch}(E_d)$. For another approach, see Ran (2).)

A hypersurface $F \in H^0(\mathbb{P}^n, \mathcal{O}(d))$ determines a section of E_d, whose zeros consist of all (P, L) such that L has contact of order at least $2n-1$ with F at P. The weighted number of such (P, L) is therefore

$$c_{2n-1}(E_d) = \int_I \prod_{i=0}^{2n-2} ((d-2i)\, c_1(\mathcal{O}(1)) + i\, \sigma_1)\,.$$

Since $p_*(c(\mathcal{O}(1))^{i+1})$ is the ith Segre class of S, i.e., σ_i, the right side may be evaluated by Schubert calculus. For example, if $n = 2$, one gets $3d(d-2)$; and if $n = 3$, it gives $5d(d-4)(7d-12)$.

15.3 Riemann-Roch Without Denominators

If E is a vector bundle on a non-singular variety X, the total Chern class $c(E) = 1 + c_1(E) + \ldots$ is an element of the multiplicative group

$$A^\times(X) = \left\{ \sum_{i \geq 0} a_i \,\middle|\, a_i \in A^i X,\ a_0 = 1 \right\}.$$

By the Whitney sum formula, the total Chern class defines a homomorphism

$$c: K(X) \to A^\times(X)$$

from the additive group $K(X)$ to the multiplicative group $A^\times(X)$. If $E.$ is a complex of vector bundles, or any indexed collection of vector bundles on X, we write

$$c(E.) = \prod_i c(E_i)^{(-1)^i}$$

for the image of $\sum (-1)^i [E_i]$ by c.

Let $f: X \to Y$ be a closed imbedding of non-singular varieties, of codimension d, with normal bundle N, and let E be a vector bundle on X. Then $f_*[E] \in K(Y)$, and the object is to find a formula for $c(f_*[E])$ in $A^\times Y$.

To solve this, we first consider the model situation, as in § 15.1: $Y = P(N \oplus 1)$, $X = Z(s)$, s the canonical section of the universal quotient bundle Q, p the projection from Y to X. Then

$$f_*[E] = \sum_{i=0}^{d} (-1)^i [\Lambda^i Q^\vee \otimes p^* E],$$

so

$$c(f_*[E]) = c(\Lambda^\cdot Q^\vee \otimes p^* E) = \prod_{i=0}^{d} c(\Lambda^i Q^\vee \otimes p^* E)^{(-1)^i}.$$

Lemma 15.3. *Fix positive integers d, e. There is a unique power series $P(T_1, \ldots, T_d, U_1, \ldots, U_e)$ with integer coefficients such that for all vector bundles D, E of ranks d, e on any variety V,*

$$c(\Lambda^\cdot D^\vee \otimes E) - 1 = c_d(D) \cdot P(D, E)$$

where $P(D, E)$ denotes $P(c_1(D), \ldots, c_d(D), c_1(E), \ldots, c_e(E))$.

Proof. Let $x_1, \ldots, x_d$ be Chern roots for D, $y_1, \ldots, y_e$ Chern roots for E. Then by Remark 3.2.3

$$c(\Lambda^\cdot D^\vee \otimes E) - 1 = \prod_{p=0}^{d} \prod_{j=1}^{e} \prod_{i_1 < \ldots < i_p} (1 + y_j - x_{i_1} - \ldots - x_{i_p})^{(-1)^p} - 1.$$

It suffices to verify that the right side of this equation is divisible by x_1, for then, by symmetry, it is divisible by all the x_i, and hence by the product, which is $c_d(D)$. If x_1 is set equal to zero, each term in the product with $i_1 > 1$ and given p cancels a term with $i_1 = 1$ and p replaced by $p + 1$. □

Returning to the model, let $e = \operatorname{rank} E$, and define P as in the lemma for this d, e. Then

$$c(f_*[E]) = 1 + c_d(Q) \cdot P(Q, p^* E).$$

Since $c_d(Q) = f_*[X]$, as in § 15.1 this gives

$$c(f_*[E]) = 1 + f_*(P(N, E)).$$

Theorem 15.3. *Let $f: X \to Y$ be a closed imbedding of non-singular varieties, of codimension d, with normal bundle N. Let E be a vector bundle of rank e on X. Then*

$$c(f_* E) = 1 + f_*(P(N, E)),$$

with P as defined in Lemma 15.3.

Proof. The proof is identical with the proof given for Theorem 15.2 for the case of a closed imbedding, with the substitution of total Chern class c for the Chern character ch throughout. □

Example 15.3.1. Let P be defined as in Lemma 15.3, for fixed d, e, and let P_q be the term of isobaric weight q of P. For $\alpha \in K(Y)$, let $c_q(\alpha)$ denote the component of $c(\alpha)$ in $A^q Y$. For $f: X \to Y$, E as in Theorem 15.3, and $q > 0$,

$$c_q(f_*[E]) = f_*(P_{q-d}(N, E)).$$

In particular, $c_q(f_*[E]) = 0$ for $0 < q < d$. Since $P_0(T_1, \ldots, T_d, U_1, \ldots, U_e) = (-1)^{d-1}(d-1)!\, e$ (cf Example 15.1.2(c)),

$$c_d(f_*[E]) = (-1)^{d-1}(d-1)!\, e\, [X].$$

In particular,

$$c_d(f_*[\mathcal{O}_X]) = (-1)^{d-1}(d-1)!\, [X].$$

Example 15.3.2 (Kleiman (3) § 5). Let X be non-singular over an algebraically closed field, $i: X \to \mathbb{P}^m$ a locally closed imbedding, $h = c_1(f^*\mathcal{O}(1))$. For any $\alpha \in A^p X$, there is a vector bundle E on X and integer n so that

$$(p-1)!\, \alpha = c_p(E) - n h^p.$$

In addition one may assume $c_i(E) = n_i h^i$ for $i < p$, n_i some integers, $\operatorname{rank}(E) \leqq \dim(X)$, and $E \otimes \mathcal{O}(-1)$ generated by its sections.

Kleiman uses this result, together with the geometry of Schubert cycles, to show that $(p-1)!\, \alpha$ is smoothable, i.e., rationally equivalent to a cycle $\sum n_i [V_i]$ with all V_i non-singular, provided $p > (\dim(X) - 2)/2$. Earlier, in characteristic zero, Hironaka (1) had shown that α itself can be smoothed if $\dim(\alpha) \leqq 3$ and $p > (\dim(X) + 2)/2$. The impossibility of smoothing cycles in general was shown by Hartshorne, Rees, and Thomas (1), although the question of smoothability with rational coefficients remains open. For a discussion of this problem see Hartshorne (4).

Example 15.3.3. Let D, E be vector bundles of ranks d, e, on a scheme X. Let Q be the universal quotient bundle on $P(D \oplus 1)$, p the projection from $P(D \oplus 1)$ to X. Then for all $\alpha \in A_* X$,

$$p_*(c(\Lambda^{\cdot} Q^{\vee} \otimes p^* E)) \cap p^* \alpha) = P(D, E) \cap \alpha$$

where P is defined in Lemma 15.3. This formula also determines P. (Assume $\alpha = [X]$, X a variety. Let f be the canonical zero section. Then

$$c(\Lambda^{\cdot} Q^{\vee} \otimes p^* E) = c_d(Q) \cdot P(Q, p^* E) = f_*(P(f^* Q, f^* p^* E) \cap [X])$$

as in § 15.1.) For other descriptions of P, bringing in λ-rings, see [SGA 6] Exp. 0.II.5 and V.6, and Jouanolou (2).

Example 15.3.4. When $d=1$, and $D=L^\vee$, the polynomial $P(D,E)$ of Lemma 15.3 is determined by the identity

$$c(E\otimes L)\,P(D,E)=(c(E\otimes L)-c(E))/c_1(L)$$
$$=\sum_{p=0}^{e-1}\sum_{i=0}^{p}\binom{e-i}{p+1-i}c_i(E)\,c_1(L)^{p-i}$$

(Example 3.2.2). One may obtain a closed expression for $P(D,E)$ by writing out $c(E\otimes L)^{-1}$ (cf. Example 3.1.1).

Example 15.3.5 (cf. Mumford (7)). Let $E=1$ be a trivial line bundle. If $d=1$ or 2, then the power series of Lemma 15.3 is

$$P(D,1)=(-1)^{d-1}c(D^\vee)^{-1}.$$

It follows that if $f\colon X\to Y$ is a closed imbedding of non-singular varieties of codimension $d=1$ or $d=2$, with normal bundle N, then

$$c(f_*\mathcal{O}_X)=1+(-1)^{d-1}f_*(c(N^\vee)^{-1}).$$

If $d=3$, $P(D,1)=c(D^\vee)^{-1}(2-c_1(D))(1-c_1(D))^{-1}$.

Example 15.3.6. Let X be a non-singular n-dimensional variety, and set $F^pX=F_{n-p}K\circ X$ (Example 15.1.5). If $\alpha\in F^pX$, then $c_i(\alpha)=0$ for $0<i<p$. The mapping $\alpha\to c_p(\alpha)$ determines a homomorphism

$$c_p\colon F^pX/F^{p+1}X\to A^pX.$$

The composites $c_p\circ\varphi$ and $\varphi\circ c_p$, which are endomorphisms of A^pX and $F^pX/F^{p+1}X$ respectively, are both multiplication by $(-1)^{p-1}(p-1)!$. (Argue as in Example 15.2.6, using Example 15.3.1. Note that Theorem 15.3 extends to arbitrary regular imbeddings, as in Example 15.2.15.) In particular

$$\begin{aligned}\operatorname{rank}&\colon F^0X/F^1X\xrightarrow{\sim}\mathbb{Z}=A^0X\\ \det&\colon F^1X/F^2X\xrightarrow{\sim}\operatorname{Pic}(X)=A^1X\\ c_2&\colon F^2X/F^3X\xrightarrow{\sim}A^2X.\end{aligned}$$

Thus if X is a surface, $A(X)\xrightarrow{\sim}\operatorname{Gr}K(X)$. For affine varieties, there are some stronger results, cf. Kumar-Murthy (1).

15.4 Blowing up Chern Classes

For any non-singular variety X, write $c(X)$ for $c(T_X)$. Consider a blow-up diagram (§ 6.7)

$$\begin{array}{ccc}\tilde{X} & \xrightarrow{j} & \tilde{Y}\\ {\scriptstyle g}\downarrow & & \downarrow{\scriptstyle f}\\ X & \xrightarrow[i]{} & Y\end{array}$$

with Y, X, and therefore $\tilde{Y}$ and $\tilde{X}$ non-singular. Our object is to compare $c(\tilde{Y})$ with $f^* c(Y)$. Let N be the normal bundle to X in Y, of rank d, and identify $\tilde{X}$ with $P(N)$, so $N_{\tilde{X}}\tilde{Y}$ is $\mathcal{O}_N(-1)$. Let F be the universal quotient bundle on $P(N)$; let $\mathcal{O}(1) = \mathcal{O}_N(1)$.

Lemma 15.4. *There are exact sequences*

(i) $0 \to \mathcal{O}(-1) \to g^* N \to F \to 0$

(ii) $0 \to \mathcal{O} \to g^* N \otimes \mathcal{O}(1) \to T_{\tilde{X}} \to g^* T_X \to 0$

(iii) $0 \to T_{\tilde{X}} \to j^* T_{\tilde{Y}} \to \mathcal{O}(-1) \to 0$

(iv) $0 \to T_{\tilde{Y}} \to f^* T_Y \to j_*(F) \to 0$.

The first three are exact sequences of vector bundles on $\tilde{X}$, the fourth an exact sequence of sheaves on $\tilde{Y}$.

Proof. (i) is the universal exact sequence on the projective bundle $P(N)$. (ii) follows from the exact sequence giving the relative tangent bundle of a projective bundle (Appendix B.5.8):

$$0 \to \mathcal{O} \to g^* N \otimes \mathcal{O}(1) \to T_g \to 0,$$

and the identification of T_g with the kernel of $dg: T_{\tilde{X}} \to g^* T_X$. (iii) is the usual relation between tangent and normal bundles (Appendix B.7.2). For (iv), note that

$$df: T_{\tilde{Y}} \to f^* T_Y$$

is an isomorphism off $\tilde{X}$, so is a monomorphism of sheaves. To prove (iv) it suffices to map $j^* f^* T_Y = g^* i^* T_Y$ to F so that

$$j^* T_{\tilde{Y}} \to g^* i^* T_Y \to F \to 0$$

is exact on $\tilde{X}$. Consider the diagram

$$\begin{array}{ccccccc} 0 \to & T_{\tilde{X}} & \to & j^* T_{\tilde{Y}} & \to & N_{\tilde{X}}\tilde{Y} & \to 0 \\ & \downarrow & & \downarrow & & \downarrow & \\ 0 \to & g^* T_X & \to & g^* i^* T_Y & \to & g^* N & \to 0 \end{array}$$

with exact rows. The first vertical map is surjective by (ii), the third injective, with cokernel F by (i). The desired map is the composite

$$g^* i^* T_Y \to g^* N \to F,$$

and the required exactness is a simple diagram chase. □

We apply Riemann-Roch without denominators to the vector bundle F on $\tilde{X}$, and the inclusion $j: \tilde{X} \to \tilde{Y}$:

$$c(j_* F) - 1 = j_*(P(\mathcal{O}(-1), F)).$$

By (iv) of the lemma,

$$c(j_* F) = f^* c(Y)/c(\tilde{Y}).$$

Therefore

(1) $$f^* c(Y) - c(\tilde{Y}) = c(\tilde{Y}) \cdot j_*(P(\mathcal{O}(-1), F)) = j_*(j^* c(\tilde{Y})\, P(\mathcal{O}(-1), F)).$$

By (iii) and (ii) of the lemma,

$$j^* c(\tilde{Y}) = c(\tilde{X})\, c(\mathcal{O}(-1)) = g^* c(X)\, c(g^* N \otimes \mathcal{O}(1))\, c(\mathcal{O}(-1))\,. \tag{2}$$

From the definition of $P(\mathcal{O}(-1), F)$,

$$\frac{c(F)}{c(F \otimes \mathcal{O}(1))} - 1 = c_1(\mathcal{O}(-1)) \cdot P(\mathcal{O}(-1), F)\,,$$

or

$$c(F \otimes \mathcal{O}(1)) \cdot P(\mathcal{O}(-1), F) = \frac{c(F \otimes \mathcal{O}(1)) - c(F)}{c_1(\mathcal{O}(1))}\,, \tag{3}$$

where the right side denotes the result of formally dividing by $c_1(\mathcal{O}(1))$ (cf. Example 15.3.4). By (i) of the Lemma, and the result of tensoring (i) by $\mathcal{O}(1)$,

$$\begin{aligned} c(F) &= c(g^* N)/c(\mathcal{O}(-1))\,, \\ c(F \otimes \mathcal{O}(1)) &= c(g^* N \otimes \mathcal{O}(1))\,. \end{aligned} \tag{4}$$

Combining (2), (3), and (4), we have

$$j^* c(\tilde{Y})\, P(\mathcal{O}(-1), F) = g^* c(X)\, \frac{c(\mathcal{O}(-1))\, c(g^* N \otimes \mathcal{O}(1)) - c(g^* N)}{c_1(\mathcal{O}(1))}\,. \tag{5}$$

Theorem 15.4. *With the above notation, and $\zeta = c_1(\mathcal{O}(1))$,*

$$c(\tilde{Y}) - f^* c(Y) = j_*(g^* c(X) \cdot \alpha)\,,$$

where

$$\alpha = \frac{1}{\zeta}\left[\sum_{i=0}^{d} g^* c_{d-i}(N) - (1-\zeta) \sum_{i=0}^{d} (1+\zeta)^i g^* c_{d-i}(N)\right].$$

In this expression, the term in brackets is expanded as a polynomial in ζ, and α is the polynomial one obtains after formally dividing by ζ.

Proof. This follows from (1) and (5) and the identity

$$c(g^* N \otimes \mathcal{O}(1)) = \sum_{i=0}^{d} c(\mathcal{O}(1))^i g^* c_{d-i}(N)$$

of Remark 3.2.3 (b). □

Example 15.4.1. The formula for α may be written explicitly as follows:

$$\alpha = \sum_{j=0}^{d} \sum_{k=0}^{d-j} \left(\binom{d-j}{j} - \binom{d-j}{k+1}\right) \zeta^k g^* c_j(N)$$

where $\binom{p}{q}$ is defined to be zero if $p < q$. With this formula for α, and $(-1)^i c_i(V)$ for the ith *canonical class* of a non-singular variety V, Theorem 15.4 takes on a form similar to that conjectured by Todd (7).

Example 15.4.2. Suppose $c_k(N) = i^* c_k$, for some classes $c_k \in A^k Y$. For example, if N is the restriction of a vector bundle E on Y, then $c_k = c_k(E)$ will do. Let $\eta = c_1(\mathcal{O}_{\tilde{Y}}(\tilde{X}))$. Then

$$c(\tilde{Y}) - f^* c(Y) = j_*(g^* c(X)) \cdot \beta$$

where

$$\beta = \frac{1}{\eta}\left[(1+\eta)\sum_{i=0}^{d}(1-\eta)^i f^* c_{d-i} - \sum_{i=0}^{d} f^* c_{d-i}\right].$$

(Since $j^*\eta = -\zeta$, it follows that $j^*\beta = \alpha$ with α as in the theorem. Apply the projection formula.) Using $c(X) = i^* c(Y)\,(\sum c_k)^{-1}$, and $j_*(1) = \eta$, this can be rewritten:

(a) $$c(\tilde{Y}) = f^* c(Y)\left(\sum_{k=0}^{d} f^* c_k\right)^{-1}(1+\eta)\sum_{i=0}^{d}(1-\eta)^i f^* c_{d-i}.$$

If $c(N) = 1$, then

(b) $$c(\tilde{Y}) = f^* c(Y)(1+\eta)(1-\eta)^d.$$

This holds in particular when X is a point. Then $f^*\alpha\cdot\eta = j_*(j^* f^*\alpha) = j_*(g^* i^*\alpha) = 0$ for $\alpha \in A^p X$, $p > 0$, so

(c) $$c(\tilde{Y}) = f^* c(Y) + (1+\eta)(1-\eta)^d - 1.$$

Example 15.4.3. Equating terms of degree 1,

$$c_1(\tilde{Y}) - f^* c_1(Y) = j_*(1-d) = (1-d)\,[\tilde{X}],$$

which is the usual formula relating the canonical divisors on Y and $\tilde{Y}$:

$$K_{\tilde{Y}} = f^* K_Y + (d-1)\,[\tilde{X}].$$

For terms of degree 2,

$$c_2(\tilde{Y}) - f^* c_2(Y) = -j_*\left((d-1)\,g^* c_1(X) + \frac{d(d-3)}{2}\zeta + (d-2)\,g^* c_1(N)\right).$$

If $d = 2$,

$$c_2(\tilde{Y}) - f^* c_2(Y) = -j_* g^* c_1(X) - [\tilde{X}]\cdot[\tilde{X}] = f^* i_*[X] - f^* c_1(Y)\cdot[\tilde{X}].$$

(For the second equation, use Proposition 6.7.)

Example 15.4.4. Let X be a non-singular projective surface, and suppose a Lefschetz pencil of curves is constructed on X, i.e., all curves in the pencil are non-singular curves, of genus g, except for d curves, which each have one ordinary node; at each of the a base points the curves are assumed to be non-singular and meet transversally. The *Zeuthen-Segre invariant* I is defined by

$$I = d - 4g - a.$$

Then $\chi(X) = \int c_2(X) = I + 4$. (If $\tilde{X}$ is the blow-up of X at the base points, then $\chi(\tilde{X}) = \chi(X) + a$ by Example 15.4.2(c). The pencil is given by a morphism from $\tilde{X}$ to $\mathbb{P}^1$, and $d = \chi(\tilde{X}) - 2(2-2g)$ by Example 14.1.5.) In particular, I is independent of the choice of pencil.

With the notation of Example 15.2.2, define $p^{(1)} = (K\cdot K) + 1$, and $p_a = p_g - q$. Then equation (*) of Example 15.2.2 is equivalent to Noether's formula (cf. Noether (1)):

$$p^{(1)} + I = 12 p_a + 9.$$

Example 15.4.5. Atiyah and Hirzebruch (3) proved GRR for closed imbeddings of complex manifolds. The proof given here can also be used in the analytic case. Similarly, the formula for blowing up Chern classes (Theorem 15.4) is also valid, with the same proof, for X and Y arbitrary complex manifolds, and Chern classes in singular cohomology with integer coefficients.

Notes and References

For more than a century, the Riemann-Roch problem has stimulated the development of intersection theory and the search for invariants of algebraic varieties. This began with Riemann's inequality

$$\dim H^0(X, \mathcal{O}_X(D)) \geqq \deg(D) + 1 - g$$

for a divisor D on a non-singular projective curve of genus g, and an identification of the error term as $\dim H^0(X, \mathcal{O}_X(K - D))$ by Roch. For surfaces, analogues of Riemann's inequality were developed primarily by Noether, Castelnuovo, and Severi (cf. Severi (13), (20), Zariski (4)) while Zeuthen and Segre (cf. Example 15.4.4) generalized the notion of genus. For a sketch of this early history see Zariski (1). The arithmetic genus was studied quite generally by Severi (3), (20).

The notion of canonical divisor was generalized from curves to surfaces by Noether. Severi (6) was the first to find an analogue in codimension greater than one, when he defined a canonical zero-cycle on a surface, as the zero set of a holomorphic one-form, for surfaces which have such one-forms. There followed a succession of papers by Segre, Todd, and Eger, constructing canonical classes in all dimensions on arbitrary non-singular projective varieties; the ith canonical class of these authors is

$$c_i(T_X^\vee) = (-1)^i c_i(T_X) = (-1)^i c_i(X),$$

although this was not proved until the machinery of Chern classes was highly developed (Nakano (1)). By ingenious calculations, Todd (4) found formulas for the arithmetic genus in terms of canonical classes (Corollary 15.2.2), and proved them in many cases.

There was also much work by Segre and Todd on the formula relating canonical classes of a variety and its monoidal transforms. Formulas were proved in many cases, and Todd (7) conjectured a general formula. The article of Todd (8) may be recommended for the early history of canonical classes.

Two developments were vital for the modern solution of these problems. One was the introduction of the methods of sheaf theory, primarily by Kodaira and Spencer for complex manifolds, and Serre for algebraic varieties. The second was the development of characteristic classes of vector bundles in topology, by Whitney, Stiefel, Pontrjagin, and Chern. (It is interesting that the study of canonical classes in algebraic geometry and Stiefel-Whitney classes in topology began almost simultaneously, with many common geometric constructions, involving singularities of projections in ambient projective or Euclidean spaces; apparently it was some time before there was any contact between these schools.)

The general formula (Corollary 15.2.1) for the Euler characteristic of a vector bundle on a non-singular complex projective variety was given and proved by Hirzebruch (1), using Thom's results in cobordism. The third edition of Hirzebruch's book contains useful historical notes at the end of chapters, and an appendix by Schwarzenberger with applications and more recent history. For surfaces and threefolds, other proofs of Corollary 15.2.2 have been give by Piene (5) and Piene and Ronga (1).

Grothendieck's transformation of the Riemann-Roch problem (cf. Borel-Serre (1), [SGA 6] Exp. 0) has influenced a wide spectrum of mathematics. Grothendieck groups and K-theory are two legacies of this work. The generalization from giving a formula on individual varieties to a formula for morphisms between varieties, and the construction of a natural transformation of functors, has had profound consequences. At the same time, and using similar sheaf-theoretic and geometric constructions, Washnitzer (2) gave an axiomatic characterization of the arithmetic genus (cf. Example 15.2.10). Some recent applications of GRR may be found in Harris-Mumford (1), and Arbarello-Cornalba-Griffiths-Harris (1).

For complex projective varieties, by GAGA theorems of Serre (3), Riemann-Roch for algebraic sheaves and morphisms is equivalent to corresponding formulas for analytic sheaves and morphisms. A generalization of GRR for imbeddings of complex manifolds has been given by Atiyah and Hirzebruch (3), where the values are taken in topological K-theory. Another far-reaching generalization of the Hirzebruch formula is the index theorem of Atiyah and Singer (1). Obrian, Toledo and Tong (3) have generalized GRR to complex manifolds, with values in the cohomology $H^*(X, \Omega_X^*)$.

The proof of GRR given in this chapter is a simplification of ideas in Baum-Fulton-MacPherson (1), (2), using the deformation to the normal bundle. The generalization to non-projective varieties, over arbitrary ground fields, depends first on having an adequate intersection theory in that generality. In characteristic zero, one may then use Chow's lemma and Hironaka's resolution of singularities to conclude the proof (cf. Example 15.2.9) – an observation I owe to Kleiman and Gillet. In all characteristics, GRR for non-singular varieties can be deduced from the full singular Riemann-Roch theorem (see Chap. 18).

Riemann-Roch without denominators (Theorem 15.3) was conjectured by Grothendieck and proved by Jouanolou (2). The general formula for blowing up Chern classes was deduced from GRR by Porteous (1), after conceptual simplification by Van de Ven (1); Porteous's proof was only valid modulo torsion until Riemann-Roch without denominators was proved. Lascu and Scott (1), (2) gave a proof without using Riemann-Roch. Mumford, Jouanolou, Lascu and Scott used the space which we constructed in Chap. 5 as the total space of the deformation to the normal bundle; the further simplifications may be attributed to realizing the full potential of the deformation itself.

It is possible to give more conceptual proofs of GRR for the case of a projection $Y \times \mathbb{P}^m \to Y$, although with some loss in brevity. See Baum-Fulton-MacPherson (2) App. 3, or Obrian-Toledo-Tong (3).

Hirzebruch pointed out the application of the geometry of the deformation to the normal bundle to signatures in Example 15.2.9. Example 15.2.18 is J. Harris's solution to a question of M. Green. Most of the other examples are standard consequences of Riemann-Roch, at least in the case of projective varieties over an algebraically closed field.

Chapter 16. Correspondences

Summary

A correspondence from X to Y, denoted $\alpha : X \vdash Y$, is a subvariety, cycle, or equivalence class of cycles on $X \times Y$. The graph of a morphism, or the closure of the graph of a rational map, are basic examples, but more general correspondences have played an important role in the development of algebraic geometry. On complete non-singular varieties correspondences have a product $\beta \circ \alpha$, and a correspondence $\alpha : X \vdash Y$ determines homomorphisms α_* from $A(X)$ to $A(Y)$, and α^* from $A(Y)$ to $A(X)$, these notions generalizing composition, push-forward, and pull-back for morphisms. The basic algebra of correspondences is deduced easily from the general theory of Chap. 8.

If $X = Y$ has dimension n, and T is an n-dimensional correspondence, then the degree of the intersection class $T \cdot \Delta$ of T with the diagonal is the virtual number of fixed points of T. In case there are non-isolated fixed points, the excess intersection formulas can be applied. (If $T = V \times W$, with V, W subvarieties of X, $T \cdot \Delta = V \cdot W$ is the intersection class studied in Chap. 8.) When one has explicit formulas for the equivalence class of $[T]$ or of $[\Delta]$ on $X \times X$, fixed point formulas for $T \cdot \Delta$ can be deduced.

Notation. Unless otherwise stated, all ambient varieties $X, Y, Z, \ldots$ in this chapter are assumed to be complete and non-singular, i.e., proper and smooth over the given ground field.

16.1 Algebra of Correspondences

Definition 16.1.1. A *correspondence* from a variety X to a variety Y is a cycle, or an equivalence class of cycles, on $X \times Y$. We shall write $\alpha : X \vdash Y$ to denote that α is a correspondence from X to Y.

If $\alpha : X \vdash Y$, $\beta : Y \vdash Z$, with X, Y, Z non-singular, the *product* (or *composite*) correspondence, denoted $\beta \circ \alpha$, is the correspondence from X to Z defined by the formula

$$\beta \circ \alpha = p_{XZ*}(p_{XY}{}^* \alpha \cdot p_{YZ}{}^* \beta) .$$

Here p_{XY}, p_{YZ}, p_{XZ} denote the projections from $X\times Y\times Z$ to $X\times Y$, $Y\times Z$, $X\times Z$, respectively. The product $p_{XY}{}^*\alpha\cdot p_{YZ}{}^*\beta$ is the intersection product on the non-singular variety $X\times Y\times Z$. If α and β are cycles and $p_{XY}{}^*\alpha$ meets $p_{YZ}{}^*\beta$ properly, then $\beta\circ\alpha$ is a well-defined cycle on $X\times Z$; in general $\beta\circ\alpha$ is defined up to rational equivalence. This product defines a bilinear homomorphism

$$A(X\times Y)\otimes A(Y\times Z)\xrightarrow{\circ}A(X\times Z).$$

A correspondence $\alpha: X\vdash Y$ has a *transpose* $\alpha': Y\vdash X$ defined by

$$\alpha'=\tau_*(\alpha)$$

where $\tau: X\times Y\to Y\times X$ reverses the factors, i.e., $\tau(P,Q)=(Q,P)$.

An *irreducible* correspondence from X to Y is a subvariety V of $X\times Y$, identified with its cycle $[V]$. Any morphism $f: X\to Y$ determines an irreducible correspondence Γ_f from X to Y, given by the graph imbedding of X in $X\times Y$.

Proposition 16.1.1. *Let* $\alpha: X\vdash Y$, $\beta: Y\vdash Z$.

(a) *If* $\gamma: Z\vdash W$, *then* $\gamma\circ(\beta\circ\alpha)=(\gamma\circ\beta)\circ\alpha$.
(b) $(\beta\circ\alpha)'=\alpha'\circ\beta'$, *and* $(\alpha')'=\alpha$.
(c) (i) *If* $\beta=\Gamma_g$, *then* $\beta\circ\alpha=(1_X\times g)_*(\alpha)$.
(ii) *If* $\alpha=\Gamma_f$, *then* $\beta\circ\alpha=(f\times 1_Z)^*(\beta)$.
(iii) *If* $\alpha=\Gamma_f$, $\beta=\Gamma_g$, *then* $\beta\circ\alpha=\Gamma_{gf}$.

Proof. (a) Denote by p_{XW}^{XZW} the projection from $X\times Z\times W$ to $X\times W$, and similarly for other projections; for projections from $X\times Y\times Z\times W$, the superscript $XYZW$ is omitted. By Proposition 1.7, we have the formula

$$(*)\qquad (p_{XZ}^{XZW})^*\circ(p_{XZ}^{XYZ})_*=(p_{XZW})_*\circ(p_{XYZ})^*$$

as homomorphisms from $A(X\times Y\times Z)$ to $A(X\times Z\times W)$. Then we have

$$\begin{aligned}\gamma\circ(\beta\circ\alpha)&=p_{XW*}^{XZW}(p_{XZ}^{XZW*}(p_{XZ*}^{XYZ}(p_{XY}^{XYZ*}\alpha\cdot p_{YZ}^{XYZ*}\beta))\cdot p_{ZW}^{XZW*}\gamma)\\&=p_{XW*}^{XZW}(p_{XZW*}(p_{XYZ}{}^*(p_{XY}^{XYZ*}\alpha\cdot p_{YZ}^{XYZ*}\beta))\cdot p_{ZW}^{XZW*}\gamma)\\&=p_{XW*}^{XZW}(p_{XZW*}((p_{XY}{}^*\alpha\cdot p_{YZ}{}^*\beta)\cdot p_{XZW}{}^*p_{ZW}^{XZW*}\gamma))\\&=p_{XW*}((p_{XY}{}^*\alpha\cdot p_{YZ}{}^*\beta)\cdot p_{ZW}{}^*\gamma)\\&=p_{XW*}(p_{XY}{}^*\alpha\cdot(p_{YZ}{}^*\beta\cdot p_{ZW}{}^*\gamma)).\end{aligned}$$

These five equalities follow from: (i) the definition of products; (ii) formula (*); (iii) the compatibility of pull-back with intersection product, functoriality of pull-back, and the projection formula; (iv) functoriality of push-forward and pull-back; (v) associativity of intersection products. Symmetrically, the last expression is $(\gamma\circ\beta)\circ\alpha$ which concludes the proof of (a).

(b) Let $\sigma: X\times Y\times Z\to Z\times Y\times X$ map (P,Q,R) to (R,Q,P). For any S, T let τ^{ST} denote the morphism from $S\times T$ to $T\times S$ that reverses the factors. By Proposition 1.7,

$$(p_{ZY}^{ZYX})^*\circ(\tau^{YZ})_*=(\sigma)_*\circ(p_{YZ}^{XYZ})^*,\qquad(p_{YX}^{ZYX})^*\circ(\tau^{XY})_*=(\sigma)_*\circ(p_{XY}^{XYZ})^*.$$

Arguing as in (a), one has

$$\begin{aligned}\alpha' \circ \beta' &= p^{ZYX}_{ZX*}(p^{ZYX*}_{ZY}(\tau^{YZ}_{*}\beta) \cdot p^{ZYX*}_{YX}(\tau^{XY}_{*}\alpha))\\ &= p^{ZYX}_{ZX*}((\sigma_* p^{XYZ*}_{YZ}\beta) \cdot (\sigma_* p^{XYZ*}_{XY}\alpha))\\ &= p^{ZYX}_{ZX*}\sigma_*(p^{XYZ*}_{XY}\alpha \cdot p^{XYZ*}_{YZ}\beta)\\ &= \tau^{XZ}_{*} p^{XYZ}_{XZ*}(p^{XYZ*}_{XY}\alpha \cdot p^{XYZ*}_{YZ}\beta) = (\beta \circ \alpha)'.\end{aligned}$$

The third equality uses the fact that σ_* is a ring homomorphism (Example 8.3.5); the fourth follows from the identity $p^{ZYX}_{ZX}\sigma = \tau^{XZ} p^{XYZ}_{XZ}$. The other identity $(\alpha')' = \alpha$ follows from the identity $\tau^{YX}\tau^{XY} = 1_{X\times Y}$.

(c) Let $\gamma_g : Y \to Y \times Z$ be the graph of g. Then

$$\begin{aligned}\Gamma_g \circ \alpha &= p^{XYZ}_{XZ*}(p^{XYZ*}_{YZ}(\gamma_{g*}[Y]) \cdot p^{XYZ*}_{XY}(\alpha))\\ &= p^{XYZ}_{XZ*}((1_X \times \gamma_g)_*(p^{XY*}_{Y}[Y]) \cdot p^{XYZ*}_{XY}(\alpha))\\ &= p^{XYZ}_{XZ*}((1_X \times \gamma_g)_*((1_X \times \gamma_g)^* p^{XYZ*}_{XY}(\alpha)))\\ &= (1_X \times g)_*(\alpha).\end{aligned}$$

The second equation uses Proposition 1.7, the third the projection formula, and the fourth the identities

$$p^{XYZ}_{XZ}(1_X \times \gamma_g) = 1_X \times g, \qquad p^{XYZ}_{XY}(1_X \times \gamma_g) = 1_{X\times Y}.$$

Similarly for c(ii),

$$\begin{aligned}\beta \circ \Gamma_f &= p^{XYZ}_{XZ*}(p^{XYZ*}_{XY}(\gamma_{f*}[X]) \cdot p^{XYZ*}_{YZ}\beta)\\ &= p^{XYZ}_{XZ*}((\gamma_f \times 1_Z)_*([X \times Z]) \cdot p^{XYZ*}_{YZ}\beta)\\ &= p^{XYZ}_{XZ*}(\gamma_f \times 1_Z)_*((\gamma_f \times 1_Z)^* p^{XYZ*}_{YZ}\beta)\\ &= (f \times 1_Z)^*(\beta).\end{aligned}$$

For c(iii), $\Gamma_g \circ \Gamma_f = (1_X \times g)_*(\Gamma_f) = (1_X \times g)_* \gamma_{f*}[X] = \gamma_{gf*}[X] = \Gamma_{gf}$. □

Corollary 16.1.1. *For a non-singular variety X, the product $\alpha \times \beta \to \alpha \circ \beta$ makes $A(X \times X)$ into an associative ring with unit $[\Delta_X]$, and with an involution $\alpha \to \alpha'$.* □

Definition 16.1.2. For $\alpha : X \vdash Y$, define a homomorphism

$$\alpha_* : A(X) \to A(Y)$$

by the formula $\alpha_*(a) = p^{XY}_{Y*}(\alpha \cdot p^{XY*}_{X}(a))$, and a homomorphism

$$\alpha^* : A(Y) \to A(X)$$

by the formula $\alpha^*(b) = p^{XY}_{X*}(\alpha \cdot p^{XY*}_{Y}(b))$.

Proposition 16.1.2. (a) *If $\alpha : X \vdash Y$, $\beta : Y \vdash Z$, then*

$$(\beta \circ \alpha)_* = \beta_* \circ \alpha_* \text{ and } (\beta \circ \alpha)^* = \alpha^* \circ \beta^*.$$

(b) *If $\alpha : X \vdash Y$, then $(\alpha')_* = \alpha^*$.*

(c) *If $f : X \to Y$, then $(\Gamma_f)_* = f_*$* and *$(\Gamma_f)^* = f^*$.*

Proof. Let $P = \mathrm{Spec}(K)$. If $a \in A(X) = A(P \times X)$, then $\alpha_*(a)$ may be identified with $\alpha \circ a$ in $A(Z) = A(P \times Z)$. Similarly if $b \in A(Y) = A(Y \times P)$, $\alpha^*(b)$, $b \circ \alpha$. With these identifications, (a), (b) and (c) follow from corresponding parts of Proposition 16.1.1. □

Corollary 16.1.2. *For a non-singular variety X, the homomorphism*

$$A(X \times X) \to \mathrm{End}(A(X)), \quad \alpha \to \alpha_* \text{ (resp. } \alpha \to \alpha^*)$$

is a homomorphism (resp. anti-homomorphism) of rings. □

Remark 16.1. Several modifications of $A(X \times X)$ have been useful, and have been called the ring of correspondences on X:

(i) Let $C(X) = A^n(X \times X)$, where $n = \dim(X)$. This is a subring of the ring of correspondences, closed under the involution and containing the identity and graphs of morphisms, and $\alpha \in C(X)$ induces homomorphisms of $A_* X$ which preserve degrees (cf. Example 16.1.1).

(ii) One may consider $A(X \times X)/I$, where I is an ideal of degenerate correspondences (cf. Example 16.1.2).

(iii) One may replace rational equivalence by algebraic, numerical, or homological equivalence (cf. Chap. 19).

The results of this section extend readily to arbitrary smooth, proper schemes over a field, using the identity

$$A(X \times Y) = \bigoplus_{i,j} A(X_i \times Y_j)$$

where X_i, Y_j are the connected components of X and Y.

In addition, the completeness of the ambient varieties is not always necessary. For example, if α is a correspondence from X to Y, and the support of α is proper over Y, then α induces a homomorphism α_* from $A(X)$ to $A(Y)$. Indeed, as in Definition 16.1.2, if $a \in A(X)$, then $\alpha \cdot p_X^{XY*}(a)$ is represented by a well-defined class on $|\alpha|$ by § 8.1; the proper push-forward of this class is $\alpha_*(a)$.

Example 16.1.1. A correspondence $\alpha : X^n \vdash Y^m$ is homogeneous of *degree* (or codimension) p if $\alpha \in A^{m+p}(X \times Y)$. If $\alpha : X \vdash Y$ has degree p, and $\beta : Y \vdash Z$ has degree q, then $\beta \circ \alpha$ has degree $p + q$. In particular the correspondences of degree zero are closed under composition. If α has degree p, then α_* maps $A_k X$ to $A_{k-p} Y$, and α^* maps $A^k Y$ to $A^{k+p} X$. If $f : X \to Y$ is a morphism, then Γ_f has degree 0. If $\alpha : X \vdash Y$ has degree p, then α' has degree $p + m - n$; in particular, Γ_f' has degree $m - n$. (A dual definition is discussed in Example 16.1.12.)

Example 16.1.2. *Degenerate correspondences.* (a) Let $I(X, Y)$ be the subgroup of $A(X \times Y)$ generated by correspondences of the form $[V \times W]$, with V (resp. W) a subvariety of X (resp. Y). If $\alpha \in I(X, Y)$ and $\beta \in A(Y \times Z)$, then $\beta \circ \alpha \in I(X, Z)$. It follows that $I(X, X)$ is a two-sided homogeneous ideal in $A(X \times X)$, with $I(X, X)' = I(X, X)$. (If $\alpha = [V \times W]$, then $\beta \circ \alpha = [V] \times p_{Z*}^{YZ}([W \times Z] \cdot \beta)$.)

The subgroup of $A(X \times X)$ generated by correspondences of the form $[P \times X]$ and $[X \times P]$ for points $P \in X$, is likewise a two-sided ideal.

(b) Consider only n-dimensional varieties X, Y, Z. Let $J(X, Y)$ be the subgroup of $A^n(X \times Y)$ generated by irreducible correspondences $[V]$, where V is a subvariety of $X \times Y$ such that $p_X^{XY}(V) \neq X$ or $p_Y^{XY}(V) \neq Y$. If $\alpha \in J(X, Y)$ and $\beta \in A^n(Y \times Z)$, then $\beta \circ \alpha \in J(X, Z)$. Hence $J(X, X)$ is a two-sided, involutive ideal in $C(X) = A^n(X \times X)$.

(Let $\alpha = [V]$, $\beta = [W]$, V, W irreducible. If $p_X^{XY}(V) = S \subsetneq X$, then $\beta \circ \alpha$ is represented by a cycle on $S \times Z$. Suppose $p_Y^{XY}(V) = T \subsetneq Y$. If $W \not\subset T \times Z$, then $p_Z^{XYZ}((V \times Z) \cap (X \times W)) \subset p_Z^{YZ}((T \times Z) \cap W) \subsetneq Z$ by a dimenson count. Any n-cycle on $Y \times Z$ is rationally equivalent to a cycle none of whose components are contained in $T \times Z$; indeed, on any smooth variety, any cycle can be moved off any given proper subvariety, as a simple local argument shows.)

(c) If X is a curve, $J(X, X) = I(X, X) \cap C(X)$. The quotient ring $C(X)/J(X, X)$ of non-degenerate correspondences operates on the Jacobian

$$\tilde{A}_0(X) = \operatorname{Ker}(A_0(X) \xrightarrow{\deg} \mathbb{Z})$$

by the operations α_* and α^* of Definition 16.1.2.

Example 16.1.3. If α and β are correspondences from X to Y, then

$$p_*(\alpha \cdot \beta) = (\beta' \circ \alpha) \cdot \Delta_X$$

in $A(X)$, with $p: X \times Y \to X$ the projection. If X, Y, α and β are n-dimensional, then $\int \alpha \cdot \beta$ is the *virtual number of coincidences* of α and β. The virtual number of coincidences of α and β is therefore equal to the virtual number of fixed points of $\beta' \circ \alpha$, or of $\alpha' \circ \beta$. (Let $\gamma = (1_X \times \delta_Y \times 1_X)^*(\alpha \times \beta')$. Form the fibre square

$$\begin{array}{ccc} X \times Y & \xrightarrow{\varphi} & X \times Y \times X \\ {\scriptstyle p}\downarrow & & \downarrow{\scriptstyle q} \\ X & \xrightarrow[\delta_X]{} & X \times X \end{array}$$

where q is the projection. Then $\alpha \cdot \beta = \varphi^*(\gamma)$ and $\beta' \circ \alpha = q_*(\gamma)$; and $\delta_X^* q_* = p_* \varphi^*$ by Theorem 6.2.)

Example 16.1.4. In this example, X, Y, Z are irreducible of the same dimension n. the *indices* (or *degrees*) $d_1(\alpha)$ and $d_2(\alpha)$ of a correspondence $\alpha \in A^n(X \times Y)$ are defined by

$$p_{X*}^{XY}(\alpha) = d_1(\alpha) \cdot [X], \quad p_{Y*}^{XY}(\alpha) = d_2(\alpha) \cdot [Y].$$

(a) If $\beta \in A^n(Y \times Z)$, then, for $i = 1$ and 2,

$$d_i(\beta \circ \alpha) = d_i(\beta) \cdot d_i(\alpha).$$

(Use Proposition 16.1.2(a) and the identity $\alpha^*[Y] = d_1(\alpha) \cdot [X]$.)

(b) For any α, $d_1(\alpha') = d_2(\alpha)$.

(c) For any rational point P on X, $d_1(\alpha) = \int \alpha \cdot [P \times Y]$.

With $a = d_2(\alpha)$, $b = d_1(\alpha)$, α is called an *(a, b)-correspondence.*

Example 16.1.5. *Valence.* The varieties X, Y, Z are assumed to be n-dimensional. According to Severi (10), a correspondence $\alpha \in A^n(X \times Y)$ has *valence zero* if α is in the group $I(X, Y)$ of degenerate correspondences defined in Example 16.1.2(a). A correspondence $\alpha \in A^n(X \times X)$ has *valence* v if $\alpha + v\Delta$ has valence zero, where Δ is the identity correspondence. These notions are most useful if rational equivalence is replaced by algebraic or numerical equivalence, as in Remark 16.1.

(a) If α and β are correspondences on X with valences u and v, then $\beta \circ \alpha$ has valence $-uv$. ($\beta \circ \alpha - uv\Delta = (\beta + v\Delta) \circ \alpha - v(\alpha + u\Delta)$.)

(b) If α has valence v, so does α'.

(c) If α has valence v, and one knows a decomposition

$$\alpha + v\Delta \sim \sum n_i [V_i \times W_i]$$

with V_i, W_i subvarieties of X such that $\dim V_i + \dim W_i = X$, then

$$\int_{X \times X} \alpha \cdot \Delta = -v\chi + \sum n_i \int_X V_i \cdot W_i ,$$

where $\chi = \int \Delta \cdot \Delta$ is the Euler characteristic of X.

(d) Every correspondence α on $\mathbb{P}^1$ has valence zero. If α is an (a, b)-correspondence on $\mathbb{P}^1$, then

$$\int \alpha \cdot \Delta = a + b .$$

This *correspondence principle* of Chasles (1), (4) – that an (a, b) correspondence on $\mathbb{P}^1$ has $a + b$ fixed points – was one of the primary tools of classical enumerative geometry.

(e) (Chasles-Cayley-Brill-Hurwitz). If α is an (a, b)-correspondence with valence v on a curve X of genus g, then

$$\int \alpha \cdot \Delta = a + b + 2vg .$$

If $\alpha = [T]$, T irreducible, $T \neq \Delta$, then T has $a + b + 2vg$ fixed points, counting multiplicities. (Let $\alpha + v\Delta = c \times [X] + [X] \times d$. Then $\deg(c) = a + v$, $\deg(d) = b + v$. Apply (c).)

(f) If X is a projective space, Grassmann variety, or any flag variety, then every correspondence in $A^n(X \times X)$ has valence zero. (See Example 1.10.2.) On a curve of general moduli, every correspondence has a valence (Hurwitz (1)).

(g) If $i: X \to \mathbb{P}^m$ is a closed imbedding, and $\mathscr{V}$ is a purely m-dimensional subscheme of $X \times \mathbb{P}^m$, then

$$\alpha = (1_X \times i)^* [\mathscr{V}] \in A^n(X \times X)$$

is a correspondence of valence zero. If the projection from $\mathscr{V}$ to X is flat, with fibre $V(P)$ over $P \in X$, then $\alpha_*(P)$ is the intersection class of $V(P)$ with X in $\mathbb{P}^m$. With some additional hypotheses, Severi (10) calls such α a *generalized Zeuthen correspondence*.

Example 16.1.6 (cf. Severi (5), p. 174). If α is a correspondence on a curve of genus g which is a product of r correspondences α_i, with α_i an (a_i, b_i)-correspondence of valence v_i, then the virtual number of fixed points is

$$\int \alpha \cdot \Delta = \prod_{i=1}^{r} a_i + \prod_{i=1}^{r} b_i + (-1)^{r-1} 2g \prod_{i=1}^{r} v_i .$$

Example 16.1.7. Let α_1 and α_2 be correspondences on a curve of genus g, α_i an (a_i, b_i)-correspondence of valence v_i. Then the virtual number of coincidences is given by

$$\int \alpha_i \cdot \alpha_2 = a_1 b_2 + a_2 b_1 - 2g\, v_1 v_2 .$$

(This follows from Examples 16.1.3 and 16.1.5 (e).)

For enumerative applications of this and the preceeding formulas for correspondences on curves, see Severi (5) VI § 4.

Example 16.1.8 (cf. Zeuthen (1), Severi (5) § 60). Let T be an irreducible (n, n')-correspondence between curves C and C'. Let g be the genus of C, and let d be the number (suitably weighted) of points of C which correspond to fewer than n' points of C'; define g' and d' symmetrically. Then

$$d - 2n(g' - 1) = d' - 2n'(g - 1) .$$

(Apply the Riemann-Hurwitz formula (Example 3.2.20) to $\tilde{T} \to C$ and $\tilde{T} \to C'$, $\tilde{T}$ the non-singular model of T.) More generally, if D (resp. D') is the divisor of double points, and K (resp. K') a canonical divisor on C (resp. C'), then

$$T_*(K + D) = n(K' + D')$$

in $A_0(C')$. (Write $T = \beta \circ \alpha'$, $\alpha : \tilde{T} \to C$, $\beta : \tilde{T} \to C'$.)

Example 16.1.9. Let $f: X \to X$ be a morphism, P a fixed point of f, rational over the ground field. Regard $1 - df$ as an endomorphism of the tangent bundle T_X.

(a) If $\det(1 - df)_P \neq 0$, then P is an isolated fixed point of f, and

$$i((P, P), \Gamma_f \cdot \Delta; X \times X) = 1 .$$

(b) If X is a curve, and the ground field has characteristic zero, then

$$i((P, P), \Gamma_f \cdot \Delta; X \times X) = 1 + \operatorname{ord}_P(1 - df) .$$

Here $\operatorname{ord}_P(1 - df)$ is the order of vanishing at P of $1 - df$ as a section of $T_X^\vee \otimes T_X$. (If t is local coordinate at P, and $f(t) = \sum a_i t^i$, the left side is the order of vanishing of $t - f(t)$ at $t = 0$.)

Example 16.1.10. *Castelnuovo-Severi inequaltiy* (cf. Mattuck-Tate (1)).

(a) Let T be an (n, n')-correspondence between curves C and C'. Then

$$\int T \cdot T \leqq 2n\, n' .$$

(Apply the index theorem (Example 15.2.4) to the divisor $D = T - n(P \times C') - n'(C \times P')$ on the surface $C \times C'$, with ample divisor $H = C \times P' + P \times C'$.) Note that the equality occurs precisely when a multiple of T is algebraically equivalent to a degenerate correspondence, or, modulo numerical equivalence, T has valence 0.

(b) With T as above, and S an (m, m') correspondence between C and C', then

$$\left(\int S \cdot T - m\, n' - m'\, n\right)^2 \leqq 4\left(n\, n' - \tfrac{1}{2}\int T \cdot T\right) \cdot \left(m\, m' - \tfrac{1}{2}\int S \cdot S\right).$$

(Let $q(a, b) = Q(a\, S + b\, T)$, where $Q(U) = n(U) \cdot n'(U) - \frac{1}{2}\int U \cdot U$. Then q is a non-negative quadratic form, so its discriminant is non-positive.)

(c) With T as in *(a)*, and $C = C'$ a curve of genus g, then

$$(\int T \cdot \Delta - n - n')^2 \leqq 4g(n\,n' - \tfrac{1}{2}\int T \cdot T).$$

Equality holds precisely when T has a valence, modulo numerical equivalence.

(d) If, in (b), S and T are graphs of morphisms e and f from C to C', then $N = \int S \cdot T$ is the weighted number of coincidences and

$$|N - \deg(e) - \deg(f)| \leqq 2g' \sqrt{\deg(e) \cdot \deg(f)}\,,$$

where g' is the genus of C'. (By the self-intersection formula, $\int T \cdot T = \int f^* c_1(T_{C'}) = (2 - 2g')\deg(f)$, so $Q(T) = g' \deg(f)$.)

(e) In (d), if $C' = C$, $e = \text{id.}$, the weighted number N of fixed points of f satisfies

$$|N - \deg(f) - 1| \leqq 2g\sqrt{\deg(f)}\,.$$

(f) If C is defined over the field $\mathbb{F}_q$ with q elements, then the number N_q of $\mathbb{F}_q$-valued points of C satisfies the "Riemann hypothesis":

$$|N_q - q - 1| \leqq 2g\sqrt{q}\,.$$

(Apply (e) to the Frobenius morphism.)

Example 16.1.11 (cf. Colliot-Thélène and Coray (1) § 6). If X and Y are birationally equivalent complete non-singular varieties over an algebraically closed field, then $A_0 X \cong A_0 Y$. (The closure of the graph of a birational map determines a correspondence α from X to Y. To see that α_* and α'_* are inverse isomorphisms on A_0, note that $\alpha' \circ \alpha$ is the sum of the identity correspondence and correspondences whose projections are contained in proper subvarieties of X; any point on X is rationally equivalent to a zero-cyclic disjoint from a given subvariety of X.)

Example 16.1.12. *Grothendieck's motives* (cf. Manin (1), Kleiman (5), Deligne-Milne-Ogus-Shih (1)). Let $\mathscr{V}$ be the category of smooth schemes (over a fixed ground field). Define an additive category $\mathscr{CV}$ as follows. The objects of $\mathscr{CV}$ are the objects of $\mathscr{V}$; denote by $\bar{X}$ the object in $\mathscr{CV}$ determined by the scheme X. The morphisms in $\mathscr{CV}$ are defined by

$$\operatorname{Hom}(\bar{X}, \bar{Y}) = A(X \times Y)$$

with composition defined by the product of correspondences. Define a direct sum in $\mathscr{CV}$ by $\bar{X} \oplus \bar{Y} = \overline{X \amalg Y}$, and a tensor product by $\bar{X} \otimes \bar{Y} = \overline{X \times Y}$. For each morphism $f\colon Y \to X$ in $\mathscr{V}$, define $\bar{f}\colon \bar{X} \to \bar{Y}$ in $\mathscr{CV}$ by $\bar{f} = [\Gamma_f]'$; this gives a contravariant functor from $\mathscr{V}$ to $\mathscr{CV}$.

For T in $\mathscr{V}$, define $\bar{X}(\bar{T}) = \operatorname{Hom}(\bar{T}, \bar{X})$. For $\varphi \in \operatorname{Hom}(\bar{X}, \bar{Y})$, define

$$\varphi_T\colon \bar{X}(\bar{T}) \to \bar{Y}(\bar{T})$$

by $\varphi_T(g) = \varphi \circ g$. When T is a point, $\bar{X}(\bar{T}) = A(X)$, and φ_T is the homomorphim φ_* of Definition 16.1.2. If $\varphi = \bar{f}$, $\varphi_T = (1_T \times f)^*$, while if $\varphi = \bar{f}'$, $\varphi_T = (1_T \times f)_*$.

Manin's identity principle. Let $\varphi, \psi \in \operatorname{Hom}(\bar{X}, \bar{Y})$. Then the following are equivalent: (1) $\varphi = \psi$. (2) $\varphi_T = \psi_T$ for all T. (3) $\varphi_X = \psi_X$. (Indeed, $\varphi_X([\Delta_X]) = \varphi \circ 1_X = \varphi$, so (3) $\Rightarrow$ (1).)

This principle can be used to show that formulas known in rational equivalence therory A^* are also valid in a cohomology theory H^*, when one has a cycle map $A^* \to H^*$. For example, the key formula (Proposition 6.7 (a)), for non-singular varieties, can be expressed as an equality of correspondences between $\tilde{Y}$ and X; the validity of the formula, after crossing the diagram with an arbitrary T, implies its truth in H^*.

Define $\operatorname{Hom}^i(\bar{X}, \bar{Y}) = \oplus A^{n_j+i}(X_j \times Y)$, where X_j are the connected components of X, $n_j = \dim X_j$. The degrees of morphisms add under composition. Allowing only morphisms of degree zero determined a subcategory $\mathscr{CV}^0$ of $\mathscr{CV}$.

Grothendieck defines a *motif* to be a pair (X, p), $X \in \mathscr{V}$, $p \in \operatorname{Hom}(\bar{X}, \bar{X})$, with $p \circ p = p$. The motif of the variety X is the pair $(X, 1_X)$, denoted $\tilde{X}$. Define $\operatorname{Hom}((X, p), (Y, q))$ to be

$$\{\varphi \in \operatorname{Hom}^0(\bar{X}, \bar{Y}) \mid q\,\varphi = \varphi\, p\} / \{\varphi \in \operatorname{Hom}^0(\bar{X}, \bar{Y}) \mid q\,\varphi = \varphi\, p = 0\}\,.$$

With these morphisms, the category of motives, denoted $\widetilde{\mathscr{CV}}^0$, is pseudo-abelian, and the natural functor from $\mathscr{CV}^0$ to $\widetilde{\mathscr{CV}}^0$ is universal for functors to pseudo-abelian categories. (An additive category $\mathscr{D}$ is called pseudo-abelian if every $p: D \to D$ in $\mathscr{D}$ with $p \circ p = p$ has a kernel $\operatorname{Ker}(p)$, and the canonical map from $\operatorname{Ker}(p) \oplus \operatorname{Ker}(1_D - p)$ to D is an isomorphism.) The category $\widetilde{\mathscr{CV}}^0$ inherits a tensor product: $(X, p) \otimes (Y, q) = (X \times Y, p \times q)$.

The *Tate motive* L is defined to be $(\mathbb{P}^1, p)$, with $p = [\mathbb{P}^1 \times \{0\}] \in A^1(\mathbb{P}^1 \times \mathbb{P}^1)$. Then $\tilde{\mathbb{P}}^1 = 1 \oplus L$, where $1 = \tilde{P}$, $P = \operatorname{Spec}(K)$. If E is a vector bundle of rank $e+1$ on X, there is a canonical morphism of motives

$$P(E) = \bigoplus_{i=0}^{e} \tilde{X} \otimes L^{\otimes i}$$

(Manin (1) § 7). Similar formulas are valid for blow-ups (loc. cit. § 9). For a curve X, the essential part of the motif of X is determined by the Jacobian variety of X (loc. cit. § 10).

Important variations are obtained by replacing rational equivalence by algebraic, homological, or numerical equivalence, as in Remark 16.1.

Example 16.1.13. The theory of correspondences can be extended to quotient varieties, provided rational coefficients are used. If $X_i = Y_i/G_i$, Y_i non-singular, $\pi_i: Y_i \to X_i$ as in Example 8.3.12, then Cartesian products $X_1 \times \ldots \times X_r$ are identified with quotients of $Y_1 \times \ldots \times Y_r$ by $G_1 \times \ldots \times G_r$. One may construct $\alpha \circ \beta$, α_*, α^* as in this section, using the intersection product for quotient varieties described in Example 8.3.12. In particular, if $f: X_1 \to X_2$ is a morphism, one may define the pull-back

$$f^*: A_*(X_2)_{\mathbb{Q}} \to A_*(X_1)_{\mathbb{Q}}$$

to be Γ_f^*, where Γ_f is the graph of f. This is a ring homomorphism, with the usual functoriality, and projection formula for proper f; if f is the identity, so is f^*. (One way to verify this is to show that this f^* agrees with that of Example 17.4.10.)

In particular, the definition of intersection product in Example 8.3.12 is independent of presentation of the variety as a quotient variety. Keeping track of supports of correspondences, one sees likewise that the same is true for the refined intersection products, and hence for the rational intersection numbers.

Example 16.1.14. Let $\alpha : X \vdash Y$ be a correspondence, and let a be a cycle on X, b a cycle on Y. Then

(i)
$$\int_X a \cdot \alpha^*(b) = \int_Y \alpha_*(a) \cdot b .$$

(Both are equal to $\int_{X\times Y} \alpha \cdot (a \times b)$; cf. Clemens-Griffiths (1) p. 289.)

If $\alpha = [T]$ is an irreducible correspondence, and $p : T \to X$, $q : T \to Y$ are the projections, then

(ii)
$$T_*(a) = q_* p^* a , \quad T^* b = p_* q^* b .$$

Here p^* and q^* are the pullbacks of § 8.1. (If i is the imbedding of T in $X \times Y$, then $i_*(p^* a) = T \cdot (a \times Y)$; apply p^{XY}_{Y*} to prove the first statement; apply the involution $T \to T'$ to deduce the second.)

If $f : X' \to X$, $g : Y' \to Y$ are morphisms, then (cf. Lieberman (2) p. 1168)

(iii)
$$(f \times g)^*(\alpha) = \Gamma'_g \circ \alpha \circ \Gamma_f .$$

Example 16.1.15. *Lefschetz fixed point formula.* The formalism of this section is also valid for homological correspondences α between compact oriented manifolds X and Y. To avoid sign problems, consider only $\alpha \in H^n(X \times Y)$, with $n = \dim(X) = \dim(Y)$ *even;* the cohomology groups are taken with rational coefficients. Such α determines

$$\alpha^* : H^i Y \to H^i X$$

by the formalism of Definition 16.1.2, using Poincaré duality to define the push-forward p^{XY}_{X*}. By the Künneth isomorphism, this corresponds to an isomorphism

$$H^n(X \times Y) \cong \bigoplus_i \operatorname{Hom}(H^i Y, H^i X) .$$

If $X = Y$, the diagonal class Δ corresponds to the identity map. The Lefschetz fixed point formula states that for $\alpha \in H^n(X \times X)$,

$$\int_{X\times X} \alpha \cdot \Delta = \sum_i (-1)^i \operatorname{trace}(\alpha^* : H^i X \to H^i X) .$$

This gives a formula for the virtual number of fixed points of α (cf. Kleiman (2)).

If α corresponds to a morphism $f : X \to X$, one recovers a fixed point theorem for f. There are formulas for the intersection numbers occurring on the left. For example, in the differentiable case, if P is an isolated fixed point and df_P does not have 1 as an eigenvalue, then the intersection number of Γ_f and Δ at (P, P) is the sign of the determinant of $I - df_P$.

When the manifolds are algebraic, and α is the class of an algebraic cycle, it follows from the discussion in § 19.1 that the topological and algebraic calculations of $\int \alpha \cdot \Delta$ agree, as do the notions of α^*, $\alpha \circ \beta$, etc., cf. Example 19.2.7. Note however, that, in general, the right side of the fixed point formula necessarily involves non-algebraic cycles.

16.2 Irregular Fixed Points

Let T be an irreducible n-dimensional correspondence on an n-dimensional variety X, with $T \neq \Delta$. Let $X \tilde{\times} X$ be the blow-up of $X \times X$ along the diagonal Δ. The exceptional divisor E is the projective tangent bundle $P(T_X)$; let p denote the projection from E to $X = \Delta$. Let $\tilde{T}$ be the proper transform of T in $X \tilde{\times} X$, i.e., $\tilde{T}$ is the blow-up of T along $T \cap \Delta$. The points of $\tilde{T} \cap E$ are certain tangent lines to X, called *principal tangents.* Since $\tilde{T} \cap E$ is a divisor in $\tilde{T}$, it has pure dimension $n-1$.

The fibres of $\tilde{T} \cap E$ over a point P in $T \cap \Delta$ can have any dimension from 0 to $n-1$. If $\tilde{T} \cap E$ contains the full fibre $P(T_P)$, P is called a *perfect* fixed point. All isolated fixed points are perfect, but there may also be a finite number of perfect fixed points lying on larger components of $T \cap \Delta$.

The degree $\int T \cdot \Delta$ of the intersection is the *virtual number of fixed points* of T. The following proposition can be used to relate this number to the geometry of $T \cap \Delta$. Several classical formulas of this type are derived in the examples that follow.

Proposition 16.2. *Let ξ be the universal quotient bundle on the projective bundle $P(T_X)$. Then*

$$T \cdot \Delta = \{c(T_X) \cap s(T \cap \Delta, T)\}_0 = p_*(c_{n-1}(\xi) \cap [\tilde{T} \cap E])$$

in $A_0(\Delta) = A_0 X$.

Proof. The class $T \cdot \Delta$ may be calculated by intersecting $T \subset X \times X$ by $\Delta \subset X \times X$, according to the prescription of §6.1. Since $\tilde{T} \cap E$ is the projective normal cone to $T \cap \Delta$ in T, the result follows from Proposition 6.1 (a) and Example 6.1.8. □

Example 16.2.1 (cf. Pieri (1)). If $X = \mathbb{P}^n$, there is a canonical morphism

$$\varphi : E = P(T_{\mathbb{P}^n}) \to G = G_1(\mathbb{P}^n)$$

which takes a projective tangent line to the corresponding imbedded tangent line in $\mathbb{P}^n$. In fact φ extends to a morphism from $X \tilde{\times} X$ to G which takes (P, Q), $P \neq Q$, to the line between P and Q (cf. Kleiman (8) V.B). The *principal lines* at P are the images by φ of the principal tangents $\tilde{T} \cap P(T_P X)$.

Define the *indices* $I_1, \ldots, I_{n+1}$ of T by setting

$$I_{i+1} = \# \{(P, Q) \in T \mid P \in [i], Q \in [n-i]\}$$

where $[i]$ and $[n-i]$ are general linear spaces in $\mathbb{P}^n$ of the indicated dimensions. These indices are simply the bidegrees of the n-cycle $[T]$ on $\mathbb{P}^n \times \mathbb{P}^n$.

Pieri defined the *ranks* $R_1, \ldots, R_n$ of T by

$$R_{i+1} = \# \{P \in T \cap \Delta \mid P \in [n-i], \text{ and some principal line at } P \text{ meets } [i]\}.$$

Thus R_1 is the number of perfect fixed points. Since the scheme $\tilde{T} \cap E$ may not be reduced, these ranks should be interpreted with multiplicities, as follows:

Let σ_m be the class in $A^m(G)$ represented by a Schubert variety of lines that meet a linear space $[n-1-m]$. Then

$$R_{i+1} = \int p_*(\varphi^*(\sigma_{n-1-i}) \frown [\tilde{T} \bigcap E]) \cdot [n-i] .$$

Pieri's Theorem. *With this notation,*

$$(*) \qquad I_1 + \ldots + I_{n+1} = R_1 + \ldots + R_n .$$

Previously cases with $n = 1, 2$, and 3 had been given by Chasles, Zeuthen, and Schubert. (Since $[\Delta] = \sum [i] \times [n-i]$, the left side of (*) is $\int T \cdot \Delta$. Since $\sigma_m = c_m(Q)$, where Q is the universal quotient bundle on G (§ 14.7), the right side of (*) is

$$\sum_{i=0}^{n-1} \int p_*(p^* c_1(\mathcal{O}(1))^i \varphi^* c_{n-1-i}(Q) \frown [\tilde{T} \bigcap E]) .$$

By Proposition 16.2 and Remark 3.2.3 (b) it suffices to show that $\xi = \varphi^* Q \otimes p^* \mathcal{O}(1)$. Indeed, the canonical surjections $p^* \mathcal{O}(1)^{\oplus n+1} \to p^* T_X \to \xi$ realize $\xi \otimes p^* \mathcal{O}(-1)$ as a quotient of the trivial bundle of rank $n+1$, and φ is determined by the universal property that this quotient is $\varphi^*(Q)$.)

For example, if T is the closure of the graph of a projection from a subspace A^{n-r-1} to a complementary space B^r, the principal lines to $P \in B$ are the lines joining P to A. In this case $I_i = R_i = 1$ for $1 \leqq i \leqq r+1$, and $I_i = R_i = 0$ for $i > r+1$.

As an application, take $T = V \times W$. The left side of (*) is $\deg(V) \deg(W)$. Suppose V meets W in N isolated points, counted with multiplicities, and also scheme-theoretically in a non-singular variety Z of dimension $k > 0$. Then

$$N = \deg(V) \deg(W) - \sum_{i=n-k}^{n} y_{(i)}$$

where $y_{(i)}$ is defined as follows (cf. Example 12.3.6). Fix a general i-plane L and a general $(n-i)$-plane M. Then $y_{(i)}$ is the number of points P in $L \bigcap Z$ such that the span of the tangent spaces to V and W at P meet M. Previous cases of this formula had been given by Salmon and Caporali. For applications see Baker (2) Ch. II, Pt. III.

Pieri's theorem can also be deduced from the formula for $A(\mathbb{P}^n \tilde{\times} \mathbb{P}^n)$, using Proposition 6.7.

Example 16.2.2. Proposition 16.2 and its proof go through for an arbitrary T of pure dimension $m \geqq n$, giving a formula for $T \cdot \Delta$ in $A_{m-n}(T \bigcap \Delta)$.

If $X = \mathbb{P}^n$, the degree of $T \cdot \Delta$ is equal to the sum of the indices of T (bidegrees of $[T]$). In particular, every irreducible correspondence of dimension $m \geqq n$ on $\mathbb{P}^n$ must have fixed points. The degree of $T \cdot \Delta$ is also equal to the sum of the ranks of T, as defined in the preceding example. If $Z_1, \ldots, Z_r$ are the irreducible components of $T \bigcap \Delta$, then (Theorem 12.2 (b))

$$\deg(T \cdot \Delta) \geqq \sum \deg(Z_i) > 0 .$$

Example 16.2.3 (cf. Severi (1)). Let $X = G_d(\mathbb{P}^m)$, $n = (d+1)(m-d)$, and let T be an n-dimensional correspondence on X. For $0 \leqq a_0 < a_1 < \ldots < a_d \leqq m$, define an index $I(a_0, \ldots, a_d)$ of T by

$$I(a_0, \ldots, a_d) = \# \{(L, M) \in T \mid L \in [a_0, \ldots, a_d], M \in [m - a_d, \ldots, m - a_0]\}$$

with notation as in Example 14.7.16. Then

$$\int T \cdot \Delta = \sum I(a_0, \ldots, a_d),$$

the sum over all such sequences $(a_0, \ldots, a_d)$. (In fact,

$$[\Delta] = \sum (a_0, \ldots, a_d) \times (m - a_d, \ldots, m - a_0)$$

in $A_n(X \times X)$, as one sees by intersecting both sides with $(m - b_d, \ldots, m - b_0) \times (c_0, \ldots, c_d)$.) A similar formula is valid for general flag manifolds.

Example 16.2.4. If X is a surface, and T contains a curve D of fixed points, Severi (8) p. 874 gave a formula for the contribution of D to the virtual number of fixed points, so that the difference is the number N of perfect fixed points. His formula is:

$$\int T \cdot \Delta - N = v + 2 - 2\pi + \varrho - 2\omega,$$

where: v is the degree of D, i.e. $\int D \cdot D$; π is the virtual (arithmetic) genus of D; choosing a general one-dimensional linear system L of curves on X, ω is the number of points of $D \cap C$ for general C in L; and if $\tilde{D}$ is the lift of D to $P(T_X)$, i.e. the closure of the tangents to smooth points of D, then ϱ is the number of points $t \in \tilde{D}$ which are tangent to any C in L passing through $p(t)$.

This may be explained, including cases when multiplicities occur, as follows. Write

$$[\tilde{T} \cap E] = \sum m_i [P(T_{P_i} X)] + \tilde{D}$$

where the P_i are the perfect fixed points, and each component of $\tilde{D}$ maps finitely by p to a curve in X. Define D to be the curve $p_* \tilde{D}$. Then

$$\sum m_i = N.$$

By the adjunction formula (Example 15.2.2)

$$\int c_1(T_X) \cap D = v + 2 - 2\pi.$$

By Example 3.2.19, if $\mathcal{O}(1) = \mathcal{O}_{T_X}(1)$ on $E = P(T_X)$,

$$\int c_1(\mathcal{O}(1)) \cap \tilde{D} = \varrho - 2\omega.$$

Severi's formula now follows from Proposition 16.2:

$$\begin{aligned} \Delta \cdot T &= \{c(T_X) \cap s(\Delta \cap T, T)\}_0 \\ &= c_1(T_X) \cap p_*[\tilde{T} \cap E] + p_*(c_1(\mathcal{O}(1)) \cap [\tilde{T} \cap E]). \end{aligned}$$

Combining the displayed equations gives Severi's formula in case D has no multiple components, and shows how to interpret the formula in general.

Notes and References

A glance at the long encyclopedia article of Berzolari (3) impresses one with the importance of correspondences in mathematics through the early part of this century. In particular, many problems in enumerative geometry were solved by constructing appropriate correspondences between curves, and using coincidence or fixed point formulas for such correspondences (cf. Examples 16.1.5(d), (e), 16.1.6, 16.1.7).

There were many attempts to find higher dimensional analogues. Among these Pieri's theorem for correspondences on $\mathbb{P}^n$ (Example 16.2.1) stands out as a precursor of modern excess intersection theory. Zeuthen and Severi also devoted a number of papers to this (cf. Examples 16.1.5, 16.2.3, and 16.2.4). Most of the success in higher dimensions was achieved for those correspondences T for which one can write the class $[T]$ of T on $X \times X$ as a sum of exterior products of cycles on X, or on varieties X for which the diagonal has such a decomposition. In general such Künneth decompositions are only possible if one allows non-algebraic cycles, including odd-dimensional homology classes, on X. The problem of finding a general fixed point formula in this context was solved by Lefschetz (1), cf. Example 16.1.15.

For history and applications of the theory of correspondences – which we have made no attempt to repeat here – we recommend the encyclopedia article quoted above, Zariski (1) Ch. VI and App. B, Severi (5) § 6, Lefschetz (2) VIII, Conforto (1), and Baker (2) I, II. Recently correspondences have appeared in the guise of Hecke operators (cf. Shimura (2) § 7, Deligne (1)).

The material in § 16.1 is a routine application of standard intersection theory; for curves this appears in Weil (4). The deduction of the formulas of Pieri and Severi in Examples 16.2.1 and 16.2.4 from the excess intersection formula are apparently the first modern proofs to appear.

Chapter 17. Bivariant Intersection Theory

Summary

Our basic intersection construction has assigned to a regular imbedding (or l.c.i. morphism) $f: X \to Y$ of codimension d a collection of homomorphisms

$$f^!: A_k Y' \to A_{k-d} X'$$

for all $Y' \to Y$, $X' = X \times_Y Y'$, all k. In this chapter we formalize the study of such operations. For any morphism $f: X \to Y$, a *bivariant class* c in $A^p(X \xrightarrow{f} Y)$ is a collection of homomorphisms from $A_k Y'$ to $A_{k-p} X'$, for all $Y' \to Y$, all k, compatible with push-forward, pull-back, and intersection products.

The group $A^{-k}(X \to \text{pt.})$ is canonically isomorphic to $A_k(X)$. The other extreme $A^k(X \xrightarrow{\text{id}} X)$ is defined to be the *cohomology* group $A^k X$. The bivariants groups have products

$$A^p(X \xrightarrow{f} Y) \otimes A^q(Y \xrightarrow{g} Z) \to A^{p+q}(X \xrightarrow{gf} Z)$$

which specialize to give a ring structure on $A^* X$, and a cap product action of $A^* X$ on $A_* X$. If X is non-singular, $A^* X \cong A_* X$. There are also a proper push-forward and a pull-back operation for bivariant groups, generalizing the push-forward on A_* and defining a pull-back on A^*. There are compatibilities among these three operations which allow one to manipulate bivariant classes symbolically with a freedom one is accustomed to with homology and cohomology in topology.

Many constructions of previous chapters actually produce classes in appropriate bivariant groups. For example, Chern classes of vector bundles on X live in $A^* X$. Flat and l.c.i. morphisms $f: X \to Y$ determine canonical elements in $A^*(X \xrightarrow{f} Y)$, which are denoted $[f]$. An element c of $A^p(X \to Y)$ determines generalized Gysin homomorphisms

$$A_k Y \xrightarrow{c^*} A_{k-p} X \quad \text{and} \quad A^k X \xrightarrow{c_*} A^{k+p} Y$$

(for the latter f is assumed to be proper). Intersection formulas such as the excess and residual intersection formulas achieve their sharpest formulation in the bivariant language.

There is a useful criterion which implies that an operation which produces rational equivalence classes on X' from subvarieties of Y' (for all $Y' \to Y$), passes to rational equivalence and defines a bivariant class (Theorem 17.1). This will be used in the next chapter to deduce the important properties of local Chern classes.

17.1 Bivariant Rational Equivalence Classes

Definition 17.1. Let $f: X \to Y$ be a morphism. For each morphism $g : Y' \to Y$, form the fibre square

$$(*) \qquad \begin{array}{ccc} X' & \xrightarrow{f'} & Y' \\ {\scriptstyle g'}\downarrow & & \downarrow{\scriptstyle g} \\ X & \xrightarrow[f]{} & Y \end{array}$$

with induced morphisms as labelled. A *bivariant class* c in $A^p(X \xrightarrow{f} Y)$ is a collection of homomorphisms

$$c_g^{(k)} : A_k Y' \to A_{k-p} X'$$

for all $g : Y' \to Y$, and all k, compatible with proper push-forward, flat pull-back, and intersection products, i.e.:

(C_1) If $h : Y'' \to Y'$ is proper, $g : Y' \to Y$ arbitrary, and one forms the fibre diagram

$$(\text{\textasteriskcentered}) \qquad \begin{array}{ccc} X'' & \xrightarrow{f''} & Y'' \\ {\scriptstyle h'}\downarrow & & \downarrow{\scriptstyle h} \\ X' & \xrightarrow[f']{} & Y' \\ {\scriptstyle g'}\downarrow & & \downarrow{\scriptstyle g} \\ X & \xrightarrow[f]{} & Y \end{array} ,$$

then, for all $\alpha \in A_k Y''$,

$$c_g^{(k)}(h_*(\alpha)) = h'_* \, c_{gh}^{(k)}(\alpha)$$

in $A_{k-p}X'$.

(C_2) If $h : Y'' \to Y'$ is flat of relative dimension n, and $g : Y' \to Y$ is arbitrary, and one forms the fibre diagram (⁑), then, for all $\alpha \in A_k Y'$,

$$c_{gh}^{(k+n)}(h^* \alpha) = h'^* \, c_g^{(k)}(\alpha)$$

in $A_{k+n-p}(X'')$.

(C_3) If $g : Y' \to Y$, $h : Y' \to Z'$ are morphisms, and $i : Z'' \to Z'$ is a regular imbedding of codimension e, and one forms the fibre diagram

$$(⁑{*}) \qquad \begin{array}{ccccc} X'' & \xrightarrow[f'']{} & Y'' & \xrightarrow[h']{} & Z'' \\ {\scriptstyle i''}\downarrow & & \downarrow{\scriptstyle i'} & & \downarrow{\scriptstyle i} \\ X' & \xrightarrow[f']{} & Y' & \xrightarrow[h]{} & Z' \\ {\scriptstyle g'}\downarrow & & \downarrow{\scriptstyle g} & & \\ X & \xrightarrow[f]{} & Y & & \end{array} ,$$

then, for all $\alpha \in A_k Y'$,

$$i^! c_g^{(k)}(\alpha) = c_{gi'}^{(k-e)}(i^! \alpha)$$

in $A_{k-p-e}(X'')$.

Notation. The group $A^p(X \xrightarrow{f} Y)$ may be denoted simply $A^p(X \to Y)$ or $A^p(f)$. The homomorphism $c_g^{(k)} : A_k Y' \to A_{k-p} X'$ determined by an element c in $A^p(X \to Y)$ will usually be denoted simply c, with an indication of where it acts. Since these homomorphisms will be seen to generalize the cap products of previous chapters, we may also write $c \cap \alpha$ in place of $c(\alpha) = c_g^{(k)}(\alpha)$ for

$\alpha \in A_k Y'$. Denote by $A^*(X \to Y)$ or $A(X \to Y)$ the direct sum of all $A^p(X \to Y)$, $p \in \mathbb{Z}$.

Proper push-forward and flat pull-back are defined on the cycle level. One case of intersection product is defined on the cycle level, namely, when i is the imbedding of a principal Cartier divisor (Remark 2.3).

Theorem 17.1. *Let $f: X \to Y$ be given. Suppose for all $g: Y' \to Y$, and all k, there are homomorphisms*

$$c_g^{(k)}: Z_k Y' \to A_{k-d} X'$$

satisfying the formulas of (C_1) *and* (C_2) *for all $h: Y'' \to Y'$ proper and flat respectively, and satisfying the formula of* (C_3) *whenever $i: Z'' \to Z'$ is the imbedding of the point $\{0\}$ in $\mathbb{A}^1$. Then the homomorphisms $c_g^{(k)}$ pass to rational equivalence, and the resulting homomorphisms from $A_k Y'$ to $A_{k-p} X'$ determine a bivariant class in $A^p(X \to Y)$.*

Proof. We show first that c passes to rational equivalence. Since c is compatible with proper push-forward, it suffices (Proposition 1.6) to show that if V is a subvariety of $Y' \times \mathbb{P}^1$, dominant over $\mathbb{P}^1$, of dimension $k+1$, then

$$c(i_0^![V]) = c(i_\infty^![V]) .$$

Since, for $t = 0, \infty$, i_t is a composite of an imbedding of $\{0\}$ in $\mathbb{A}^1$, followed by an open imbedding of $\mathbb{A}^1$ in $\mathbb{P}^1$ (which is flat),

$$c(i_t^![V]) = i_t^! c([V])$$

with $c([V]) \in A_{k+1-p}(X' \times \mathbb{P}^1)$. But $i_0^! \alpha = i_\infty^! \alpha$ for any $\alpha \in A_*(X' \times \mathbb{P}^1)$ (Example 3.3.6), so c passes to rational equivalence.

It remains to show that c verifies (C_3), when i is an arbitrary regular imbedding. Let N be the normal bundle to Z'' in Z', $\pi: N \to Z''$ the projection, and let $M^\circ = M^\circ_{Z''} Z'$ be the deformation space for deforming $Z'' \hookrightarrow Z'$ into $Z'' \hookrightarrow N$ (§ 5.1). Let

$$N \overset{i_\infty}{\hookrightarrow} M^\circ \overset{j}{\hookleftarrow} Z' \times \mathbb{A}^1 \overset{\mathrm{pr}}{\longrightarrow} Z'$$

be the canonical inclusions and projection. Let $N_{Y''}$ be the pull-back of N to Y'', with projection $\pi_{Y''}: N_{Y''} \to Y''$, and let $M^\circ_{Y'} = M^\circ \times_{Z'} Y'$. There is a unique homomorphism σ so that the diagram

$$\text{(1)} \qquad \begin{array}{ccccc} A_{k+1}(N_{Y''}) & \xrightarrow{i_{\infty *}} & A_{k+1}(M^\circ_{Y'}) & \xrightarrow{j^*} & A_{k+1}(Y' \times \mathbb{A}^1) \to 0 \\ & \searrow^{0} & \downarrow i_\infty^* & & \uparrow \mathrm{pr}^* \\ & & A_k(N_{Y''}) & \xleftarrow[\sigma]{} & A_k(Y') \end{array}$$

commutes. Then $i^! = (\pi^*_{Y''})^{-1} \circ \sigma: A_k Y' \to A_{k-e} Y''$. This follows easily from the construction of $i^!$, as in § 6.2: if $V \subset Y'$, $j^*[M^\circ_{V \cap Y''} V] = \mathrm{pr}^*[V]$, and $i_\infty^*[M^\circ_{V \cap Y''} V] = [C_{V \cap Y''} V]$.

One has a similar diagram for the base extension $X' \to Z'$ in place of $Y' \to Z'$, and a similar description for $i^!: A_{k-p} X' \to A_{k-p-e} X''$. From the fibre diagram

$$\begin{array}{ccccc} N_{X''} & \to & N_{Y''} & \to & \infty \\ \downarrow & & \downarrow & & \downarrow i_\infty \\ M^\circ_{X'} & \to & M^\circ_{Y'} & \to & \mathbb{P}^1 \\ \downarrow & & \downarrow & & \\ X & \to & Y & & \end{array}$$

one sees as in the first step that c commutes with $i^!_\infty$. Since c commutes with $i_{\infty *}$ (by (C_1)), and with j^* and pr^* (by (C_2)), c also commutes with σ. Since c commutes with π^* (by (C_2)), c commutes with $i^! = (\pi^*)^{-1} \circ \sigma$, as required. □

Example 17.1.1. A pseudo-divisor $D = (L, Z, s)$ on a scheme X determines a bivariant class $c(D) \in A^1(Z \to X)$. If $g: X' \to X$ is a morphism, $\alpha \in A_*(X')$, then $c(D)(\alpha)$ is defined, to be $g^*(D) \cdot \alpha$ (cf. Definitions 2.2.4 and 2.3). (The main results of Chap. 2 verify the hypotheses of Theorem 17.1.)

17.2 Operations and Properties

There are three basic operations on the bivariant groups $A^*(X \to Y)$.

(P_1) *Product.* For all morphisms $f: X \to Y$, $g: Y \to Z$, and integers p, q, there is a homomorphism

$$A^p(X \xrightarrow{f} Y) \otimes A^q(Y \xrightarrow{g} Z) \to A^{p+q}(X \xrightarrow{gf} Z) .$$

The image of $c \otimes d$ is denoted $c \cdot d$. Given $Z' \to Z$, form the fibre diagram

$$(**) \qquad \begin{array}{ccccc} X' & \xrightarrow{f'} & Y' & \xrightarrow{g'} & Z' \\ \downarrow & & \downarrow & & \downarrow \\ X & \xrightarrow[f]{} & Y & \xrightarrow[g]{} & Z \end{array} .$$

If $\alpha \in A_k Z'$, then $d(\alpha) \in A_{k-q} Y'$, and $c(d(\alpha)) \in A_{k-q-p} X'$. We define $c \cdot d$ by

$$c \cdot d(\alpha) = c(d(\alpha)) .$$

(P_2) *Push-forward.* If $f: X \to Y$ is a proper morphism, $g: Y \to Z$ any morphism, and p an integer, there is a homomorphism

$$f_*: A^p(X \xrightarrow{gf} Z) \to A^p(Y \xrightarrow{g} Z) .$$

Given $Z' \to Z$, form the fibre diagram (**). If $c \in A^p(gf)$, and $\alpha \in A_k(Z')$, then $c(\alpha) \in A_{k-p}(X')$. Since f' is proper, $f'_*(c(\alpha)) \in A_{k-p}(Y')$. Define $f_*(c)$ by the formula

$$f_*(c)(\alpha) = f'_*(c(\alpha)) .$$

(P_3) *Pull-back.* Given $f: X \to Y$, $g: Y_1 \to Y$, form the fibre square

$$(*) \qquad \begin{array}{ccc} X_1 & \xrightarrow{f_1} & Y_1 \\ \downarrow & & \downarrow g \\ X & \xrightarrow[f]{} & Y \end{array} .$$

For each p there is a homomorphism

$$g^*: A^p(X \xrightarrow{f} Y) \to A^p(X_1 \xrightarrow{f_1} Y_1) .$$

Given $c \in A^p(f)$, $Y' \to Y_1$, $\alpha \in A_k(Y')$, then composing with g gives a morphism $Y' \to Y$. Therefore $c(\alpha) \in A_{k-p}(X')$, $X' = X \times_Y Y' = X_1 \times_{Y_1} Y'$. Set

$$g^*(c)(\alpha) = c(\alpha) .$$

It is very easy to verify that, in these three cases, $c \cdot d$, $f_*(c)$, and $g^*(c)$ determine bivariant classes in the appropriate groups, i.e. that $(C_1)-(C_3)$ of Definition 17.1 are verified. The following seven axioms satisfied by these three operations are similarly straightforward to verify, using the basic functorial properties of Chaps. 1 and 6.

(A_1) *Associativity of products. If $c \in A(X \to Y)$, $d \in A(Y \to Z)$, $e \in A(Z \to W)$,* then

$$(c \cdot d) \cdot e = c \cdot (d \cdot e) \in A(X \to W).$$

(A_2) *Funtoriality of push-forwards.* If $f: X \to Y$ and $g: Y \to Z$ are proper, $Z \to W$ arbitrary, and $c \in A(X \to W)$, then

$$(g f)_*(c) = g_*(f_* c) \in A(Z \to W).$$

(A_3) *Functoriality of pull-backs.* If $c \in A(X \to Y)$, $g: Y_1 \to Y$, $h: Y_2 \to Y_1$, then

$$(g\, h)^*(c) = h^* g^*(c) \in A(X_2 \to Y_2),$$

$X_2 = X \times_Y Y_2$.

(A_{12}) *Product and push-forward commute.* If $f: X \to Y$ is proper, $Y \to Z$, $Z \to W$ are arbitrary and $c \in A(X \to Z)$, $d \in A(Z \to W)$, then

$$f_*(c) \cdot d = f_*(c \cdot d) \in A(Y \to W).$$

(A_{13}) *Product and pull-back commute.* If $c \in A(X \xrightarrow{f} Y)$, $d \in A(Y \to Z)$, and $g: Z_1 \to Z$ is a morphism, form the fibre diagram

$$(**) \qquad \begin{array}{ccccc} X_1 & \xrightarrow{f'} & Y_1 & \to & Z_1 \\ \downarrow & & \downarrow{\scriptstyle g'} & & \downarrow{\scriptstyle g} \\ X & \xrightarrow{f} & Y & \to & Z \end{array} .$$

Then

$$g^*(c \cdot d) = g'^*(c) \cdot g^*(d) \in A(X_1 \to Z_1).$$

(A_{23}) *Push-forward and pull-back commute.* If $f: X \to Y$ is proper, $Y \to Z$, $g: Z_1 \to Z$, and $c \in A(X \to Z)$ are given, then, with notation as in the preceding diagram,

$$g^* f_*(c) = f'_*(g^*(c)) \in A(Y_1 \to Z_1).$$

(A_{123}) *Projection formula.* Given a diagram

$$\begin{array}{ccccc} X' & \xrightarrow{f'} & Y' & & \\ {\scriptstyle g'}\downarrow & & \downarrow{\scriptstyle g} & & \\ X & \xrightarrow{f} & Y & \xrightarrow{h} & Z \end{array}$$

with g proper, the square a fibre square, and $c \in A(X \to Y)$, $d \in A(Y' \to Z)$, then

$$c \cdot g_*(d) = g'_*(g^*(c) \cdot d) \in A(X \to Z).$$

17.3 Homology and Cohomology

Let $S = \mathrm{Spec}(K)$, K the ground field. For each p, there is a canonical homomorphism
$$\varphi : A^{-p}(X \to S) \to A_p(X)$$
taking a bivariant class c to $c([S])$.

Proposition 17.3.1. *The above homomorphisms φ are isomorphisms: $A^{-p}(X \to S) \cong A_p(X)$.*

Proof. Given $a \in A_p(X)$, define a bivariant class $\psi(a) \in A^{-p}(X \to S)$ as follows: for any morphism $Y \to S$, and $\alpha \in A_k Y$, define
$$\psi(a)(\alpha) = a \times \alpha \in A_{p+k}(X \times_S Y)$$
where $a \times \alpha$ is the exterior product (§ 1.19). Since exterior products are compatible with proper push-forward, flat pull back, and intersections (Proposition 1.10, Example 6.5.2), $\psi(a)$ is a bivariant class. Clearly $\psi(a)([S]) = a$, so $\varphi \circ \psi$ is the identity. To show that $\psi \circ \varphi$ is the identity, we must show that
$$c(\alpha) = \varphi(c) \times \alpha \in A_{k+p}(X \times_S Y)$$
for all $\alpha \in A_k Y$. By compatibility with push-forward, we may assume $\alpha = [V]$, $V = Y$ a variety of dimension k. Then $\alpha = p^*[S]$, where $p : V \to S$ is the morphism from V to S. Since c commutes with flat pull-back,
$$c(\alpha) = c(p^*[S]) = p^* c[S] = \varphi(c) \times [V],$$
as required. □

Definition 17.3. For any scheme X, and any integer p, define the pth *cohomology* group $A^p X$ by
$$A^p X = A^p(X \xrightarrow{\mathrm{id}} X).$$
Thus an element $c \in A^p X$ is a collection of homomorphisms $A_k X' \to A_{k-p} X'$, for all $X' \to X$ and all k, compatible with proper push-forward, flat pull-back and intersections ($(C_1)-(C_3)$). There is an element $1 \in A^0 X$, which acts as the identity on all $A_k X'$, such that
$$1 \cdot c = c, \quad d \cdot 1 = d$$
for all $c \in A(X \to Y)$, $d \in A(W \to X)$.

The product from the composite $X \xrightarrow{\mathrm{id}} X \xrightarrow{\mathrm{id}} X$ determines "cup" products
$$A^p X \otimes A^q X \to A^{p+q} X$$
which make $A^* X$ into an associative, graded ring with unit 1. For any $g : X_1 \to X$, the pull-back $g^* : A^* X \to A^* X_1$ is a ring homomorphism; this is functorial in g.

There are canonical homomorphisms
$$A^p X \otimes A_q X \xrightarrow{\cap} A_{q-p} X$$

taking $c \otimes \alpha$ to $c \cap \alpha = c(\alpha)$. If one identifies $A_q X$ with $A^{-q}(X \to S)$, this is the bivariant product from the composite $X \xrightarrow{\text{id}} X \to S$. This "cap" product makes $A_* X$ into a left $A^* X$-module. One has the projection formula

$$f_*(f^* \beta \cap \alpha) = \beta \cap f_*(\alpha)$$

for $f\colon X_1 \to X$, $\alpha \in A_* X_1$, $\beta \in A^* X$. (All of the above assertions are formal consequences of the seven axioms.)

Given a vector bundle E on a scheme X, and an integer p, there is a Chern class $c_p(E)$ in $A^p(X)$. The action of $c_p(E)$ on $\alpha \in A_k(X')$, $g: X' \to X$, is defined by

$$c_p(E)(\alpha) = c_p(g^* E) \overset{\prime}{\cap} \alpha\,,$$

where the right side is the class defined in § 3.2. Theorem 3.2(c), (d) and Proposition 6.3 amount to saying that $c_p(E)$ is a bivariant class. In addition, all the formal identities proved for Chern classes in § 3.2 are valid for these classes.

In fact, Chern classes commute with all bivariant classes, not just the three stated in the definition. Put another way, any operation which commutes with push-forward, pull-back, and intersections automatically commutes with Chern classes:

Proposition 17.3.2. Let $c \in A^q(X \xrightarrow{f} Y)$, $Y' \to Y$, $\alpha \in A_k(Y')$, E *a vector bundle on* Y'. *Then*

$$c(c_p(E) \cap \alpha) = c_p(f'^* E) \cap c(\alpha) \in A_{k-q-p} X'\,.$$

where $f'\colon X' = X \times_Y Y' \to Y'$ *is the morphism induced by* f.

Proof. Since Chern classes are polynomials in Segre classes, which come from operations of the form $\alpha \to p_*(c_1(\mathcal{O}(1))^i \cap p^*(\alpha))$ (cf. § 3.1), and since c commutes with p_* and p^*, one is reduced to showing that c commutes with $c_1(L)$, L a line bundle on Y'. We may also assume $\alpha = [V]$, $V = Y'$ a variety, so $L = \mathcal{O}(D)$, D a Cartier divisor on V. After replacing V by V', where $V' \to V$ is proper and birational, we may assume $D = D_1 - D_2$, D_1, D_2 effective (cf. Theorem 2.4, Case 3). Since $c_1(L) = c_1(\mathcal{O}(D_1)) - c_1(\mathcal{O}(D_2))$, we assume D is effective. Let i be the inclusion of D in V. Then $c_1(L) \cap \alpha = i_* i^!(\alpha)$, and since c commutes with i_* and $i^!$, c commutes with $c_1(L)$. □

Example 17.3.1. For X closed in Y, one may define a analogue of local cohomology by setting

$$A^p_X Y = A^p(X \to Y)\,.$$

If also Z is closed in Y, there are products

$$A^p_X Y \otimes A^q_Z Y \xrightarrow{\cup} A^{p+q}_{X \cap Z}(Y)$$

by $c \cup d = i^*(c) \cdot d$, where i is the inclusion of Z in Y. This product satisfies an obvious associativity, and refines the cup product on $A^* Y$.

Example 17.3.2. Let X be a scheme, $\pi: X' \to X$ a proper morphism such that every irreducible variety in X is the birational image of some subvariety of X'. Then

$$\pi^*: A^* X \to A^* X'$$

is injective. More generally for any $f\colon Y \to X$, π^* injects $A^*(Y \to X)$ into $A^*(X' \times_X Y \to X')$. (For any $h: V \to X$, V a variety, there is a proper birational morphism $V' \to V$ so that the composite $V' \to V \to X$ factors through X'. Then the action of A^*X on $[V']$ factors through the action of A^*X', and the action of A^*X on $[V']$ determines the action on $[V]$ by (C_1).)

Example 17.3.3. Let $f\colon X \to Y$ be a morphism. Let $m = \dim(Y)$, and let n be the largest dimension of any fibre $f^{-1}(y)$, $y \in Y$. Then

$$A^p(X \xrightarrow{f} Y) = 0 \quad \text{if} \quad p < -n \quad \text{or} \quad p > m\,.$$

In particular, for any X,

$$A^p X = 0 \quad \text{if} \quad p < 0 \quad \text{or} \quad p > \dim(X)\,.$$

(Let $c \in A^p(f)$, $p < -n$ or $p > m$. It suffices to show $c([V]) = 0$, $h: V \to Y$, V a variety. Restricting to the closure of $h(V)$, we may assume Y is a variety and h dominant. By the flattening theorem of Raynaud and Gruson (1) 5.5.2, there is a proper, birational morphism $Y' \to Y$, and a closed subscheme $V' \subset V \times_Y Y'$ such that the induced map $g: V' \to Y'$ is flat, and $V' \to V$ is birational. It suffices by (C_1) to show $c([V']) = 0$. By (C_2), $c([V']) = g^*(c([Y']))$. But $c([Y'])$ is in $A_{m-p}(X \times_Y Y')$, which is zero since $\dim(X \times_Y Y') \leqq m + n$.)

17.4 Orientations

Certain morphisms $f\colon X \to Y$ have naturally determined elements in $A(X \xrightarrow{f} Y)$, called *canonical orientations*, and denoted $[f]$:

(1) If $f\colon X \to Y$ is flat of relative dimension n, then $[f] \in A^{-n}(X \xrightarrow{f} Y)$ is defined by flat pull-back. If $g: Y' \to Y$ is a morphism, the induced morphism $f': X' \to Y'$ is flat, and for $\alpha \in A_k Y'$, we set

$$[f](\alpha) = f'^*(\alpha)$$

with f'^* the flat pull-back of § 1.7.

(2) If $f\colon X \to Y$ is a regular imbedding of codimension d, then $[f] \in A^d(X \xrightarrow{f} Y)$ is defined by the refined Gysin homomorphism. If $g: Y' \to Y$, and $\alpha \in A_k Y'$, we constructed a class $f^!(\alpha)$ in $A_{k-d}X'$ in § 6.2. Set

$$[f](\alpha) = f^!(\alpha)\,.$$

(3) More generally, if $f\colon X \to Y$ is a l.c.i. morphism which factors

$$X \xrightarrow{i} P \xrightarrow{p} Y$$

with i a regular imbedding of codimension e, p smooth of relative dimension n, set $d = e - n$, the codimension of f, and set

$$[f] = [i] \cdot [p]\,.$$

Here $[i] \in A^e(X \to P)$ by (2), $[p] \in A^{-n}(P \to Y)$ by (1), so the product is in $A^d(X \to Y)$. Equivalently, if $g: Y' \to Y$, $\alpha \in A_k Y'$, then $[f](\alpha) = f^!(\alpha)$, where $f^!$ is the refined Gysin homomorphism of § 6.6.

It follows from Proposition 6.6 that $[f]$ is independent of the factorization, and that the definitions of (1) and (3) agree if f is both flat and l.c.i. If $f: X \to Y$ and $g: Y \to Z$ are both flat, or both regular imbeddings, or both l.c.i. morphisms with compatible factorizations as in the proof of Proposition 6.6 (c), then

$$[f] \cdot [g] = [gf] \quad \text{in} \quad A(X \xrightarrow{gf} Z).$$

Proposition 17.4.1 (Excess intersection formula). *Let*

$$\begin{array}{ccc} X' & \xrightarrow{f'} & Y' \\ {\scriptstyle g'}\downarrow & & \downarrow \\ X & \xrightarrow[f]{} & Y \end{array}$$

be a fibre square, with f and f' l.c.i. morphisms of codimensions d and d'. Then

$$g^*[f] = c_e(E) \cdot [f'] \in A^d(X' \to Y'),$$

where $e = d - d'$, E the excess normal bundle.

If f is a closed imbedding, and N and N' are the normal bundles fo f and f', then $E = g'^* N/N'$. The general definition, and the proof, are given in Proposition 6.6 (c). □

Several other compatibilities of these orientations are sketched in Example 17.4.6.

Proposition 17.4.2. *Let $g: Y \to Z$ be a smooth morphism of relative dimension n, and let $[g] \in A^{-n}(Y \xrightarrow{g} Z)$ be its orientation class. Then for any morphism $f: X \to Y$, and any integer p,*

$$A^p(X \xrightarrow{f} Y) \xrightarrow{\cdot [g]} A^{p-n}(X \xrightarrow{gf} Z)$$

is an isomorphism.

Proof. Form the fibre diagram

$$\begin{array}{ccccc} X & \xrightarrow{f} & Y & & \\ {\scriptstyle \gamma}\downarrow & & \downarrow{\scriptstyle \delta} & & \\ X \times_Z Y & \xrightarrow{f'} & Y \times_Z Y & \xrightarrow{q} & Y \\ {\scriptstyle p'}\downarrow & & \downarrow{\scriptstyle p} & & \downarrow{\scriptstyle g} \\ X & \xrightarrow[f]{} & Y & \xrightarrow[g]{} & Z \end{array}$$

where δ is the diagonal imbedding, and p and q are the first and second projections. Define the inverse homomorphism

$$L: A^{p-n}(X \xrightarrow{gf} Z) \to A^p(X \xrightarrow{f} Y)$$

by $L(c) = [\gamma] \cdot g^*(c)$. Note that δ and γ are regular imbeddings of codimension n, with $f'^*[\delta] = [\gamma]$. The verification that L and multiplication by $[g]$ are inverse isomorphisms is as follows: If $c \in A^{p-n}(gf)$, then

$$L(c)\cdot[g] = [\gamma]\cdot(g^*(c)\cdot[g]) \qquad (A_1)$$
$$= [\gamma]\cdot[p']\cdot c \qquad (C_2)$$
$$= [p'\circ\gamma]\cdot c = 1\cdot c = c \qquad (A_1)\,.$$

Similarly, if $c \in A^p(f)$, then

$$L(c\cdot[g]) = f'^*[\delta]\cdot p^*(c)\cdot g^*[g] \qquad (A_{13}), (A_1)$$
$$= (p\circ\delta)^*(c)\cdot[\delta]\cdot[q] \qquad (C_3)$$
$$= c\cdot[\delta\circ q] = c\cdot 1 = c \qquad (A_1)\,. \quad \square$$

Corollary 17.4 (Poincaré duality). *Let Y be a smooth, purely n-dimensional scheme.*

(a) *The canonical homomorphisms*

$$A^p Y \xrightarrow{\cap[Y]} A_{n-p} Y$$

are isomorphisms.

(b) *The ring structure on A^*Y is compatible with that defined on A_*Y in* §8.3. *More generally, if $f: X \to Y$ is a morphism, $\beta \in A^*Y$, $\alpha \in A_*X$, then the class $f^*(\beta)\cap\alpha \in A_*X$ coincides with that constructed in* §8.3.

Proof. (a) Apply the proposition to $f = 1_Y$, $g: Y \to S$, and identify $A_{n-p}Y$ with $A^{p-n}(g)$ (Proposition 17.3.1). (b) follows from the construction of $f^*(\beta)\cap\alpha$ as $\gamma_f^*(\alpha\times\beta)$ in §8.3, with γ_f the graph of f, and the construction of the inverse isomorphism given in the proposition. $\square$

A bivariant class c in $A^p(X \xrightarrow{f} Y)$ determines *Gysin homomorphisms*

$$(G_1) \qquad c^*: A_k Y \to A_{k-p} X$$

and, if f is proper,

$$(G_2) \qquad c_*: A^k X \to A^{k+p} Y\,.$$

Define $c^*(\alpha) = c(\alpha)$, $\alpha \in A_k Y$, and $c_*(\beta) = f_*(\beta\cdot c)$, $\beta \in A^k(X)$; note that $\beta\cdot c \in A^{k+p}(X \to Y)$, so $f_*(\beta\cdot c) \in A^{k+p}(Y \to Y)$. Formal properties satisfied by Gysin homomorphisms are listed in Fulton-MacPherson (3) § 2.5.

If f is flat or l.c.i., and $[f]$ is its canonical orientation, we write f^* for $[f]^*$, and f_* for $[f]_*$.

Example 17.4.1. Consider the situation of Proposition 17.4.1.

(a) If g is proper, then for all $\alpha \in A_k Y'$

$$f^* g_*(\alpha) = g'_*(c_e(E)\cap f'^*(\alpha))$$

in $A_{k-d}X$.

(b) If f is proper, then for all $\beta \in A^p X$,

$$g^* f_* \beta = f'_*(g'^*(\beta)\cdot c_e(E))$$

in $A^{p+d}Y'$.

(These formulas, and other similar formulas when g has an orientation class, follow formally from Proposition 17.4.1 using the axioms for a bivariant theory (cf. Fulton-MacPherson (3) § 9.2.1).)

Example 17.4.2. Let $\sigma: E \to F$ be a vector bundle homomorphism on a scheme X. Let $e = \operatorname{rank} E$, $f = \operatorname{rank} F$, $k \leqq \min(e, f)$. Let $D_k(\sigma)$ be the locus where $\operatorname{rank}(\sigma) \leqq k$. Then there is a class

$$\delta_k(\sigma) \in A^{(e-k)(f-k)}(D_k(\sigma) \to X)$$

whose action on $[V]$, $h: V \to X$, is

$$\delta_k(\sigma)([V]) = \mathbb{D}_k(h^*\sigma),$$

with $\mathbb{D}_k(h^*\sigma) \in A_*(h^{-1}(D_k(\sigma)))$ constructed in § 14.4. (If s is a section of a bundle E of rank e, define

$$z(s) = s_E^*[s] \in A^e(Z(s) \to X)$$

where $s_E: X \to E$ is the zero section. Then with $\pi: G_d(E) \to X$, s_σ, η as in § 14.4, define

$$\delta_k(\sigma) = \eta_*(z(s_\sigma) \cdot [\pi]).)$$

Note that Theorem 14.4(d) is immediate from this description. The push-forward of $\delta_k(\sigma)$ by the inclusion $D_k(\sigma) \to X$ is the class

$$\Delta_{f-k}^{(e-k)}(c(F-E)) \in A^{(e-k)(f-k)}(X).$$

Example 17.4.3. If X is purely n-dimensional, $f: X \to Y$ a morphism, with Y smooth and purely m-dimensional, there is a class $[f] \in A^{m-n}(X \xrightarrow{f} Y)$ which corresponds to the element $[X]$ in $A_m(X)$ by the isomorphisms of Proposition 17.4.2 (with $Z = S$) and 17.3.1.

(a) If f is flat, or f is a l.c.i. morphism, then this class $[f]$ agrees with the classes constructed in (1) and (3).

(b) If $f: X \to Y$, $g: Y \to Z$, all pure dimensional, Y, Z smooth, then $[gf] = [f] \cdot [g]$.

(c) The Gysin homomorphism

$$f^*: A_k Y \to A_{k+n-m} X$$

determined by $[f]$ coincides with that defined in Chap. 8.

Example 17.4.4. Assume the ground field has characteristic zero. Let $f_1: X_1 \to Y$ and $f_2: X_2 \to Y$ be morphisms, $c_i \in A^{p_i}(X_i \xrightarrow{f_i} Y)$. Then

$$f_1^*(c_2) \cdot c_1 = f_2^*(c_1) \cdot c_2$$

in $A^{p_1+p_2}(X_1 \times_Y X_2 \to Y)$. In particular, for all X, A^*X is a *commutative* ring. (Use resolution of singularities, Example 17.3.2, and Corollary 17.4.) We do not know a proof of this commutativity without resolution of singularities.

Example 17.4.5. (a) A diagram

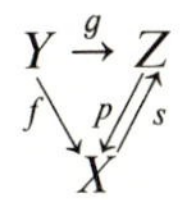

with f flat of relative dimension n, p smooth of relative dimension m, g proper, and $pg = f$, $ps = 1_X$, determines a class c in $A^d X$, $d = m - n$. Indeed

$$[f] \in A^{-n}(Y \xrightarrow{f} X), \quad [s] \in A^m(X \xrightarrow{s} Z),$$

so $g_*[f] \in A^{-n}(Z \xrightarrow{p} X)$, and we may define

$$c = [s] \cdot g_*[f] \in A^{m-n}(X \xrightarrow{ps} X) = A^d X .$$

Explicitly, if $X' \to X$, c acts on $A_k X'$ as the composite

$$A_k X' \xrightarrow{f'^*} A_{k+n} Y' \xrightarrow{g'_*} A_{k+n} Z' \xrightarrow{s^!} A_{k+n-m} X'$$

where the primes denote fibre products of the given diagram with X' over X.

If two such diagrams, over the same base X, determine classes $c_i \in A^{d_i} X$, then the fibre product diagram determines the class $c_1 \cdot c_2$ in $A^{d_1+d_2}(X)$. (This follows formally from the bivariant axioms, using the commutativity of the orientation classes $[s]$ with other bivariant classes.)

(b) Let E, F be vector bundles of ranks e, f on a scheme X, and let $Z = \operatorname{Hom}(E, F)$, u the universal bundle map on Z (§ 14.4). Then $D_k(u)$ is a subcone of Z, flat over X. A bundle homomorphism $\sigma: E \to F$ determines a section $t_\sigma: X \to Z$. The class constructed from this data by the prescription of (a) (for $Y = D_k(u)$, g the inclusion, $s = t_\sigma$) is the polynomial

$$\Delta_{f-k}^{(e-k)}(c(F-E)) \in A^{(e-k)(f-k)}(X) .$$

Similarly if $\underline{A}$ is a flag in E, the determinantal loci determine classes given by the polynomials of § 14.3. (These assertions follow readily from the constructions, cf. Example 17.4.2. These classes were used in Fulton-Lazarsfeld (3) § 3c.)

Example 17.4.6. Let $f: X \to Y$, $g: Y \to Z$ be morphisms. Assume f and g have compatible factorizations through smooth morphisms, as in § 6.6. Let $h = gf$. Assume that each of f, g, h is either flat or a l.c.i. morphism. Then in each of the following cases, $[h] = [f] \cdot [g]$:

i) f, g, and therefore h l.c.i.

ii) f, g, and therefore h flat.

iii) f, h l.c.i., g flat. (When f is a regular imbedding there is a neighborhood of $f(X)$ on which g is a l.c.i. morphism, by [EGA] IV. 19.1.5.)

iv) f l.c.i., g, h flat. (Proposition 6.5 (a)

v) g l.c.i., f, h flat. (Replacing Y by an open subscheme, g will be flat, by [EGA] IV. 2.4.6, 2.2.13.)

The remaining case, with f flat, g and h l.c.i., remains open. Is there a class of morphisms, closed under composition, containing flat morphisms, and l.c.i. morphisms, with orientations $[f]$, compatible with composition? (See Example 18.3.17.)

Example 17.4.7. Let $A \subset Y$, $B \subset Y$ be effective Cartier divisors on a scheme Y, let $D = A + B$ be the sum, and let $i: D \to Y$, $a: A \to D$, $b: B \to D$ be the inclusions. Then

(i) $[i] = a_*[i\,a] + b_*[i\,b] \in A^1(D \xrightarrow{i} Y)$.

(ii) $i^* c_1(\mathcal{O}_Y(B)) \cdot a_*[i\,a] = i^* c_1(\mathcal{O}_Y(A)) \cdot b_*[i\,b] \in A^2(D \xrightarrow{i} Y)$.

(If $g: Y' \to Y$ is any morphism, and α is a cycle on Y', then

(i)′ $(g^* D) \cdot \alpha = (g^* A) \cdot \alpha + (g^* B) \cdot \alpha$

(ii)″ $(g^* B) \cdot (g^* A) \cdot \alpha = (g^* A) \cdot (g^* B) \cdot \alpha$

by Proposition 2.3 (b) and Corollary 2.4.2. Here $g^* D$, $g^* A$, $g^* B$ are the pull-back pseudo-divisors on Y'.)

Example 17.4.8. Localized top Chern classes are multiplicative, in the following sense. Let s_i be sections of vector bundles E_i of ranks e_i on X, defining bivariant classes

$$z(s_i) \in A^{e_i}_{Z(s_i)} X = A^{e_i}(Z(s_i) \to X)$$

(see Examples 17.3.1 and 17.4.2). Let $E = E_1 \oplus E_2$, $s = s_1 \oplus s_2$. Then

$$z(s) = z(s_1) \cup z(s_2) \in A^{e_1+e_2}_{Z(s)} X .$$

Example 17.4.9. The canonical homomorphism

$$\operatorname{Pic}(X) \to A^1 X , \quad L \to c_1(L)$$

need not be injective. (Take X to be a singular curve, and use Example 17.3.2.) In particular, if X is quasi-projective, the canonical homomorphism

$$\varinjlim A^* Y \to A^* X$$

need not be injective; the limit is over all $X \to Y$, Y non-singular quasi-projective (cf. Example 8.3.13). Mumford (7) has considered the image of $\varinjlim A^* Y$ in $A^* X$ as a cohomology theory with some of the concrete advantages of the former, and formal properties of the latter.

Example 17.4.10. If $X = Y/G$ is a quotient variety as in Example 8.3.12, then the canonical homomorphism

$$A^* X_{\mathbb{Q}} \xrightarrow{\cap [X]} A_* X_{\mathbb{Q}}$$

is an isomorphism of rings. This shows in particular that the ring structure on $A_* X_{\mathbb{Q}}$ is independent of Y, and constructs pull-back ring homomorphism for arbitrary morphisms of such varieties. (Let $\pi^* : A_* X_{\mathbb{Q}} \to (A_* Y_{\mathbb{Q}})^G$ be the isomorphism of Example 1.7.6. If $c \in A^* X_{\mathbb{Q}}$, let $\bar{c} = c \cap [X]$. For V a variety, $f: V \to X$ a morphism, let $\eta: W \to V$ be a finite surjective morphism of varieties, and $\tilde{f}: W \to Y$ a morphism so that $\pi \tilde{f} = f \eta$. Then

$$c \cap [V] = \deg(W/V)^{-1} \eta_*([W] \cdot_{\tilde{f}} \pi^* \bar{c}) .$$

Conversely, given $\bar{c} \in A_* X_{\mathbb{Q}}$, this formula defines a class in $A_* V_{\mathbb{Q}}$, independent of choice of W; one may use Theorem 17.1 to show that this construction determines an element c of $A^* X_{\mathbb{Q}}$.)

Similarly for any $f: X' \to X$, $A(X' \to X)_{\mathbb{Q}} \cong A_* X'_{\mathbb{Q}}$. This may be used to show that the refined intersection products (and hence intersection numbers) of Example 8.3.12 are independent of the isomorphism $X \cong Y/G$.

17.5 Monoidal Transforms

Let X be a regularly imbedded closed subscheme of a scheme Y. Let $f\colon \tilde{Y} \to Y$ be the blow-up of Y along X, and form the fibre square

$$\begin{array}{ccc} \tilde{X} & \xrightarrow{j} & \tilde{Y} \\ {\scriptstyle g}\downarrow & & \downarrow{\scriptstyle f} \\ X & \xrightarrow[i]{} & Y \,. \end{array}$$

We assume that there is a surjection of a locally free sheaf E on Y onto the ideal sheaf $\mathscr{I}(X)$ of X; such always exists, for example, if Y is imbeddable in a smooth scheme. Then f factors into a regular closed imbedding in $P(E^{\vee})$ followed by the projection, so f is a factorable l.c.i. morphism, of relative dimension 0. Therefore f has an orientation class $[f] \in A^0(\tilde{Y} \to Y)$.

Proposition 17.5 (a) *With the above notation*

$$f_*[f] = 1 \in A^0(Y)\,.$$

(b) *Let $h\colon Y' \to Y$ be any morphism, and form the fibre square*

$$(*) \qquad \begin{array}{ccc} \tilde{Y}' & \xrightarrow{\tilde{h}} & \tilde{Y} \\ {\scriptstyle f'}\downarrow & & \downarrow{\scriptstyle f} \\ Y' & \xrightarrow{h} & Y \end{array}$$

Then $f^*\colon A^p(Y' \xrightarrow{h} Y) \to A^p(\tilde{Y}' \xrightarrow{\tilde{h}} \tilde{Y})$ is a split monomorphism, with inverse $c \to f_*(c \cdot [f])$.

Proof (cf. Proposition 6.7 (b) and Example 6.7.1).

(a) Let $h\colon Y' \to Y$, $\alpha \in A_k Y'$. Forming the fibre square (*) of (b), we must show that $f'_*([f]^*\alpha) = \alpha$. We may assume $\alpha = [V]$, V a variety, and, by covariance, $V = Y'$. Let $X' = h^{-1}(X)$, $\tilde{X}' = \tilde{h}^{-1}(\tilde{X})$, and let $g'\colon \tilde{X} \to X'$, $j'\colon \tilde{X}' \to \tilde{Y}'$ be the induced morphisms.

If $h(V) \subset X$, then by Proposition 17.4.1,

$$[f]^*(\alpha) = c_{d-1}(F) \cdot [g]^*(\alpha) = c_{d-1}(F) \cdot g'^*(\alpha)$$

with F the universal quotient bundle on $\tilde{X} = P(N)$, g'^* the flat pull-back. And

$$g'_*(c_{d-1}(F) \cdot g'^*(\alpha)) = \alpha$$

by Example 3.3.3, which proves (a) in this case.

If $h(V) \not\subset X$, let $\tilde{V} \subset \tilde{Y}'$ be the blow-up of V along X'. Then

$$[f]^*(\alpha) = [\tilde{V}] + j'_*(\beta)$$

for some $\beta \in A_k(\tilde{X}')$; an explicit formula for β is given in Example 6.7.1. Therefore

$$f'_*([f]^*(\alpha)) = f'_*[\tilde{V}] + f'_* j'_*(\beta) = [V] + i'_* g'_*(\beta) = [V]\,,$$

since $g'_*\beta \in A_k(X')$, and $\dim X' < \dim V = k$.

(b) This follows from (a) and the identity (Axiom A_{123})

$$f_*(f^*(c)\cdot[f]) = c\cdot f_*[f]. \quad \square$$

Example 17.5.1. (a) If $\pi: E \to Y$ is a vector bundle, $f: X \to Y$ any morphism, then

$$\pi^*: A^p(X \to Y) \to A^p(f^*E \to E)$$

is an isomorphism.

(b) With E, f as in (a), $r = \text{rank}(E)$,

$$A^p(P(f^*E) \to P(E)) \cong \bigoplus_{i=0}^{r-1} A^{p-i}(X \to Y).$$

In particular, $A^p(P(E)) \cong \bigoplus_{i=0}^{r-1} A^{p-i}(Y)$.

(c) Let (*) be a blow-up diagram as in § 6.7. Let $Y' \to Y$ be any morphism. Then there are split exact sequences

$$0 \to A^{p-d}(X' \to X) \to A^{p-1}(\tilde{X}' \to \tilde{X}) \oplus A^p(Y' \to Y) \to A^p(\tilde{Y}' \to \tilde{Y}) \to 0$$

where $X' = X \times_Y Y'$, etc. In particular,

$$0 \to A^{p-d}X \to A^{p-1}\tilde{X} \oplus A^pY \to A^p\tilde{Y} \to 0.$$

(The maps and proofs are parallel to those in § 3.3 and § 6.7.)

17.6 Residual Intersection Theorem

Consider a diagram

$$\begin{array}{ccccc} & & R & & \\ & & \downarrow b & & \\ D & \xrightarrow{a} & X' & \xrightarrow{j} & Y' \\ & & \downarrow g & & \downarrow f \\ & & X & \xrightarrow[i]{} & Y \end{array}$$

with the square a fibre square, and i, j, a, b closed imbeddings. Assume

(i) $j\,a$ imbeds D as a Cartier divisor on Y', and R is the residual scheme to D in X'.

(ii) i and $j\,b$ are regular imbeddings of codimensions d and e, respectively.

Set

$$c(i/j\,a) = c_{d-1}((g\,a)^* N_X Y - (j\,a)^* \mathscr{O}_{Y'}(D)),$$
$$c(i/jb; D) = c_{d-e}((gb)^* N_X Y \otimes (jb)^* \mathscr{O}_{Y'}(-D) - N_R Y'),$$

classes in $A^{d-1}D$ and $A^{d-e}R$ respectively.

Theorem 17.6. *With the preceding notation,*

$$f^*[i] = a_*(c(i/j\,a)\cdot[j\,a]) + b_*(c(i/jb; D)\cdot[jb])$$

in $A^d(X' \xrightarrow{j} Y')$.

Proof. We first assume that $e=1$, so X' is the sum of Cartier divisors D and R on Y'. Let $\delta=c_1(j^*\mathcal{O}_{Y'}(D))$, $\varrho=c_1(j^*\mathcal{O}_{Y'}(R))$ in $A^1(X')$, and let $\alpha=a_*[j\,a]$, $\beta=b_*[j\,b]$ in $A^1(X'\xrightarrow{j}Y')$. Then (Example 17.4.7)

(i) $[j]=\alpha+\beta$

(ii) $\varrho\cdot\alpha=\delta\cdot\beta$.

From (ii) it follows by induction on q that

$$\text{(iii)}\qquad (\delta+\varrho)^q\cdot(\alpha+\beta)=\delta^q\cdot\alpha+\sum_{s+t=q}\binom{q+1}{s}\delta^s\varrho^t\cdot\beta\,.$$

Let $c_p=c_p(g^*N_XY)$. By Proposition 17.4.1,

$$\begin{aligned}\text{(iv)}\quad f^*[i]&=c_{d-1}(g^*N_XY-j^*\mathcal{O}_{Y'}(D+R))\cdot[j]\\ &=\sum_{p+q=d-1}(-1)^q c_p(\delta+\varrho)^q\cdot(\alpha+\beta)\\ &=\sum_{p+q=d-1}(-1)^q c_p\delta^q\cdot\alpha+\sum_{p+s+t=d-1}(-1)^{s+t}\binom{s+t+1}{s}c_p\delta^s\varrho^t\cdot\beta\,.\end{aligned}$$

Writing out $c(i/j\,a)$, one has

$$\text{(v)}\quad a_*(c(i/j\,a)\cdot[j\,a])=\sum_{p+q=d-1}(-1)^q c_p\,\delta^q\cdot\alpha\,.$$

Similarly, and using Example 3.2.2,

$$\begin{aligned}\text{(vi)}\quad b_*(c(i/j\,b;D)\cdot[j\,b])&=\sum_{m+t=d-1}c_m(g^*N_XY\otimes\mathcal{O}(-D))(-\varrho)^t\cdot\beta\\ &=\sum_{p+s+t=d-1}(-1)^{s+t}\binom{d-p}{s}c_p\,\delta^s\varrho^t\cdot\beta\,.\end{aligned}$$

Comparing (iv), (v), and (vi) gives the required equation.

If $e>1$, let $\pi:\tilde{Y}'\to Y'$ be the blow-up of Y' along R. Put a $\sim$ over the symbols for subschemes of Y', and morphisms between them, to denote their inverse images, and induced morphisms, in $\tilde{Y}'$. By Proposition 17.5, it suffices to show that both sides of the equation of the theorem become equal after applying π^*. Thus the case $e>1$ is reduced to the case $e=1$, once it is verified that the three terms in the equation pull back to the corresponding three terms in $A^d(\tilde{j})$. This is obvious for the first two. For the third, let $\eta:\tilde{R}\to R$ be the induced morphism. Let

$$E=(g\,b)^*N_XY\otimes(j\,b)^*\mathcal{O}_{Y'}(-D)\,.$$

Since N_RY' is a sub-bundle of E (Example 9.2.2), and $N_{\tilde{R}}\tilde{Y}'$ is a sub-bundle of η^*N_RY', the Whitney sum formula gives

$$\text{(vii)}\quad c_{d-1}(\eta^*E/N_{\tilde{R}}\tilde{Y}')=c_{d-e}(\eta^*E/\eta^*N_RY')\,c_{e-1}(\eta^*N_RY'/N_{\tilde{R}}\tilde{Y}').$$

By Proposition 17.4.1,

$$\text{(viii)}\qquad \pi^*[j\,b]=c_{e-1}(\eta^*N_RY'/N_{\tilde{R}}\tilde{Y}')\cdot[\tilde{j}\,\tilde{b}]\,.$$

From (vii) and (viii),

$$\pi^*(c(i/j\,b;D)\cdot[j\,b])=c(i/\tilde{j}\,\tilde{b};\tilde{D})\,[\tilde{j}\,\tilde{b}]\,,$$

which concludes the proof, if $e>1$.

Finally, if $e=0$, i.e. $R=X'=Y'$, the required equation amounts to the identity

$$c_d(N)=c_{d-1}(N-L)\,c_1(L)+c_d(N\otimes L^\vee)$$

in A^dX', $N=g^*N_XY$, $L=\mathscr{O}_{Y'}(D)$. This is a simple formal calculation, using Example 3.2.2. □

This residual intersection theorem implies many of the previous intersection formulas. Besides those in Chap. 9, it also implies such basic facts as the functoriality theorem of Chap. 6 (cf. Example 17.6.3). Generalizations are given in the examples.

Example 17.6.1. Consider the situation of Theorem 17.6.

(a) If f is proper, and $\alpha\in A_kY'$, then

$$i^*f_*(\alpha)=(g\,a)_*(c(i/j\,a)(j\,a)^*\alpha)+(g\,b)_*(c(i/j\,b;\,D)(j\,b)^*\alpha)$$

in $A_{k-d}(X)$.

(b) For all $\beta\in A^pX$,

$$f^*i_*(\alpha)=(j\,a)_*(c(i/j\,a)(g\,a)^*\beta)+(j\,b)_*(c(i/j\,b;\,D)(g\,b)^*\beta)$$

in $A^{p+d}Y'$.

Example 17.6.2 (Kleiman (12) 3.6). There is a useful generalization of the residual intersection theorem. Consider a diagram labelled as in this section, with the square a fibre square, and a a closed imbedding. Assume

(i) $R=\operatorname{Proj}(\operatorname{Sym}(\mathscr{I}))$, where $\mathscr{I}$ is the ideal sheaf of D in X', and $b:R\to X'$ is the projection.

(ii) i and $j\,b$ are l.c.i. morphisms of codimension d.

(iii) $j\,a$ is a l.c.i. morphism of codimension d'.

Assume also that all schemes appearing can be imbedded in smooth schemes (weaker assumptions suffice to factor appropriate maps through smooth maps). Then

$$f^*[i]=a_*(c(i/j\,a)\,[j\,a])+b_*[j\,b]$$

in $A^d(X'\to Y')$, where

$$c(i/j\,a)=c_{d-d'}((g\,a)^*N_i-N_{ja})\,,$$

N_i and N_{ja} the virtual normal bundles to i and ja. (Factor i into a closed imbedding $X\hookrightarrow P$ followed by a smooth morphism $P\to Y$. Let $P'=P\times_YY'$. Let $\tilde{P}'$ be the blow-up of P' along D. Then R is the residual scheme to the exceptional divisor in $\tilde{P}'$, and Theorem 17.6 applies.)

Example 17.6.3. The residual intersection theorem can be used to prove the functoriality theorem of § 6.5: If $f:X\to Y$, $g:Y\to Z$ are regular imbeddings, then $[g\,f]=[f]\cdot[g]$. (Let $\pi:\tilde{Z}\to Z$ be the blow-up of Z along X, with exceptional divisor $\tilde{X}$. The residual scheme R to $\tilde{X}$ in $\tilde{Y}=\pi^{-1}(Y)$ is the blow-up of Y along X (Appendix B.6.10). It suffices to show that $\pi^*[g\,f]=\eta^*[f]\cdot\pi^*[g]$, where $\eta:\tilde{Y}\to Y$ is the induced morphism. Use Theorem 17.6 to calculate $\pi^*[g]$, and Proposition 17.4.1 for $\pi^*[g\,f]$.)

Example 17.6.4. There is a useful generalization of the orientation class of a regular imbedding. If $f\colon Y \to Y$ is a closed imbedding, and

$$C_X Y \subset E$$

is a closed imbedding of the normal cone in a vector bundle E of rank e on Y, this data determines a bivariant class c in $A^e(X \xrightarrow{f} Y)$. Given $g: Y' \to Y$, let $X' = g^{-1}(X)$, $h: X' \to X$ induced. Then

$$C_{X'} Y' \subset X' \times_X (C_X Y) \subset h^* E .$$

Define $c_g^{(k)}$ to be the composite

$$A_k Y' \xrightarrow{\sigma} A_k(C_{X'} Y') \xrightarrow{\text{incl}_*} A_k(h^* E) \xrightarrow{s_E!} A_{k-e} X'$$

where σ is the specialization map (§ 5.2), s_E the zero section of E. Alternatively,

$$c_g^{(k)}([V]) = \{c(h^* E) \cap s(X' \cap V, V)\}_{k-e} .$$

(That (C_1) and (C_2) are satisfied follows from Proposition 4.2; (C_3) is deduced as in Theorem 17.1.)

(a) If f is a regular imbedding of codimension d, then

$$c = c_{e-d}(E/N_X Y) \cdot [f] .$$

(b) If $g: Y' \to Y$ is a morphism, then $g^*(c)$ is the class determined by the canonical imbedding of $C_{X'} Y'$ in $h^* E$.

Example 17.6.5. Consider a residual intersection diagram as in this section, but with no assumptions on the imbedding $j\,b$ of R in Y'. With $E = (g\,b)^* N_X Y \otimes (j\,b)^* \mathcal{O}_{Y'}(-D)$, there is a canonical imbedding of $C_R Y'$ in E (Example 9.2.2). By the preceding example, this determines a bivariant class

$$\mathbf{r} \in A^d(R \xrightarrow{jb} Y') .$$

Then

$$f^*[i] = a_*(c(i/j\,a) \cdot [j\,a]) + b_*(\mathbf{r}) .$$

When $j\,b$ is a regular imbedding, this is equivalent to Theorem 17.6. (To show both sides have the same effect on $[V]$, $h: V \to Y'$ any morphism, apply Theorem 9.2 if $h(V) \not\subset D$. If $h(V) \not\subset R$, let $\tilde{V}$ be the blow-up of V along $h^{-1}(R)$, $\tilde{Y}'$ the blow-up of Y' along R. It suffices to show both sides have the same effect on $[\tilde{V}]$; but $\tilde{V}$ maps to $\tilde{Y}'$, and the case $e=1$ of Theorem 17.6 applies. Finally if $h(V) \subset D \cap R$, it is a formal calculation, as in the case $e=0$ of Theorem 17.6.)

Example 17.6.6. Consider a diagram

$$\begin{array}{ccccc} D & \xrightarrow{a} & X' & \xrightarrow{j} & Y' \\ & & {\scriptstyle g}\downarrow & & \downarrow{\scriptstyle f} \\ & & X & \xrightarrow[i]{} & Y \end{array}$$

with the square a fibre square, a a closed imbedding. Assume i and $j\,a$ are regular imbeddings of codimension d and d'. Assume Y' can be imbedded in a smooth scheme. Let R be the residual scheme to D in X', i.e., the ideal sheaf of

R in X' is the annihilator of the ideal sheaf of D in X'. Let b be the inclusion of R in X'. Define

$$c(i/j\,a) = c_{d-d'}((g\,a)^* N_i - N_{j\,a}),$$

where N_i and N_{ja} are the normal bundles to i and $j\,a$. There is a canonically defined class $\mathbf{r}$ in $A^d(R \xrightarrow{jb} Y')$ with

$$f^*[i] = a_*(c(i/j\,a) \cdot [j\,a]) + b_*(\mathbf{r}),$$

in $A^d(X' \xrightarrow{j} Y')$. (Let $\pi: \bar{Y}' \to Y'$ be the blow-up of Y' along D, with exceptional divisor $\bar{D}$, $\bar{X}' = \pi^{-1}(X')$. Let $\bar{R}$ be the residual scheme to $\bar{D}$ in $\bar{X}'$, $\eta: \bar{R} \to R$ the induced morphism. In the previous example a class $\bar{\mathbf{r}} \in A^d(\bar{R} \to \bar{Y}')$ was constructed. Then $[\pi] \in A^0(\bar{Y} \to Y')$, and one may set

$$\mathbf{r} = \eta_*(\bar{\mathbf{r}} \cdot [\pi]) \in A^d(R \to Y').)$$

Notes and References

The source and reference for most of this chapter is Fulton-MacPherson (3). There the reader can find examples of the utility of bivariant theories in areas other than intersection theory or algebraic geometry.

Previously a cohomology theory had been developed for quasi-projective schemes (cf. Example 8.3.13), which was adequate for formulating and extending Grothendieck's Riemann-Roch theorem to singular quasi-projective (Baum-Fulton-MacPherson (1)). However, this cohomology theory lacks many of the formal properties one would like, such as a Gysin push-forward for proper l.c.i. morphisms. The coarser cohomology theory associated with the operational bivariant theory (§ 17.3) has these formal properties, and may be used on arbitrary algebraic schemes. For a recent application see Mumford (7).

Motivation for developing a pair of theories, both "homology" and "cohomology" came primarily from topology. For a time most topologists had regarded cohomology as the proper object of study on singular spaces, superior to homology – just as most algebraic geometers regarded the "Cartier divisor" as a replacement for the notion of a "Weil divisor". Since Sullivan, MacPherson and others discovered that important invariants of singular spaces could lie in homology instead of cohomology, a more balanced view has been achieved.

More recently we have seen that these two theories are not rich enough for the study of singular varieties. Bivariant theory is one extension of homology-cohomology, particularly useful for functorial and formal properties. Another theory, called intersection homology, has been developed by Goresky and MacPherson (1), (2). This had led to a deep insight into the geometry of singular spaces. The functorial properties of intersection homology, and the place of algebraic cycles in the theory, is not yet clear, however.

I have had useful conversations with D. Gabber and S. Kleiman on some topics in this chapter, particularly related to the still unsettled question of which maps have orientations for rational equivalence theory (cf. Example 17.4.6). Kleiman (12) has developed some variations on bivariant theory, and a generalization of the residual intersection theorem (cf. Example 17.6.2).

The proof of Proposition 17.5 corrects an error in Fulton-MacPherson (3)9.2.2.

Chapter 18. Riemann-Roch for Singular Varieties

Summary

The basic tool for a general Riemann-Roch theorem is MacPherson's graph construction, applied to a complex $E.$ of vector bundles on a scheme Y, exact off a closed subset X. This produces a localized Chern character[1] $\mathrm{ch}_X^Y(E.)$ which lives in the bivariant group $A(X \to Y)_{\mathbb{Q}}$. For each class $\alpha \in A_* Y$, this gives a class

$$\mathrm{ch}_X^Y(E.) \cap \alpha \in A_* X_{\mathbb{Q}}$$

whose image in $A_* Y_{\mathbb{Q}}$ is $\sum (-1)^i \mathrm{ch}(E_i) \cap \alpha$. The properties needed for Riemann-Roch, in particular the invariance under rational deformation, follow from the bivariant nature of $\mathrm{ch}_X^Y(E.)$.

The general Riemann-Roch theorem constructs homomorphisms

$$\tau_X : K_\circ X \to A_* X_{\mathbb{Q}},$$

covariant for proper morphisms, such that $\tau_X(\beta \otimes \alpha) = \mathrm{ch}(\beta) \cap \tau_X(\alpha)$ for $\beta \in K^\circ X$, $\alpha \in K_\circ X$. If X is imbedded in a non-singular variety M, and a coherent sheaf $\mathscr{F}$ is resolved by a complex of vector bundles $E.$ on M, then

$$\tau_X(\mathscr{F}) = \mathrm{ch}_X^M(E.) \cap \mathrm{Td}(M).$$

where $\mathrm{Td}(M) = \mathrm{td}(T_M) \cap [M]$. Such τ_X is constructed for quasi-projective schemes in the second section. The extension to arbitrary algebraic schemes, using Chow's lemma, is carried out in the last section. As a corollary one has the GRR formula

$$f_*(\mathrm{ch}(\alpha) \cdot \mathrm{td}(T_X)) = \mathrm{ch}(f_* \alpha) \cdot \mathrm{td}(T_Y),$$

for $f: X \to Y$ proper. X, Y arbitrary non-singular varieties, $\alpha \in K^\circ X$. In the singular case, there are refinements for $f: X \to Y$ a l.c.i. morphism.

---

1 In topology, the Chern character of such a complex lives in $H^*(Y, Y-X; \mathbb{Q})$; capping with this class determines homomorphisms from $H_*(Y; \mathbb{Q})$ to $H_*(X; \mathbb{Q})$. The bivariant class $\mathrm{ch}_X^Y(E.)$ is an analogue for rational equivalence.

18.1 Graph Construction

Let X be a closed subscheme of a scheme Y, and let $E.$ be a complex of vector bundles on Y which is exact off X:

$$0 = E_{n+1} \to E_n \xrightarrow{d_n} E_{n-1} \to \ldots \to E_{m+1} \xrightarrow{d_{m+1}} E_m \to E_{m-1} = 0,$$

with d_i homomorphisms of locally free sheaves, $d_i \circ d_{i+1} = 0$, and $\operatorname{Im}(d_{i+1}) = \operatorname{Ker}(d_i)$ on $Y - X$. Let $e_i = \operatorname{rank} E_i$. Let $G_i = \operatorname{Grass}_{e_i}(E_i \oplus E_{i-1})$, and set

(1) $$G = G_n \times_Y G_{n-1} \times_Y \ldots \times_Y G_m.$$

Let ξ_i be the tautological bundle of rank e_i on G_i, and set

(2) $$\xi = \sum_{i=m}^{n} (-1)^i [(\mathrm{pr}_i)^* (\xi_i)] \in K^\circ(G),$$

$\mathrm{pr}_i : G \to G_i$ the projection. For each $y \in Y$, and each λ in the ground field K, the graph of $\lambda d_i(y)$ is an e_i-dimensional subspace $\Gamma(\lambda d_i(y))$ of $E_i(y) \oplus E_{i-1}(y)$. This determines a morphism

(3) $$Y \times \mathbb{A}^1 \to G_i, \qquad (y, \lambda) \to \Gamma(\lambda d_i(y))$$

such that the pull-back of ξ_i is the graph of $\lambda d_i(y)$ at $(y, \lambda) \in Y \times \mathbb{A}^1$. This determines a closed imbedding

(4) $$Y \times \mathbb{A}^1 \xrightarrow{\varphi} G \times \mathbb{A}^1, \qquad (y, \lambda) \to \left(\prod_i \Gamma(\lambda d_i(y)), \lambda\right).$$

(To make this pointwise description scheme-theoretic, see Example 18.1.1.)

Define integers k_i by setting $k_i = 0$ if $i \geqq n$, and requiring that

(5) $$k_i + k_{i-1} = e_i$$

for all i. Thus $k_i = e_{i+1} - e_{i+2} + \ldots \pm e_n$. We assume that these k_i are non-negative. This is the case if $E.$ is exact at any point $y \in Y$ (e.g. if $Y - X$ is non-empty), for then $k_i = \dim(\operatorname{Ker}(d_i(y)))$. Let $H_i = \operatorname{Grass}_{k_i}(E_i)$, and set

(6) $$H = H_n \times_Y H_{n-1} \times \ldots \times_Y H_m.$$

On $Y - X$, $\operatorname{Ker}(d_i)$ is a subbundle of E_i of rank k_i. This determines a section of H_i on $Y - X$, and hence a closed imbedding

(7) $$Y - X \xrightarrow{\psi} H^\circ,$$

where H° is the restriction of H over $Y - X$. There is a canonical closed imbedding

(8) $$H \xrightarrow{\iota} G$$

which takes a collection of k_i-planes L_i in E_i to the collection of e_i-planes $L_i \oplus L_{i-1}$ in $E_i \oplus E_{i-1}$. We identify H as a closed subscheme of G via ι. Note that

(9) $$\iota^*(\xi) = 0 \quad \text{in} \quad K^\circ(H).$$

Indeed, if ζ_i is the tautological bundle on H_i,

$$\imath^*(\xi) = \sum_{i=m}^{n} (-1)^i (\mathrm{pr}_i^*(\zeta_i) + \mathrm{pr}_{i-1}^*(\zeta_{i-1})) = 0\,,$$

where pr_i is the projection from H to H_i.

The inclusions in (4), (7), and (8) combine to form a fundamental (non-commutative!) diagram

$$\begin{array}{ccccc} Y\times\mathbb{A}^1 & & \xrightarrow{\quad\varphi\quad} & G\times\mathbb{A}^1 & \subset G\times\mathbb{P}^1 \\ \cup & & & & \uparrow{\scriptstyle \imath\times 1} \\ (Y-X)\times\mathbb{A}^1 & \subset (Y-X)\times\mathbb{P}^1 & \xrightarrow[\psi\times 1]{} & H^\circ\times\mathbb{P}^1 & \subset H\times\mathbb{P}^1\,. \end{array} \tag{10}$$

Let $\pi: G\to Y$ be the projection, $G_X = \pi^{-1}(X)$, and let $\eta: G_X\to X$ be the restriction. Given a subvariety V of Y, the *localized Chern character*

$$\mathrm{ch}_X^Y(E.)\cap[V]\in A_*X_{\mathbb{Q}} \tag{11}$$

may be defined as follows. Let W be the closure of $\varphi(V\times\mathbb{A}^1)$. Let

$$Z_\infty = i_\infty^*[W] \tag{12}$$

where $i_\infty: G = G\times\{\infty\}\hookrightarrow G\times\mathbb{P}^1$. Then Z_∞ is a k-cycle on G. One shows (Lemma 18.1) that one irreducible component of Z_∞ is a variety $\tilde{V}$ projecting birationally onto V, and that

$$Z = Z_\infty - [\tilde{V}] \tag{13}$$

is a k-cycle on G_X. Set

$$\mathrm{ch}_X^Y(E.)\cap[V] = \eta_*(\mathrm{ch}(\xi)\cap[Z])\,. \tag{14}$$

It is useful, however, to have a definition with more flexibility. Given a cycle α on Y, let α° denote the restriction of α to $Y-X$.

(i) Choose a cycle α' on $G\times\mathbb{P}^1$ which restricts to $\varphi_*(\alpha\times[\mathbb{A}^1])$ on $G\times\mathbb{A}^1$.

(ii) Choose a cycle α'' on $H\times\mathbb{P}^1$ which restricts to $\psi_*(\alpha^\circ)\times[\mathbb{P}^1]$ on $H^\circ\times\mathbb{P}^1$.

Then $\alpha'-\alpha''$ is a cycle on $G\times\mathbb{P}^1$. Set

$$\gamma = i_\infty^*(\alpha'-\alpha'')\in Z_*(G)\,. \tag{15}$$

Note that γ is well-defined as a *cycle* on G since the normal bundle to i_∞ is trivial (Remark 2.3). In other words, γ is the specialization cycle $(\alpha'-\alpha'')_\infty$ discussed in § 10.1.

Lemma 18.1. (a) *γ is a cycle on G_X.*

(b) *Another choice of α' does not change γ.*

(c) *Another choice of α'' changes γ to $\gamma+\beta$, where β is a cycle on $H_X = G_X\cap H$.*

Proof. Another choice of α' is of the form $\alpha'+\varrho$, ϱ a cycle on $G\times\{\infty\}$. Then $i_\infty^*(\varrho)=0$, which proves (b). Another choice of α'' is of the form $\alpha''+\varrho$, ϱ a cycle on $H_X\times\mathbb{P}^1$. Then $i_\infty^*(\varrho)$ is a cycle on H_X, which proves (c).

To prove (a), since constructions of γ commute with restrictions to open subschemes, we may assume $X=\emptyset$, in which case we must show, for some choices of α' and α'', that $\gamma=0$. Thus we may assume $E.$ is exact on all of Y. Let $K_i=\operatorname{Ker}(d_i)$. Define a rank e_i subbundle L_i of the pull-back of $E_i \oplus E_{i-1}$ to $Y\times \mathbb{P}^1$, whose fibre over $(y, (\lambda_0:\lambda_1))$ is

$$\{(v_i, v_{i-1}) \in E_i(y) \oplus K_{i-1}(y) \mid \lambda_0 v_{i-1} = \lambda_1 d_i v_i\} .$$

This determines morphisms $Y\times \mathbb{P}^1 \to G_i$, and hence a closed imbedding

$$\bar{\varphi}: Y\times \mathbb{P}^1 \to G\times \mathbb{P}^1 .$$

Away from $\infty=(0:1)$, L_i is the graph of λd_i, so $\bar{\varphi}$ extends the imbedding φ of (4). At ∞, $L_i=K_i\oplus K_{i-1}$, so, over ∞, $\bar{\varphi}$ is the imbedding $\iota\circ\psi$ of (7) and (8).

Now given α on Y, let $\alpha'=\bar{\varphi}_*(\alpha\times[\mathbb{P}^1])$, and let $\alpha''=\psi_*(\alpha)\times[\mathbb{P}^1]$. Then

$$\gamma = i_\infty^*\alpha' - i_\infty^*\alpha'' = \psi_*(\alpha) - \psi_*(\alpha) = 0 . \quad \square$$

Definition 18.1. Given $E.$ on Y, exact off X, and a cycle α on Y, define

$$\operatorname{ch}_X^Y(E.)\cap\alpha = \eta_*(\operatorname{ch}(\xi)\cap\gamma) \tag{16}$$

in $A_*X_{\mathbb{Q}}$. Here γ is a cycle on G_X defined by (15) and Lemma 18.1(a), η projects G_X to X, and ξ is the virtual bundle defined in (2). By Lemma 18.1, γ is unique up to adding a cycle in H_X. Since ξ restricts to 0 on H, $\operatorname{ch}(\xi)\cap\gamma$ is independent of choice of γ, so $\operatorname{ch}_X^Y(E.)\cap\alpha$ is well-defined.

More generally, if $g: Y'\to Y$ is an arbitrary morphism, set $X'=g^{-1}(X)$, and define, for $\alpha\in Z_*Y'$,

$$\operatorname{ch}_X^Y(E.)\cap\alpha = \operatorname{ch}_{X'}^{Y'}(g^*E.)\cap\alpha \in A_*X'_{\mathbb{Q}} . \tag{17}$$

Let $A(X\to Y)_{\mathbb{Q}} = \oplus A^p(X\to Y)\otimes\mathbb{Q}$. Equivalently, $A(X\to Y)_{\mathbb{Q}}$ is the bivariant group defined in § 17.1, but using cycles and cycle classes with rational coefficients.

Theorem 18.1. *The operation* $\alpha\to\operatorname{ch}_X^Y(E.)\cap\alpha$ *determines a bivariant class, denoted* $\operatorname{ch}_X^Y(E.)$, *in* $A(X\to Y)_{\mathbb{Q}}$.

Proof. We show that the operation defined by (17) satisfies the conditions of Theorem 17.1. To show the commutativity with proper push-forward, it suffices to show that if $g: Y'\to Y$ is proper, and $h: X'\to X$ is the induced morphism, and α is a cycle on Y', then

$$h_*(\operatorname{ch}_{X'}^{Y'}(g^*E.)\cap\alpha) = \operatorname{ch}_X^Y(E.)\cap g_*\alpha$$

in $A_*X_{\mathbb{Q}}$. Note that the fundamental diagram for $g^*E.$ is obtained from the fundamental diagram (10) for $E.$ by the base extension $g: Y'\to Y$. To avoid cumbersome notation, we use g_* and g^* for push-forward and pull-back for all morphisms (except h) induced by this base extension. If α' and α'' are choices satisfying (i) and (ii) for α, then $g_*\alpha'$ and $g_*\alpha''$ are choices for $g_*\alpha$ which satisfy (i) and (ii), since push-forward commutes with restriction. Similarly if $\gamma=i_\infty^*(\alpha'-\alpha'')$, since g_* commutes with i_∞^* (Theorem 6.2(a), cf. Proposition 1.4),

$$g_*(\gamma) = i_\infty^*(g_*\alpha' - g_*\alpha'')$$

is a valid choice for the γ-cycle for $g_* \alpha$. Therefore

$$\begin{aligned} \mathrm{ch}_X^Y(E.) \cap g_* \alpha &= \eta_*(\mathrm{ch}(\xi) \cap g_* \gamma) = \eta_* g_*(\mathrm{ch}(g^* \xi) \cap \gamma) \\ &= h_* \eta'_*(\mathrm{ch}(g^* \xi) \cap \gamma) = h_* \mathrm{ch}_{X'}^{Y'}(g^*(E.) \cap \alpha) . \end{aligned}$$

Here $\eta' : G_{X'} \to X'$ is the map induced by η.

The proof that $\mathrm{ch}_X^Y(E.)$ commutes with flat pullback is entirely analogous, and left to the reader.

Finally, let $D \subset Y$ be a principal divisor, i.e., $D = h^{-1}(0)$ for some morphism $h : Y \to \mathbb{A}^1$. Let j be the imbedding of 0 in $\mathbb{A}^1$. To complete the proof, we must show that

$$j^!(\mathrm{ch}_X^Y(E.) \cap \alpha) = \mathrm{ch}_{D \cap X}^D(E.|_D) \cap j^! \alpha$$

in $A_*(D \cap X)_{\mathbb{Q}}$. The proof is again similar. If α' and α'' are choices for α, and $\gamma = i_\infty^*(\alpha' - \alpha'')$, then $j^! \alpha'$ and $j^! \alpha''$ are cycles on $G_D \times \mathbb{P}^1$ and $H_D \times \mathbb{P}^1$ which satisfy (i) and (ii) with respect to the cycle $j^!(\alpha)$; again, this follows from the fact that $j^!$ commutes with restriction to open subschemes. Therefore

$$i_\infty^! j^!(\alpha' - \alpha'') = i_\infty^* j^!(\alpha') - i_\infty^! j^!(\alpha'')$$

is a valid choice for the γ-cycle for $j^!(\alpha)$. Now by the fundamental commutativity for divisors (Theorem 6.4),

$$i_\infty^! j^!(\alpha' - \alpha'') = j^! i_\infty^!(\alpha' - \alpha'') = j^!(\gamma) .$$

Therefore, if η' denotes the projection from $G_{D \cap X}$ to $D \cap X$,

$$\begin{aligned} &\mathrm{ch}_{D \cap X}^D(E.|_D) \cap j^! \alpha = \eta'_*(\mathrm{ch}(\xi|_{G_{D \cap X}}) \cap j^! \gamma) \\ &= \eta'_* j^!(\mathrm{ch}(\xi) \cap \gamma) = j^! \eta_*(\mathrm{ch}(\xi) \cap \gamma) = j^!(\mathrm{ch}_X^Y(E.) \cap \alpha) . \quad \square \end{aligned}$$

In particular, the homomorphism $\alpha \to \mathrm{ch}_X^Y(E.) \cap \alpha$ passes to rational equivalence, defining

$$\mathrm{ch}_X^Y(E.) \cap _ : A_* Y_{\mathbb{Q}} \to A_* X_{\mathbb{Q}} . \tag{18}$$

In addition, if $g : Y' \to Y$ is a regular imbedding or a l.c.i. morphism, and $X' = g^{-1}(X)$, then, for all $\alpha \in A_* Y$,

$$g^!(\mathrm{ch}_X^Y(E.) \cap \alpha) = \mathrm{ch}_{X'}^{Y'}(g^* E.) \cap g^* \alpha \tag{19}$$

in $A_* X_{\mathbb{Q}}$.

Corollary 18.1.1 (Homotopy). *Let $X \subset Y$, and let $E.$ be a complex of vector bundles on $Y \times \mathbb{A}^1$, exact off $X \times \mathbb{A}^1$. For each rational point $t \in \mathbb{A}^1$, let $E._t$ be the induced complex on $Y = Y \times \{t\}$. Then for any $\alpha \in A_* Y$,*

$$\mathrm{ch}_X^Y(E._1) \cap \alpha = \mathrm{ch}_X^Y(E._0) \cap \alpha \in A_* X_{\mathbb{Q}} .$$

Proof. $\mathrm{ch}_X^Y(E._t) \cap \alpha = i_t^*(\mathrm{ch}_{X \times \mathbb{A}^1}^{Y \times \mathbb{A}^1}(E.) \cap (\alpha \times [\mathbb{A}^1]))$ by (19), where i_t imbeds X in $X \times \mathbb{A}^1$ at t. Since all homomorphisms

$$i_t^* : A_*(X \times \mathbb{A}^1) \to A_*(X)$$

are the same (Corollary 6.5), the corollary follows. $\square$

Proposition 18.1. *Let i be the inclusion of X in Y.*

(a) *For all* $\alpha \in A_* Y$,

$$i_*(\mathrm{ch}_X^Y(E.) \cap \alpha) = \sum_{i=m}^{n} (-1)^i \mathrm{ch}(E_i) \cap \alpha$$

in $A_* Y_{\mathbb{Q}}$.

(b) *Let* $0 \to E.^{(1)} \to E.^{(2)} \to E.^{(3)} \to 0$ *be an exact sequence of complexes of vector bundles on Y, each exact off X. Then*

$$\mathrm{ch}_X^Y(E.^{(2)}) = \mathrm{ch}_X^Y(E.^{(1)}) + \mathrm{ch}_X^Y(E.^{(3)}) .$$

(c) *Let F be a vector bundle on Y. Then*

$$\mathrm{ch}_X^Y(F \otimes E.) = \mathrm{ch}(i^* F) \cdot \mathrm{ch}_X^Y(E.) .$$

Proof. (a) Since the localized Chern character commutes with i_*, we may assume $X = Y$. In the choice of α' and α'', we may take $\alpha'' = 0$. Then

$$\mathrm{ch}(\xi) \cap \gamma = \mathrm{ch}(\xi) \cap i_\infty^* \alpha' = \mathrm{ch}(\xi) \cap i_0^* \alpha' ,$$

where $i_0 : G = G \times \{0\} \hookrightarrow G \times \mathbb{P}^1$ (cf. Example 2.6.6). If $\varphi_0 : Y \to G$ is the restriction of φ to $Y = Y \times \{0\}$, then $i_0^* \alpha' = \varphi_{0*}(\alpha)$ by condition (i) determining α'. Since φ_0 is the section corresponding to the complex with zero boundary maps, the graph of $0\, d_i$ is $E_i \oplus 0$, so ξ_i pulls back to E_i; therefore $\varphi_0^* \xi = \sum (-1)^i [E_i]$. Hence

$$\mathrm{ch}(\xi) \cap i_0^* \alpha' = \varphi_{0*}(\sum (-1)^i \mathrm{ch}(E_i) \cap \alpha) .$$

Applying η_* yields (1), since $\eta \varphi_0 = \mathrm{id}_Y$.

(b) We first deform $E.^{(2)}$ into the direct sum $E.^{(1)} \oplus E.^{(3)}$. By Corollary 18.1.1, it will then suffice to consider this split case. Let β be the given map from $E.^{(2)}$ to $E.^{(3)}$. Define a family of vector bundle surjections

$$h_t : E.^{(2)} \oplus E.^{(3)} \to E.^{(3)}$$

parametrized by $t \in \mathbb{A}^1$, by $h_t(v_2, v_3) = \beta(v_2) - t\, v_3$. Then $\mathrm{Ker}(h_0) = E.^{(1)} \oplus E.^{(3)}$, and $\mathrm{Ker}(h_1) = E.^{(2)}$.

In the split case, denote by superscripts ${}^{(i)}$ the spaces, bundles, and maps constructed for $E.^{(i)}$, $i = 1, 2, 3$. There is a canonical imbedding of $G^{(1)} \times_Y G^{(3)}$ in $G^{(2)}$ such that $\xi^{(2)}$ restricts to $p^* \xi^{(1)} + q^* \xi^{(3)}$, where p and q are the projections from $G^{(1)} \times_Y G^{(3)}$ to $G^{(1)}$ and $G^{(3)}$. The imbeddings $\varphi^{(i)}$ factor

$$\begin{array}{ccccc} & & & \overset{p\times 1}{\nearrow} & G^{(1)} \times \mathbb{P}^1 \\ Y \times \mathbb{A}^1 & \hookrightarrow & G^{(1)} \times_Y G^{(3)} \times \mathbb{P}^1 & \hookrightarrow & G^{(2)} \times \mathbb{P}^1 \\ & & & \underset{q\times 1}{\searrow} & G^{(3)} \times \mathbb{P}^1 . \end{array}$$

There are analogous factorings for the $H^{(i)}$ and $\psi^{(i)}$. Therefore one may choose the cycles α', α'' for $E.^{(2)}$ to be cycles on $G^{(1)} \times_Y G^{(3)} \times \mathbb{P}^1$. Then the images of α', α'' by $(p \times 1)_*$ and $(q \times 1)_*$ are legitimate choices for $E.^{(1)}$ and $E.^{(3)}$. Note also that $\pi^{(2)} = \pi^{(1)} \circ p = \pi^{(3)} \circ q$. Therefore if $\gamma = i_\infty^*(\alpha' - \alpha'')$, then

$$\begin{aligned} \mathrm{ch}_X^Y(E.^{(2)}) \cap \alpha &= \eta_*^{(2)}(\mathrm{ch}(p^* \xi^{(1)} + q^* \xi^{(3)}) \cap \gamma) \\ &= \eta_*^{(1)}(\mathrm{ch}(\xi^{(1)}) \cap p_* \gamma) + \eta_*^{(3)}(\mathrm{ch}(\xi^{(3)}) \cap q_* \gamma) \\ &= \mathrm{ch}_X^Y(E.^{(1)}) \cap \alpha + \mathrm{ch}_X^Y(E.^{(3)}) \cap \alpha . \end{aligned}$$

(c) If G is constructed for $E.$, $\tilde{G}$. for $F \otimes E.$, there is a canonical imbedding of G in $\tilde{G}$ so that ξ restricts to $F \otimes \xi$. The proof concludes as in (b) (cf. Baum-Fulton-MacPherson (1)II.2.3). □

Corollary 18.1.2. *Let $i: X \to Y$, $j: Y \to Z$ be closed imbeddings, where j is a regular imbedding with normal bundle N. Let $\mathscr{F}$ be a coherent sheaf on X, and suppose $\mathscr{F}$ has a resolution by a finite complex of locally free sheaves $E.$ (resp. $F.$) on Y (resp. Z). Then*

$$\operatorname{ch}_X^Z(F.) = \operatorname{ch}_X^Y(E.) \cdot \operatorname{td}(N)^{-1} \cdot [j]$$

in $A(X \to Z)_{\mathbb{Q}}$.

Proof. Note that by (b) of the proposition, $\operatorname{ch}_X^Y(E.)$ and $\operatorname{ch}_X^Z(F.)$ are independent of the resolutions. We must show that both sides of the displayed equation have the same effect on α, for any cycle α on any scheme V, $g: V \to Z$ any morphism.

Assume first that $Z = P(N \oplus 1)$, j is the imbedding by the zero section, and $V = Z$. Let p be the bundle projection from Z to Y, and assume also that $\alpha = p^* \beta$ for some $\beta \in A_* Y$. Let Q be the universal quotient bundle on $P(N \oplus 1)$. Then (cf. § 15.1) $\Lambda^{\bullet} Q^{\vee} \otimes p^* E.$ is a resolution of $\mathscr{F}$ on Z. Let $\tilde{X} = p^{-1}(X)$, $q: \tilde{X} \to X$ the projection induced by p. By compatibility of local Chern character and push-forward, since j imbeds X in $\tilde{X}$,

$$\text{(1)} \qquad \operatorname{ch}_X^Z(\Lambda^{\bullet} Q^{\vee} \otimes p^* E.) \cap \alpha = q_*(\operatorname{ch}_{\tilde{X}}^Z(\Lambda^{\bullet} Q^{\vee} \otimes p^* E.) \cap \alpha)\,.$$

Since $p^* E.$ is exact off $\tilde{X}$, $\Lambda^{\bullet} Q^{\vee} \otimes p^* E.$ is homotopic, as a complex on Z exact off $\tilde{X}$, to the complex with the same vector bundles, but using the zero boundary map on $\Lambda^{\bullet} Q^{\vee}$. Let $\tilde{Q}$ be the restriction of Q to $\tilde{X}$. By Corollary 18.1.1 and Proposition 18.1 (c),

$$\text{(2)} \qquad \operatorname{ch}_{\tilde{X}}^Z(\Lambda^{\bullet} Q^{\vee} \otimes p^* E.) \cap \alpha = \operatorname{ch}(\Lambda^{\bullet} \tilde{Q}^{\vee}) \cap (\operatorname{ch}_{\tilde{X}}^Z(p^* E.) \cap \alpha)\,.$$

Let k be the zero section imbedding of X in $\tilde{X}$. Since $\operatorname{ch}(\Lambda^{\bullet} \tilde{Q}^{\vee}) = c_d(\tilde{Q}) \cdot \operatorname{td}(\tilde{Q})^{-1}$, $d = \operatorname{rank} N$, and $c_d(\tilde{Q}) \cdot \zeta = k_*(k^* \zeta)$ for all ζ, and $k^* \tilde{Q} = i^* N$, (2) can be rewritten:

$$\text{(3)} \qquad \operatorname{ch}_{\tilde{X}}^Z(\Lambda^{\bullet} Q^{\vee} \otimes p^* E.) \cap \alpha = k_*(\operatorname{td}(i^* N)^{-1} \cap k^* q^* (\operatorname{ch}_X^Y(E.) \cap \beta))\,,$$

using the commutativity of local Chern character and flat pull-back. Applying q_*, and noting that $q\,k = \mathrm{id}_X$, (1) and (3) yield

$$\text{(4)} \qquad \operatorname{ch}_X^Z(\Lambda^{\bullet} Q^{\vee} \otimes p^* E.) \cap \alpha = \operatorname{td}(i^* N)^{-1} \cap (\operatorname{ch}_X^Y(E.) \cap \beta)\,,$$

which is the required equation in this case.

For the general case, deform the imbedding j into the imbedding in the normal bundle N, by the construction of § 5.1:

$$J: Y \times \mathbb{P}^1 \hookrightarrow M_Y Z\,.$$

Let $\mathrm{pr}: Y \times \mathbb{P}^1 \to Y$ be the projection, and let $\tilde{F}.$ be a resolution of $\mathrm{pr}^* \mathscr{F}$ on $M_Y Z$ (which exists since J is a regular imbedding). Let i_0 and i_∞ be the imbeddings of Z and $P(N \oplus 1)$ in $M_Y Z$ at 0 and ∞. Then $i_0^* \tilde{\mathscr{F}}.$ (resp. $i_\infty^* \tilde{\mathscr{F}}.$) is a resolution of $\mathscr{F}$ on Z (resp. $P(N \oplus 1)$).

We may assume V is a variety and $\alpha = [V]$. Let $W = g^{-1}(Y)$, $h: W \to Y$ the induced morphism, and consider the induced morphism $M_W V \to M_Y Z$ of deformation spaces. Then, since local Chern character commutes with specialization,

$$\text{(5)} \qquad \mathrm{ch}_X^Z(F.) \cap \alpha = i_0^! (\mathrm{ch}_{X \times \mathbb{P}^1}^{M_Y Z}(\tilde{F}.) \cap [M_W V]) .$$

As in the proof of Corollary 18.1.1, we may replace $i_0^!$ by $i_\infty^!$ on the right side of (5), which yields:

$$\text{(6)} \qquad \mathrm{ch}_X^Z(F.) \cap \alpha = \mathrm{ch}_X^{P(N \oplus 1)}(i_\infty^* \tilde{F}.) \cap [P(C_W V \oplus 1)] .$$

By construction, $[P(C_W V \oplus 1)] = \bar{p}^* j^! [V]$, where $\bar{p}$ is the projection from $P(h^* N \oplus 1)$ to W. Since $i_\infty^* \tilde{F}.$ can be replaced by $\Lambda^{\bullet} Q^{\vee} \otimes p^* E.$, the required equation

$$\mathrm{ch}_X^Z(F.) \cap \alpha = \mathrm{td}(i^* N)^{-1} \cap (\mathrm{ch}_X^Y(E.) \cap j^! \alpha$$

follows from (6) and (4). □

Example 18.1.1. For any vector bundles E, F of ranks e, f on a scheme Y, there is an open imbedding

$$\text{(1)} \qquad \mathrm{Hom}(E, F) \to \mathrm{Grass}_e(E \oplus F)$$

which, over geometric points $y \in Y$, takes $\sigma(y): E(y) \to F(y)$ to the graph of $\sigma(y)$. (Indeed, if Y is affine, $\mathrm{Hom}(E, F)$ is one of the basic sets in an affine open covering of the Grassman bundle (cf. [EGA] I.9.7).

If $\sigma: E \to F$ is a vector bundle homomorphism, there is a homomorphism

$$\text{(2)} \qquad Y \times \mathbb{A}^1 \to \mathrm{Hom}(E, F)$$

which, on geometric points, takes (y, λ) to $\lambda\,\sigma(y)$. (Let s_σ be the section of $\mathrm{Hom}(E, F)$ corresponding to σ. Then

$$Y \times \mathbb{A}^1 \xrightarrow[s_\sigma \times 1]{} \mathrm{Hom}(E, F) \times \mathbb{A}^1 \xrightarrow[\mu]{} \mathrm{Hom}(E, F) ,$$

where μ is scalar multiplication on the bundle $\mathrm{Hom}(E, F)$.) The composite of (2) and (1):

$$\text{(3)} \qquad Y \times \mathbb{A}^1 \to \mathrm{Grass}_e(E \oplus F)$$

takes the geometric point (y, λ) to the graph of $\lambda\,\sigma(y)$.

The other constructions of this section extend similarly to arbitrary schemes.

Example 18.1.2. Let $E.$ be a complex of vector bundles on Y, exact off X. Let $E.[p]$ be the complex obtained by translating $E.$ p steps to the left: $(E[p])_i = E_{i-p}$. Then

$$\text{(a)} \qquad \mathrm{ch}_X^Y(E.[p]) = (-1)^p \, \mathrm{ch}_X^Y(E.) .$$

Let $E.^{\vee}$ be the dual complex: $(E.^{\vee})_n = E_{-n}^{\vee}$, with n^{th} boundary map $d_{-n+1}^{\vee}$. Let $\mathrm{ch}_i(E.)$ denote the component of $\mathrm{ch}_X^Y(E.)$ in $A^i(X \to Y)_{\mathbb{Q}}$. Then

$$\text{(b)} \qquad \mathrm{ch}_i(E.^{\vee}) = (-1)^i \, \mathrm{ch}_i(E.) .$$

(For (a), the geometry of the construction is the same for $E.$ and for $E.[p]$; only ξ changes to $(-1)^p\xi$. For (b), using the duality of Grassmannians (Example 14.6.5), the Grassmann bundles may also be identified, and ξ corresponds to $\xi^\vee$.)

Example 18.1.3. With the same notation as in this section, define

$$c(\xi) = 1 + c_1(\xi) + c_2(\xi) + \ldots = \prod_{i=m}^{n} c(\xi_i)^{(-1)^i}.$$

and define

$$c_i(E.) \cap \alpha = \eta_*(c_i(\xi) \cap \gamma).$$

Then for all $i > 0$, $c_i(E.)$ determines an element in $A^i(X \to Y)$, called the ith *localized Chern class* of E. There are analogues of Proposition 18.1 and Corollary 18.1.2, "without denominators", for these classes.

Example 18.1.4. Let $E.$, $F.$ be complexes of vector bundles on Y, both exact off X. Let $\sigma: E. \to F.$ be a quasi-isomorphism, i.e., σ is a homomorphism of complexes which induces an isomorphism of homology sheaves. Then

$$\mathrm{ch}_X^Y E. = \mathrm{ch}_X^Y F.,$$

and similarly for local Chern classes. (There is an exact sequence

$$0 \to F. \to G. \to E.[1] \to 0$$

where $G.$ is the mapping cone, which is a complex exact on all of Y.)

Example 18.1.5. Let $E.$ and $F.$ be complexes of vector bundles on Y, exact off X and Z respectively. Then $E.\otimes F.$ is exact off $X \cap Z$, and, with the notation of Example 17.3.1,

$$\mathrm{ch}_{X\cap Z}^Y(E.\otimes F.) = \mathrm{ch}_X^Y(E.) \cup \mathrm{ch}_Z^Y(F.).$$

(It would be interesting to find a direct proof of this multiplicativity property, as in Proposition 18.1; in topology this has been done by Iversen (3). The formula may be deduced from the formula of Example 18.3.12, the surjectivity of the Riemann-Roch map τ_Y, and the spectral sequence

$$\mathscr{H}_p(E.\otimes \mathscr{H}_q(F.)\otimes \mathscr{G}) \Rightarrow \mathscr{H}_{p+q}(E.\otimes F.\otimes \mathscr{G})$$

for a coherent sheaf $\mathscr{G}$ on Y.)

Example 18.1.6. *MacPherson's graph construction for vector bundle homomorphisms.* Let $\sigma: E \to F$ be a homomorphism of vector bundles of ranks e, f on a variety Y. Let $G = \mathrm{Grass}_e(E \oplus F)$, with universal bundle ζ of rank e on G. There is a canonical imbedding

$$\varphi: Y \times \mathbb{A}^1 \to G \times \mathbb{P}^1$$

taking (y, λ) to (graph of $\lambda\,\sigma(y)$, $(1:\lambda)$). Let W be the closure of the image of φ, and let

$$Z_\infty = i_\infty^*[W] = \sum n_i [V_i],$$

an n-cycle on G, $n = \dim(Y)$. Let $\eta_i: V_i \to Y$ be the maps induced by projection.

(a) For any characteristic class $c\,l$ (i.e., polynomial in Chern classes),

$$c\,l(E) \cap [Y] = \sum n_i\, \eta_{i_*}(c\,l(\zeta) \cap [V_i]) .$$

(b) If σ has rank k off a proper closed subset X of Y, then there is one component $[\tilde{Y}]$ appearing in Z_∞ with multiplicity one, such that $\tilde{Y}$ projects birationally to Y, and $\tilde{Y}$ lies in

$$H = \mathrm{Grass}_{e-k}(E) \times_Y \mathrm{Grass}_k(F) \hookrightarrow G .$$

The other components V_i all project into X.

(c) If $k = e = f$, then $\tilde{Y} = Y$, and

$$c\,l(E) \cap [Y] - c\,l(F) \cap [Y] = \sum_{V_i \to X} n_i\, \eta_{i*}(c\,l(\zeta) \cap [V_i]) .$$

This is a special case of the construction of this section for the two-term complex $E \to F$.

(d) If E is a trivial line bundle, so σ is a section of F, and $X = Z(s)$, then $G = P(F \oplus 1)$, $H = P(F)$, and

$$Z_\infty = [\tilde{Y}] + [P(C \oplus 1)] ,$$

where $\tilde{Y}$ is the blow-up of Y along X, and C is the normal cone to X in Y.

(e) If σ is sufficiently generic, then $Z_\infty = \sum [V_i]$, where V_i projects onto the locus where rank $\sigma \leqq i$. In general, Z_∞ contains much of the usable information about how σ degenerates.

(f) If $f: Y_1 \to Y_2$ is a morphism of non-singular varieties, the graph construction may be applied to $df: T_{Y_1} \to f^* T_{Y_2}$. The components of Z_∞ will lie over various singularity loci of f. For example, if Y_1 is the blow-up of Y_2 along a smooth subvariety X_2, and X_1 is the exceptional divisor, then

$$Z_\infty = [Y_1] + [P(L \oplus 1)]$$

where $L = \mathscr{O}_N(1)$ on $X_1 = P(N)$, $N = N_{X_2} Y_2$. The formula for blowing up Chern classes (§ 15.4) may also be derived from this.

On the other hand, if f is a branched covering, the graph construction produces a decomposition

$$c(T_{Y_1}) = f^* c(T_{Y_2}) + \beta ,$$

for a class β on the ramification locus of f.

Example 18.1.7. In the situation of Corollary 18.1.1, the classes $\mathrm{ch}_X^Y(E_{.t}) \in A(X \to Y)_{\mathbb{Q}}$ are independent of $t \in \mathbb{A}^1$. (See Example 17.5.1.)

18.2 Riemann-Roch for Quasi-projective Schemes

In this section we work in the category of schemes X which are quasi-projective over a fixed non-singular base variety S. Such schemes are equipped with a morphism to S, which factors

$$X \xrightarrow{i} U \xrightarrow{j} P(E) \xrightarrow{p} S ,$$

with i a closed imbedding, j an open imbedding, E a vector bundle on S, p the projection. The case $S = \operatorname{Spec}(K)$, K the ground field is paramount; the extra generality will be used in the extension to the non-projective case (§ 18.3).

If M is smooth over S, let $T_{M/S}$ denote the relative tangent bundle of M over S, and set

$$\operatorname{Td}(M) = \operatorname{td}(T_{M/S}) \in A(M)_{\mathbb{Q}}.$$

For any closed imbedding $i: X \to M$, M smooth over S, and any coherent sheaf $\mathscr{F}$ on X, set

$$\operatorname{ch}_X^M(\mathscr{F}) = \operatorname{ch}_X^M(E.) \in A(X \to M)_{\mathbb{Q}}$$

where $E.$ is any resolution of $i_*(\mathscr{F})$ by locally free sheaves E_i on M. It follows from Proposition 18.1 (b) that this class is independent of choice of resolution, and that ch_X^M determines a homomorphism

$$\operatorname{ch}_X^M: K_\circ X \to A(X \to M)_{\mathbb{Q}}.$$

Define a homomorphism

$$\tau_X^M: K_\circ X \to A_* X_{\mathbb{Q}}$$

by the formula

$$\tau_X^M(\alpha) = \operatorname{ch}_X^M(E.) \cap \operatorname{Td}(M),$$

where $\operatorname{Td}(M) = \operatorname{td}(T_{M/S}) \cap [M]$. Identifying $A(X \to M)_{\mathbb{Q}}$ with $A.X_{\mathbb{Q}}$, this may also be written $\tau_X^M(\alpha) = \operatorname{td}(T_{M/S}) \cap (\operatorname{ch}_X^M(E.) \cap [M])$.

Theorem 18.2. *The homomorphism τ_X^M is independent of the imbedding of X in M. Denote by* τ or τ_X,

$$\tau_X: K_\circ X \to A_* X_{\mathbb{Q}}$$

the homomorphism τ_X^M for any such imbedding. Then the following properties hold:

(1) (Covariance). *If $f: X \to Y$ is proper, then*

$$\begin{array}{ccc} K_\circ X & \xrightarrow{\tau_X} & A_* X_{\mathbb{Q}} \\ f_* \downarrow & & \downarrow f_* \\ K_\circ Y & \xrightarrow{\tau_Y} & A_* Y_{\mathbb{Q}} \end{array}$$

commutes.

(2) (Module). *If $\alpha \in K_\circ X$, $\beta \in K^\circ X$, then*

$$\tau_X(\beta \otimes \alpha) = \operatorname{ch}(\beta) \cap \tau_X(\alpha).$$

(3) (Local complete intersections). *If $f: X \to Y$ is a l.c.i. morphism, with virtual tangent bundle T_f, then the diagrams*

$$\begin{array}{ccc} K^\circ X & \xrightarrow{\operatorname{ch}} & A^* X_{\mathbb{Q}} \\ f_* \downarrow & & \downarrow f_*(\operatorname{td}(T_f)\cdot_) \\ K^\circ Y & \xrightarrow{\operatorname{ch}} & A^* Y_{\mathbb{Q}} \end{array} \qquad \begin{array}{ccc} K_\circ Y & \xrightarrow{\tau_Y} & A_* Y_{\mathbb{Q}} \\ f^* \downarrow & & \downarrow \operatorname{td}(T_f)\cdot f^* \\ K_\circ X & \xrightarrow{\tau_X} & A_* X_{\mathbb{Q}} \end{array}$$

commute. For the first diagram, f is assumed proper. The Gysin homomorphisms f_*, f^* are those constructed in § 6.2 and Example 15.1.8. The virtual bundle T_f is defined in Appendix B.7.6.

Proof. The proof is divided into ten steps.

Step 1. If $i: X \to M$ is a closed imbedding, and $j: M \to P$ an open or closed imbedding, with M and P smooth over S, then $\tau_X^M = \tau_X^P$. This follows from the definition if j is an open imbedding, and from Corollary 18.1.2 if j is a closed imbedding.

Step 2. If $i: X \to M$ is a closed imbedding, $\beta \in K^\circ M$, $\alpha \in K_\circ X$, then

$$\tau_X^M(i^*\beta \otimes \alpha) = \mathrm{ch}(i^*\beta) \cap \tau_X^M(\alpha).$$

This follows from Proposition 18.1 (c).

Step 3. If $f: X \to Y$ is a closed imbedding, and $j: Y \to M$ is a closed imbedding, M smooth over S, then $f_* \tau_X^M(\alpha) = \tau_Y^M(f_* \alpha)$ for all $\alpha \in K_\circ X$. This is a consequence of the construction of τ_X^M.

Step 4. Let $i: Y \to M$ be a closed imbedding, M smooth over S, and let $p: P \to M$ be a smooth morphism. Form the fibre square

$$\begin{array}{ccc} X & \xrightarrow{j} & P \\ {\scriptstyle f}\downarrow & & \downarrow{\scriptstyle p} \\ Y & \xrightarrow[i]{} & M \end{array} .$$

Then for all $\alpha \in K_\circ Y$,

$$\tau_X^P f^* \alpha = \mathrm{td}(T_f) \cdot f^* \tau_Y^M(\alpha).$$

This follows from the commutativity of local Chern character with flat pullback (Theorem 18.1).

Step 5. With notation as in Step 4, assume $P = P(E)$, E a vector bundle on M, p the projection. Then for all $\alpha \in K_\circ X$,

$$\tau_Y^M(f_* \alpha) = f_* \tau_X^P(\alpha).$$

It suffices to prove this for $\alpha = \mathscr{O}(n) \otimes f^* \beta$, $\beta \in K_\circ Y$, since any class in $K_\circ X$ is a sum of such classes (cf. Example 15.1.1). Then

$$f_* \tau_X^P(\alpha) = f_* (\mathrm{ch}(\mathscr{O}(n)) \cdot \mathrm{td}(T_f) \cap f^* \tau_Y^M(\beta))$$

by Steps 2 and 4. By Example 15.2.17, since $\mathrm{td}(T_f) = \mathrm{td}(f^*(E|_Y) \otimes \mathscr{O}(1))$, this may be rewritten:

$$f_* \tau_X^P(\alpha) = i^* \mathrm{ch}(\mathrm{Sym}^n E^\vee) \cap \tau_Y^M(\beta).$$

On the other hand, $f_* \alpha = \mathrm{Sym}^n(i^* E^\vee) \otimes \beta$, so

$$\tau_Y^M(f_* \alpha) = i^* \mathrm{ch}(\mathrm{Sym}^n E^\vee) \cap \tau_Y^M(\beta),$$

by Step 2 again.

Step 6. τ_X^M is independent of M. By Step 1, it suffices to consider closed imbeddings $i: X \to U$, U open in $P(E)$, E a vector bundle on S. Suppose $j: X \to V \subset P(F)$ is another such imbedding. Let $P = P(E)$, $Q = P(F)$. There are induced imbeddings

$$\begin{array}{ccc} X & \xrightarrow{k} & X \times_S Q \\ {\scriptstyle \delta}\downarrow & & \downarrow \\ U \times_S V & \to & U \times_S Q \end{array}$$

with $k=(1,j)$, $\delta=(i,j)$. Let $Z=X\times_S Q$, p the projection from Z to X. Consider the diagram

$$\begin{array}{ccc} K_\circ X & \xrightarrow{\tau_X^{U\times_S V}} & A_* X_{\mathbb{Q}} \\ {\scriptstyle k_*}\downarrow & & \downarrow{\scriptstyle k_*} \\ K_\circ Z & \xrightarrow{\tau_Z^{U\times_S Q}} & A_* Z_{\mathbb{Q}} \\ {\scriptstyle p_*}\downarrow & & \downarrow{\scriptstyle p_*} \\ K_\circ X & \xrightarrow{\tau_X^{U}} & A_* X_{\mathbb{Q}} \end{array}.$$

The upper square commutes by Steps 1 and 3, the lower square by Step 5. Since $p_* k_* = (p\,k)_* = \mathrm{id}$, $\tau_X^U = \tau_X^{U\times_S V}$; and $\tau_X^{U\times_S V} = \tau_X^V$ by symmetry.

Step 7. The covariance of τ follows from Steps 3 and 5. Indeed, if $X \hookrightarrow U \subset P(E) \to S$, E a bundle, and $Y \hookrightarrow M \xrightarrow{q} S$, with q smooth, one has a diagram

$$\begin{array}{ccccc} X & \xrightarrow{i} & P(q^* E|_Y) & \hookrightarrow & P(q^* E) \\ & {\scriptstyle f}\searrow & {\scriptstyle p}\downarrow & & \downarrow \\ & & Y & \hookrightarrow & M \end{array},$$

and one may use Step 3 for i, Step 5 for p.

Step 8. The module property follows from Step 2 and the following lemma.

Lemma 18.2. *Let E be a vector bundle on a scheme X. Then there is a closed imbedding $i: X \to M$, with M smooth over S, and a vector bundle $\tilde{E}$ on M such that $i^* \tilde{E} \cong E$.*

Proof. Choose a closed imbedding j of X in U, with U smooth over S. Choose a vector bundle F on U and a surjection of sheaves $F \to j_*(E)$ (Appendix B.8.1). Let $e = \mathrm{rank}(E)$, and let M be the Grassmannian of rank e quotients of F, $\pi: M \to U$ the projection. Then there is a morphism $i: X \to M$ with $\pi \circ i = j$, such that E is the pull-back of the universal quotient bundle on M. □

Step 9. To verify that the first diagram in (3) commutes, when f is a proper l.c.i. morphism, factor f as in Step 7. For the projection, the calculation is the same formal calculation done in Step 5. For the case of a regular imbedding, the proof is essentially the same as that given in § 15.2 in the non-singular case. Indeed, a stronger result is given in Corollary 18.1.2.

Step 10. For the second diagram in (3), again factor f as in Step 7:

$$X \hookrightarrow U \times_S Y \hookrightarrow P(q^* E|_Y) \xrightarrow{p} Y.$$

For p, the assertion was proved in Step 4, and for open imbeddings it follows from Step 1. It therefore suffices to prove that

$$(*) \qquad f^* \tau_Y(\alpha) = \mathrm{td}(N) \cap \tau_X f^* \alpha$$

for $f: X \to Y$ a (closed) regular imbedding, with normal bundle N.

We first verify (*) when $Y = N$, and f is the zero section imbedding of X in N. Let p be the projection of N to X. Since N is the restriction of a bundle on a smooth variety containing X, it follows from Step 4 that

$$(*') \qquad \tau_N(p^*\beta) = \mathrm{td}(p^*N) \cap p^*\tau_X(\beta) = p^*(\mathrm{td}(N) \cap \tau_X(\beta))$$

for all $\beta \in K_\circ X$. Since $p^*: K_\circ X \to K_\circ N$ is surjective, we may write $\alpha = p^*\beta$, $\beta = f^*\alpha$. Since f^* and p^* are inverse isomorphisms between A_*X and A_*N, (*) follows from (*').

We next verify (*) when $X = \mathrm{div}(t)$ is a principal Cartier divisor on Y. Then t determines a morphism $p: Y \to \mathbb{A}^1_S$, with $X = p^{-1}(0)$. Choose an imbedding $i: Y \to M$, M smooth over S. Then $j = (i, p)$ imbeds Y in $M \times_S \mathbb{A}^1_S = \mathbb{A}^1_M$, and one has a fibre square

$$\begin{array}{ccc} X & \xrightarrow{f} & Y \\ k\downarrow & & \downarrow j \\ M & \xrightarrow[g]{} & \mathbb{A}^1_M \end{array}$$

where g is the zero section, and $k = i \circ f$. It suffices to prove (*) for $\alpha = [\mathcal{O}_V]$, V a subvariety of Y. If $V \subset X$, both sides of (*) are zero, since $f^* \circ f_*$ is zero on $K_\circ X$ and on A_*X, and $\tau_Y \circ f_* = f_* \circ \tau_X$. If $V \not\subset X$, let $W = V \cap X$, so $f^*\alpha = [\mathcal{O}_W]$. Let $E.$ be a resolution of $j_*\mathcal{O}_V$ by locally free sheaves on $\mathbb{A}^1_M$. Then $g^*E.$ resolves $k_*\mathcal{O}_W$ on M (Lemma A.4.2). Now

$$f^*\tau_Y(\mathcal{O}_V) = f^*(\mathrm{ch}_Y^{\mathbb{A}^1_M}(E.) \cap \mathrm{Td}(\mathbb{A}^1_M)) = \mathrm{ch}_X^M(g^*E.) \cap \mathrm{Td}(M)$$

by formula (19) of § 18.1. But this last term is $\tau_X(\mathcal{O}_W)$, which is the required equation (*) since the normal bundle to X in Y is trivial.

The general case of (*) follows from the preceding two cases and a deformation to the normal bundle. As for rational equivalence (§ 5.2) there is a specialization homomorphism

$$\sigma: K_\circ Y \to K_\circ N$$

defined to make the diagram

$$\begin{array}{ccccc} K_\circ N & \xrightarrow{i_*} K_\circ M^\circ & \xrightarrow{j^*} & K_\circ(Y \times \mathbb{A}^1) & \to 0 \\ & i^* \downarrow & & \uparrow \mathrm{pr}^* & \\ & K_\circ N & \xleftarrow[\sigma]{} & K_\circ(Y) & \end{array}$$

commute. In addition,

$$f^* = \bar{f}^* \circ \sigma: K_\circ Y \to K_\circ X$$

where $\bar{f}$ is the zero-section imbedding of X in N. (To see this, consider the deformation diagram

$$\begin{array}{ccccc} X & \xrightarrow{i_0} & X \times \mathbb{P}^1 & \xleftarrow{i_\infty} & X \\ f\downarrow & & \downarrow F & & \downarrow \bar{f} \\ Y & \xrightarrow[k]{} & M^\circ & \xleftarrow[i]{} & N \end{array} .$$

Let $\alpha \in K_\circ Y$, $\mathrm{pr}^*\alpha = j^*\beta$, $\beta \in K_\circ M^\circ$. Then

$$\bar{f}^*\sigma^*(\alpha) = \bar{f}^* i^*\beta = i_\infty^* F^*\beta = i_0^* F^*\beta = f^*k^*\beta = f^*\alpha .)$$

Now τ commutes with the specialization maps σ, since τ commutes with pr^* (Step 4), j^* (Step 1), and i^* (since i imbeds N as a principal Cartier divisor on

M°). Therefore

$$f^* \tau_Y(\alpha) = \bar{f}^* \sigma \tau_Y(\alpha) = \bar{f}^* \tau_N(\sigma(\alpha))$$
$$= \operatorname{td}(N) \cap \tau_X(\bar{f}^* \sigma(\alpha)) = \operatorname{td}(N) \cap \tau_X(f^* \alpha),$$

which concludes the proof. □

Example 18.2.1. Let $f: X \to Y$ be a regular imbedding of codimension d, with normal bundle N, and let E be a vector bundle of rank e on X. Define $P(N, E) \in A^* X$ by the prescription of Lemma 15.3. Then

$$c(f_* E) = 1 + f_*(P(N, E))$$

in $A^* Y$. (The proof is the same as in Corollary 18.1.2, cf. Example 18.1.3.)

Example 18.2.2. *Lefschetz-Riemann-Roch.* There are equivariant analogues of the Riemann-Roch theorems, at least for automorphisms of finite order prime to the characteristic. For a singular version see Baum-Fulton-Quart (1). The formalism and use of deformation to normal bundle are completely analogous. There is an analogous formula for the Frobenius endomorphism in characteristic p (cf. Fulton (5)).

18.3 Riemann-Roch for Algebraic Schemes

In this section we work in the category of algebraic schemes (locally of finite type and separated) over a fixed base field.

Theorem 18.3. *For all schemes X there is a homomorphism*

$$\tau = \tau_X : K_\circ X \to A_* X_{\mathbb{Q}}$$

satisfying:

(1) (Covariance). *If $f: X \to Y$ is proper, $\alpha \in K_\circ X$, then $f_* \tau_X(\alpha) = \tau_Y f_*(\alpha)$.*

(2) (Module). *If $\alpha \in K_\circ X$, $\beta \in K^\circ X$, then*

$$\tau_X(\beta \otimes \alpha) = \operatorname{ch}(\beta) \cap \tau_X(\alpha).$$

(3) *If $i: X \to M$ is a closed imbedding in a smooth M, and $E.$ is a resolution of $i_* \mathscr{F}$ by locally free sheaves on M, then*

$$\tau_X(\mathscr{F}) = \operatorname{ch}_X^M(E.) \cap \operatorname{Td}(M) = \operatorname{td}(i^* T_M) \cap (\operatorname{ch}_X^M(E.) \cap [M]).$$

(4) *Let $f: X \to Y$ be a l.c.i. morphism. Assume that there are closed imbeddings $X \subset M$, $Y \subset P$ in smooth schemes M, P. Then*

$$\tau_X f^* \alpha = \operatorname{td}(T_f) \cdot f^* \tau_Y(\alpha)$$

for all $\alpha \in K_\circ Y$.

(5) (Top term). *If V is a closed subvariety of X, with* dim $V = n$, *then*

$$\tau_X(\mathscr{O}_V) = [V] + \text{terms of dimension} < n.$$

In addition, τ is uniquely determined by properties (1), (4) *(for open imbeddings of quasi-projective schemes), and* (5) *(for $V = X = \mathbb{P}^n$).*

The proof will be given at the end of this section. We first record some corollaries. For any scheme X, we define the *Todd class* of X, denoted $\mathrm{Td}(X)$, by the formula

$$\mathrm{Td}(X) = \tau_X(\mathcal{O}_X) \in A_*(X)_{\mathbb{Q}}.$$

Corollary 18.3.1. (a) *If X is complete, then for any vector bundle E on X,*

$$\chi(X, E) = \int_X \mathrm{ch}(E) \cap \mathrm{Td}(X).$$

In particular,

$$\chi(X, \mathcal{O}_X) = \int_X \mathrm{Td}(X).$$

(b) *If X is a l.c.i. scheme which is imbeddable in a smooth scheme, then*

$$\mathrm{Td}(X) = \mathrm{td}(T_X) \cap [X].$$

Here T_X is the virtual tangent bundle to X. This holds, in particular, for all smooth schemes X.

(c) *Let $f: X \to Y$ be a proper morphism. Let $\beta \in K^\circ X$. Assume there is an element $f_*(\beta)$ in $K^\circ Y$ such that*

$$f_*(\beta \otimes \mathcal{O}_X) = f_*(\beta) \otimes \mathcal{O}_Y \in K_\circ Y.$$

For example, if f is a l.c.i. morphism, or more generally, any perfect morphism (cf. Example 15.1.8*), there is a canonical such element $f_*(\beta)$. Then*

$$f_*(\mathrm{ch}(\beta) \cap \mathrm{Td}(X)) = \mathrm{ch}(f_*\beta) \cap \mathrm{Td}(Y).$$

In particular, if X and Y are smooth, then

$$f_*(\mathrm{ch}(\beta) \cdot \mathrm{td}(T_X)) = \mathrm{ch}(f_*\beta) \cdot \mathrm{td}(T_Y).$$

Proof. (a) follows from the covariance for the mapping from X to a point, and the module property (2). If X is smooth, (b) follows from property (3) of the theorem, imbedding X in itself to calculate $\tau_X(\mathcal{O}_X)$. If X is a l.c.i. scheme, and $f: X \to Y$ an imbedding of X in a smooth Y, then by property (4),

$$\tau(\mathcal{O}_X) = \mathrm{td}(N_f)^{-1} f^*(\mathrm{td}(T_Y) \cap [Y]) = \mathrm{td}(T_X) \cap [X].$$

(c) follows from the covariance and module properties (1) and (2):

$$\begin{aligned} f_*(\mathrm{ch}(\beta) \cap \tau_X(\mathcal{O}_X)) &= f_*(\tau_X(\beta \otimes \mathcal{O}_X)) \\ &= \tau_Y(f_*(\beta \otimes \mathcal{O}_X)) = \tau_Y(f_*(\beta) \otimes \mathcal{O}_Y) \\ &= \mathrm{ch}(f_*(\beta)) \cap \tau_Y(\mathcal{O}_Y). \quad \square \end{aligned}$$

Corollary 18.3.2. *For all X, the homomorphism τ_X induces an isomorphism*

$$(\tau_X)_{\mathbb{Q}}: K_\circ X_{\mathbb{Q}} \xrightarrow{\sim} A_* X_{\mathbb{Q}}.$$

where $K_\circ X_{\mathbb{Q}} = (K_\circ X) \otimes_{\mathbb{Z}} \mathbb{Q}$.

Proof. In fact, if

$$\varphi: A_* X \to \mathrm{Gr}\, K_\circ X$$

is the canonical surjection (Example 15.1.5) the composite $\varphi_{\mathbb{Q}} \circ \tau$ induces an isomorphism on the associated graded groups $\mathrm{Gr}\, K_\circ X_{\mathbb{Q}}$, as follows from Property (5). □

In particular, if X is non-singular, the Chern character determines an isomorphism

$$\mathrm{ch}_{\mathbb{Q}}: K^\circ X_{\mathbb{Q}} \xrightarrow{\cong} A^* X_{\mathbb{Q}}.$$

The proof of Theorem 18.3 will depend on the following two propositions. If $i: X \to M$ is a closed imbedding of a scheme X in a smooth scheme M, and $\mathscr{F}$ is a coherent sheaf on X, we define $\mathrm{ch}_X^M(\mathscr{F}) \in A_* X_{\mathbb{Q}}$ by

$$\mathrm{ch}_X^M(\mathscr{F}) = \mathrm{ch}_X^M(E.) \cap [M]$$

where $E.$ is a resolution of $i_*\mathscr{F}$ on M. This determines a homomorphism ch_X^M from $K_\circ X$ to $A_* X_{\mathbb{Q}}$.

Proposition 18.3.1. *Let $X \overset{i}{\hookrightarrow} M$, $X \overset{j}{\hookrightarrow} P$ be closed imbeddings of a scheme X in smooth schemes M and P. Then*

$$\mathrm{td}(T_M) \cap \mathrm{ch}_X^M(\alpha) = \mathrm{td}(T_P) \cap \mathrm{ch}_X^P(\alpha)$$

in $A_* X_{\mathbb{Q}}$, for all $\alpha \in K_\circ X$.

Proof. It suffices to show that

$$\text{(i)} \qquad \mathrm{td}(T_P) \cap \mathrm{ch}_X^P(\mathscr{F}) = \mathrm{td}(T_{M\times P}) \cap \mathrm{ch}_X^{M\times P}(\mathscr{F}),$$

where $X \subset M \times P$ by the diagonal, and $\mathscr{F}$ is a coherent sheaf on X. Consider the diagram

$$\begin{array}{ccccc} X & \xrightarrow{h} & M \times X & \xrightarrow{k} & M \times P \\ & & {\scriptstyle q}\downarrow & & \downarrow{\scriptstyle p} \\ & & X & \xrightarrow[j]{} & P \end{array}$$

with $h = (i, 1_X)$, $k = 1_M \times j$, p, q the projections. Note that h is a regular imbedding with normal bundle $i^*(T_M)$. Let $E.$ be a resolution of $j_*\mathscr{F}$ by locally free sheaves on P. Then $p^* E.$ is a resolution of $q^*\mathscr{F}$ on $M \times P$, since p is flat; therefore by Theorem 18.1,

$$\text{(ii)} \qquad p^* \mathrm{ch}_X^P(\mathscr{F}) = \mathrm{ch}_{M\times X}^{M\times P}(q^*\mathscr{F}).$$

Now apply Theorem 18.2(3) to the regular imbedding h, regarded as a morphism of schemes which are projective *over the base scheme* $S = M \times P$. This gives

$$\text{(iii)} \qquad \mathrm{ch}_X^{M\times P}(h^*\beta) = \mathrm{td}(i^* T_M)^{-1} \cap h^* \mathrm{ch}_{M\times X}^{M\times P}(\beta)$$

for any $\beta \in K_\circ(M \times X)$. Apply this formula to $\beta = [q^*\mathscr{F}]$. Then $h^*\beta = (q\,h)^*[\mathscr{F}] = [\mathscr{F}]$, so by (ii) and (iii),

$$\text{(iv)} \qquad \mathrm{ch}_X^{M\times P}(\mathscr{F}) = \mathrm{td}(i^* T_M)^{-1} \cap h^* q^* \mathrm{ch}_X^P(\mathscr{F}) = \mathrm{td}(i^* T_M)^{-1} \cap \mathrm{ch}_X^P(\mathscr{F}).$$

Multiplying by $\mathrm{td}(T_{M\times P})$, one achieves (i), as required. □

Proposition 18.3.2. *Let*

$$\begin{array}{ccc} Y' & \xrightarrow{j} & X' \\ {\scriptstyle q}\downarrow & & \downarrow{\scriptstyle p} \\ Y & \xrightarrow[i]{} & X \end{array}$$

be a fibre square, with i a closed imbedding, and p projective, such that p induces an isomorphism of $X'-Y'$ onto $X-Y$. Then the sequence

$$K_\circ Y' \xrightarrow{a} K_\circ Y \oplus K_\circ X' \xrightarrow{b} K_\circ X \to 0$$

is exact, where $a(\alpha)=(q_*\alpha, -j_*\alpha)$, $b(\alpha,\beta)=i_*\alpha+p_*\beta$.

Proof. Let $U=X-Y$, $U'=X'-Y'$, r the induced isomorphism from U' to U. Then there is a commutative diagram with exact rows

$$\begin{array}{ccccccccc} K_1 U' & \to & K_\circ Y' & \xrightarrow{j_*} & K_\circ X' & \to & K_\circ U' & \to & 0 \\ \wr\wr\downarrow{\scriptstyle r_*} & & \downarrow{\scriptstyle q_*} & & \downarrow{\scriptstyle p_*} & & \wr\wr\downarrow{\scriptstyle r_*} & & \\ K_1 U & \to & K_\circ Y & \xrightarrow[i_*]{} & K_\circ X & \to & K_\circ U & \to & 0, \end{array}$$

where K_1 is Quillen's first higher K-group (Quillen (2)§ 7 Prop. 3.2, Fulton-Gillet (1)). Then the conclusion follows from a simple diagram chase. □

The homomorphism $\tau_X: K_\circ X \to A_* X_{\mathbb{Q}}$ has been constructed in § 18.2 for all schemes X which are quasi-projective over $S=\operatorname{Spec}(K)$, K the ground field. To extend this to arbitrary schemes X, we introduce the notion of a Chow envelope.

Definition 18.3. An *envelope* of a scheme X is a proper morphism $p: X' \to X$ such that for every subvariety V of X, there is a subvariety V' of X' such that p maps V' birationally onto V. If X' is quasi-projective over $\operatorname{Spec}(K)$, we call $p: X' \to X$ a *Chow envelope.*

We will use the following elementary lemma.

Lemma 18.3. (1) *If $p: X' \to X$ and $q: X'' \to X'$ are envelopes, then $pq: X'' \to X$ is an envelope.*

(2) *If $p: X' \to X$ is an envelope, and $f: Y \to X$ is an arbitrary morphism, then the fibre product*

$$p \times_X Y \;:\; X' \times_X Y \to Y$$

is an envelope.

(3) *For any scheme X, there is a Chow envelope $p: X' \to X$, and a closed subscheme $Y \subset X$, with $X-Y$ dense in X, such that p maps $X'-p^{-1}(Y)$ isomorphically onto $X-Y$.*

(4) *If $p_i: X_i' \to X$ are envelopes, $i=1,2$, then there is a Chow envelope $p: X' \to X$, with morphisms $q_i: X' \to X_i'$ such that $p=p_i \circ q_i$ for $i=1,2$.*

(5) *For any morphism $f: Y \to X$, and any Chow envelope $p: X' \to X$, there is a Chow envelope $q: Y' \to Y$ and a morphism $f': Y' \to X'$ such that $pf'=fq$. If f is proper, one may achieve this with f' proper.*

(6) *If $p: X' \to X$ is an envelope, then the homomorphisms*

$$p_*: K_\circ X' \to K_\circ X, \quad p_*: A_* X' \to A_* X$$

are surjective.

Proof. (1) is immediate. For (2), if $V \subset Y$, let W be the closure of $f(V)$, and let p map $W' \subset X'$ birationally onto W. There are open sets $U \subset W$, $U' \subset W'$ such that p maps U' isomorphically onto U. Then the closure of $U' \times_U V$ in $W' \times_W V$ is a subvariety of $X' \times_X Y$ which maps birationally onto V.

For (3), by Chow's lemma (cf. [EGA]II.5.6), there is a proper morphism $p_1: X_1' \to X$, with X_1' quasi-projective over S, and a closed $Y \subset X$ with $X - Y$ dense in X, so that p_1 restricts to an isomorphism over $X - Y$. By Noetherian induction, there is a Chow envelope $p_2: X_2' \to Y$. Then

$$X' = X_1' \coprod X_2' \xrightarrow{p_1 \coprod p_2} X$$

is the desired Chow envelope for X.

In the situation of (4), choose a Chow envelope

$$X' \to X_1' \times_X X_2'.$$

Then by (1) and (2), the composite $X' \to X$ is also a Chow envelope. In (5), choose any Chow envelope

$$Y' \to Y \times_X X'.$$

Finally, if $F_i K_\circ X$ denotes the subgroup of $K_\circ X$ generated by sheaves whose support has dimension at most i, then $F_i K_\circ X$ is generated by $[\mathcal{O}_V]$, where V ranges over subvarieties of dimension at most i. If p maps V' birationally onto V, then

$$p_*[\mathcal{O}_{V'}] = [\mathcal{O}_V] + \alpha$$

with $\alpha \in F_{i-1} K_\circ X$, and (6) follows by Noetherian induction. The case for A_* is trivial since p_* is surjective on the cycle level. □

We can now complete the proof of the Riemann-Roch theorem. For all quasi-projective schemes X over $S = \mathrm{Spec}(K)$, we denote by τ_X the homomorphism which was constructed in § 18.2. The extension to general X will be carried out in several steps.

Step 1. If $p: X' \to X$ is a Chow envelope, there is at most one homomorphism $\tau: K_\circ X \to A_* X_{\mathbb{Q}}$ so that the diagram

$$(*) \qquad \begin{array}{ccc} K_\circ X' & \xrightarrow{\tau_{X'}} & A_* X'_{\mathbb{Q}} \\ p_* \downarrow & & \downarrow p_* \\ K_\circ X & \xrightarrow{\tau} & A_* X_{\mathbb{Q}} \end{array}$$

commutes. Thus there can be at most one way to extend from quasi-projective to general schemes, preserving the covariance property. This is a consequence of Lemma 18.3(6). If (*) commutes, we say τ is *compatible* with p.

Step 2. Suppose $\tau: K_\circ X \to A_* X_{\mathbb{Q}}$ is a homomorphism, compatible with a Chow envelope $p: X' \to X$. Then for any proper morphism $f: Y \to X$, with Y quasi-projective, the diagram

$$\begin{array}{ccc} K_\circ Y & \xrightarrow{\tau_Y} & A_* Y_{\mathbb{Q}} \\ f_* \downarrow & & \downarrow f_* \\ K_\circ X & \xrightarrow{\tau} & A_* X_{\mathbb{Q}} \end{array}$$

commutes. In particular, the homomorphism τ, if it exists, does not depend on the Chow envelope used to construct it. To see this, choose

$$\begin{array}{ccc} Y' & \xrightarrow{f'} & X' \\ {\scriptstyle q}\downarrow & & \downarrow{\scriptstyle p} \\ Y & \xrightarrow[f]{} & X \end{array}$$

as in Lemma 18.3(5). The covariance with respect to f then follows from the known covariance with respect to q and f' (§ 18.2), the assumed covariance with respect to p, and the surjectivity of q_*. For if $\alpha \in K_\circ Y$, let $\alpha = q_* \alpha'$, $\alpha' \in K_\circ Y'$; then

$$\begin{aligned} f_* \tau_Y(q_* \alpha') &= f_* q_* \tau_{Y'}(\alpha') = p_* f'_* \tau_{Y'}(\alpha') \\ &= p_* \tau_{X'}(f'_* \alpha') = \tau\, p_* f'_* \alpha' = \tau f_*(q_* \alpha') \,. \end{aligned}$$

Step 3. We prove the existence of τ by induction on the dimension of X. Given X, choose a Chow envelope $p: X' \to X$, with $Y \subset X$ as in Lemma 18.3(3). Form the fibre square

$$\begin{array}{ccc} Y' & \xrightarrow{j} & X' \\ {\scriptstyle q}\downarrow & & \downarrow{\scriptstyle p} \\ Y & \xrightarrow[i]{} & X \,. \end{array}$$

Since j is a closed imbedding, Y' is quasi-projective. By Noetherian induction, there is a homomorphism $\tau_Y: K_\circ Y \to A_* Y_{\mathbb{Q}}$, compatible with the Chow envelope $p: Y' \to Y$. Consider the diagram

$$\begin{array}{ccccccc} K_\circ Y' & \xrightarrow{a} & K_\circ Y \oplus K_\circ X' & \xrightarrow{b} & K_\circ X & \to & 0 \\ {\scriptstyle \tau_{Y'}}\downarrow & & \downarrow{\scriptstyle \tau_Y \oplus \tau_{X'}} & & & & \\ A_* Y'_{\mathbb{Q}} & \xrightarrow[a]{} & A_* Y_{\mathbb{Q}} \oplus A_* X'_{\mathbb{Q}} & \xrightarrow[b]{} & A_* X_{\mathbb{Q}} & \to & 0 \end{array}$$

where a and b are defined as in Proposition 18.3.2. By assumption, and the known covariance with respect to the morphism j of quasi-projective schemes, the square in this diagram commutes. By Proposition 18.3.2 the top row is exact; in fact, the bottom row is also exact (Example 1.8.1), but this is not needed. Therefore there is a unique homomorphism $\tau: K_\circ X \to A_* X_{\mathbb{Q}}$ making the diagram commute. In particular, τ is compatible with p.

Step 4. By steps 2 and 3, we have, for all X, a homomorphism

$$\tau_X: K_\circ X \to A_* X_{\mathbb{Q}}$$

compatible with any Chow envelope $X' \to X$. We prove the covariance property. If $f: Y \to X$ is proper, choose p, q, f' as in Lemma 18.3(5). The equality $f_* \tau_Y(\alpha) = \tau_X f_*(\alpha)$ follows exactly as in Step 2.

Step 5. For the module property, if $\beta \in K^\circ X$, $\alpha \in K_\circ X$, choose a Chow envelope $p: X' \to X$, and choose $\alpha' \in K_\circ X'$ with $p_* \alpha' = \alpha$. Then

$$\begin{aligned} \tau_X(\beta \otimes \alpha) &= \tau_X p_*(p^* \beta \otimes \alpha') = p_* \tau_{X'}(p^* \beta \otimes \alpha') \\ &= p_* \operatorname{ch}(p^* \beta) \cap \tau_{X'}(\alpha') = \operatorname{ch}(\beta) \cap p_* \tau_{X'}(\alpha') = \operatorname{ch}(\beta) \cap \tau_X(\alpha) \end{aligned}$$

by the known result for X' (Theorem 18.2), and the projection formula.

Step 6. To prove Property (3), let $i: X \to M$ be a closed imbedding, M smooth, and let $p: X' \to X$ be a Chow envelope. Let $j: X' \to U$ be a closed imbedding, U open in some projective space. Then $k=(ip,j)$ is a closed imbedding of X' in $M\times U$. Let $\alpha'\in K_\circ X'$, $\alpha=p_*\alpha'$, and set

$$\tilde{\tau}_{X'}(\alpha')=\mathrm{td}\,(k^*T_{M\times U})\cap \mathrm{ch}_{X'}^{M\times U}(\alpha')\,,$$
$$\tilde{\tau}_X(\alpha)=\mathrm{td}\,(i^*T_M)\cap \mathrm{ch}_X^M(\alpha)\,.$$

By Theorem 18.2, applied to the category of schemes which are quasi-projective over M,

$$p_*\tilde{\tau}_{X'}(\alpha')=\tilde{\tau}_X(p_*\alpha')=\tilde{\tau}_X(\alpha)\,.$$

(To be precise, Theorem 18.2 gives this equality before multiplying both sides by $\mathrm{td}\,(i^*T_M)$.) But by Proposition 18.3.1, $\tilde{\tau}_{X'}(\alpha')$ agrees with the class obtained by imbedding X' in U, i.e.

$$\tilde{\tau}_{X'}(\alpha')=\tau_{X'}(\alpha')\,.$$

Therefore $\tilde{\tau}_X=p_*\circ\tau_{X'}$, and by Step 1, $\tilde{\tau}_X=\tau_X$, as required.

Step 7. Let $f:X\to Y$ be a l.c.i. morphism as in Property (4). One obtains a commutative diagram

$$\begin{array}{ccccc} X & \overset{h}{\hookrightarrow} & M\times Y & \overset{k}{\hookrightarrow} & M\times P \\ & {\scriptstyle f}\searrow & \downarrow{\scriptstyle q} & & \downarrow{\scriptstyle p} \\ & & Y & \underset{j}{\hookrightarrow} & P\,. \end{array}$$

For the projection q, the proof of (4) is the same as in Step 4 of the proof of Theorem 18.2. For the regular imbedding h, (4) follows from Theorem 18.2(3), regarding X and $M\times Y$ as schemes quasi-projective over the smooth base $M\times P$.

Step 8. To prove Property (5), by covariance one may assume $V=X$, and by taking a Chow envelope, that X is quasi-projective. Since τ is compatible with restriction to open subvarieties, it suffices to prove it when X is projective. Choose a finite surjective morphism $f:X\to\mathbb{P}^n$ of degree d. Then $f_*[X]=d[\mathbb{P}^n]$, and

$$f_*[\mathscr{O}_X]=d[\mathscr{O}_{\mathbb{P}^n}]+\alpha\,,$$

$\alpha\in F_{n-1}K_\circ\mathbb{P}^n$. Then by covariance

$$f_*\tau_X(\mathscr{O}_X)=d\tau_{\mathbb{P}^n}(\mathscr{O}_{\mathbb{P}^n})+\tau_{\mathbb{P}^n}(\alpha)=d[\mathbb{P}^n]+\beta\,,$$

where β is a class of dimension $<n$; this uses the truth of (5) for $\mathbb{P}^n$, which follows from Property (3), and the covariance of τ with respect to inclusions of subvarieties of $\mathbb{P}^n$. If $m[X]$ is the top term of $\tau_X(\mathscr{O}_X)$, then

$$f_*\tau_X(\mathscr{O}_X)=dm[\mathbb{P}^n]+\beta'$$

with $\dim(\beta')<n$. Comparing, one has the required conclusion that $m=1$.

As for uniqueness, suppose τ' also satisfies the indicated properties. Since $\tau_{\mathbb{Q}}$ is an isomorphism (Corollary 18.3.2), one may consider the composite $T=\tau'_{\mathbb{Q}}\circ\tau_{\mathbb{Q}}^{-1}$. Then T is a transformation from $A_{*\mathbb{Q}}$ to itself, which is covariant for proper morphisms, contravariant for open imbeddings of quasi-projective schemes, and $T[\mathbb{P}^n]=[\mathbb{P}^n]+\beta_n$ for some β_n of dimension $<n$. Therefore by Example 1.9.6, T is the identity, as required.

Example 18.3.1. *Cartesian products.* Let X, Y be algebraic schemes. Then the diagram

$$\begin{array}{ccc} K_\circ X \otimes K_\circ Y & \xrightarrow{\tau_X \otimes \tau_Y} & A_* X_{\mathbb{Q}} \otimes A_* Y_{\mathbb{Q}} \\ \times \downarrow & & \downarrow \times \\ K_\circ (X \times Y) & \xrightarrow[\tau_{X \times Y}]{} & A_* (X \times Y)_{\mathbb{Q}} \end{array}$$

commutes. In particular,

$$\mathrm{Td}(X \times Y) = \mathrm{Td}(X) \times \mathrm{Td}(Y) .$$

(If X and Y are quasi-projective, this is proved in Baum-Fulton-MacPherson (1)III.2. The general case follows by taking Chow envelopes $X' \to X$, $Y' \to Y$. This is equivalent to the multiplicativity of the localized Chern character (Example 18.1.5).)

Example 18.3.2. *Uniqueness.* On the category of complete schemes, τ is uniquely determined by the covariance property (1), the module property (2), and the normalization that $\tau(\mathcal{O}_P) = [P]$ when $P = \mathrm{Spec}(K)$ is a point. (See Baum-Fulton-MacPherson (1)III.2.) On the category of all algebraic schemes, adding the property that τ is compatible with open imbeddings determines τ uniquely.

Example 18.3.3. If $\pi : \tilde{X} \to X$ is a proper birational morphism, isomorphic off $Z \subset X$, then

$$\mathrm{Td}(X) = \pi_* \mathrm{Td}(\tilde{X}) + \alpha$$

where α is a class supported in Z. In particular

$$\mathrm{Td}_k(X) = \pi_* \mathrm{Td}_k(\tilde{X}) \in A_k(X)_{\mathbb{Q}}$$

for $k > \dim Z$.

For example, if X is an n-dimensional variety with a finite number of singularities, and $\tilde{X}$ is a resolution of singularities,

$$\mathrm{Td}_k(X) = \pi_* (\mathrm{td}_{n-k}(T_{\tilde{X}}) \cap [\tilde{X}])$$

for $k > 0$, and $\mathrm{Td}_0(X) = \pi_* (\mathrm{td}_n(T_{\tilde{X}}) \cap [\tilde{X}]) + \sum n_P[P]$, with the sum over all singular points P in X, and

$$n_P = \sum_{i=1}^{n-1} (-1)^{i-1} l(R^i \pi_* \mathcal{O}_{\tilde{X}})_P - l(\pi_* \mathcal{O}_{\tilde{X}} / \mathcal{O}_X)_P$$

where $l(\mathcal{F})_P$ denotes the length of the stalk of the sheaf $\mathcal{F}$ at P. When X is normal the last term is zero.

Example 18.3.4. (a) If X is a projective curve, and $\pi : \tilde{X} \to X$ is a resolution of singularities, then

$$\mathrm{Td}(X) = [X] + \tfrac{1}{2} \pi_* (c_1(T_{\tilde{X}}) \cap [\tilde{X}]) - \sum \delta_P[P] ,$$

where $\delta_P = l(\pi_* \mathcal{O}_{\tilde{X}} / \mathcal{O}_X)_P$, the sum over singular points P of X. In particular, $\chi(X, \mathcal{O}_X) = \int_X \mathrm{Td}_0(X) = 1 - g(\tilde{X}) - \sum \delta_P$. If E is a vector bundle of rank r on X, then

$$\chi(X, E) = \int_X c_1(E) + r\, \chi(X, \mathcal{O}_X) .$$

(b) If X is a normal surface, and $\pi : \tilde{X} \to X$ a resolution of singularities, then

$$\operatorname{Td}(X) = [X] + \frac{1}{2}\pi_*(\tilde{c}_1) + \frac{1}{12}\pi_*(\tilde{c}_1^2 + \tilde{c}_2) + \sum n_P[P],$$

where $\tilde{c}_i = c_i(T_{\tilde{X}}) \cap [\tilde{X}]$, $n_P = l(R^1\pi_*\mathcal{O}_{\tilde{X}})_P$. In particular,

$$\chi(X, \mathcal{O}_X) = \frac{1}{12}\int_{\tilde{X}} \tilde{c}_1^2 + \tilde{c}_2 + \sum n_P.$$

If E is a vector bundle of rank r on X, then

$$\chi(X, E) = \tfrac{1}{2}\int_X c_1(E)^2 - 2c_2(E) + \tfrac{1}{2}\int_X c_1(E)\,\pi_*\tilde{c}_1 + r\,\chi(X, \mathcal{O}_X).$$

Example 18.3.5. Let $f: X \to Y$ be a regular imbedding of codimension d, with normal bundle N. Assume Y can be imbedded in a non-singular variety. Then

$$\operatorname{Td}(X) = \operatorname{td}(N)^{-1} \cap f^*\operatorname{Td}(Y).$$

If X is a complete intersection of hypersurfaces $D_1, \ldots, D_d$ on Y, then

$$\operatorname{td}(N)^{-1} = \prod_{i=1}^{d}(1 - \exp(-x_i)/x_i),$$

where $x_i = c_1(f^*\mathcal{O}_Y(D_i))$, and

$$\chi(X, \mathcal{O}_X) = \int_Y \prod_{i=1}^{d}(1 - \exp(-D_i)) \cap \operatorname{Td}(Y).$$

In particular, if X is a Cartier divisor on Y, then

$$\chi(X, \mathcal{O}_X) = \sum_{k \geqq 1}(-1)^{k-1}/k!\int_Y X^k \cdot \operatorname{Td}_k(Y).$$

When Y is smooth of dimension n, $\operatorname{Td}_k(Y) = \operatorname{td}_{n-k}(T_Y) \cap [Y]$.

There are singular projective varieties X such that $\operatorname{Td}(X)$ does not lift to cohomology. In Baum-Fulton-MacPherson (1)IV.6 a normal projective 3-fold X is constructed with one singular point, such that the image of $\operatorname{Td}_2(X)$ in $H_4(X; \mathbb{Q})$ cannot be written in the form $a \cap [X]$, for any $a \in H^2(X; \mathbb{Q})$.

Example 18.3.6. Let X be an n-dimensional complete scheme, $\mathcal{F}$ a coherent sheaf on X, $L_1, \ldots, L_r$ line bundles on X, with $x_i = c_1(L_i)$. Then

$$\begin{aligned}\chi(X, \mathcal{F} \otimes L_1^{\otimes d_1} \otimes \ldots \otimes L_r^{\otimes d_r}) &= \int_X e^{d_1x_1 + \ldots + d_rx_r} \cap \tau_X(\mathcal{F}) \\ &= \sum_{k=0}^{n}\frac{1}{k!}\int_X (d_1x_1 + \ldots + d_rx_r)^k \cap \tau_{X,k}(\mathcal{F})\end{aligned}$$

where $\tau_{X,k}(\mathcal{F})$ is the component of $\tau_X(\mathcal{F})$ in $A_kX_{\mathbb{Q}}$. This is a polynomial of degree n in $d_1, \ldots, d_r$ whose top degree term is

$$\frac{1}{n!}\int_X (d_1x_1 + \ldots + d_rx_r)^n \cap [X].$$

This gives a concrete form to polynomials considered by Snapper (1); for related results see Kleiman (1) and Mumford (6).

Example 18.3.7. (a) Let E be a vector bundle of rank n on a complete scheme X of dimension $\leqq n$. Then

$$\sum_{i=0}^{n} (-1)^i \chi(X, \Lambda^i E^\vee) = \int_X c_n(E) .$$

(By Hirzebruch-Riemann-Roch and Example 3.2.5,

$$\sum (-1)^i \chi(X, \Lambda^i E^\vee) = \int_X \operatorname{ch}(\sum (-1)^i \Lambda^i E^\vee) \cap \tau(X) = \int_X c_n(E) \operatorname{td}(E)^{-1} \cap \tau(X) .$$

The conclusion follows since $\tau(X)$ and $[X]$ agree in dimension n.)

(b) If E has a regular section s, with a finite zero scheme $Z(s)$, the equation in (a) reduces to

$$\chi(Z(s), \mathcal{O}_{Z(s)}) = \deg [Z(s)]$$

(cf. § 14.1). If $E \otimes L$ has a regular section, for some line bundle L – for example, if X is a projective variety – then, as R. Lazarsfeld points out, (a) can be deduced directly from (b).

(c) If X is a non-singular complete variety, then

$$\sum_{p,q=0}^{n} (-1)^{p+q} \dim H^p(X, \Omega_X^q) = \int_X c_n(T_X) ,$$

the topological Euler characteristic of X.

Example 18.3.8. *Flat families.* Let T be non-singular, $X \to T$ a quasi-projective morphism. For $t \in T$, let X_t be the fibre over t, and let

$$i_t^! : A_* X \to A_* X_t$$

be the specialization map (§ 10.1). Let $\mathcal{F}$ be a coherent sheaf on X which is flat over T. Then

$$i_t^! \tau_X(\mathcal{F}) = \tau_{X_t}(\mathcal{F}_t) ,$$

where $\mathcal{F}_t$ is the fibre of $\mathcal{F}$ over t. (Use Theorem 18.1 and the fact that the normal bundle to t in T is trivial). This generalizes (and reproves) the fact that $\chi(X_t, \mathcal{F}_t)$ is a locally constant function of t. In particular, if X is flat over T, then

$$i_t^* \operatorname{Td}(X) = \operatorname{Td}(X_t) .$$

This is a special case of the theorem discussed in Example 18.3.16.

Example 18.3.9 Let $f: X \to Y$ be a finite étale morphism. Then

$$\tau_X \circ f^* = f^* \circ \tau_Y .$$

(Taking a Chow envelope $Y' \to Y$, one is reduced to the case where Y and X are quasi-projective. This is special case of Theorem 18.2(3).) It follows that

$$f^* \operatorname{Td}(Y) = \operatorname{Td}(X) .$$

Therefore $f_* \operatorname{Td}(X) = \deg(f) \cdot \operatorname{Td}(Y)$. In particular, if X and Y are complete, then

$$\chi(X, \mathcal{O}_X) = \deg(f) \, \chi(Y, \mathcal{O}_Y) .$$

(S. Kleiman has given an elementary proof of the above equation for the arithmetic genus.)

Example 18.3.10. Let $f: X \to Y$ be a proper l.c.i. morphism. Assume that f factors into a regular closed imbedding $X \hookrightarrow P$ followed by a smooth proper morphism $P \to Y$. Then for all $\beta \in K^\circ X$,

$$\operatorname{ch}(f_* \beta) = f_*(\operatorname{td}(T_f) \cdot \operatorname{ch}(\beta))$$

in $A^* Y_{\mathbb{Q}}$. (It suffices to prove this when f is a regular imbedding or f is smooth. The former case is covered by Theorem 18.2. If f is smooth, we must show both sides have the same effect on $\alpha \in A_* Y'_{\mathbb{Q}}$, $Y' \to Y$, Y' quasi-projective. Write $\alpha = \tau_{Y'}(\gamma)$, $\gamma \in K_\circ Y'_{\mathbb{Q}}$. The desired equation then follows formally by applying Theorem 18.3(4) to the morphism $X \times_Y Y' \to Y'$.)

Example 18.3.11. Let $\mathscr{F}$ be a coherent sheaf on a scheme X with $\dim \operatorname{Supp}(\mathscr{F}) \leqq n$. Define the n-cycle of $\mathscr{F}$,

$$Z_n(\mathscr{F}) = \sum_{\dim V = n} l_V(\mathscr{F})[V]$$

where the sum is over all n-dimensional subvarieties V in the support of $\mathscr{F}$, and $l_V(\mathscr{F})$ is the length of the stalk of $\mathscr{F}$ over the local ring of X at V. Then

$$\tau_X(\mathscr{F}) = Z_n(\mathscr{F}) + \text{terms of dimension} < n\,.$$

If $F_n K_\circ X$ is the subgroup of $K_\circ X$ generated by sheaves whose support has dimension $\leqq n$, then assigning $Z_n(\mathscr{F})$ to $\mathscr{F}$ determines a well-defined homomorphism

$$\psi : F_n K_\circ X \to A_n X_{\mathbb{Q}}\,.$$

(In fact, mod $F_{n-1} K_\circ X$, ψ is the associated graded homomorphism of τ_X.) Since generators of $F_n K_\circ X$ may be related by sheaves of higher dimensions, this fact is not obvious from the definitions. Note that one must tensor by $\mathbb{Q}$ for this to be well-defined.

Example 18.3.12. *A Riemann-Roch formula.* Let $E.$ be a complex of vector bundles on a scheme Y, which is exact off a closed subscheme X. Let $\mathscr{F}$ be a coherent sheaf on Y. Then the homology sheaves $\mathscr{H}_i(E. \otimes \mathscr{F})$ of the complex $E. \otimes_{\mathscr{O}_Y} \mathscr{F}$ are supported on X, so define classes $[\mathscr{H}_i(E. \otimes \mathscr{F})]$ in $K_\circ X$. Then one has

$$(*) \qquad \sum (-1)^i \tau_X[\mathscr{H}_i(E. \otimes \mathscr{F})] = \operatorname{ch}_X^Y(E.) \cap \tau_Y(\mathscr{F})$$

in $A_* X_{\mathbb{Q}}$. If Y is smooth, $\mathscr{F} = \mathscr{O}_Y$, and $E.$ resolves a sheaf on X, this is Property (3) of Theorem 18.3. Property (4) may also be deduced from (*) (cf. Example 18.3.10). Conversely, with resolution of singularities one may deduce (*) from the Riemann-Roch theorem. Note that if $X = Y$ (*) follows from the module property (2). We sketch the general proof of (*).

For fixed $E.$, both sides of (*) are additive in $\mathscr{F}$, so define homomorphisms from $K_\circ X$ to $A_* X_{\mathbb{Q}}$. Let $\tau_X(\mathscr{H}(E. \otimes \alpha))$ denote the value of the left side on the element $\alpha \in K_\circ Y$. Then (*) is equivalent to

$$(**) \qquad \tau_X(\mathscr{H}(E. \otimes \alpha)) = \operatorname{ch}_X^Y(E.) \cap \tau_Y(\alpha)$$

for all $\alpha \in K_\circ Y$. The proof will be given in four steps.

1. Suppose $f: Y' \to Y$ is proper, $X' = f^{-1}(X)$, $g: X' \to X$ induced, and $\alpha' \in K_\circ Y'$. Suppose

$$\tau_{X'}(\mathscr{H}(f^* E. \otimes \alpha')) = \mathrm{ch}_{X'}^{Y'}(f^* E.) \cap \tau_{Y'}(\alpha')$$

in $A_* X'_{\mathbb{Q}}$. Then (**) holds for $\alpha = f_* \alpha'$. (Apply g_* to both sides, and use the covariance of τ. One also needs the formula

$$g_*(\mathscr{H}(f^* E. \otimes \alpha')) = \mathscr{H}(E. \otimes f_* \alpha')$$

coming from a spectral sequence $\mathscr{H}^p(E. \otimes R^q f_* \mathscr{F}') \Rightarrow R^{p+q} f_*(f^* E. \otimes \mathscr{F}')$.)

2. If $0 \to E'. \to E. \to E''. \to 0$ is an exact sequence of complexes of vector bundles on Y exact off X, and (**) holds for $E'.$ and $E''.$, then (**) holds for $E.$ (Proposition 18.1(b)).

3. Suppose Y is a quasi-projective variety, and $E.$ is an elementary complex, i.e. there is an n so that $E_i = 0$ for $i \neq n, n-1$, and E_n and E_{n-1} are line bundles. Suppose also that $\mathscr{F} = \mathscr{O}_Y$. Then (*) is true. (After translating and tensoring by a line bundle, one may assume $n = 1$, $E_0 = \mathscr{O}_Y$, $E_1 = \mathscr{O}(-D)$, D an effective divisor on Y, $E_1 \to E_0$ the natural inclusion. Then (*) follows from Theorem 18.2(3) and Corollary 18.1.2.)

4. For the general proof, by Step 1 and Chow's lemma, we may assume Y is a quasi-projective variety. We induct on the sum of the ranks of the bundles E_i; let $E.$ be the complex

$$0 \to E_n \xrightarrow{d_n} E_{n-1} \to \dots \to E_0 \to 0$$

with $E_n \neq 0$. We may assume $X \neq Y$. Since the projection maps $K_\circ(P(E_n))$ onto $K_\circ Y$, by Step 1 we may assume E_n contains a sub-line-bundle L. In this situation, we induct on the dimension of the support of $\mathscr{F}$. We may assume $\mathscr{F} = \mathscr{O}_Y$. By Steps 1 and 2 and the inductive assumptions, it suffices to find a proper birational $f: Y' \to Y$ so that $f^* E.$ contains an elementary subcomplex $E'.$, exact off $f^{-1}(X)$.

Let $\mathbb{P} = P(E_{n-1})$, $p: \mathbb{P} \to Y$ the projection, $\mathscr{O}(-1)$ and Q the universal sub and quotient bundles of $p^* E_{n-1}$. Let $\mathbb{P}' \subset \mathbb{P}$ be the scheme-theoretic intersection of the loci where the bundle maps

$$p^* L \xrightarrow{p^* d_n} p^* E_{n-1} \to Q \quad \text{and} \quad \mathscr{O}(-1) \to p^* E_{n-1} \to p^* E_{n-2}$$

vanish. The restriction of $p^* E.$ to $\mathbb{P}'$ has an elementary subcomplex $F.$, with F_n and F_{n-1} the restrictions of $p^* L$ and $\mathscr{O}(-1)$ to $\mathbb{P}'$. On $Y - X$, d_n is injective, so $d_n(L(y))$ is a line in $E_{n-1}(y)$ for $y \in Y - X$. This determines a section $s: Y - X \to \mathbb{P}'$ such that $s^* F.$ is exact. The closure Y' of $s(Y - X)$ in $\mathbb{P}'$ is the required variety, $f: Y' \to Y$ is induced by projection, and $E'. = F. | Y'$.

Example 18.3.13. (a) Let $f: X \to Y$ be a regular imbedding of codimension d, with normal bundle N, and assume Y can be imbedded in a smooth scheme. Then for any coherent sheaf $\mathscr{F}$ on Y,

$$\sum (-1)^i \tau_X(\mathrm{Tor}_i^Y(\mathscr{O}_X, \mathscr{F})) = \mathrm{td}(N)^{-1} \cap f^* \tau_Y(\mathscr{F}).$$

If $\dim \mathrm{Supp}(\mathscr{F}) = n$, equating top terms gives

$$\sum (-1)^i Z_{n-d}(\mathrm{Tor}_i^Y(\mathscr{O}_X, \mathscr{F})) = f^* Z_n(\mathscr{F}).$$

(Apply Theorem 18.3(4) or Example 18.3.12.)

(b) Let X be a non-singular variety, let V, W be closed subvarieties of X. Then

$$[V]\cdot[W]=\sum(-1)^i Z_m(\mathrm{Tor}_i^X(\mathcal{O}_V,\mathcal{O}_W))$$

in $A_m(V\cap W)_{\mathbb{Q}}$, where $m=\dim V+\dim W-\dim X$. (Apply (a) to the diagonal imbedding of X in $X\times X$, $\mathcal{F}=\mathcal{O}_{V\times W}$.)

For a discussion of the role of Tor in defining intersection products, see § 20.4.

Example 18.3.14. Suppose $E.$ is a complex of vector bundles on a scheme Y, exact off a *point* X. For a coherent sheaf $\mathcal{F}$ on Y,

$$\mathcal{H}(E.\otimes\mathcal{F})=\sum(-1)^i\,\mathrm{length}\,(\mathcal{H}_i(E.\otimes\mathcal{F}))$$

in $K_\circ X=\mathbb{Z}$. Let $\mathrm{ch}_i(E.)\in A^i(X\to Y)_{\mathbb{Q}}$ denote the term of degree i of $\mathrm{ch}_X^Y(E.)$. Let $n=\dim Y$, and assume that

$$(*)\qquad\qquad \mathrm{ch}_i(E.)=0\quad\text{ for }\quad i<n\,.$$

Let $E.^\vee$ be the dual complex of vector bundles. Then

(a) $\mathcal{H}(E.\otimes\mathcal{F})\ =0$ if $\dim(\mathrm{Supp}\,\mathcal{F})<n$

(b) $\mathcal{H}(E.^\vee\otimes\mathcal{F})=(-1)^n\mathcal{H}(E.\otimes\mathcal{F})$.

(Use Examples 18.3.12 and 18.1.2.) The condition (*) holds whenever $E.$ is the pull-back of some complex of vector bundles on a non-singular variety. However, Dutta, Hochster and McLaughlin (1) have recently produced a complex $E.$ on a three-dimensional variety, resolving a module of finite length, so that $\mathcal{H}(E.\otimes\mathcal{F})$ takes on both positive and negative values, for $\mathcal{F}$ the structure sheaf of appropriate surfaces in Y; thus $\mathrm{ch}_2(E.)\neq 0$. For a discussion of relations between Riemann-Roch and local algebra, see Szpiro (1).

Example 18.3.15. Let $f\colon X\to Y$ be a closed imbedding such that $\mathcal{O}_X$ can be resolved by a complex of vector bundles on Y. Therefore f is perfect (cf. Example 15.1.8), and $\mathcal{H}(E.\otimes\alpha)=f^*\alpha$, where $f^*: K_\circ Y\to K_\circ X$ is the Gysin map. In this case Example 18.3.12 reads

$$\tau_X(f^*\alpha)=\mathrm{ch}_X^Y(E.)\cap\tau_Y(\alpha)\,.$$

In particular, if f is a regular imbedding, then (Corollary 18.1.2)

$$\tau_X(f^*\alpha)=\ \mathrm{td}\,(N)^{-1}f^*\tau_Y(\alpha)\,,$$

with N the normal bundle to X in Y. This reproves Theorem 18.3(4). (Gillet reports that, using twisted complexes as in Toledo and Tong (1), the existence of global resolutions $E.$ is not needed for this conclusion.)

Example 18.3.16. *Bivariant Riemann-Roch.* All schemes are assumed to be quasi-projective over a fixed non-singular base S. For a morphism $f\colon X\to Y$, define

$$K(X\xrightarrow{f}Y)$$

to be the Grothendieck group of *f-perfect* complexes on X; if one factors f into a closed imbedding $i: X\to M$ followed by a smooth $p: M\to Y$, a complex $\mathcal{A}^\bullet$ of sheaves on X is f-perfect if $i_*(\mathcal{A}^\bullet)$ is quasi-isomorphic to a bounded complex

$E^\bullet$ of locally free sheaves, called a resolution of $\mathscr{A}^\bullet$ on M. There is a natural notion of push-forward (for proper morphisms), pull-back (for Tor-independent squares), and product, making K into a bivariant theory (cf. Fulton-MacPherson (3)). If f is factored as above, then $K(X \xrightarrow{f} Y) = K(X \xrightarrow{i} M)$.

In particular, if Y is smooth, $K(X \to Y) = K_\circ X$, while if $Y = X$, $f = \text{id}$, $K(X \to X) = K^\circ X$.

Define a homomorphism

$$\tau_f \colon K(X \xrightarrow{f} Y) \to A(X \xrightarrow{f} Y)_{\mathbb{Q}}$$

as follows. Factor f as above. Given an f-perfect complex $\mathscr{A}^\bullet$ on X, resolve it by $E^\bullet$ on M, and set

$$\tau_f[\mathscr{A}^\bullet] = \text{td}(i^* T_{M/Y}) \cdot \text{ch}_X^M(E^\bullet).$$

where $\text{ch}_X^M(E^\bullet) \in A(X \to M)_{\mathbb{Q}} = A(X \to Y)_{\mathbb{Q}}$.

Theorem. *The homomorphism τ is independent of factorization, and compatible with push-forward, pull-back, and product.*

(Except for the compatibility with products, the proof is much the same as for Theorem 18.2. For products, the essential point is the following. Let

$$X \xrightarrow{f} Y \xrightarrow{g} Z$$

be closed imbeddings, $\mathscr{A}^\bullet$ f-perfect, $\mathscr{B}^\bullet$ g-perfect. Let $F^\bullet$ resolve $\mathscr{A}^\bullet$ on Y; let $G^\bullet$ resolve $\mathscr{B}^\bullet$ on Z. Assume $\mathscr{B}^\bullet$ is a complex of locally free sheaves on Y (not necessarily bounded above). Then the product $[\mathscr{A}^\bullet] \cdot [\mathscr{B}^\bullet]$ is represented by the complex $\mathscr{A}^\bullet \otimes_{\mathscr{O}_X} f^* \mathscr{B}^\bullet$. Let $E^\bullet$ be a resolution of $\mathscr{A}^\bullet \otimes_{\mathscr{O}_X} f^* \mathscr{B}^\bullet$ on Z. Compatibility of τ with products asserts that for all $Z' \to Z$, $\alpha \in A_* Z'_{\mathbb{Q}}$,

$$(*) \qquad \text{ch}_X^Z(E^\bullet) \cap \alpha = \text{ch}_X^Y(F^\bullet) \cap (\text{ch}_Y^Z(G^\bullet) \cap \alpha).$$

At first glance, this appears to be difficult, since we do not know how $E^\bullet$ relates *algebraically* to $F^\bullet$ and $G^\bullet$; the essence of Fulton-MacPherson (3)PRR was to prove such a *topological* relation, for complex varieties. There is a simple subterfuge, however. An element in $K(X \xrightarrow{f} Y)$ determines a homomorphism from $K_\circ Y'$ to $K_\circ(X \times_Y Y')$ for any $h: Y' \to Y$: if $\mathscr{A}^\bullet$ is an f-perfect complex on X, and $\mathscr{F}$ is a coherent sheaf on Y', then $[\mathscr{A}^\bullet]$ takes $[\mathscr{F}]$ to

$$[\mathscr{A}^\bullet](\mathscr{F}) = \sum (-1)^i [\text{Tor}_i^Y(\mathscr{A}^\bullet, \mathscr{F})].$$

If f is a closed imbedding, and $E.$ a resolution of $\mathscr{A}^\bullet$ on Y, then

$$[\mathscr{A}^\bullet](\mathscr{F}) = \sum (-1)^i [\mathscr{H}^i(g^* E^\bullet \otimes \mathscr{F})].$$

It follows from Example 18.3.12 that

$$\tau_{X'}([\mathscr{A}^\bullet](\mathscr{F})) = \tau_f[\mathscr{A}^\bullet] \cap \tau_Y(\mathscr{F}).$$

The analogue of (*) for $\mathscr{F}$ in $K_\circ Z'$, $Z' \to Z$:

$$[\mathscr{A}^\bullet \otimes f^* \mathscr{B}^\bullet](\mathscr{F}) = [\mathscr{A}^\bullet]([\mathscr{B}^\bullet](\mathscr{F}))$$

follows from the Tor spectral sequence. Then (*) follows, since $\tau_{Z'} \otimes \mathbb{Q}$ is surjective.)

Example 18.3.17. Let $f: X \to Y$ be a perfect morphism of pure-dimensional quasi-projective schemes, and let $d = \dim Y - \dim X$. Define

$$[f] \in A^d(X \xrightarrow{f} Y)_{\mathbb{Q}}$$

to be the term of degree d in $\tau_f(\mathcal{O}_X)$, with τ_f as in the preceding example.

(a) If f is flat, l.c.i., or Y is smooth, then $[f]$ is the image of the integral class in $A^d(X \to Y)$ defined in § 17.4. The terms of degree less than d in $\tau_f(\mathcal{O}_X)$ are all zero.

(b) If $f: X \to Y$, $g: Y \to Z$ are morphisms such that f and g can each be factored into a sequence of morphisms which are flat, l.c.i., or morphisms to smooth schemes. Then

$$[g f] = [f] \cdot [g] \in A(X \to Z)_{\mathbb{Q}}.$$

It is not known if this holds with integer coefficients (cf. Example 17.4.6). For a general perfect morphism f, it is not known if the terms of degree less than d in $\tau_f(\mathcal{O}_X)$ vanish, or if the term of degree d comes from a class with integer coefficients. For f a closed imbedding, and $E.$ a resolution of $\mathcal{O}_X$ on Y, this asks for vanishing and integrality for the local classes $\mathrm{ch}_i(E.)$, which seems doubtful, cf. Example 18.3.14.

Example 18.3.18. For complex algebraic schemes, there are homomorphisms

$$K_\circ(X) \to K_\circ^{\mathrm{top}}(X)$$

where $K_\circ^{\mathrm{top}}$ denotes the homology theory (with locally closed supports) of topological K-theory. This homomorphism satisfies covariance, module and other properties analogous to those in Theorem 18.3; if X is non-singular, the image of $\mathcal{O}_X$ is the orientation class of X. If

$$\mathrm{ch}: K_\circ^{\mathrm{top}} X \to H_*(X; \mathbb{Q})$$

is the Chern character in homology, the composite

$$K_\circ X \to H_*(X; \mathbb{Q})$$

satisfies the same properties and formulas as τ. (In the quasi-projective case, the map is constructed in Baum-Fulton-MacPherson (2). The extension to non-projective schemes proceeds as in this section.)

Example 18.3.19. *Riemann-Roch and Grothendieck duality.* Let $\mathscr{A}^\bullet$ be a bounded complex of sheaves on a scheme X such that the cohomology sheaves $\mathscr{H}^i(\mathscr{A}^\bullet)$ are coherent. Assume that X can be imbedded in a smooth scheme. Let

$$D(\mathscr{A}^\bullet) = R\,Hom(\mathscr{A}^\bullet, \omega_X^\bullet)$$

with $\omega_X^\bullet$ a dualizing complex on X (cf. Hartshorne (2)). Then for all k,

$$\sum (-1)^i \tau_k(\mathscr{H}^i(\mathscr{A}^\bullet)) = (-1)^k \sum (-1)^i \tau_k(\mathscr{H}^i(D\mathscr{A}^\bullet))$$

in $A_k(X)_{\mathbb{Q}}$. If X is Cohen-Macaulay, with dualizing sheaf ω_X, and E is locally free, then

$$\tau_k(E \otimes \mathcal{O}_X) = (-1)^{n-k} \tau_k(E^\vee \otimes \omega_X),$$

$n = \dim X$. When $k = 0$, this yields

$$\chi(X, E) = (-1)^n \chi(X, E^\vee \otimes \omega_X),$$

which also follows from Serre-Grothendieck duality. (If $X \subset M$, M smooth, and $E^\bullet$ resolves $\mathscr{A}^\bullet$ on M, then $E^{\bullet\vee}$ resolves $D(\mathscr{A}^\bullet)$, cf. Fulton-MacPherson (3)§ 7.)

Example 18.3.20. For an algebraic scheme X, let $K_\circ^c X$ be the Grothendieck group of sheaves on X whose supports are complete. There is a homomorphism

$$\tau : K_\circ^c X \to A_*^c X_{\mathbb{Q}},$$

with $A_*^c X$ as in Example 10.2.8, covariant for arbitrary morphisms, and satisfying analogues of the properties of Theorem 18.3. In particular, τ induces isomorphisms $K_\circ^c X_{\mathbb{Q}} \cong A_*^c X_{\mathbb{Q}}$.

Notes and References

The localized Chern character of § 18.1 was developed in Baum-Fulton-MacPherson (1), modelled on an earlier graph construction of MacPherson (cf. Example 18.1.6). The fact that $\mathrm{ch}_X^Y(E.)$ determines an element in the bivariant group $A(X \to Y)_{\mathbb{Q}}$ is new here. This strengthens the properties proved in (*loc. cit.*) as well as simplifying the proofs; in addition, no separate argument is needed for base fields which are not algebraically closed. A topological construction of local Chern classes, with values in $H_X^* Y = H^*(Y, Y - X)$, has been given by Iversen (3). An important precedent had been given by Atiyah and Hirzebruch (2), (3). A version of Riemann-Roch for singular curves may be found in Serre (5), and for normal surfaces, in Zariski (3).

The Riemann-Roch theorem of § 18.2 was first given in Baum-Fulton-MacPherson (1), with the exception of Theorem 18.2(3). The first part of (3) is analogous to the RR theorem for l.c.i. morphisms proved earlier in [SGA 6]. In the SGA 6 version, values were taken in graded K-groups. Using deformation to the normal bundle, the proof of the SGA 6 theorem may also be simplified as in Chap. 15. The second formula of Theorem 18.2(3) was conjectured in Baum-Fulton-MacPherson (1), and proved by Verdier (5). The proofs in § 18.2 follow these sources, with modifications for use in the non-projective case.

The extension to arbitrary algebraic schemes (§ 18.3) follows Fulton-Gillet (1).

In the complex case one also wants a Riemann-Roch theorem to give a transformation from algebraic geometry to topology. The most natural such transformation is from algebraic K-theory to topological K-theory. Such was constructed in the non-singular case in Atiyah-Hirzebruch (3), and in the quasi-projective case in Baum-Fulton-MacPherson (2). Using the methods of § 18.3, this extends to arbitrary algebraic $\mathbb{C}$-schemes (cf. Example 18.3.18).

Composing with the homology Chern character gives a homomorphism

$$K_\circ X \to H_*(X; \mathbb{Q})$$

satisfying the properties of Theorem 18.3.

For arbitrary complex analytic spaces such homomorphisms have not yet been constructed. For complex manifolds X, Obrian, Toledo and Tong (3) have constructed a homomorphism with values in $H^*(X, \Omega_X^*)$, and proved the GRR formula in this context.

The Riemann-Roch formula of Example 18.3.12 has not appeared before, although special cases were known from the singular Riemann-Roch theorem. This was motivated by joint speculation with MacPherson, Peskine, and Szpiro on the relations of Riemann-Roch with conjectures in local algebra. The note of Peskine-Szpiro (2) provided initial evidence that such relations exist.

Chapter 19. Algebraic, Homological, and Numerical Equivalence

Summary

Each k-dimensional complex variety V has a cycle class $cl(V)$ in $H_{2k}V$, where H_* denotes homology with locally finite supports (Borel-Moore homology). If V is a subvariety of an n-dimensional complex manifold X, then $H_{2k}(V) \cong H^{2n-2k}(X, X-V)$. The resulting homomorphism from cycles to homology passes to algebraic equivalence. There results in particular a cycle map

$$cl: A_*X \to H_*X$$

for complex schemes X, which is covariant for proper morphisms, and compatible with Chern classes of vector bundles.

If V and W are subvarieties of dimensions k and l of a non-singular n-dimensional variety X, a refined topological intersection product $cl(V) \cdot cl(W)$ is constructed in $H_{2m}(V \cap W)$, $m = k + l - n$. If $cl^X V$ is the class in $H^{2n-2k}(X, X-V)$ dual to $cl(V)$, and similarly for $cl^X(W)$, then $cl(V) \cdot cl(W)$ is defined to be the class dual to

$$cl^X(V) \cup cl^X(W) \in H^{2n-2m}(X, X - V \cap W).$$

We show that the cycle map takes the refined intersection $V \cdot W \in A_m(V \cap W)$ of Chap. 8 to the class $cl(V) \cdot cl(W)$. In particular, cl is a ring homomorphism from A^*X to H^*X. More generally, if $i: X \to Y$ is a regular imbedding of codimension d, the cycle classes of the refined products $i^!(\alpha)$ of Chap. 6 are given by cap product with an orientation class $u_{X,Y}$ in $H^{2d}(Y, Y-X)$.

In the final section we discuss what is known about algebraic, homological, and numerical equivalence on non-singular projective varieties. Only a few salient facts are mentioned which relate most directly to other chapters, and few proofs are included. Together with the examples, this may serve as an introduction to the literature on the transcendental theory of algebraic cycles.

Notation. Unless otherwise stated, all schemes in this chapter are assumed to be complex algebraic schemes which admit a closed imbedding into some non-singular complex variety. All topological spaces will be locally compact Hausdorff spaces which admit a closed imbedding into some Euclidean space. As in preceding chapters, a k-cycle on X is a formal sum of algebraic subvarieties of X.

19.1 Cycle Map

Even when one is primarily interested in projective varieties, it is useful, as we have seen, to have rational equivalence groups defined for varieties which may not be complete, so that every variety has a fundamental class, with restriction homomorphisms for open imbeddings, etc. The analogous homology theory for locally compact topological spaces is the Borel-Moore homology, which we denote by H_*. Geometrically, Borel-Moore homology may be thought of as constructed from possibly infinite singular chains with locally finite support. We sketch here the main features of Borel-Moore homology needed for present purposes; for details, see Example 19.1.1 and the references cited there.

Borel-Moore homology can be defined using singular cohomology. If a space X is imbedded as a closed subspace of $\mathbb{R}^n$, then

$$H_i X \cong H^{n-i}(\mathbb{R}^n, \mathbb{R}^n - X) \tag{1}$$

where the group on the right is relative singular cohomology with integer coefficients. More generally, if X is a closed subspace of a space Y, there are *cap products*

$$H^j(Y, Y-X) \otimes H_k Y \xrightarrow{\cap} H_{k-j} X. \tag{2}$$

If Y is an oriented, connected, real n-manifold, then $H_n Y$ is freely generated by a *fundamental class* μ_Y, and capping with μ_Y determines an isomorphism

$$H^{n-i}(Y, Y-X) \underset{\cong}{\xrightarrow{\cap \mu_Y}} H_i X. \tag{3}$$

When $Y = \mathbb{R}^n$, this is the isomorphism of (1). When $X = Y$, (3) is the Poincaré duality isomorphism of $H^{n-i} Y$ with $H_i Y$.

If $f: X \to Y$ is a proper morphism, there are covariant homomorphisms

$$f_*: H_i X \to H_i Y. \tag{4}$$

If $j: U \to Y$ is an open imbedding, there are contravariant restriction homomorphisms

$$j^*: H_i Y \to H_i U. \tag{5}$$

If X is the complement of U in Y, $i: X \to Y$ the closed imbedding, there is a long exact sequence

$$\ldots \to H_{i+1} U \to H_i X \xrightarrow{i_*} H_i Y \xrightarrow{j^*} H_i U \to H_{i-1} X \to \ldots. \tag{6}$$

If X is a disjoint union of a finite number of spaces $X_1, \ldots, X_m$, then

$$H_i X = H_i X_1 \oplus \ldots \oplus H_i X_m. \tag{7}$$

There are evident compatibilities among these homomorphisms, which will be used without comment. One such is the equation

$$(u \cup v) \cap a = j^*(u) \cap (v \cap a) \tag{8}$$

in $H_*(X \cap W)$, where $X \hookrightarrow Y$, $j: W \to Y$ are closed imbeddings, $u \in H^*(Y, Y-X)$, $v \in H^*(Y, Y-W)$, $a \in H_* Y$.

Lemma 19.1.1. *If X is an n-dimensional complex scheme, then $H_i X = 0$ for $i > 2n$, and $H_{2n} X$ is a free abelian group with one generator for each irreducible component* of X.

Proof. Let S be the singular locus of X, $U = X - S$. For U this follows from (7) and the Poincaré isomorphism (3), since a complex manifold has a canonical orientation. Assuming the assertion for S by induction on the dimension, one concludes by the exact sequence (6). □

The generator of $H_{2n} X$ corresponding to an n-dimensional irreducible component X_i will be denoted $cl(X_i)$; it is determined by the fact that it restricts to the fundamental class μ_{U_i} for any connected open subset U_i of the non-singular locus of X_i. More generally, if V is a k-dimensional closed subvariety of a scheme X, define the *cycle class* $cl_X(V)$ by

$$cl_X(V) = i_* \, cl(V) \in H_{2k} X,$$

where i is the inclusion of V in X. When no confusion is likely, we may write $cl(V)$ in place of $cl_X(V)$.

Lemma 19.1.2. *Let $f: V \to W$ be a proper, surjective morphism of varieties. Then*

$$f_* \, cl(V) = \deg(V/W) \cdot cl(W).$$

Proof. Let $n = \dim V$. If $\dim W < n$, the lemma is obvious since $H_{2n} W = 0$. If $\dim W = n$, let U be a small open topological ball contained in the smooth part of W, such that $f^{-1}(U)$ is the disjoint union of $d = \deg(V/W)$ balls $U_1, \ldots, U_d$, each of which is mapped homeomorphically by f to U. The lemma follows from the commutativity of the diagram

$$\begin{array}{ccc} H_{2n} V & \xrightarrow{\text{res}} & H_{2n}(f^{-1} U) = \oplus H_{2n} U_i \\ {\scriptstyle f_*}\downarrow & & \downarrow{\scriptstyle (f|U)_*} \\ H_{2n} W & \xrightarrow{\text{res}} & H_{2n} U \end{array}$$

and the fact that the restrictions $H_{2n} W \to H_{2n} U$ and $H_{2n} V \to H_{2n} U_i$ are isomorphisms, as are the induced maps from $H_{2n} U_i$ to $H_{2n} U$. □

For any X, there is a homomorphism denoted cl, or cl_X,

$$cl: Z_k X \to H_{2k} X$$

from the algebraic k-cycles to the Borel-Moore homology, which takes $\sum n_i [V_i]$ to $\sum n_i \, cl_X(V_i)$. It follows from Lemma 19.1.2 and the definitions in Chap. 1 that cl commutes with push-forward for proper morphisms, and with restriction to open subschemes.

If $M \subset P$ is a closed imbedding of complex manifolds of codimension d, there is a canonical *orientation class* $u_{M,P}$,

$$u_{M,P} \in H^{2d}(P, P - M).$$

In case P is a complex vector bundle over M with M imbedded in P by the zero section, $u_{M,P}$ is the Thom class of the bundle. If $f: P' \to P$ is transversal to M, and $M' = f^{-1} M$, then $f^* u_{M,P} = u_{M',P'}$. In particular, if N is the normal bundle to M in P, and $f: N \to P$ is a tubular neighborhood, $u_{M,P}$ is determined by the fact that $f^* u_{M,P}$ is the Thom class of N. If M and P are oriented, then

(9) $$u_{M,P} \cap \mu_P = \mu_M .$$

Lemma 19.1.3. *Let X be an n-dimensional variety, T a non-singular curve, $f: X \to T$ a surjective morphism, $t \in T$, $X_t = f^{-1}(t)$. Then*

$$f^*(u_{t,T}) \cap c\,l_X(X) = c\,l_{X_t}[X_t]$$

in $H_{2n-2}(X_t)$.

Proof. Since $c\,l$ and specialization commute with proper push-forward (Proposition 10.1), we may replace X by its normalization, i.e., we may assume X is normal. It suffices to show that both sides of the required equation restrict to the same class in $H_{2n-2}(U \cap X_t)$, where U is any open subset of X which meets X_t inside a given irreducible component of X_t. Thus we may assume T is the unit disc in $\mathbb{C}$, $t = 0$, and that X is non-singular, with $f = g^m$, where $g = 0$ defines a non-singular hypersurface $Y \subset X$. Consider the commutative diagram

$$\begin{array}{ccc} & \overset{g}{\nearrow} & (T, T - \{0\}) \\ (X, X - X_t) & & \downarrow \eta \\ & \underset{f}{\searrow} & (T, T - \{0\}) \end{array}$$

where $\eta(z) = z^m$. Let $u = u_{t,T}$, the canonical generator of $H^2(T, T - \{0\})$. Then $\eta^* u = m \cdot u$, and

$$\begin{aligned} f^*(u) \cap c\,l(X) &= m \cdot g^*(u) \cap c\,l(X) = m \cdot u_{Y,X} \cap \mu_X \\ &= m \cdot \mu_Y = m \cdot c\,l(Y) = c\,l[X_t]. \quad \square \end{aligned}$$

Proposition 19.1.1. *If a cycle α on a complex scheme X is algebraically equivalent to zero, then $c\,l(\alpha) = 0$ in $H_* X$.*

Proof. It suffices to show that

$$c\,l[V_{t_0}] = c\,l[V_{t_1}]$$

in $H_* V$ where V is a variety, $f: V \to T$ dominant morphism to a non-singular curve T, $t_0, t_1 \in T$ (cf. Example 10.3.2). By Lemma 19.1.3,

$$c\,l(V_t) = f^*(\bar{u}_{t,T}) \cap c\,l(V)$$

in $H_* V$, where $\bar{u}_{t,T}$ is the image of $u_{t,T}$ in $H^2 T$. Since T is connected, $\bar{u}_{t,T}$ is independent of t. Therefore $c\,l(V_t)$ is independent of t, as required. $\square$

It follows that the cycle map passes to algebraic equivalence, defining homomorphisms

$$c\,l: B_k X = Z_k X / \mathrm{Alg}_k X \to H_{2k} X$$

which are covariant for proper morphisms. The composite

$$A_* X \to B_* X \to H_* X$$

is also called the cycle map, and denoted $c\,l$ or $c\,l_X$.

A complex vector bundle E on a space X has Chern classes $c_i(E) \in H^{2i}(X)$, satisfying formulas analogous to those proved in § 3.2.

Proposition 19.1.2. *If E is an algebraic vector bundle on a scheme X, α a k-cycle on X, then*

$$cl(c_i(E) \cap \alpha) = c_i(E) \cap cl(\alpha)$$

in $H_{2k-2i}X$, for all i.

Proof. By the projection formula, it suffices to find a proper morphism $f: X' \to X$ and $\alpha' \in A_* X'$ with $f_* \alpha' = \alpha$, so that the formula holds for the bundle $f^* E$ and the cycle α' on α. Taking $X' \to X$ to be a composite of projective bundles (cf. Example 3.3.3), one may assume E is filtered, with line bundle quotients. By the Whitney formula, one may therefore assume E is a line bundle, and $i = 1$. As usual, we may assume $\alpha = [V]$, $V = X$ a variety, $E = \mathcal{O}(D)$, D a Cartier divisor. Blowing up denominators as in Case 3 of the proof of Theorem 2.4, and using the additivity of the first Chern class, we may assume D is effective. Normalizing, X may be assumed to be normal. Let s be the section of E whose zero is D. If u is the Thom class of E in $H^2(E, E - \{0\})$, then $s^*(u) \in H^2(X, X - D)$ is a localized first Chern class of E, i.e., the image of $s^*(u)$ in $H^2 X$ is $c_1(E)$. The same argument as in Lemma 19.1.3 shows that

$$s^*(u) \cap cl(X) = cl[D]$$

in $H_*(D)$. Since $[D] = c_1(E) \cap [X]$, the conclusion follows. □

Definition 19.1. Define $\mathrm{Hom}_k X$ to be the group of k-cycles whose homology class is zero, i.e.

$$\mathrm{Hom}_k X = \mathrm{Ker}(cl: Z_k X \to H_{2k} X) .$$

Denote by $\mathrm{Hom}_k^{\mathbb{Q}} X$ the cycles for which some non-trivial multiple is in $\mathrm{Hom}_k X$, i.e., $\mathrm{Hom}_k^{\mathbb{Q}} X$ is the kernel of the cycle map from $Z_k X$ to $H_{2k}(X; \mathbb{Q})$. Two cycles are *homologous* if their difference is in $\mathrm{Hom}_*^{\mathbb{Q}} X$ (or in $\mathrm{Hom}_* X$, according to some authors).

For any complete scheme X, we say that a k-cycle α is *numerically equivalent* to *zero* if

$$\int_X P \cap \alpha = 0$$

for all polynomials P in Chern classes of vector bundles on X. When X is non-singular, this is equivalent to requiring $\int_X \beta \cdot \alpha = 0$ for all $\beta \in A^k X$ (cf. Example 19.1.5). Let $\mathrm{Num}_k X$ be the group of k-cycles numerically equivalent to zero on X, and let $N_k X = Z_k X / \mathrm{Num}_k X$.

Let $\mathrm{Alg}_*^{\mathbb{Q}} X$ (resp. $\mathrm{Rat}_*^{\mathbb{Q}} X$) be the group of cycles some non-zero multiple of which is in $\mathrm{Alg}_* X$ (resp. $\mathrm{Rat}_* X$). By the results of this section, for any complete complex scheme X, there are inclusions

$$\begin{array}{ccccccccc} \mathrm{Rat}_* X & \subset & \mathrm{Alg}_* X & \subset & \mathrm{Hom}_* X & & & & \\ \cap & & \cap & & \cap & & & & \\ \mathrm{Rat}_*^{\mathbb{Q}} X & \subset & \mathrm{Alg}_*^{\mathbb{Q}} X & \subset & \mathrm{Hom}_*^{\mathbb{Q}} X & \subset & \mathrm{Num}_* X & \subset & Z_* X . \end{array}$$

Example 19.1.1. *On Borel-Moore homology.* For a space X which admits a closed imbedding into a Euclidean space $\mathbb{R}^n$, one may define the Borel-Moore homology groups $H_i(X)$ by

$$H_i(X) = H^{n-i}(\mathbb{R}^n, \mathbb{R}^n - X)$$

where the groups on the right are relative singular, or Čech (cf. Dold (1) VIII.6.12) cohomology groups. Most of the basic properties needed in § 19.1 were verified more generally in Fulton-MacPherson (3), which we cite here as [BT]. For example the fact that this definition is independent of the imbedding is proved in [BT] 3.1.5, the covariance for proper maps in [BT] 3.1.8, the cap product (equation (2) of § 19.1) in [BT] 3.1.7, the duality isomorphism (equation (3)) in [BT] 4.1.3.

If $i: X \to Y$ is a closed imbedding, $j: U \to Y$ the complementary open set, the long exact sequence (b) of § 19.1 is the cohomology exact sequence of the triple $(\mathbb{R}^n, \mathbb{R}^n - X, \mathbb{R}^n - Y)$, for some closed imbedding of Y in $\mathbb{R}^n$. Equation (7) follows from the definition and excision; equation (8) is a special case of [BT] 2.2 (A1); equation (9) follows from [BT] 4.1.3.

If the one-point compatification $X^c = X \cup \{*\}$ of X is a CW complex, one may imbed X^c in the n-sphere S^n, and then

$$H_i(X) = H^{n-i}(S^n - \{*\}, S^n - X^c) \cong H_i(X^c, \{*\}),$$

which identifies $H_i X$ with homology with closed or locally finite supports, i.e. Borel-Moore homology (cf. Borel-Moore (1) Thm. 3.8 and § 5). Borel and Moore give a sheaf-theoretic construction of homology groups $H_i X$ for any locally compact space, from which the basic properties can also be derived. For the analogous l-adic theory for algebraic varieties, see Deligne (2) and Laumon (1).

Example 19.1.2 (cf. Bloch-Ogus (1)). Let X be a complete scheme over an algebraically closed field. Then

$$\mathrm{Alg}_* X/\mathrm{Rat}_* X$$

is a divisible group. (The group is generated by classes in $A_* X$ of the form $p_* q^*(\beta)$ for V a subvariety of $X \times C$, C a complete non-singular curve, β a zero-cycle of degree zero on C, p, q the projections from V to X and C. Since $\mathrm{Alg}_0 C/\mathrm{Rat}_0 C = J(C)$ is divisible (Example 1.6.6), for any N we may write $\beta = N\beta'$, so $\alpha = N\, p_* q^* \beta'$, as required.

Example 19.1.3. For any complex algebraic scheme X, the groups $H_* X$ are finitely generated. It follows that $Z_* X/\mathrm{Hom}_* X$ is finitely generated. If X is complete, the quotient group $Z_* X/\mathrm{Num}_* X$ is a finitely generated free abelian group, and $\mathrm{Hom}_*^{\tau} X/\mathrm{Hom}_* X$ is finite. (Using the long exact sequence (6) and Noetherian induction, one is reduced to the case where X is projective. For a simple proof of the triangulability of projective varieties, see Hironaka (2).)

Example 19.1.4. For any complete scheme X over any algebraically closed field, $N_* X = Z_* X/\mathrm{Num}_* X$ is a finitely generated free abelian group. (There is a pairing $N' \otimes Z_* X \to \mathbb{Z}$ for an abelian group N', so that $\mathrm{Num}_* X$ is the annihilator of N'. Thus $N = N_* X$ is torsion-free. Using the cycle map to étale homology, it follows that $N \otimes \mathbb{Q}_l$ is finitely generated. Let $x_1, \ldots, x_n \in N$ give a $\mathbb{Q}$-basis for $N \otimes \mathbb{Q}$. Choose $y_i \in N'$ so that $y_i \cdot x_i = m_i \neq 0$, and $y_i \cdot x_j = 0$ for $i \neq j$. Then $N \subset \sum \mathbb{Z}(x_i/m_i) \subset N \otimes \mathbb{Q}$, so N is finitely generated.)

Example 19.1.5. (a) If X is a non-singular, complete, n-dimensional variety over a field, then a k-cycle α on X is numerically equivalent to zero in the sense of this section if and only if $\int_X \alpha \cdot \beta = 0$ for all $(n-k)$-cycles β on X. (By Riemann-Roch (§ 18.3), the Chern character determines an isomorphism form $K^\circ X_{\mathbb{Q}}$ to $A(X)_{\mathbb{Q}}$.)

(b) If X is a complete scheme which can be imbedded in a non-singular variety, then the following are equivalent:

(i) α is numerically equivalent to zero.

(ii) $\int_X \operatorname{ch}(\xi) \cap \alpha = 0$ for all $\xi \in K^\circ X$.

(iii) $\int_X f^*\beta \cap \alpha = \int_X \alpha \cdot_f \beta = 0$ for all $f\colon X \to Y$, Y non-singular, $\beta \in A^k Y$.

(Use Lemma 18.2 with the argument in (a).)

Example 19.1.6. Let $f\colon X \to Y$ be a morphism of complete schemes over an arbitrary field. Then $f_*(\mathrm{Num}_* X) \subset \mathrm{Num}_* Y$, so f induces a homomorphism $f_*\colon N_* X \to N_* Y$, making N_* a covariant functor.

If X and Y are non-singular, $f^*(\mathrm{Num}_* Y) \subset \mathrm{Num}_* X$, and f induces a homomorphism $f^*\colon N^* Y \to N^* X$ of graded rings. (Both follow from the projection formula, using Example 19.1.5) in the second case.) More generally, for any correspondence $\alpha\colon X \vdash Y$, α_* and α^* induce homomorphisms between $N_* X$ and $N_* Y$.

Example 19.1.7. *MacPherson Chern classes* (MacPherson (1)). For any complex scheme X, and any subvariety $V \subset X$, the function which assigns to P the local Euler obstruction $\mathrm{Eu}_P V$ (cf. Example 4.2.9 (a)) is a constructable function $\mathrm{Eu}_- V$ on X. The mapping

$$\sum n_i [V_i] \to \sum n_i \,\mathrm{Eu}_- V_i$$

defines an isomorphism Φ from the group of cycles $Z_* X$ to the group of constructable functions $F_* X$.

Define $c_M(\sum n_i [V_i])$ to be $\sum n_i c_M(V_i)$ with $c_M(V_i)$ the Mather Chern class of V_i (Example 4.2.9 (b)). Define

$$c_*\colon F_* X \to A_* X$$

by $c_* = c_M \circ \Phi^{-1}$. The theorem of MacPherson (1) is that the composite

$$F_* X \xrightarrow{c_*} A_* X \xrightarrow{cl} H_*(X; \mathbb{Z})$$

commutes with push-forward for proper morphisms; for $f\colon X \to Y$, 1_V the characteristic function of a subvariety V of X, $f_*(1_V)(y)$ is defined to be the topological Euler characteristic of $V \cap f^{-1}(y)$. The MacPherson Chern class $c_* X$ of X is defined to be the class $c_*(1_X)$. For X complete, mapping X to a point shows that MacPherson's Chern class satisfies

$$\int_X c_*(X) = \chi(X),$$

the topological Euler characteristic of X.

In fact, the proof of MacPherson (1) shows that $c_*: F_* \to A_*$ commutes with proper push-forward. Verdier (6) has shown that these Chern classes are compatible with specialization. Relations with other intersection operations are discussed by Sabbah (1).

Note: Three other natural definitions of Chern classes of singular varieties. all agreeing with $c(T_X) \cap [X]$ for X non-singular, have been discussed (Examples 4.2.6 (a), (c), and 4.2.9 (b)). Few functorial properties are known for these other classes, however.

Example 19.1.8. *Cohomology operations.* Let $S^\cdot: H^* \to H^*$ be a cohomology operation, where H^* is cohomology with $\mathbb{Z}/p\,\mathbb{Z}$ coefficients, and let w be the corresponding characteristic class:

$$w(E) = \varphi^{-1} S^\cdot \varphi(1)$$

where $\varphi: H^* X \to H^*(E, E - \{0\})$ is the Thom isomorphism for an H-oriented bundle E. For example, if $S^\cdot$ is the Steenrod square, w is the Stiefel-Whitney class (cf. Milnor-Stasheff (1), Atiyah-Hirzebruch (1).) Assume that for any non-singular projective algebraic (resp. complex analytic) manifold M, $w(T_M) \cap \mu_M$ is represented by an algebraic (resp. analytic) cycle. For example, if $S^\cdot$ is the Steenrod square, w is the reduction of the total Chern class mod 2, so this is clear. Then $S^\cdot$ preserves algebraic (resp. analytic) cycles.

More generally, for any complete algebraic scheme (resp. compact analytic space) X, the corresponding homology operation

$$S_\cdot: H_* X \to H_* X$$

preserves algebraic (resp. analytic) cycles. (If V is a subvariety of X, by Hironaka there is a proper morphism $\pi: M \to V$, M non-singular, with $\pi_*(\mu_M) = cl[V]$. Therefore

$$S_\cdot(cl(V)) = \pi_* S_\cdot(\mu_M) = \pi_*(w(T_M)^{-1} \cap \mu_M)$$

is algebraic.)

The case of Steenrod squares and reduced power operations on non-singular varieties was proved by Kawai (1), by a more complicated argument. Note in particular that Sq^i vanishes on algebraic cycles if i is odd. Related results were given by Atiyah and Hirzebruch (1).

We do not know if Sq lifts to an operation on the group $A_* X \otimes (\mathbb{Z}/2\mathbb{Z})$. The above proof shows what Sq must be. The question is whether $\pi_*(c(T_M)^{-1} \cap [M])$ is independent of the resolution.

Example 19.1.9. There are exterior products

$$H_i X \otimes H_j Y \xrightarrow{\times} H_{i+j}(X \times Y)$$

for Borel-Moore homology. If α and β are algebraic cycles on X and Y, then

$$cl_{X \times Y}(\alpha \times \beta) = cl_X(\alpha) \times cl_Y(\beta)\,.$$

(Since the products are compatible with restriction, this follows from the fact that $\mu_{M \times N} = \mu_M \times \mu_N$ for manifolds M, N.)

Example 19.1.10. *Refined topological intersections.* For X a closed subspace of Y, $c \in H^*(Y, Y-X)$, and $f\colon Y' \to Y$, $a \in H_*(Y')$, define $c \cap a$ in $H_*(X')$, $X' = f^{-1}(X)$, by setting

$$c \cap a = f^*(c) \cap a,$$

where the cap product on the right is the cap product (2).

Let M be an oriented real n-manifold, $A \subset M$ a closed subspace, $a \in H_k A$. Define $c\,l^M(a) \in H^{n-k}(M, M-A)$ to be the class dual to a, i.e.,

$$c\,l^M(a) \cap \mu_M = a.$$

If also $B \subset M$, $b \in H_l B$, define

$$a \cdot b = (c\,l^M(a) \cup c\,l^M(b)) \cap \mu_M = c\,l^M(a) \cap b$$

in $H_{k+l-n}(A \cap B)$. If $u_\delta \in H^n(M \times M, M \times M - \Delta)$ is the orientation class of the diagonal, then

$$a \cdot b = u_\delta \cap (a \times b).$$

In addition, $b \cdot a = (-1)^{(n-k)(n-l)} a \cdot b$.

Example 19.1.11. There is an important class of schemes X for which the cycle map

$$c\,l_X\colon A_* X \to H_* X$$

is an isomorpism. Such varieties can have no odd-dimensional homology, so the only curves for which this holds are $\mathbb{P}^1$ and $\mathbb{A}^1$.

(a) If $Y \subset X$ is a closed subscheme, $U = X - Y$, and $c\,l_Y$ and $c\,l_U$ are isomorphisms, then $c\,l_X$ is an isomorphism. (Use the compatible exact sequences of § 1.8 and § 19.1 (6).)

(b) If X has a cellular decomposition in the sense of Example 1.9.1, then $c\,l_X$ is an isomorphism.

(c) If $X = G/B$, with B a Borel subgroup of a reductive group G, the Bruhat decomposition (cf. Borel (1) IV.14) is cellular in this sense, so $c\,l_X$ is an isomorphism.

(d) If X is a projective space, Grassmannian, or arbitrary flag manifold, then $c\,l_X$ is an isomorphism. More generally, if X is a flag bundle over a scheme Y, then $c\,l_X$ is an isomorphism if and only if $c\,l_Y$ is an isomorphism.

Example 19.1.12. If $H_*^c X$ denotes singular homology (homology with compact supports), and $A_*^c X$ is the group defined in Example 10.2.8, there are homomorphisms

$$c\,l\colon A_k^c X \to H_{2k}^c X,$$

covariant for arbitrary morphisms.

19.2 Algebraic and Topological Intersections

Every regular (closed) imbedding $i\colon X \to Y$ of codimension d has an *orientation class*

$$u_i = u_{X,Y} \in H^{2d}(Y, Y-X).$$

We describe this class in three important cases, which suffice for many applications, and then give the general construction:

(i) If Y is a vector bundle on X, and i is the zero-section imbedding, then $u_{X,Y}$ is the Thom class of the bundle.

(ii) If X and Y are non-singular, $u_{X,Y}$ is the class defined in § 19.1.

(iii) If there is a vector bundle E of rank d on Y and a section s of E so that X is the zero-scheme of s, then $u_{X,Y} = s^*(u)$, where u is the Thom class of E.

(iv) In general, choose a vector bundle E of rank $e \geqq d$ on Y and a section s of E so that X is the zero-scheme of s (Appendix B.8.2). The normal bundle N to X in Y is a sub-bundle of the restriction $E|_X$ of the E to X. Choose a (classical) neighborhood V of X in Y, and a complex topological sub-bundle C of $E|_V$ so that $C|_X \oplus N = E|_X$. Let Q be the quotient bundle of $E|_V$ by C, and let $\bar{s}$ be the section of Q induced by s. Shrinking V if necessary, $\bar{s}$ maps the pair $(V, V-X)$ to $(Q, Q-\{0\})$, with $\{0\}$ the zero-section of Q. Let u be the Thom class of Q. Then set

$$u_{X,Y} = \bar{s}^*(u) \in H^{2d}(V, V-X) = H^{2d}(Y, Y-X).$$

It may be verified (Baum-Fulton-MacPherson (1) IV.4) that this class is independent of choices, and agrees with the classes considered in (i)–(iii).

Lemma 19.2 (a) *Consider a fibre square*

$$\begin{array}{ccc} X' & \xrightarrow{i'} & Y' \\ {\scriptstyle f'}\downarrow & & \downarrow{\scriptstyle f} \\ X & \xrightarrow[i]{} & Y \end{array}$$

with i and i' regular imbeddings of codimension d. Then

$$f^*(u_i) = u_{i'} \in H^{2d}(Y', Y'-X').$$

(b) *If, in (a), f and f' are also regular imbeddings of codimension e, then*

$$u_i \cup u_f = u_f \cup u_i = u_{fi'} = u_{f'i}$$

in $H^{2d+2e}(Y, Y-X')$.

(c) *Let T be a non-singular, connected curve, X a scheme, $i_t: X \to X \times T$ the imbedding $i_t(x) = (x, t)$. Then for all $a \in H_k(X \times T)$,*

$$u_{i_t} \cap a \in H_{k-2}(X)$$

is independent of t in T.

(d) *Let $i: X \to N$ be the zero-section of a vector bundle N of rank d on X. Then for all k-cycles α on N,*

$$u_i \cap c\,l_N(\alpha) = c\,l_X(i^*\alpha)$$

in $H_{2k-2d}(X)$.

Proof. Properties (a), (b) and (c) follow readily from the constructions of the orientation classes. For (d), by Theorem 3.3 (a) one may assume $\alpha = [\pi^{-1}V]$, V a subvariety of X, π the projection from N to X. Using (a), we may assume $V = X$. The required formula is then $u_i \cap \mu_N = \mu_X$; this is verified by restricting to a non-singular open subset of X, in which case equation (9) of § 19.1 suffices. □

Theorem 19.2. *Consider a fibre square*

$$\begin{array}{ccc} X' & \xrightarrow{f'} & Y' \\ {\scriptstyle g'}\downarrow & & \downarrow{\scriptstyle g} \\ X & \xrightarrow[f]{} & Y \end{array}$$

with f a regular imbedding of codimension d. Then for all k-cycles α on Y',

$$cl_{X'}(f^!\alpha) = g^*(u_{X,Y}) \cap cl_{Y'}(\alpha)$$

in $H_{2k-2d}X'$. Here $f^!\alpha$ is the class in $A_{k-d}(X')$ constructed in § 6.2.

Proof. We may assume $\alpha = [V]$, with $V = Y'$ a variety. Construct, from § 5.1, the diagram for deformation to the normal bundle N of X in Y:

$$\begin{array}{ccccc} X & \xrightarrow[f]{} & N & \rightarrow & \{\infty\} \\ {\scriptstyle i_\infty}\downarrow & & \downarrow{\scriptstyle j_\infty} & & \downarrow \\ X \times \mathbb{P}^1 & \xrightarrow[F]{} & M^\circ & \rightarrow & \mathbb{P}^1 \\ {\scriptstyle i_0}\uparrow & & \uparrow{\scriptstyle j_0} & & \uparrow \\ X & \xrightarrow[f]{} & Y & \rightarrow & \{0\} \end{array} .$$

Here $M^\circ = M_X^\circ Y$ is the blow-up of $Y \times \mathbb{P}^1$ along $X \times \{\infty\}$, with the proper transform $\tilde{Y}$ of $Y \times \{\infty\}$ omitted. Factoring g into a closed imbedding followed by a smooth morphism, one is reduced to the case where g is a closed imbedding. Then $M^{\circ\prime} = M_{X'}^\circ Y'$ is a subvariety of M°, imbedded by a morphism $G: M^{\circ\prime} \to M^\circ$. By Lemma 19.1.3,

$$cl(Y') = G^*(u_{j_0}) \cap cl(M^{\circ\prime})$$
$$cl[C'] = G^*(u_{j_\infty}) \cap cl(M^{\circ\prime}) ,$$

where $C' = (M^{\circ\prime})_\infty$ is the normal cone to X' in Y'. By Lemma 19.2(b) and equation (8) of § 19.1,

$$g^*(u_f) \cap cl(Y') = g^*(u_f) \cap (G^*(u_{j_0}) \cap cl(M^{\circ\prime}))$$
$$= (g' \times 1)^*(u_{i_0}) \cap (G^*(u_F) \cap cl(M^{\circ\prime})) .$$

By Lemma 19.2(c), we may replace u_{i_0} by u_{i_∞} in the last expression. Reversing the argument at ∞,

$$(g' \times 1)^*(u_{i_\infty}) \cap (G^*(u_F) \cap cl(M^{\circ\prime})) = \bar{g}^*(u_{\bar{f}}) \cap cl[C'] ,$$

where $\bar{g}$ is the imbedding of C' in N. By Lemma 19.2(d),

$$\bar{g}^*(u_{\bar{f}}) \cap cl[C'] = cl_{X'}(\bar{f}^*[C']) .$$

This concludes the proof, since $f^![Y']$ is defined to be $\bar{f}^*[C']$. □

Let y be an l-cycle on an n-dimensional non-singular variety Y, with support $|y| \subset Y$. Define

$$cl^Y(y) \in H^{2n-2l}(Y, Y - |y|)$$

to be the class such that $cl^Y(y) \cap \mu_Y = cl(y)$ in $H_{2l}(|y|)$ (cf. equation (3) of § 19.1). If $f: X \to Y$ is a morphism, X any scheme, x a k-cycle on X, then we have constructed a refined intersection class $x \cdot_f y$ in $A_m(Z)$, where $Z = |x| \cap f^{-1}(|y|)$, $m = k + l - n$ (§ 8.1).

Proposition 19.2. *Let f' be the morphism from $|x|$ to Y induced by f. Then*

$$c\,l_Z(x \cdot_f y) = f'^*(c\,l^Y(y)) \cap c\,l_{|x|}(x)\,.$$

in $H_{2m}(Z)$.

Proof. Note that $f'^* c\,l^Y(y) \in H^*(|x|, |x| - Z)$, so the cap product on the right lands in $H_* Z$. As usual, we may assume $x = [V]$, $V = |x| = X$ a variety, so $f' = f$. Apply Theorem 19.2 to the fibre square

$$\begin{array}{ccc} Z & \rightarrow & X \times |y| \\ \downarrow & & \downarrow g \\ X & \underset{\gamma}{\rightarrow} & X \times Y \end{array}$$

with γ the graph of f, g the inclusion. This yields:

$$c\,l_Z(x \cdot_f y) = g^*(u_\gamma) \cap c\,l(X \times y)\,.$$

Now if p is the projection from $X \times Y$ to Y,

$$c\,l(X \times y) = p^*(c\,l^Y(y)) \cap c\,l(X \times Y)\,.$$

By the commutativity of even dimensional cohomology classes, and equation (8) of § 19.1,

$$\begin{aligned} g^*(u_\gamma) \cap (p^*(c\,l^Y(y)) \cap c\,l(X \times Y)) &= f^*(c\,l^Y(y)) \cap (u_\gamma \cap c\,l(X \times Y)) \\ &= f^*(c\,l^Y(y)) \cap c\,l(X)\,, \end{aligned}$$

the last step by Theorem 19.2 again. Combining the displayed equations concludes the proof. □

Let X be a fixed non-singular n-dimensional variety, V, W subvarieties of X of dimensions k, l. We have constructed $V \cdot W$ in $A_m(V \cap W)$, $m = k + l - n$, in § 8.1. A corresponding refined class is defined topologically as follows. Since

$$c\,l^X(V) \in H^{2n-2k}(X, X - V)\,, \qquad c\,l^X(W) \subset H^{2n-2l}(X, X - W)\,,$$

we have $c\,l^X(V) \cup c\,l^X(W) \in H^{4n-2k-2l}(X, X - V \cap W)$. Set

$$c\,l(V) \cdot c\,l(W) = (c\,l^X(V) \cup c\,l^X(W)) \cap \mu_X$$

in $H_{2m}(V \cap W)$ (cf. Example 19.1.10).

Corollary 19.2. (a) *With the above notation,*

$$c\,l(V \cdot W) = c\,l(V) \cdot c\,l(W)$$

in $H_{2m}(V \cap W)$.

(b) *The mapping $c\,l : A^* X \rightarrow H^* X$ is a homomorphism of graded rings, contravariant for morphisms of non-singular varieties.*

Proof. For (a), apply Proposition 19.2 to the inclusion f of V in X, $x=[V]$, $y=[W]$. This yields

$$\begin{aligned} cl_{V\cap W}(V\cdot W) &= f^* cl^X(W) \cap cl_V(V) \\ &= f^* cl^X(W) \cap (cl^X(V) \cap \mu_X) \\ &= (cl^X(W) \cup cl^X(V)) \cap \mu_X, \end{aligned}$$

using equation (8) of § 19.1, as required. The fact that cl commutes with f^* for $f: X \to Y$, is a special case of Proposition 19.2, with $x=[X]$. □

For X quasi-projective and non-singular, the fact that cl is a ring homomorphism may also be deduced from the moving lemma, using the strong form stated in Example 11.4.2.

Analogous results hold for quasi-projective schemes over an arbitrary algebraically closed field, with values in étale homology and cohomology. For details, as well as a sheaf-theoretic approach in the classical case, see Exposés VI–IX in the seminar of Douady-Verdier (1). This approach yields several important generalizations, including: (i) a flat pull-back homomorphism in homology, compatible with that defined in § 1.7 for cycles; (ii) cycle classes for varieties not necessarily imbeddable in smooth varieties, and for general complex analytic spaces. Gillet (3) constructs the classes $u_{X,Y}$ for a regular imbedding $X \hookrightarrow Y$ without assuming Y can be imbedded in a smooth variety.

Example 19.2.1 (cf. Baum-Fulton-MacPherson (1) IV.4, and Fulton-MacPherson (3)). For any l.c.i. morphism $f: X \to Y$ of relative dimension d there are Gysin maps

$$f^*: H_i Y \to H_{i-2d} X.$$

These are functorial, and compatible with the cycle class, i.e. $f^* cl_Y(\alpha) = cl_X f^* \alpha$ for any cycle α on Y. (Imbed X in a smooth M, so f factors into an imbedding i into $Y \times M$ followed by the projection p to Y. Define $f^* = i^* p^*$, where $p^*(\alpha) = \alpha \times \mu_M$, and $i^*(\beta) = u_i \cap \beta$.) If f is also proper, there are also functorial Gysin maps $f_*: H^i X \to H^{i+2d} Y$.

Example 19.2.2. *Excess intersection formula* (cf. Fulton-MacPherson (3)). Given a fibre square

$$\begin{array}{ccc} X' & \xrightarrow{i'} & Y' \\ {\scriptstyle g}\downarrow & & \downarrow{\scriptstyle f} \\ X & \xrightarrow[i]{} & Y \end{array}$$

with i (resp. i') a regular imbedding of codimension d (resp. d'), with excess bundle E (§ 6.3), then

$$f^* u_{X,Y} = c_e(E) \cdot u_{X',Y'}$$

in $H^{2d}(Y', Y'-X')$, $e=d-d'$. The product on the right is defined as follows. If $c \in H^* X'$, extend c to a class $\tilde{c} \in H^* U'$ for some topological neighborhood U' of X' in Y'. For u in $H^*(Y', Y'-X')$, identified with $H^*(U', U'-X')$ by excision, $c \cdot u$ is defined to be $\tilde{c} \cup u$.

Theorem 19.2 follows from this formula, since by blowing up we may assume $\alpha=[V]$, $V=Y'$, and either X' is a divisor on Y' or $X'=Y'$.

Example 19.2.3. Let $f: X \to Y$ be a l.c.i. morphism of complete schemes which are imbeddable in non-singular varieties, over an arbitrary field. Then $f^*(\mathrm{Num}_* Y) \subset \mathrm{Num}_* X$, so N_* is a contravariant functor for l.c.i. morphisms. (If $\alpha \in \mathrm{Num}_* Y$, write the image of α in $A_* Y_{\mathbb{Q}}$ as $\tau_Y(\theta)$, $\theta \in K_\circ Y$, and use Theorem 18.3(4).)

Example 19.2.4. *Linking numbers.* Let V^a, W^b be subvarieties of a non-singular complex variety X^n, of the indicated dimensions, meeting properly at a point P in X; so $n = a + b$. Identify (complex analytically) a neighborhood of P in X with a neighborhood of 0 in $\mathbb{C}^n$. If S is a small $(2n-1)$-sphere about P, then S meets V in a real topological $(2a-1)$-cycle, which we denoted by A. In fact, if D is the closed ball around P with boundary S, then $D \cap V$ is homeomorphic to the cone over A with vertex P. Similarly, S meets W in a topological $(2b-1)$-cycle B.

The linking number $L(A, B)$ of A and B in S may be defined as follows. Choose a $2b$-chain B' in S whose boundary is B, so B' determines a class $\{B'\}$ in $H_{2b}(S, B)$. Let $cl^S(A) \in H^{2b}(S, S - A)$ be the class dual to the class of A in $H_{2a-1}(A)$ (cf. Example 19.1.10). Then the relative cap product $cl^S(A) \cap \{B'\}$ lies in $H_0(A, A \cap B) = H_0 A$, since $A \cap B = \emptyset$, and we may set

$$L(A, B) = \deg(cl^S(A) \cap \{B'\}) .$$

Proposition. $L(A, B) = i(P, V \cdot W; X)$.

(Since $cl^X(V) \in H^{2b}(X, X - V)$ restricts to $cl^S(A)$,

$$\text{(i)} \qquad L(A, B) = \deg(cl^X(V) \cap \{B'\}).$$

Let $W_D = W \cap D$. Then W_D determines a chain on D whose boundary is B, so a class $\{W_D\}$ in $H_{2b}(W_D, B)$, and

$$\text{(ii)} \qquad i(P, V \cdot W; X) = \deg(cl^X(V) \cap \{W_D\})$$

where $cl^X(V) \cap \{W_D\} \in H_0(V \cap W_D, V \cap B) = H_0(P)$. This follows from Corollary 19.2 and the fact that $\{W_D\} = cl^X(W) \cap \{D\}$, where $\{D\}$ is the canonical generator of $H_{2n}(D, S)$.

Now B' and W_D both have boundary B, so $B' - W_D$ is a cycle on D. Since $H_{2b} D = 0$, $\{B'\}$ and $\{W_D\}$ have the same image in $H_{2b}(D, B)$. The zero cycles appearing in (i) and (ii) therefore have the same image in $H_0(V \cap D, V \cap B) = H_0(V \cap D)$, so they have the same degree.)

The case of two curves on a smooth surface is discussed by Martinelli (1) and Reeve (1)

Example 19.2.5. *Analytic spaces.* Much of the intersection theory developed in this text for algebraic schemes can be extended to analytic spaces. (One should use projective bundles and cones, rather than affine cones, to avoid arguments using closures of locally closed subvarieties.) For spaces imbeddable in complex manifolds, the methods of this chapter apply. For example, if V and W are irreducible analytic subspaces of a complex manifold X, the intersection class $cl(V) \cdot cl(W)$ can be constructed in $H_*(V \cap W)$; if V and W meet properly, this determines intersection numbers, and therefore $V \cdot W$ as an analytic cycle on X. By the methods of Chap. 6, $cl(V) \cdot cl(W)$ is always represented by an analytic cycle on $V \bigcap W$ (cf. King (3)).

For divisors, the results of Chap. 2 extend without change. Such a refined intersection theory is useful in studying families of compact analytic (or algebraic) families parametrized by a disk (cf. Persson (1)).

The multiplicity of an analytic variety at a point, as in Chap. 4, is the degree of the projective tangent cone to the variety at the point. This multiplicity can also be realized analytically as the Lelong number of the variety at the point (cf. Harvey (1), Griffiths-Harris (1) 3.2). More generally, for any closed imbedding $X \subset Y$ of analytic spaces, there is a Segre class $s(X, Y)$ in $H_*(X)$.

If $D_1, \ldots, D_n$ are divisors meeting properly at a point P in an n-dimensional manifold X, and f_i is a local equation for D_i at P, then (Griffiths-Harris (1) 5.2, cf. Picard (1))

$$i(P, D_1 \cdot \ldots \cdot D_n; X) = \mathrm{Res}_0\left(\frac{df_1}{f_1} \wedge \ldots \wedge \frac{df_n}{f_n}\right).$$

For an n-dimensional subvariety X of $\mathbb{P}^m$, with the Fubini-Study metric,

$$\mathrm{Volume}(X) = \deg(X) \cdot \mathrm{Volume}(L)$$

with L an n-dimensional linear subspace of $\mathbb{P}^m$. For this and other relations between volumes and intersection theory, see the discussion of Wirtinger's theorem in Mumford (5) and Griffiths-Harris (1).

Example 19.2.6. If a vector bundle E of rank r on a scheme X has a section s, and Z is the zero-scheme of s, then the top Chern class $c_r(E)$ in $H^{2r}(X)$ comes from a well-determined class in $H^{2r}(X, X-Z)$, namely the class $s^*(u)$, where $u \in H^{2r}(E, E-\{0\})$ is the Thom class of E. More generally, if Z is closed in X, a nowhere vanishing section of E on $X-Z$ determines such a localization of $c_r(E)$ (cf. Atiyah-Bott-Shapiro (1) II).

A complex $E.$ of vector bundles on a space Y exact of a closed subspace X, determines a local Chern character $\mathrm{ch}_X^Y(E.)$ in $H^*(Y, Y-X; \mathbb{Q})$ (cf. Atiyah-Hirzebruch (2), Iversen (3)). Similarly, it is only necessary to have the bundles E_i on Y, and the exact complex on $Y-X$, to construct such a class (Atiyah-Bott-Shapiro (1) II, cf. Baum-Fulton-MacPherson (2) App. 1). This Chern character is compatible with that of § 18.1, i.e.,

$$cl_X(\mathrm{ch}_X^Y(E.) \cap \alpha) = \mathrm{ch}_X^Y(E.) \cap cl_Y(\alpha)$$

in $H_*(X; \mathbb{Q})$, for any $\alpha \in A_* Y$.

Example 19.2.7 A *topological correspondence* $\theta: X \vdash Y$ between compact oriented manifolds X and Y is a class $\theta \in H^*(X \times Y)$. The formalism of Chap. 16 goes through without change. For example, θ determines homomorphisms $\theta_*: H_* X \to H_* Y$ by $\theta_*(a) = q_*(\alpha \cdot p^* a)$, where p and q are the projections from $X \times Y$ to X and Y, and similarly for θ^*, and for composites of correspondences.

If X and Y are non-singular complete varieties, and $\alpha \in A^*(X \times Y)$ is an algebraic correspondence, then $cl(\alpha)$ is a topological correspondence, and if $a \in A_*(X)$, then

$$cl(\alpha)_*(cl_X(a)) = cl_Y(\alpha_*(a)).$$

(This follows from Lemma 19.1.2 and Corollary 19.2.)

19.3 Equivalence on Non-singular Varieties

In this section X will be a non-singular, complex projective variety of dimension n. The groups of cycles will be indexed by superscripts denoting codimension. Our man purpose is to sketch what is known about relations among the groups defined generally in § 19.1:

$$\begin{array}{ccccccc} \mathrm{Rat}^p X & \subset & \mathrm{Alg}^p X & \subset & \mathrm{Hom}^p X & & \\ & & \cap & & \cap & & \\ & & \mathrm{Alg}^p_\tau X & \subset & \mathrm{Hom}^p_\tau X & \subset & \mathrm{Num}^p X \subset Z^p X. \end{array}$$

19.3.1. *Divisors.* For $p = 1$, the situation is quite well understood. In this case:

(i) $\mathrm{Alg}^1 X = \mathrm{Hom}^1 X$.
(ii) $\mathrm{Alg}^1_\tau X = \mathrm{Hom}^1_\tau X = \mathrm{Num}^1 X$.
(iii) The Neron-Severi group $NS(X) = Z^1 X/\mathrm{Alg}^1 X$ is finitely generated.
(iv) The Picard variety $\mathrm{Pic}^0 X = \mathrm{Alg}^1 X/\mathrm{Rat}^1 X$ has the structure of an abelian variety whose dimension is $\dim H^1(X, \mathscr{O}_X)$.
(v) $\mathrm{Alg}^1_\tau X/\mathrm{Alg}^1 X = H^2(X, \mathbb{Z})_{\mathrm{tors}}$, a finite group.

The basis for these facts is the cohomology sequence of the exact sequence of analytic sheaves

$$0 \to \mathbb{Z} \to \mathscr{O}_X^{an} \xrightarrow{e} \mathscr{O}_X^{an*} \to 0,$$

where $e(f) = \exp(2\pi i f)$, and the GAGA theorems that $H^i(X, \mathscr{O}_X^{an}) = H^i(X, \mathscr{O}_X)$, $H^1(X, \mathscr{O}_X^{an*}) = H^1(X, \mathscr{O}_X^*) = \mathrm{Pic}(X)$. This gives an exact sequence

$$H^1(X, \mathbb{Z}) \to H^1(X, \mathscr{O}_X) \to \mathrm{Pic}(X) \xrightarrow{\delta} H^2(X, \mathbb{Z}) \to H^2(X, \mathscr{O}_X).$$

The boundary map δ is the first Chern class, i.e. $\delta(\mathscr{O}(D)) = c_1(\mathscr{O}(D)) = cl[D]$. The Picard variety $\mathrm{Pic}^0(X)$ is the quotient of $H^1(X, \mathscr{O}_X)$ by the image of $H^1(X, \mathbb{Z})$, which is a lattice in $H^1(X, \mathscr{O}_X)$. Assertions (i)–(v) follow from these facts.

19.3.2. $Z^* X/\mathrm{Hom}^* X$.

In general, since $H^* X$ is finitely generated (for example, from the fact that X is a compact manifold), $Z^* X/\mathrm{Hom}^* X$ is finitely generated. Therefore

(i) $Z^p X/\mathrm{Num}^p X$ is a finitely generated free abelian group.
(ii) $\mathrm{Hom}^p_\tau X/\mathrm{Hom}^p X$ is a finite group.

It is not known if homological and numerical equivalence coincide for all p, i.e.

(iii) Is $\mathrm{Hom}_\tau^p X = \mathrm{Num}^p X$?

For $p = 0, n$, this is trivially true, and for $p = 1$ it is true by 19.3.1. Lieberman (1) has verified the equality for $p = 2$ and $p = n - 2$, and for all p when X is an abelian variety; in general the question is closely related to Grothendieck's "standard conjectures" (cf. Example 19.3.14, Grothendieck (6), Kleiman (2)).

19.3.3. $\mathrm{Hom}_\tau^* X/\mathrm{Alg}_\tau^* X$.

Griffiths (2) showed that $\mathrm{Alg}_\tau^p X$ can be unequal to $\mathrm{Hom}_\tau^p X$ if $1 < p < n$. Griffiths constructed a 3-fold with

$$\mathrm{Hom}_\tau^2 X/\mathrm{Alg}_\tau^2 X \text{ infinite.}$$

Ceresa (1) has shown that if X is the Jacobian of a general curve C of genus $n \geqq 3$, φ the endomorphism of X which is multiplication by -1, and $w_p \in A^p X$ is the class of the image of $C^{(n-p)}$ in X, then

$$\varphi^*(w_p) - w_p$$

is not algebraically equivalent to zero, if $1 < p < n$, although it is homologous to zero, since φ^* is the identity on $H^* X$. Other naturally occurring examples have been given by Ceresa and Collino (1).

Recently Clemens (1) has shown that there are 3-folds X with $\mathrm{Hom}_\tau^2 X/\mathrm{Alg}_\tau^2 X$ *not* finitely generated.

It is not known if $\mathrm{Alg}_\tau^p X/\mathrm{Alg}^p X$ is always finite.

19.3.4. $\mathrm{Alg}^* X/\mathrm{Rat}^* X$.

For $p > 1$, the structure of the groups $\mathrm{Alg}^p X/\mathrm{Rat}^p X$ has been determined in only a few cases. In several cases this group has been given the structure of an abelian variety, usually by relating these groups to Jacobians of certain curves associated to X. For examples of this see Murre (2), Bloch-Murre (1), and Beauville (1).

In general, however, for $p \geqq 2$, the groups $\mathrm{Alg}^p X/\mathrm{Rat}^p X$ can be larger than any abelian variety, as shown by examples of Mumford, discussed next.

19.3.5. *Zero-cycles.* For $p = n$, $\mathrm{Alg}^n X = \mathrm{Num}^n X$, and $Z^n X/\mathrm{Alg}^n Z = \mathbb{Z}$. The interesting group is the group

$$\tilde{A}_0 X = \mathrm{Alg}^n X/\mathrm{Rat}^n X$$

of zero-cycles of degree zero, modulo rational equivalence.

Mumford (3) showed that when X is a surface with geometric genus $p_g > 0$, then $\tilde{A}_0 X$ cannot be given a structure of an abelian variety, if one requires that the canonical maps

$$S^m X \to \tilde{A}_0 X, \quad P_1 + \ldots + P_m \to \sum [P_i] - m[P_0]$$

for P_0 a base point, be morphisms of algebraic varieties. Indeed, the fibres of this map act as if $\tilde{A}_0 X$ is infinite dimensional.

Bloch has conjectured that $\tilde{A}_0 X$ is finite dimensional when $p_g = 0$. For a discussion of evidence for this, see Bloch-Kas-Lieberman (1) and Bloch (4). For relations with vector bundles, see Murthy-Swan (1).

For a general X, Roĭtman (2) has shown that the canonical map (Example 1.4.4)

$$\theta : \tilde{A}_0 X \to \mathrm{Alb}(X)$$

to the Albanese variety induces an isomorphism on the torsion points.

19.3.6. *Hodge theory.* With complex coefficients, the cohomology of X has a canonical Hodge decomposition

$$H^k(X; \mathbb{C}) = \bigoplus_{p+q=k} H^{p,q}(X)$$

with $\overline{H^{q,p}}(X) = H^{p,q}(X) \cong H^q(X, \Omega_X^p)$. If α is an algebraic cycle of codimension p on X, then the image of $cl(\alpha)$ in $H^{2p}(X; \mathbb{C})$ lies in $H^{p,p}(X)$. This follows from the fact that, if $z_1, \ldots, z_n$ are local holomorphic coordinates on X, a form with more than $n-p$ dz's or $d\bar{z}$'s must vanish on the non-singular locus of a subvariety of codimension p.

For $p=1$, the Lefschetz-Hodge theorem asserts that every class in $H^{1,1}(X)$ which is in the image of $H^2(X, \mathbb{Z})$ is algebraic. This is proved by a direct calculation (cf. Griffiths-Harris (1) p. 163). For $p > 1$, the celebrated Hodge conjecture asks if every class in $H^{p,p}X \cap H^{2p}(X; \mathbb{Q})$ is a rational combination of algebraic cycles.

A principal aim of the theory of intermediate jacobians has been to study algebraic cycles by means of a complex torus $J^p(X)$ and an *Abel-Jacobi* map

$$\theta : \mathrm{Hom}^p X / \mathrm{Rat}^p X \to J^p X$$

generalizing the map $\tilde{A}_0 X \to \mathrm{Alb}(X)$ for 0-cycles. Griffiths, modifying a construction of Weil, achieves this by defining

$$J^p X = H^{2p-1}(X; \mathbb{R}) / H_{\mathbb{Z}}^{2p-1} = H^{2n-2p+1}(X; \mathbb{R})^* / H_{2n-2p+1}^{\mathbb{Z}},$$

where $H_{\mathbb{Z}}^i$ (resp. $H_i^{\mathbb{Z}}$) denotes the image of $H^i(X; \mathbb{Z})$ (resp. $H_i(X; \mathbb{Z})$) in $H^i(X; \mathbb{R})$ (resp. $H_i(X; \mathbb{R})$), and the $*$ denotes the dual of a real vector space. Set

$$F^p = H^{p,p-1} \oplus H^{p+1,p-2} \oplus \ldots \oplus H^{2p-1,0}.$$

Then

$$J^p X = H^{2p-1}(X; \mathbb{C}) / F^p + H_{\mathbb{Z}}^{2p-1} = (F^{n-p+1})^{\vee} / H_{2n-2p+1}^{\mathbb{Z}}$$

where $\vee$ denotes the complex dual. These descriptions put a complex structure on $J^p X$.

Given a cycle $\alpha \in \mathrm{Hom}^p X$, write α as the boundary of a $(2n-2p+1)$-chain (or current) Γ, and define $\theta(\alpha)$ to be the functional

$$\omega \to \int_\Gamma \omega, \qquad \omega \in F^{n-p+1}.$$

These complex tori vary holomorphically if X varies in a holomorphic family $\mathscr{X} \to T$. If α is a cycle of codimension p on the total space $\mathscr{X}$, whose specializations α_t are homologous to zero on X_t, one may study how $\theta(\alpha_t)$ varies with t. For some of the successful applications of this theme, see Griffiths (2), (3), Clemens-Griffiths (1), Zucker (1), Carlson-Green-Griffiths-Harris (1).

The Hodge conjecture is also closely related to Grothendieck's "standard conjectures" for algebraic cycles (Grothendieck (6), Kleiman (2)). For some recent progress on the Hodge conjecture see Deligne-Milne-Ogus-Shih (1).

Example 19.3.1. Let X be a non-singular projective surface. The group

$$N(X) = Z^1 X/\mathrm{Num}^1 X = Z^1 X/\mathrm{Alg}^1_\tau X$$

is a free abelian group of rank $\varrho = \operatorname{rank} NS(X) = \dim H^{1,1}(X)$. This group plays an important role in the study of curves on X (cf. Mumford (2), Beauville (2)).

(a) By the Hodge index theorem (Example 15.2.4), the intersection pairing

$$N(X) \otimes N(X) \to \mathbb{Z}$$

given by $C, D \to \int C \cdot D$, is a perfect pairing over $\mathbb{Q}$, of index $2-\varrho$. Equivalently, if C, D are divisors on X, with $\int C^2 > 0$ and $\int C \cdot D < 0$, then $\int D^2 \leqq 0$, with equality if and only if a multiple of D is algebraically equivalent to zero.

(b) If $\tilde{X}$ is the blow-up of X at a point P, then

$$N(\tilde{X}) = N(X) \oplus \mathbb{Z}\, e$$

with $e \cdot e = -1$, and e perpendicular to $N(X)$. (See Proposition 6.7; e is the class of the exceptional divisor.)

(c) If $X = P(E)$, where E is a vector bundle of rank 2 on a curve Y, then $\varrho = 2$ and $N(X)$ has a basis x, y with $x^2 = 0$, $x \cdot y = 1$, $y^2 = -c$, where $c = \int c_1(E)$.

(d) (Segre (2)). For any divisor D on X,

$$\int D^2 \leqq (\deg(D))^2/\deg(X)$$

where $\deg(D) = \int D \cdot H$, $\deg(X) = \int H^2$, for any ample divisor H on X. Equality holds when D and H are dependent in $N(X)$. (Apply (a) to $\deg(X) \cdot D - \deg(D) \cdot H$.)

Example 19.3.2. The index of X^n is the index of the symmetric bilinear form on $H^n(X; \mathbb{R})$ given by the cup product (intersection pairing). When $n = 2$, since $H^{1,1}$ is all algebraic, and the pairing puts $H^{2,0}$ and $H^{0,2}$ into duality, the index theorem of Example 15.2.4 is equivalent to the equation

$$\operatorname{Index}(X) = 2 + 2h^{0,2} - h^{1,1},$$

where $h^{p,q} = \dim H^{p,q}$. Equivalently, with $c_i = c_i(T_X)$,

$$\operatorname{Index}(X) = \tfrac{1}{3} \int_X c_1^2 - 2c_2 .$$

($\int c_2 = \sum (-1)^{p+q} h^{p,q}$ and $\int c_1^2 + c_2 = 12(1 - h^{0,1} + h^{0,2})$ by Examples 18.3.7 and (15.2.2.)

The Hodge index theorem also has implications for algebraic cycles if $n > 2$. For example, if $n = 2p$ is even, and α is a p-cycle on X such that $H \cdot \alpha$ is homologous to zero for an ample divisor H, then

$$(-1)^p \int \alpha \cdot \alpha \geqq 0$$

with equality if and only if α is homologous to zero. For $n > 2$, only the transcendental proof is known. For generalizations to $p < 2n$, see Kleiman (2).

Example 19.3.3. Let X be a projective variety of dimension $n \geqq 2$ over an algebraically closed field, H an ample divisor on X. For a Cartier divisor D on X, the following are equivalent:

(i) $\int_X D \cdot \alpha = 0$ for all 1-cycles α on X.

(ii) $\mathscr{O}(m\,D)$ is algebraically equivalent to zero in $\mathrm{Pic}(X)$, for some $m \neq 0$.

(iii) $\int_X D \cdot H^{n-1} = \int_X D^2 \cdot H^{n-2} = 0$.

(iv) $\chi(X, \mathscr{F} \otimes \mathscr{O}(D)) = \chi(X, \mathscr{F})$ for all coherent sheaves $\mathscr{F}$ on X.

In particular, D is numerically equivalent to zero if and only if the restriction of D to a hyperplane section of X is numerically equivalent to zero. (For non-singular complex varieties, (iii) $\Rightarrow$ (i) follows from the Hard Lefshetz theorem and the index theorem on surfaces. (i) $\Leftrightarrow$ (iv) also follows from the singular Riemann-Roch theorem. For details see Matsusaka (1) and Kleiman [SGA 6] XIII.4.6.)

If X is normal and $n \geqq 3$, Weil (5) (cf. [SGA 6] XIII.3) shows that D is linearly equivalent to zero if and only if the restriction of D to a generic hyperplane section is linearly equivalent to zero. A similar result holds for $D \in \mathrm{Rat}_1^\tau X$ if $n = 2$.

There are few results of this kind known for cycles of codimension greater than 1. Note that by Griffiths' example a 1-cycle α on a 3-fold X can be homologous to zero, so the restriction of α to all surfaces in X will be algebraically equivalent to zero, without any multiple of α being algebraically equivalent to zero.

Example 19.3.4. If D is a Cartier divisor on a complex manifold X, corresponding to a line bundle L with a section s, then $cl\ (D) = c_1(L) \in H^{1,1}(X)$ is represented by the form $-\dfrac{1}{2\pi i}\theta$, where θ is the curvature form of any connection for L; equivalently, $\theta = \bar{\partial}\partial \log|s|^2$, where the norm is taken with respect to a Hermitian metric on L. (Cf. Griffiths-Harris (1) p. 141.)

Weil (3) showed that, if X is projective, D is homologous to zero if and only if D is the residue of a closed meromorhpic 1-form on X. For extensions of this result to 0-cycles and cycles of arbitrary dimension see Griffiths (5) and Coleff-Herrera-Lieberman (1).

Example 19.3.5. For $p > 1$, using a construction of Serre, Atiyah and Hirzebruch (2) give examples of classes of finite order in $H^{2p}(X, \mathbb{Z})$ which are not algebraic.

Example 19.3.6. If $\mathscr{H}^p$ is the sheaf associated to the pre-sheaf $U \to H^p(U, \mathbb{Z})$, in the Zariski topology, on a non-singular complex variety X, Bloch and Ogus (1) show that there is a natural isomorphism

$$Z^p(X)/\mathrm{Alg}^p X \cong H^p(X, \mathscr{H}^p)\,.$$

Example 19.3.7. The canonical decomposition

$$A^p(X \times \mathbb{P}^m) \cong \bigoplus_{i=0}^{r} A^{p-i}(X),$$

where $r = \min(p, m)$ (Theorem 3.3) induces corresponding isomorphisms on the subgroups G^*/Rat^*, where G^* denotes any of the groups Alg^*, Hom^*, Num^*, Alg^*_τ, or Hom^*_τ. In particular, examples of X and p where any of the quotient groups is large can be crossed with $\mathbb{P}^m$ to find similar examples with larger codimensions and dimensions.

Example 19.3.8 (Roĭtman (2)4.2). If an n-dimensional variety X is an irreducible component of the intersection of r hypersurfaces of degrees $d_1, \ldots, d_r$ in $\mathbb{P}^{n+r}$, with $\sum d_i \leqq n + r$, then $\tilde{A}_0 X = 0$. If, however, X is a non-singular complete intersection of r hypersurfaces, with $\sum d_i > n + r$, $n \geqq 2$, then $\tilde{A}_0 X$ is infinite dimensional.

Example 19.3.9. If X is non-singular, the images of $\mathrm{Alg}^* X$, $\mathrm{Hom}^* X$, $\mathrm{Num}^* X$, $\mathrm{Alg}^*_\tau X$, and $\mathrm{Hom}^*_\tau X$ in $A^* X$ are all ideals. The quotient of $Z^* X$ modulo each of these subgroups is a graded ring, contravariant for morphisms of non-singular varieties.

Example 19.3.10. If X is a non-singular complete intersection in $\mathbb{P}^m$, then $H^i X = H^i \mathbb{P}^m$ for $i < n$, so $Z^p X / \mathrm{Hom}^p X = \mathbb{Z}$ for $p < n/2$. Hartshorne (4) asks if $\mathrm{Hom}^p X = \mathrm{Alg}^p X = \mathrm{Rat}^p X$ for $p < n/2$.

Example 19.3.11. Several equivalence relations lying between $\mathrm{Rat}^* X$ and $\mathrm{Alg}^* X$ have been defined, usually involving families of cycles parametrized by abelian varieties. For discussions of some of these, we refer to Samuel (4), Weil (5), Griffiths (3), and Kleiman (4).

Example 19.3.12. If X is a complex scheme which is imbeddable in a non-singular scheme, and one defines

$$A^* X = \varinjlim A^* Y$$

the direct limit taken over all $f: X \to Y$, Y non-singular (cf. Example 8.3.13), there is an induced homomorphism of graded rings

$$cl: A^* X \to H^*(X).$$

If $A^* X$ is the bivariant cohomology ring defined in § 17, however, we do not know if there is such a map.

Example 19.3.13. If $f: X \to Y$ is holomorphic, then the Abel-Jacobi map θ commutes with f_* and f^*. Equivalently, if $T: Y \vdash X$ is an algebraic correspondence, then for all cycles α on Y

$$cl(T)_* \, \theta(\alpha) = \theta(T_* \alpha)$$

(cf. Example 19.2.7). See Griffiths (3) for a discussion of this fact, and King (1) for the technique involved.

The fact that θ vanishes on $\mathrm{Rat}^* X$ may be seen as a special case of this formula (cf. Lieberman (2)). Indeed, a cycle in $\mathrm{Rat}^* X$ is of the form $T_*(\alpha)$, for

$T: \mathbb{P}^1 \vdash X$, α a cycle of degree zero on $\mathbb{P}^1$. Then $\theta(T_* \alpha) = T_* \theta(\alpha) = 0$ since $\theta(\alpha) \in J^1 \mathbb{P}^1 = \mathrm{Pic}^\circ(\mathbb{P}^1) = 0$.

Example 19.3.14. If $H^{n-p,n-p}(X) \bigcap H^{2n-2p}(X; \mathbb{Q})$ consists of rational linear combinations of algebraic cycles, (i.e., the Hodge conjecture holds for p-cycles on X), then $\mathrm{Hom}^p_\tau X = \mathrm{Num}^p X$.

Notes and References

Several references for the transcendetal theory of algebraic cycles have been included in the text and examples. For general surveys, we recommend Griffiths (3), (4), Griffiths-Harris (1), Carlson-Green-Griffiths-Harris (1), Hartshorne (4), and Kleiman (2).

Although the use of topology and homology on algebraic varieties can be traced back to Riemann and Poincaré, it was Lefschetz who developed a simplicial theory for intersecting homology classes on an ordinary manifold, and applied it to the classes of algebraic subvarieties of complex projective manifold (cf. Lefschetz (2)). This theory depended on the triangulability of complex varieties (cf. Hironaka (2)). Borel and Haefliger (1) used relative cohomology and Borel-Moore homology to construct the cycle map and give a modern proof that algebraic and topological intersections agree. Refinements were given by Atiyah and Hirzebruch (2) and Douady (1).

Sections 19.1 and 19.2 extend these results to refined intersections on nonsingular varieties, and to local complete intersections on possibly singular varieties. These results stem from Baum-Fulton-MacPherson (1), Fulton-MacPherson (1), (3), Deligne (2), and the sources cited at the end of § 19.2.

Many of the above sources construct intersections on complex analytic as well as algebraic varieties. Others have used the machinery of differential forms, residues, currents and kernel functions to give more or less explicit analytic formulas for cycle classes. In particular, King (3) gave an analytic version of the excess intersection formula, some time before the formula appeared in modern algebraic geometry. Other references for the analytic approach are Draper (1), King (1), (2), Raisonnier (1), and Toledo and Tong (2).

Borel-Moore homology is the natural homology theory for use on noncompact algebraic or analytic varieties. In addition to the original source, Borel-Moore (1), one may refer to Verdier (1), Douady-Verdier (1), or Iversen (4) for the general sheaf-theoretic development. Since this theory also arises naturally in the intersection homology theory of Goresky and MacPherson (1), (2), we look forward to its becoming more familiar – perhaps appearing in standard topology texts.

Many of the results discussed in this chapter are valid for varieties over arbitrary algebraically closed fields, using étale in place of classical homology and cohomology. For this we refer to Deligne (2), and to articles by Laumon and Verdier in the seminar of Douady-Verdier (1). Grothendieck and Katz have extended Griffiths' example (§ 19.3.3) to characteristic p (Deligne-Katz (1)XX). The basic facts on divisors (§ 19.3.1) were proved in the abstract case by Matsusaka and Weil (see Example 19.3.3 for references). Bloch (3) and Milne (1) have extended Roĭtman's theorem (§ 19.3.5) to characteristic p

For discussion of relations of algebraic cycles to K-theory and number theory, as well as much more on zero-cycles, we recommend the lectures of Bloch (4).

Chapter 20. Generalizations

Summary

Much of the intersection theory developed in this text is valid for more general schemes than algebraic schemes over a field. A convenient category, sufficient for applications envisaged at present, is the category of schemes X of finite type over a regular base scheme S. Using an appropriate definition of relative dimension, one has a notion of k-cycle on X, and a graded group $A_*(X)$ of rational equivalence classes, satisfying the main functorial properties of Chaps. 1–6. The Riemann-Roch theorem also holds; in particular

$$A_*(X) \otimes \mathbb{Q} \cong K_\circ(X) \otimes \mathbb{Q} .$$

The main missing ingredient in such generality is an exterior product

$$A_k(X) \otimes A_l(Y) \to A_{k+l}(X \times_S Y) .$$

If S is one-dimensional, however, there is such a product. In particular, if X is smooth over S, then $A_*(X)$ has a natural ring structure.

When $S = \operatorname{Spec}(R)$, R a discrete valuation ring, and X is a scheme over S, with general fibre X° and special fibre $\bar{X}$, there are specialization maps

$$\sigma : A_k(X^\circ) \to A_k(\bar{X}) ,$$

which are compatible with all our intersection operations. If X is smooth over S, σ is a homomorphism of rings.

For proper intersections on a regular scheme, Serre has defined intersection numbers using Tor. For smooth schemes over a field, these numbers agree with those in § 8.2. Indeed, even for improper intersections, a Riemann-Roch construction shows how to recover intersection classes from Tor, at least with rational coefficients.

Although higher K-theory is outside the scope of this work, the chapter concludes with a brief discussion of Bloch's formula

$$A^p X = H^p(X, \mathscr{K}_p) .$$

20.1 Schemes Over a Regular Base Scheme

In this section S denotes an arbitrary regular scheme, i.e., S is a Noetherian scheme all of whose local rings are regular local rings. All schemes X will be of finite type and separated over S. No assumption of a ground field is made; in particular, S could be $\operatorname{Spec}(\Lambda)$, with Λ: (i) the ring of integers in a number field on any Dedekind domain, (ii) a p-adic ring of integers or any discrete valuation ring; (iii) any regular ring.

We introduce a notion of relative dimension for schemes over S, which shares many properties with absolute dimension over a ground field. For $p: X \to S$ a scheme over S, and $V \subset X$ a closed integral subscheme of X, define

$$\dim_S V = \text{tr.deg.}(R(V)/R(T)) - \operatorname{codim}(T, S)$$

where T is the closure of $p(V)$ in S, and $R(V)$, $R(T)$ are the function fields. If $v \in X$ is the generic point of V, and $t = p(v)$, then

$$\dim_S V = \text{tr.deg.}(\varkappa(v)/\varkappa(t)) - \dim(\mathcal{O}_{t,S}) .$$

Note that in case S is an algebraic scheme over a field, $\dim_S V = \dim(V) - \dim(S)$.

Lemma 20.1. (1) *If V° is a non-empty open subscheme of V, then* $\dim_S V^\circ = \dim_S V$.

(2) *If $W \subset V$ is a closed integral subscheme of V, then*

$$\dim_S V = \dim_S W + \operatorname{codim}(W, V) .$$

(3) *If $f: V \to W$ is a dominant morphism of integral schemes over S, then*

$$\dim_S V = \dim_S W + \text{tr.deg.}(R(V)/R(W)) .$$

In particular, $\dim_S W \leqq \dim_S V$, *with equality if and only if $R(V)$ is a finite extension of $R(W)$.*

Proof. (1) and (3) are immediate from the definition. For (2), let T and U be the closures of the images of V and W in S. Since S is universally catenary, we have the dimension formula [EGA]IV.5.6.5:

$$\text{tr.deg.}(R(V)/R(T)) + \operatorname{codim}(U, T) = \text{tr.deg.}(R(W)/R(U)) + \operatorname{codim}(W, V) ,$$

which is equivalent to (2). □

Using this relative notion of dimension, a k-cycle on a scheme X can be defined, as in § 1.3, to be a formal sum $\sum n_i [V_i]$, with V_i closed integral subschemes of X such that $\dim_S V_i = k$. The notion of rational equivalence is defined as in § 1.3. Denote by $A_k X$ or $A_k(X/S)$ the group of rational equivalence classes of k-cycle. Note that the group

$$A_* X = \bigoplus_k A_k(X/S)$$

does not depend on S; only the grading depends on the dimension relative to S.

One has the covariance of A_k for proper morphisms, contravariance for flat morphisms (of some relative dimension), and the exact sequence of § 1.8. (For Proposition 1.4(b) the proof appealing to [EGA]III must be used.) The alternate definition of $A_* X$ (§ 1.6) is also valid provided one uses $X \times_S \mathbb{P}^1_S$ in place of $X \times \mathbb{P}^1$. However, one does *not* in general have exterior products as in § 1.10, since subvarieties of X need not be flat over S.

With these foundations, the rest of § 2 – § 6 carries over without essential change. As above, appearances of $X \times \mathbb{P}^n$ or $X \times \mathbb{A}^n$ should be replaced by $X \times_S \mathbb{P}^n_S$ or $X \times_S \mathbb{A}^n_S$, and any other mention of exterior products should be omitted. Section 7.1 goes over as stated, but § 7.2 (where finiteness of a normalization is appealed to) does not.

In particular, one has Chern class operators

$$c_i(E) \cap _ : A_k X \to A_{k-i} X$$

for E a vector bundle on X, and refined Gysin homomorphisms:

$$f^! : A_k Y' \to A_{k-d}(X \times_Y Y')$$

for $f: X \to Y$ a (factorable) l.c.i. morphism of codimension d, $Y' \to Y$ arbitrary, satisfying the fundamental properties of § 3 and § 6. If Y is a scheme which is smooth over S of relative dimension n, the diagonal is a regular imbedding, so one has Gysin homomorphisms

$$\delta^* : A_k(Y \times_S Y) \to A_{k-n}(Y).$$

Lacking exterior products $A_* Y \otimes A_* Y \to A_*(Y \times_S Y)$, however, this does not determine products for arbitrary cycle classes on Y (cf. Example 20.1.1).

The residual intersection theorem (§ 9) generalizes without change. Many of the other results of § 9 – § 13 have analogues in this generality; we leave this to the interested reader. The formulas for degeneracy loci in § 14 generalize without change. In addition, the Riemann-Roch theorems in § 15 and § 18 are valid, with the same proofs, in this generality. In particular, this implies that

$$K_\circ X_{\mathbb{Q}} \cong A_* X_{\mathbb{Q}} \cong \operatorname{Gr} K_\circ X_{\mathbb{Q}}$$

where $\operatorname{Gr} K_\circ X$ is constructed from the filtration of $K_\circ X$ using relative dimension over S.

The definition and formalism (§ 17) of bivariant theories $A(X \to Y)$ extends to schemes over S; only Proposition 17.3.1 and Corollary 17.4, where exterior products were used, must be omitted. In particular, one has a contravariant functor A^* from schemes over S to rings.

Example 20.1.1. (a) If $Y \to S$ is smooth of relative dimension n, then, with the notation of Chap. 17,

$$A^p Y \cong A^{p-n}(Y \to S).$$

For any $f: X \to Y$, $A^p(X \to Y) \cong A^{p-n}(X \to S)$.

(b) If $V \subset Y$ is a closed subscheme of Y which is flat over S of relative dimension m, then V determines an element c_V in $A^{-m}(Y \to S)$, and hence a class in $A^{n-m} Y$ if Y is as in (a). In particular, c_V determines homomorphisms

$$f^*(c_V) \cap _ : A_k X \to A_{k+m-n} X$$

for any $f: X \to Y$. (For proofs of (a) and (b) see Proposition 17.4.2.)

(c) If S is smooth over a ground field, then the canonical homomorphism from $A^*(Y \to S)$ to $A_* Y$ is an isomorphism. On the other hand, if one knows that this homomorphism is an isomorphism, and Y is smooth over S, then $A_* Y \cong A^* Y$ has an intersection product (cf. Example 20.2.3).

Example 20.1.2. If $f: X \to Y$ is a morphism of smooth S-schemes, of relative dimensions n and m over S, there is a pull-back

$$f^*: A_k(Y/S) \to A_{k+n-m}(X/S)$$

defined as the composite $\gamma_f^* \circ p^*$, $\gamma_f: X \to X \times_S Y$ the graph of f, $p: X \times_S Y \to Y$ the (flat) projection (for this X need only be flat over S). The construction of the double point class $\mathbb{D}(f)$, and the proof of the double point formula in § 9.3, generalize without essential change to this setting.

Example 20.1.3. For an arbitrary Noetherian scheme X, let $Z.X$ be the free abelian group on the subvarieties (closed integral subschemes) V of X. For $r \in R(V)^*$, let $[\operatorname{div}(r)] = \sum \operatorname{ord}_W(r)[W]$, the sum over all subvarieties W of V of codimension one. Define $A.X$ to be $Z.X/\sim$, where $\sim$ is the subgroup generated by all such $[\operatorname{div}(r)]$. If $f: X \to Y$ is proper, define $f_*: Z.X \to Z.Y$ by

$$f_*[V] = \deg(V/f(V)) \cdot [f(V)]$$

where $\deg(V/f(V)) = [R(V): R(f(V))]$ if this number is finite, and $\deg(V/f(V)) = 0$ otherwise. Then f_* passes to rational equivalence, making $A.$ a covariant functor for proper morphisms. Similarly, $A.$ is contravariant for flat morphisms.

Using infinite cycles $\sum n_V[V]$ which are locally finite (i.e., each point has a neighborhood U such that only a finite number of V with $n_V \neq 0$ meet U), the definition extends to arbitrary locally Noetherian schemes.

For X locally of finite type over a universally catenary base scheme S, the definition of relative dimension given in this section determines a grading for $A.(X)$. We do not know if the Riemann-Roch theorem extends to this generality.

Example 20.1.4. The preceding construction applies in particular to the case $X = \operatorname{Spec}(R)$, R a local Noetherian ring. If R is the local ring of a variety Y at y, $A_* X$ has been called the local Chow group of Y at y. Equivalently, $A_* X = \varinjlim A_*(U)$, where the limit is over all open subschemes U of Y which contain y. The case when Y is the cone over a smooth projective variety, with vertex y, is discussed by Grothendieck in [SGA 6] Exp. XIV.8.

If R is regular, it follows from Riemann-Roch that $A_*(X)_{\mathbb{Q}} = \mathbb{Q}$, generated by $[X]$.

A related, but different, construction of rational equivalence for locally Cohen-Macaulay rings has been studied by Claborn and Fossum (1).

Example 20.1.5. If the ordinary dimension of integral schemes is used, none of the three assertions of Lemma 20.1 is true. However, the reader who wishes may replace each $\dim_S V$ by $\dim_S V + \dim(S)$, with only notational changes; in § 20.3 specialization will then change dimension of cycle classes.

20.2 Schemes Over a Dedekind Domain

In this section S is a one-dimensional regular scheme. The important case is when $S = \mathrm{Spec}(A)$, A a Dedekind domain. All schemes are of finite type and separated over S. Any variety (integral scheme) V is either flat over S, or maps to a closed point P in S. Let X and Y be schemes over S, $V \subset X$, $W \subset Y$ subvarieties (closed integral subschemes). Define a *product* cycle $[V] \times_S [W]$ on $X \times_S Y$ by

$$[V] \times_S [W] = \begin{cases} [V \times_S W] & \text{if } V \text{ or } W \text{ is flat over } S \\ 0 & \text{otherwise}. \end{cases}$$

If $\dim_S V = k$, $\dim_S W = l$, then $[V] \times_S [W]$ is a $(k+l)$-cycle.

Proposition 20.2. *The above product passes to rational equivalence, defining an exterior product*

$$A_k(X/S) \otimes A_l(Y/S) \xrightarrow{\times} A_{k+l}(X \times_S Y/S).$$

Proof. If α is rationally equivalent to 0 on X, $W \subset Y$ a subvariety, we must show that $\alpha \times_S [W] \sim 0$. If W is flat over S, the induced map $p: X \times_S W \to X$ is flat, and

$$\alpha \times_S [W] = p^*(\alpha) \sim 0$$

since rational equivalence pulls back by flat maps. If W maps to P in S, let i_P be the imbedding of P in S. Then

$$\alpha \times_S [W] = i_P^!(\alpha) \times_P [W] \sim 0$$

since $i_P^!$ preserves rational equivalence, as does exterior product over P (§ 1.10). □

From the construction one sees that this product satisfies the usual commutative and associative properties for exterior products. For schemes which are smooth over S, the material of Chap. 8 extends without change. If Y is smooth of relative dimension n over S, set

$$A^p(Y/S) = A_{n-p}(Y/S).$$

There is an internal product

$$A^p(Y/S) \otimes A^q(Y/S) \xrightarrow{\cdot} A^{p+q}(Y/S)$$

defined by $\alpha \cdot \beta = \delta^*(\alpha \times_S \beta)$, with $\delta: Y \to Y \times_S Y$ the diagonal – a regular imbedding of codimension n. This makes $A^*(Y/S)$ into a commutative, graded ring with unit $1 = [Y]$. Given $f: X \to Y$, $A_*(X/S)$ is a module over $A^*(Y/S)$ by defining

$$f^*(\beta) \cap \alpha = \gamma_f^*(\alpha \times_S \beta),$$

$\gamma_f: X \to X \times_S Y$ the graph of f. In particular, if X is smooth, defining $f^*(\beta) = \gamma_f^*([X] \times_S \beta)$ defines a pull-back

$$f^*: A^p(Y/S) \to A^p(X/S),$$

making A^* a contravariant functor from smooth schemes over S to rings.

Example 20.2.1. *Refined products.* Let $f: X \to Y$, Y smooth of relative dimension n over S, and let $X' \to X$, $Y' \to Y$ be given morphisms. If $\alpha \in A_k(X'/S)$, $\beta \in A_l(Y'/S)$, define a refined product

$$\alpha \cdot_f \beta \in A_{k+l-n}(X' \times_Y Y'/S)$$

by $\alpha \cdot_f \beta = \gamma_f^!(\alpha \times_S \beta)$. For example, if X' and Y' are the supports of cycles α and β on X and Y, then $\alpha \cdot_f \beta$ is a class in $A_*(|\alpha| \cap f^{-1}(|\beta|))$ which maps to $f^*(\beta) \cap \alpha$ in $A_* X$. The basic properties are given in Proposition 8.1.1.

Example 20.2.2. *Intersection multiplicity.* If V, W are subvarieties of a scheme Y which is smooth over S, and P is a proper component of $V \cap W$, i.e.

$$\dim_S P = \dim_S V + \dim_S W - \dim_S Y,$$

then the coefficient of $[P]$ in $[V] \cdot [W]$ is a positive integer, denoted $i(P, V \cdot W; Y)$. The discussion of Chap. 7 carries over to this intersection multiplicity. (If neither V nor W is flat over S, no component of $V \cap W$ can be proper in this sense. Otherwise $[V \times_S W]$ is a positive cycle on $Y \times_S Y$, and the intersection with the diagonal proceeds as in § 8.2.)

Example 20.2.3. For any X over S, the canonical homomorphism

$$A^{-k}(X \to S) \to A_k(X/S),$$

$c \to c([S])$, is an isomorphism. (An element α in $A_k(X/S)$ determines homomorphisms

$$A_q(Y/S) \to A_{q+k}(X \times_S Y/S)$$

by $\beta \to \alpha \times_S \beta$. Clearly this operator takes $[S]$ to α. To show that the maps are isomorphisms, it suffices to show an element c in $A^{-k}(X \to S)$ vanishes if $c([S]) = 0$. If V is integral, and $V \to S$ is flat, then $c([V]) = 0$ by property (C_2) of § 17.1. If V maps to $P \in S$, i_P the inclusion of P in S, then c acts on $[V]$ through $i_P^*(c) \in A^{-k}(X_P \to P) = A_k(X_P)$; but $i_P^*(c)[P] = i_P^*(c[S]) = 0$, so $i_P^* c = 0$ by Proposition 17.3.1.)

If Y is smooth of relative dimension n over S, then (Example 20.1.1)

$$A^p Y \cong A^{p-n}(Y \to S) = A_{n-p}(Y/S).$$

20.3 Specialization

Let S, $\bar{S}$ be regular schemes, $i: \bar{S} \to S$ a closed regular imbedding of codimension d. Assume that the normal bundle to $\bar{S}$ in S in trivial, or at least that its top Chern class is zero. Let $S^\circ = S - \bar{S}$, j the inclusion of S° in S.

For any scheme X over S, set $\bar{X} = X \times_S \bar{S}$, $X^\circ = X \times_S S^\circ$. Note that

$$A_k(\bar{X}/\bar{S}) = A_{k-d}(\bar{X}/S) \quad \text{and} \quad A_k(X^\circ/S^\circ) = A_k(X^\circ/S)$$

by our conventions on relative dimensions. From § 1.8 we have an exact sequence

$$A_k(\bar{X}/S) \xrightarrow{i_*} A_k(X/S) \xrightarrow{j^*} A_k(X^\circ/S^\circ) \to 0,$$

where i_* and j^* denote push-forward and pull-back induced by inclusions of $\bar{X}$ and X° in X. From § 6.2 there is a Gysin homomorphism

$$i^!: A_k(X/S) \to A_{k-d}(\bar{X}/S) = A_k(\bar{X}/\bar{S}).$$

Since $i^! i_* = 0$ (Theorem 6.3), there is a unique homomorphism

$$\sigma: A_k(X^\circ/S^\circ) \to A_k(\bar{X}/\bar{S})$$

such that $\sigma(j^*\alpha) = i^!(\alpha)$ for all $\alpha \in A_k(X/S)$. This homomorphism σ is called the *specialization* map.

For any morphism $f: X \to Y$ of schemes over S, let $\bar{f}: \bar{X} \to \bar{Y}$, $f^\circ: X^\circ \to Y^\circ$ denote the induced morphisms.

Proposition 20.3. (a) *If $f: X \to Y$ is proper, and $\alpha \in A_* X^\circ$, then $\bar{f}_* \sigma(\alpha) = \sigma f^\circ_*(\alpha)$ in $A_* \bar{Y}$.*

(b) *If $f: X \to Y$ is flat, or a regular imbedding, and $\alpha \in A_* Y^\circ$, then $\bar{f}^* \sigma(\alpha) = \sigma f^{\circ *}(\alpha)$ in $A_* \bar{X}$.*

Proof. These follow from the commutativity of $i^!$ and j^* with proper push-forward, flat pull-back, and intersection products (§ 6). □

Suppose $S = \operatorname{Spec}(R)$, with R a discrete valuation ring, with residue field $\varkappa$ and quotient field K, and $\bar{S} = \operatorname{Spec}(\varkappa)$, $S^\circ = \operatorname{Spec}(K)$. For a scheme X over S, $\bar{X}$ and X° are algebraic schemes over $\varkappa$ and K, and $A_k(\bar{X}) = A_k(\bar{X}/\bar{S})$, $A_k(X^\circ) = A_k(X^\circ/S^\circ)$. In this case σ has a simple description on the cycle level: If V° is a subvariety of X°, and V is the closure of V° in X, then V is flat over S, and

$$\sigma[V^\circ] = [\bar{V}].$$

Corollary 20.3. *If X is smooth over S, S as above, then the specialization map*

$$\sigma: A_*(X^\circ) \to A_*(\bar{X})$$

is a ring homomorphism, i.e., σ preserves the intersection product.

Proof. Given subvarieties V°, W° of X°, let V, W be their closures in X. Then $V \times_S W$ is a flat subscheme of $X \times_S X$, restricting to $V^\circ \times_{S^\circ} W^\circ$ over S°. Applying (b) of the proposition to the regular imbedding $\delta: X \to X \times_S X$,

$$\begin{aligned}\sigma([V^\circ] \cdot [W^\circ]) &= \sigma\, \delta^{\circ *}[V^\circ \times_{S^\circ} W^\circ] = \bar{\delta}^* \sigma[V^\circ \times_{S^\circ} W^\circ] \\ &= \bar{\delta}^*([\bar{V} \times_{\bar{S}} \bar{W}]) = [\bar{V}] \cdot [\bar{W}],\end{aligned}$$

as required. □

In a similar way, all our other intersection operations are compatible with specialization.

Example 20.3.1. If Y is smooth over $S = \operatorname{Spec}(R)$, R a discrete valuation ring, then the homomorphisms in the commutative diagram

$$\begin{array}{ccc} & A^*(Y/S) & \\ {\scriptstyle j^*}\swarrow & & \searrow{\scriptstyle i^*} \\ A^*(Y^\circ) & \xrightarrow{\sigma} & A^*(\bar{Y}) \end{array}$$

are all homomorphisms of graded rings.

Example 20.3.2. Let $S = \operatorname{Spec}(R)$, R a discrete valuation ring, $f: X \to Y$ a morphism of S-schemes, with Y smooth over S. Let Z be a closed subscheme of X. Suppose α° and β° are cycles on X° and Y°, with $|\alpha| \cap (f^\circ)^{-1}|\beta^\circ| \subset Z^\circ$. Then

$$\sigma(\alpha^\circ \cdot_{f^\circ} \beta^\circ) = \sigma(\alpha^\circ) \cdot_{\bar{f}} \sigma(\beta^\circ)$$

in $A_*(\bar{Z})$. (This is a refinement of Corollary 20.3, with a similar proof.)

Example 20.3.3. Let X be a scheme over S, $D_1, \ldots, D_n$ Cartier divisors, or more generally, pseudo-divisors (cf. § 2.2) on X. Let V be a subscheme of X, flat over S of relative dimension n. Assume that $\bigcap D_i \bigcap V$ is proper over S. Let D_i°, $\bar{D}_i$ denote the restrictions of D_i to X°, $\bar{X}$. Then

$$\int_{\bigcap D_i^\circ \bigcap V^\circ} D_1^\circ \cdot \ldots \cdot D_n^\circ \cdot [V^\circ] = \int_{\bigcap \bar{D}_i \bigcap \bar{V}} \bar{D}_1 \cdot \ldots \cdot \bar{D}_n \cdot [\bar{V}].$$

Indeed, for any pseudo-divisor D and any cycle α on X°, $\sigma(D^\circ \cdot \alpha) = \bar{D} \cdot \sigma(\alpha)$.

Example 20.3.4. The Riemann-Roch homomorphism commutes with specialization map $\sigma: K_\circ(X^\circ) \to K_\circ(\bar{X})$, defined as with rational equivalence, provided the normal bundle to $\bar{S}$ in S is trivial. Then the diagram

$$\begin{array}{ccc} K_\circ(X^\circ) & \xrightarrow{\tau_{X^\circ}} & A_*(X^\circ/S^\circ) \\ \sigma\downarrow & & \downarrow\sigma \\ K_\circ(\bar{X}) & \xrightarrow{\tau_{\bar{X}}} & A_*(\bar{X}/\bar{S}) \end{array}$$

commutes. In particular, if $\mathscr{F}$ is a coherent sheaf on X which is flat over S, then $\sigma(\tau_{X^\circ}(\mathscr{F}^\circ)) = \tau_{\bar{X}}(\bar{\mathscr{F}})$; this generalizes the equality of Euler characteristics $\chi(X^\circ, \mathscr{F}^\circ) = \chi(\bar{X}, \bar{\mathscr{F}})$, when X is proper over S.

Example 20.3.5 (cf. [SGA6]X App. and Fulton (2)§ 4). Let R be a complete discrete valuation ring with quotient field K and residue field $\varkappa$. Let $\bar{K}$ and $\bar{\varkappa}$ be algebraic closures of K and $\varkappa$. For a scheme X over R there are specialization homomorphisms

$$\sigma: A_*(X \otimes_R \bar{K}) \to A_*(X \otimes_R \bar{\varkappa})$$

compatible with our intersection operations. (For all finite extensions R' of R in $\bar{K}$, with quotient field K' and residue field $\varkappa'$, we have specialization maps

$$A_*(X \otimes_R K') \to A_*(X \otimes_R \varkappa'),$$

constructed from the R'-scheme $X \otimes_R R'$. Passing to the direct limit over R' gives σ.)

If X is smooth over R, one has a commutative diagram

$$\begin{array}{ccc} A^i(X \otimes_R \bar{K}) & \xrightarrow{\quad\sigma\quad} & A^i(X \otimes_R \bar{\varkappa}) \\ \downarrow cl & & \downarrow cl \\ H^{2i}(X \otimes_R \bar{K}, \mathbb{Z}_l(i)) & \cong & H^{2i}(X \otimes_R \bar{\varkappa}, \mathbb{Z}_l(i)). \end{array}$$

where $l \neq \operatorname{char}(\varkappa)$, and $H^{2i}(\ , \mathbb{Z}_l(i))$ is l-adic étale cohomology.

Example 20.3.6. Let X be smooth and proper over $\operatorname{Spec}(R)$, with R, $\bar{K}$, $\bar{\varkappa}$ as in the preceding example. Then

$$\operatorname{rank} NS(X \otimes_R \bar{K}) \leqq \operatorname{rank} NS(X \otimes_R \bar{\varkappa}).$$

(If $\alpha_1, \ldots, \alpha_\varrho$ is a basis for divisors modulo numerical equivalence on $X \otimes_R \bar{K}$, there are 1-cycles $\beta_1, \ldots, \beta_\varrho$ on $X \otimes_R \bar{K}$ with $\det(\alpha_i \cdot \beta_j) \neq 0$. Then

$$\det(\sigma(\alpha_i) \cdot \sigma(\beta_j)) = \det(\alpha_i \cdot \beta_j) \neq 0 ,$$

so $\sigma(\alpha_1), \ldots, \sigma(\alpha_\varrho)$ are independent.)

20.4 Tor and Intersection Products

Let X be a regular scheme, V and W subvarieties (closed integral subschemes) of X. The sheaves $\mathrm{Tor}_i^{\mathcal{O}_X}(\mathcal{O}_V, \mathcal{O}_W)$ are supported on $V \cap W$, so determine elements of the Grothendieck group $K_\circ(V \cap W)$ of coherent sheaves on $V \cap W$ (cf. Example 15.1.7 (b)). Set

$$\mathrm{Tor}^X(V, W) = \sum_{i=0}^{\dim(X)} (-1)^i [\mathrm{Tor}_i^{\mathcal{O}_X}(\mathcal{O}_V, \mathcal{O}_W)]$$

in $K_\circ(V \cap W)$.

Let P be an irreducible component of $V \cap W$, with its reduced structure. Any element α in $K_\circ(V \cap W)$ may be written in the form

$$\alpha = m_P(\alpha)[\mathcal{O}_P] + \alpha'$$

where $m_P(\alpha)$ is a well-defined integer, and α' is a linear combination of sheaves whose supports do not contain P. Serre (4)V.6 has proved that

$$\mathrm{codim}(P, X) \leqq \mathrm{codim}(V, X) + \mathrm{codim}(W, X) .$$

If P is a *proper* component, i.e., equality holds in this inequality, Serre defines an intersection number $i(P, V \cdot W; X)$ by the formula

$$i(P, V \cdot W; X) = m_P(\mathrm{Tor}^X(V, W)) .$$

This intersection number is obviously commutative in V and W, and the expected associativity formula follows from a Tor spectral sequence. Serre's conjecture that this intersection number is always positive is one of the outstanding open questions in commutative algebra (cf. Example 20.4.2).

If X is smooth over a field, this definition agrees with that given in Chap. 8. For then one may reduce to the diagonal, where both terms agree with the multiplicity obtained from a Koszul complex (cf. Example 7.1.2 and Serre (4) V.C).

More generally, with no assumption on the properness of the intersection, the refined intersection class, at least with rational coefficients, is determined by Tor. Let X be smooth over a field, V, W subvarieties of X. Set

$$m = \dim V + \dim W - \dim X ,$$

so the refined intersection class $V \cdot W$ is in $A_m(V \cap W)$ (§ 8.1). Let

$$\tau : K_\circ(V \cap W) \to A_*(V \cap W)_{\mathbb{Q}}$$

be the Riemann-Roch map (§ 18). Then (Example 18.3.13 (b))

$$\tau(\mathrm{Tor}^X(V, W)) = V \cdot W + \text{terms of dim} < m .$$

Example 20.4.1 (Fulton-MacPherson (1) § 6). With notation as in the last paragraph of this section,

$$\mathrm{Tor}^X(V, W) \in F_m K_\circ(V \cap W) .$$

This is not known for X an arbitrary regular scheme. Indeed, Serre also conjectured that

$$m_P(\mathrm{Tor}^X(V, W)) = 0$$

for P any irreducible component of $V \cap W$ of dimension greater than m; this conjecture too remains open.

Example 20.4.2. Let X be a regular scheme, P an irreducible component of $V \cap W$. Let A be the local ring of X at P. Serre (4) p. V-15 proved the positivity of $m_P(\mathrm{Tor}^X(V, W))$ in the case of proper intersections, and its vanishing in the improper case, under the assumptions that all localizations of A at prime ideals are either equicharacteristic or unramified. This problem is also discussed in Nastold (1) and Malliavin (1).

For A a regular local ring, Serre's question is equivalent to the positivity or vanishing of

$$\sum (-1)^i l_A(\mathrm{Tor}_i^A(M, N))$$

for finitely generated A-modules M, N with $l_A(M \otimes_A N) < \infty$. The hope that this might hold for arbitrary local ring A, if M or N had a finite free resolution, has been dashed by the recent example of Dutta-Hochster-McLaughlin (1). Note that, since the completion $\hat{A}$ of A is flat over A, it suffices to consider the question for complete regular local rings.

Example 20.4.3. Let X be regular, P a proper component of $V \cap W$. Assume (with no loss of generality, replacing X by an open subscheme) that P is regular, and $V \cap W = P$ (set-theoretically). Let $\pi: \tilde{X} \to X$ be the blow-up of X along P, E the exceptional divisor, $\eta: E \to P$ the induced morphism. Let $\tilde{V}$, $\tilde{W}$ be the proper transforms of V and W in $\tilde{X}$. Then

$$(*) \qquad i(P, V \cdot W; X) = e_P(V)\, e_P(W) + m_P(\eta_*(\mathrm{Tor}^{\tilde{X}}(\tilde{V}, \tilde{W}))) .$$

It is natural to conjecture that, as in the geometric case (Theorem 12.4), the second term is always non-negative, so that

$$i(P, V \cdot W; X) \geqq e_P(V)\, e_P(W) .$$

(The multiplicities $e_P(V)$, $e_P(W)$ are defined as in the geometric case. The proof of (*) given in § 12.5 works equally well with no base field. Note that the proof of the non-negativity of the second term used reduction to the diagonal, so does not work on general schemes).

A particular case of (*) is the formula

$$i(P, V \cdot W; X) = e_P(V)\, e_P(W)$$

in case $\tilde{V} \cap \tilde{W} \cap E = P(C_P V) \cap P(C_P W)$ is empty. This had been proved by Tennison (1).

Serre's definition extends to the proper intersection of more than two varieties, and (*) and the other results of this section extend readily to this generality.

Example 20.4.4. For V, W subvarieties of an arbitrary scheme X, P a component of $V \cap W$, one may form the formal power series

$$F(T) = \sum_{i=0}^{\infty} m_P(\mathrm{Tor}_i^X(V, W))\, T^i$$

in $\mathbb{Z}[[T]]$. If X is regular, and P proper, then $F(-1)$ is the intersection number $i(P, V \cdot W; X)$. Can $F(-1)$ be defined more generally, say by analytic continuation? Is this related to fractional intersection multiplicities (cf. Example 7.1.16)?

20.5 Higher K-theory

This section sketches the relation of Quillen's higher K-theory to cycles and rational equivalence. For details we refer to the basic paper of Quillen (2), and to Grayson (1), (2) and Gillet (1).

For a commutative ring R, Quillen (2) has constructed a sequence of higher K-groups $K_p(R)$, with $K_0(R)$ the Grothendieck group of projective R-modules (locally free sheaves on Spec(R). If R is local, $K_1(R)$ is the group R^* of units in R. For a scheme X, the presheaf

$$U \to K_p(\Gamma(U, \mathcal{O}_X))$$

determines a sheaf, denoted $\mathcal{K}_p$, on X. The sheaf $\mathcal{K}_0$ is the constant sheaf $\mathbb{Z}$, and $\mathcal{K}_1 \cong \mathcal{O}_X^*$.

If X is regular, there are complexes

$$(*)_p \qquad 0 \to \mathcal{K}_p \to \mathcal{G}_p^0 \xrightarrow{d} \mathcal{G}_p^1 \to \ldots \to \mathcal{G}_p^{p-1} \xrightarrow{d} \mathcal{G}_p^p \to 0$$

with

$$\mathcal{G}_p^i = \bigoplus_{\mathrm{codim}(V)=i} (\varphi_V)_*(K_{p-i}(R(V))) .$$

Here the sum is over closed integral subschemes V of X, φ_V is the inclusion of V in X, and $K_j(R(V))$ is the jth higher K-group of the function field $R(V)$, regarded as a constant sheaf on V. In particular, $\Gamma(X, \mathcal{G}_p^p)$ is the group of cycles of codimension p on X, and

$$\Gamma(X, \mathcal{G}_p^{p-1}) = \bigoplus_{\mathrm{codim}(V)=p-1} R(V)^* .$$

The boundary map from the complex $(*)_p$: $\Gamma(X, \mathcal{G}_p^{p-1}) \to \Gamma(X, \mathcal{G}_p^p)$ takes $r \in R(V)^*$ to $[\mathrm{div}(r)]$. The cokernel of this map is therefore the group of rational equivalence classes of codimension p on X, as defined in § 1.3.

Gersten has conjectured that $(*)_p$ is exact for any regular scheme X. Since the sheaves $\mathscr{G}_p^i$ are flasque, Gersten's conjecture implies that $H^i(X, \mathscr{K}_p)$ is the ith cohomology group of the complex $\Gamma(X, \mathscr{G}_p^*)$. Quillen (2) has proved Gersten's conjecture when X is of finite type over a field. This gives the following theorem, first discovered and verified in some cases by S. Bloch.

Theorem. *If X is a regular scheme of finite type over a field, of pure dimension n, then*

$$H^p(X, \mathscr{K}_p) \cong A_{n-p}(X) .$$

This theorem generalizes the isomorphisms $H^0(X, \mathbb{Z}) \cong A_n X$, $H^1(X, \mathscr{O}_X^*) \cong \operatorname{Pic}(X) \cong A_{n-1}(X)$. There are products

$$\mathscr{K}_i \otimes \mathscr{K}_j \to \mathscr{K}_{i+j}$$

which make $H^*(X, \mathscr{K}_*)$ into a ring. When X is smooth over a field, Grayson (1) has shown that this product agrees with the product defined in Chap. 8; for the general construction see Gillet (1).

For possibility singular X, Gillet (3) has constructed Chern classes of vector bundles in $H^p(X, \mathscr{K}_p)$. More generally, he defines, for Y a closed subscheme of X,

$$A_Y^p X = H_Y^p(X, \mathscr{K}_p)$$

where the subscript Y on the right denotes cohomology with supports in Y. There are cap products

$$A_Y^p X \otimes A_q X \to A_{q-p} Y .$$

A complex of vector bundles on X, exact off Y, has local Chern classes in $A_Y^* X$. This capacity for localization is one significant advantage of the higher K-theory realization of cycle classes.

Higher K-theory has also led to insight into the group $\tilde{A}_0(X)$ of classes of 0-cycles of degree zero on a non-singular projective variety X over an algebraically closed field. Bloch (3) has used higher K-theory to prove a theorem of Roĭtman (2) that the canonical map from $\tilde{A}_0(X)$ to the Albanese of X is an isomorphism on torsion prime to the characteristic. Using $\mathscr{K}_2$, Bloch has found evidence to support the conjecture that $\tilde{A}_0(X)$ is finite dimensional when X is a surface with $p_g = 0$. Higher K-theory has also been used to suggest a theory of infinitesimal variation of Chow groups, cf. Bloch (2) and Stienstra (1). For other relations between cycles and K-theory the lectures of Bloch (4) are recommended. Some calculations on singular varieties have been made by Collino (2) and Srinivas (1).

Notes and References

Specialization of cycles and proper intersections was studied by Samuel (3) in the geometric case, and by Shimura (1) over arbitrary discrete valuation rings. The general problem of showing that rational equivalence is preserved under specialization was discussed by Grothendieck [SGA6]X App., and solved in Fulton (2)4. Grothendieck also asked for the construction of a product on A_*X for X smooth over a discrete valuation ring; this is a special case of Theorem 20.2, which is new here. For arithmetic surfaces see Lichtenbaum (1).

The construction of rational equivalence groups of cycles on Noetherian schemes, covariant for proper morphisms, appears first in Fulton (2); previous theories using graded K-groups were discussed in [SGA6]. The definition given in § 20.1, with the notion of relative dimension, is new here; it enables the usual constructions and proofs over a field to extend to schemes over a regular base. Kleiman (12) is one who has advocated doing intersection theory over general base schemes whenever possible. Until interesting applications are in sight, however, the gain in generality must be weighed against the loss of simplicity.

For schemes smooth over a ring of integers in a number field. Arakelov (1) has proposed a definition of intersection product which brings in the infinite primes, as one would like for applications to number theory, cf. Neron (1). Such a ring should surject onto the ring constructed in § 20.2. Arakelov even conjectured a Riemann-Roch theorem in this context. Recent progress on this has been reported by G. Faltings and L. Szpiro.

The definition of intersection product using Tor, discussed in § 20.4, was originated by Serre (4), for the case of proper intersections. The generalization to improper intersections of algebraic schemes using Riemann-Roch is from Fulton-MacPherson (1); the numerical consequence of this formula, in case the expected dimension is zero, had been proved by de Boer (1).

The Riemann-Roch theorem for schemes of finite type over a regular base, and the consequence that $A_*X_{\mathbb{Q}} \cong K_\circ X_{\mathbb{Q}}$, is new here. P. Wagreich (unpublished, 1966) had initiated a study of relative cycles and Riemann-Roch over a base scheme, although his results were quite different from ours.

The formula relating rational equivalence to higher K-theory was discovered by Bloch (1), and established by Quillen (2) when he proved a conjecture of Gersten. The fact that $H^p(X, \mathscr{K}_p)$ makes sense for an arbitrary scheme X was another early indication that one could construct a rational equivalence ring for schemes other than smooth quasi-projective varieties over a field. The material in § 20.5 is based on lectures of Bloch, Gillet, and Grayson at the Institute for Advanced Study in 1981-82.

Appendix A. Algebra

Summary

The main concepts needed from commutative algebra are set forth in this appendix, and the facts needed for the basic theorems are proved. Some further useful results are stated, with references to the literature for proofs.

Notation. All *rings* will be commutative, Noetherian rings with unit. The localization of a ring A (resp. module M) at a prime ideal p is denoted A_p (resp. M_p). If L is a finite extension of a field K, $[L:K]$ denotes the degree of the field extension.

A.1 Length

Definition A.1. For any finitely generated A-module M there is a chain of submodules

(*) $$M = M_0 \supsetneqq M_1 \supsetneqq \ldots \supsetneqq M_r = 0$$

with $M_{i-1}/M_i \cong A/p_i$, p_i a prime ideal in A (Bourbaki (1) IV. § 1 Th. 1). If the p_i which occur in such a chain are all maximal ideals, M is said to have *finite length.* Equivalently, the localization M_p is non-zero for only a finite number of prime ideals p, all necessarily maximal. In this case, the length r of a chain (*) is independent of the chain, and is the *length* of M, denoted $l_A(M)$. Note that

$$l_A(M) = l_{A/I}(M)$$

for any ideal I such that $IM = 0$.

Lemma A.1.1. *If* $0 \to M' \to M \to M'' \to 0$ *is an exact sequence of A-modules, and two of the modules have finite length, then the third module also has finite length, and*

$$l_A(M) = l_A(M') + l_A(M'') .$$

More generally, if

$$0 \to M_t \to M_{t-1} \to \ldots \to M_0 \to 0$$

is an exact sequence of modules of finite length, then

$$\sum_{i=0}^{t} (-1)^i \, l_A(M_i) = 0 .$$

Proof. Chains (*) for M' and M'' of length r' and r'' determine one of length $r'+r''$ for M. The second assertion follows by breaking the long exact sequence into consecutive short exact sequences. □

Lemma A.1.2. *If M has finite length, then*

$$l_A(M) = \sum_p l_{A_p}(M_p)\,,$$

the sum taken over the prime ideals p in A.

Proof. By localizing a chain (*) at a maximal ideal p, one sees that A/p occurs as a factor $l_{A_p}(M_p)$ times. □

Lemma A.1.3. *Let $A \to B$ be a local homomorphism of local rings. Let d be the degree of the residue field extension. A non-zero B-module M has finite length over A if and only if $d < \infty$ and M has finite length over B, in which case*

$$l_A(M) = d \cdot l_B(M)\,.$$

Proof. Since both sides are additive on exact sequences, one is reduced to the case $M = B/q$, q the maximal ideal of B. If p is the maximal ideal of A, then

$$l_A(M) = l_{A/p}(B/q) = d\,,$$

since length and vector space dimension agree over a field. □

Example A.1.1. Let A be a local ring with a subring k which maps isomorphically onto the residue field of A. Then

$$l_A(M) = l_k(M) = \dim_k(M)$$

for any A-module M.

A.2 Herbrand Quotients

Definition A.2. Let $\varphi: M \to M$ be an A-linear homomorphism of a finitely generated A-module. Let

$$M_\varphi = \text{Coker}(\varphi) = M/\text{Image}(\varphi)$$
$${}_\varphi M = \text{Kernel}(\varphi)\,.$$

We say that $e_A(\varphi, M)$ is defined if M_φ and ${}_\varphi M$ both have finite length over A. Equivalently, the localization

$$\varphi_p: M_p \to M_p$$

is an isomorphism for all but a finite number of prime (necessarily maximal) ideals. Then set

$$e_A(\varphi, M) = l_A(M_\varphi) - l_A({}_\varphi M)\,.$$

In case φ is multiplication by an element a in A, we write $e_A(a, M)$ for $e_A(\varphi, M)$. Note that

$$e_A(\varphi, M) = e_{A/I}(\varphi, M)$$

for any ideal I with $IM = 0$.

For applications in the text, only the case where a is a non-zero-divisor on M is needed, in which case $e_A(a, M) = l_A(M/aM)$. As the following lemmas show, the general case is more natural, from an algebraic point of view.

Lemma A.2.1. *If $l_A(M) < \infty$, then*

$$e_A(\varphi, M) = 0$$

for all $\varphi: M \to M$.

Proof. This follows from Lemma A.1.1 and the exact sequence

$$0 \to {}_{\varphi}M \to M \xrightarrow{\varphi} M \to M_{\varphi} \to 0 . \quad \square$$

Lemma A.2.2. *If $e_A(\varphi, M)$ is defined, then*

$$e_A(\varphi, M) = \sum_{p} e_{A_p}(\varphi_p, M_p) ,$$

the sum taken over all prime ideals p in A.

Proof. This follows from Lemma A.1.2 and the fact that the construction of kernels and cokernels commutes with localization. $\square$

Lemma A.2.3. *Let $A \to B$ be a local homomorphism of local rings, d the degree of the residue field extension. Let $\varphi: M \to M$ be a B-linear endomorphism of a B-module M. If $d < \infty$, then $e_A(\varphi, M)$ is defined if and only if $e_B(\varphi, M)$ is defined, and*

$$e_A(\varphi, M) = d \cdot e_B(\varphi, M) .$$

Proof. This follows from Lemma A.1.3. $\square$

Lemma A.2.4. Given a commutative diagram

$$\begin{array}{ccccccccc} 0 & \to & M' & \xrightarrow{\alpha} & M & \xrightarrow{\beta} & M'' & \to & 0 \\ & & \downarrow{\scriptstyle \varphi'} & & \downarrow{\scriptstyle \varphi} & & \downarrow{\scriptstyle \varphi''} & & \\ 0 & \to & M' & \xrightarrow[\gamma]{} & M & \xrightarrow[\delta]{} & M'' & \to & 0 \end{array}$$

with exact rows, if two of $e_A(\varphi', M')$, $e_A(\varphi, M)$, and $e_A(\varphi'', M'')$ are defined, then so is the third, and

$$e_A(\varphi, M) = e_A(\varphi', M') + e_A(\varphi'', M'') .$$

Proof. By the snake lemma, the above diagram determines an exact sequence

$$0 \to {}_{\varphi'}M' \to {}_{\varphi}M \to {}_{\varphi''}M'' \to M'_{\varphi'} \to M_{\varphi} \to M''_{\varphi''} \to 0 ,$$

and the result follows from Lemma A.1.1. $\square$

Lemma A.2.5. *Let φ and ψ be A-linear endomorphisms of M. If two of $e_A(\varphi, M)$, $e_A(\psi, M)$, and $e_A(\varphi\psi, M)$ are defined, so is the third, and*

$$e_A(\varphi\psi, M) = e_A(\varphi, M) + e_A(\psi, M) .$$

Proof. This follows from Lemma A.1.1 and the exact sequence

$$0 \to {}_{\psi}M \to {}_{\varphi\psi}M \xrightarrow{\psi} {}_{\varphi}M \to M_{\psi} \xrightarrow{\varphi} M_{\varphi\psi} \to M_{\varphi} \to 0\,.$$

The maps labelled ψ and φ are induced by ψ and φ, the others by the identity map on M. □

Lemma A.2.6. *If M is a finitely generated free A-module, $\varphi: M \to M$ an A-linear endomorphism, then $e_A(\varphi, M)$ is defined if and only if $e_A(\det(\varphi), A)$ is defined, and*

$$e_A(\varphi, M) = e_A(\det(\varphi), A)\,.$$

Proof. (See also Example A.2.5.) Let $M = A^n$, and regard φ as an $n \times n$ matrix. Note that φ_p is an isomorphism if and only if $\det(\varphi) \notin p$, which proves that both sides of the equation are simultaneously defined.

In case φ is a triangular matrix, the formula follows by induction on n from Lemma A.2.4. Indeed, there are direct summands M' of M mapped to themselves by φ, and the induced maps φ' on M', φ'' on M/M' are given by smaller triangular matrices. Another easy case is when φ is an isomorphism, in which case both sides of the formula are zero.

We prove the lemma first under the assumption that A is a one-dimensional domain – the only case needed in the book. In this case the terms are defined whenever $\det(\varphi) \neq 0$, in which case φ is one-to-one, as follows from the identity

$$\text{Adjoint}(\varphi) \cdot \varphi = \det(\varphi) \cdot I\,,$$

where I is the identity matrix. The required equation is then

$$(*) \qquad l_A(\text{Coker}(\varphi)) = l_A(A/\det(\varphi)\,A)\,.$$

Let K be the quotient field of A. Define a homomorphism

$$h: \text{GL}(n, K) \to \mathbb{Z}$$

by the formula

$$h(\psi) = e_A(a\,\psi, A^n) - e_A(\det(a\,\psi), A)\,,$$

where a is any non-zero element in A such that $a\,\psi$ has entries in A. To see that h is well-defined, let b be another such element in A. Then

$$e_A(a\,b\,\psi, A^n) = e_A(a\,I, A^n) + e_A(b\,\psi, A^n)$$
$$e_A(a\,b\,\psi, A^n) = e_A(b\,I, A^n) + e_A(a\,\psi, A^n)$$

by Lemma A.2.5. Similarly

$$e_A(\det(a\,b\,\psi), A) = e_A(a^n, A) + e_A(\det(b\,\psi), A)$$
$$e_A(\det(a\,b\,\psi), A) = e_A(b^n, A) + e_A(\det(a\,\psi), A).$$

Since the formula of the lemma is obvious for scalar matrices, the four displayed equations combine to give

$$e_A(a\,\psi, A^n) - e_A(\det(a\,\psi), A) = e_A(b\,\psi, A^n) - e_A(\det(b\,\psi), A)$$

as required. Another application of Lemma A.2.5 shows that h is a homomorphism.

To complete the proof, we must show that h is identically zero. Since any element in $\mathrm{GL}(n, K)$ is a product of elementary matrices, it suffices to see that h vanishes on these. Indeed, from the definition of h and the case considered at the beginning of the proof, h vanishes on any triangular matrix in $GL_n(K)$, and any permutation matrix is an isomorphism over A, which was also considered initially.

For the general case of the lemma, by Lemma A.2.2, we may assume A is local, with maximal ideal m. We may assume $\det(\varphi)$ is not a unit, and that $A/\det(\varphi) A$ has finite length, and therefore that A is one-dimensional. (If A is zero-dimensional, both sides are zero). Let I be the ideal of elements in A which are annihilated by some power of the maximal ideal. Since φ induces endomorphisms of IM and M/IM, and IM has finite length, we are reduced by Lemma A.2.4 to the case where $I = 0$. Hence m does not consist of zero divisors. Let K be the total ring of fractions of A. Define h from $\mathrm{GL}(n, K)$ to $\mathbb{Z}$ by the same formula as before, but requiring a to be a non-zero-divisor in A. Since K is Artinian, it is a product of local Artinian rings. It follows, by standard elementary row operations, that every element in $\mathrm{GL}(n, K)$ is a product of triangular matrices, and the proof concludes as before. □

Lemma A.2.7. *Assume A is a one-dimensional local ring, and let $p_1, \ldots, p_t$ be the minimal prime ideals of A. Let M be a finitely generated A-module, and let a be an element of A not in any p_i. Then*

$$e_A(a, M) = \sum_{i=1}^{t} l_{A_{p_i}}(M_{p_i}) \cdot e_A(a, A/p_i A) = \sum_{i=1}^{t} l_{A_{p_i}}(M_{p_i}) \cdot l_A(A/p_i + aA) .$$

Proof. The second equality follows from the fact that $a \notin p_i$ and p_i is prime. For the first, since both sides are additive for exact sequences of A-modules, we may assume $M = A/p$, with p a prime ideal in A (cf. Example A.2.2 (iii)). If p is maximal, then $M_{p_i} = 0$, and $e_A(a, M) = 0$ by Lemma A.2.1. If p is minimal, then $l_{A_p}(M_p) = 1$, and the localizations of M at the other minimal primes are zero, from which the required formula is clear. □

Lemma A.2.8. *Let φ and ψ be two commuting endomorphisms of M, both assumed to be injective. Let $\bar{\varphi}$ (resp. $\bar{\psi}$) be the endomorphism of M_ψ (resp. M_φ) induced by φ (resp. ψ). Then $e_A(\bar{\varphi}, M_\psi)$ is defined if and only if $e_A(\bar{\psi}, M_\varphi)$ is defined, and*

$$e_A(\bar{\varphi}, M_\psi) = e_A(\bar{\psi}, M_\varphi) .$$

Proof. There are canonical isomorphisms of $(M_\psi)_{\bar{\varphi}}$ with $(M_\varphi)_{\bar{\psi}}$ and of ${}_{\bar{\varphi}}(M_\psi)$ with ${}_{\bar{\psi}}(M_\varphi)$, from which the assertion is obvious. (See Example A.5.2 for a generalization to the case when φ and ψ are not injective.) □

Example A.2.1. Let T be an indeterminant, F, G polynomials in $A[T]$, with F monic of degree n. Let $B = A[T]/(F)$, a free A-module of rank n. Let φ_G be the endomorphism of B induced by multiplication by G. Let $R = \det(\varphi_G) \in A$.

(i) R is the resultant of F and G. (See Van der Waerden (4) for properties of resultants. The assertion in (i) is formal. It therefore suffices to verify it when G factors completely. The case where $\deg(G) \leqq 1$ is straightforward.)

(ii) If A is a domain, it follows from Lemma A.2.6 that

$$l_A(A[T]/(F, G)) = l_A(A/(R)) .$$

Example A.2.2 Let M be a finitely generated A-module, φ an A-linear endomorphism of M.

(i) If φ is surjective, then φ is one-to-one.

(ii) If Coker (φ) has finite length, then $e_A(\varphi, M)$ is defined, and

$$0 \leqq e_A(\varphi, M) \leqq l_A(\text{Coker}(\varphi)) .$$

(Let $N_r = \{m \in M \mid \varphi^r(m) = 0\}$. Take n so $N_n = \bigcup_{r \geqq 0} N_r$. Then $e_A(\varphi, M) = l_A(M/N_n + \varphi(M))$ by Lemmas A.2.4 and A.2.1.)

(iii) With the notation of Lemma A.2.4, if $e_A(\varphi, M)$ is defined, then $e_A(\varphi', M')$ and $e_A(\varphi'', M'')$ are defined.

Example A.2.3. Let M be a finitely generated free A-module, φ an A-linear endomorphism of M. Then φ is one-to-one if and only if $\det(\varphi)$ is a non-zero-divisor in A, and in this case,

$$l_A(M/\varphi(M)) = l_A(A/\det(\varphi)A) .$$

Example A.2.4. For any homomorphism $\varphi: M \to N$ of finitely generated A-modules, define

$$e_A(\varphi) = l_A(\text{Coker}(\varphi)) - l_A(\text{Ker}(\varphi))$$

where these lengths are finite. Lemmas A.2.1–2.4 have generalizations to this setting. If M and N are projective A-modules of rank n, then $e_A(\varphi) = e_A(\Lambda^n \varphi)$, where $\Lambda^n \varphi$ is the induced homomorphism on top exterior powers.

Example A.2.5. If A is a principal ideal domain, Lemma A.2.6 follows from the theory of elementary divisors. If A is a one-dimensional domain whose integral closure A' is a finite A-module, the assertion for A follows from that for A' and Lemma A.2.3. This case suffices for the application in Chap. 1.

A.3 Order Functions

Definition A.3. Let A be a one-dimensional domain, with quotient field K. For any non-zero element a in A, set

$$\text{ord}_A(a) = l_A(A/a\,A) = e_A(a, A) .$$

For any non-zero element r in K, set

$$\text{ord}_A(r) = \text{ord}_A(a) - \text{ord}_A(b)$$

for any a, b in A with $r = a/b$. If also $r = a'/b'$, then $a'\,b = a\,b'$, so

$$\text{ord}_A(a') + \text{ord}_A(b) = \text{ord}_A(a) + \text{ord}_A(b')$$

by Lemma A.2.5, so $\operatorname{ord}_A(r)$ is well-defined. Another application of Lemma A.2.5 shows that

$$\operatorname{ord}_A: K^* \to \mathbb{Z}$$

is a homomorphism.

Lemma A.3. *Let A be a one-dimensional domain with quotient field K. Let $\varphi: M \to M$ be an endomorphism of a finitely generated A-module, and let φ_K be the induced endomorphism of $M_K = M \otimes_A K$. If* $\det(\varphi_K) \neq 0$, *then*

$$e_A(\varphi, M) = \operatorname{ord}_A(\det(\varphi_K)) .$$

Proof. By Lemmas A.2.1 and 2.4, we may replace M by M/M', where M' is the torsion submodule of M; i.e., we may assume M imbeds in $M \otimes_A K$. Choosing a basis for $M \otimes_A K$ from elements in M, one constructs a free submodule F of M with $F \otimes_A K = M \otimes_A K$. Choose a common denominator a for a matrix for φ, so that $a\,\varphi(F) \subset F$. Since $(M/F) \otimes_A K = 0$, M/F has finite length, so

$$e_A(a\,\varphi, M) = e_A(a\,\varphi, F)$$

by Lemmas A.2.1 and 2.4 again. Similarly $e_A(a, M) = e_A(a, F)$. Then, using Lemma A.2.5,

$$e_A(a\,\varphi, M) = e_A(a, M) + e_A(\varphi, M) = e_A(a, F) + e_A(\varphi, M) = \operatorname{ord}_A(a^n) + e_A(\varphi, M),$$

where n is the rank of F. By Lemma A.2.6,

$$e_A(a\,\varphi, F) = \operatorname{ord}_A(\det(a\,\varphi)) = \operatorname{ord}_A(a^n) + \operatorname{ord}_A(\det(\varphi_K)) .$$

Comparing these equations gives the lemma. □

Example A.3.1. Let A be a one-dimensional local domain, with quotient field K. Assume that the integral closure of A in K is a finitely generated A-module. Then for any $r \in K^*$,

$$\operatorname{ord}_A(r) = \sum_R \operatorname{ord}_R(r) \cdot [R/m_R : A/m_A]$$

where the sum is over all discrete valuation rings R of K which dominate A, i.e. $R \supset A$ and the maximal ideal m_R of R contains the maximal ideal m_A of A. (If B is the integral closure of A in K, then $l_A(B/A) < \infty$, so $l_A(A/a\,A) = l_A(B/a\,B)$ for $a \in A$. The discrete valuation rings R are the localizations of B at maximal ideals, so Lemmas A.1.2 and 1.3 apply.)

Example A.3.2. If A is a discrete valuation ring, and $r \in A$, then the order of a is the minimum integer n such that $a \in m^n$, m the maximal ideal of A. For general A such a minimality definition gives a non-additive order function (cf. Example 1.2.4).

Example A.3.3. Let A be a discrete valuation ring with fraction field K, L a finite extension of K of degree n, B the integral closure of A in L. If B is a finitely generated A-module, then (Lemma A.3) for $a \in A$,

$$l_A(B/a\,B) = n \cdot \operatorname{ord}_A(a) .$$

If a is a uniformizing parameter in A, this yields the familiar formula "$\sum e_i f_i = n$".

A.4 Flatness

Definition A.4. A homomorphism $A \to B$ of rings is *flat* if every exact sequence of A-modules remains exact after tensoring over A with B. In particular, for any A-algebra A', the base extension $A' \to A' \otimes_A B$ is then flat.

Lemma A.4.1. *Assume that $A \to B$ is a flat, local homomorphism. Then the induced mapping from* $\operatorname{Spec}(B)$ *to* $\operatorname{Spec}(A)$ *is surjective. If A and B are zero-dimensional (Artinian), then*

$$l_B(B) = l_A(A) \cdot l_B(B/mB)$$

where m is the maximal ideal in A.

Proof. For the first statement, we must show that a prime ideal p of A is the restriction of a prime ideal in B. Since B/pB is flat over A/p, we may assume $p = 0$. All non-zero elements a in A are then non-zero-divisors; by flatness, multiplication by a remains injective on B. A prime ideal in B not containing the images of any non-zero elements in A then restricts to the zero ideal in A, as required.

For the second statement, take a chain

$$A = I_0 \supset I_1 \supset \ldots \supset I_r = 0$$

of ideals in A with $I_{i-1}/I_i \cong A/m$. Then

$$B = I_0 B \supset I_1 B \supset \ldots \supset I_r B = 0$$

with

$$I_{i-1}B/I_iB \cong (I_{i-1}/I_i) \otimes_A B \cong B/mB\,,$$

so $l_B(B) = r \cdot l_A(B/mB)$ by Lemma A.1.1. □

Lemma A.4.2. *Let $0 \to M_n \to M_{n-1} \to \ldots \to M_1 \to 0$ be an exact sequence of flat A-modules. Then for any A-module N, the sequence*

$$0 \to M_n \otimes_A N \to M_{n-1} \otimes_A N \to \ldots \to M_1 \otimes_A N \to 0$$

is exact.

Proof. Let $M.$ be the given complex. We show $H_i(M. \otimes_A N) = 0$ by induction on i, the assertion being obvious for $i \leqq 0$. Map a free A-module F onto N, and let N' be the kernel. There arises an exact sequence of complexes

$$0 \to M. \otimes_A N' \to M. \otimes_A F \to M. \otimes_A N \to 0\,.$$

Since $M. \otimes_A F$ is exact, the long exact homology sequence gives isomorphisms

$$H_i(M. \otimes_A N) \cong H_{i-1}(M. \otimes_A N')\,.$$

The inductive assumption, applied to N', concludes the proof. □

A.5 Koszul Complexes

Definition A.5. Let E be a finitely generated A-module, and let $s \in E^\vee$, i.e.,

$$s : E \to A$$

is an A-linear homomorphism. Let $I(s)$ be the image of s, an ideal in A. Define

$$d_k : \Lambda^k E \to \Lambda^{k-1} E$$

by the formula

$$d_k(e_1 \wedge \ldots \wedge e_k) = \sum_{i=1}^{k} (-1)^{i+1} s(e_i)\, e_1 \wedge \ldots \wedge \hat{e}_i \wedge \ldots \wedge e_k .$$

One checks that $d_k\, d_{k+1} = 0$, so one has a complex of A-modules, called a *Koszul complex;* we denote it $\Lambda^\bullet(s)$. Let $H_k(s)$ be the kth homology group of this complex:

$$H_k(s) = \operatorname{Ker}(d_k)/\operatorname{Im}(d_{k+1}) .$$

In particular,

$$H_0(s) = A/I(s) .$$

In case E is free, we say that s is a *regular* section of $E^\vee$, or s is A-regular, if $H_k(s) = 0$ for all $k > 0$.

If all $H_k(s)$ have finite length, we say that $\chi_A(s)$ is defined, and set

$$\chi_A(s) = \sum_{k \geqq 0} (-1)^k l_A(H_k(s)) .$$

Lemma A.5.1. *Let $E = F \oplus A$, $s = t \oplus u$, $t \in F^\vee$, $u \in A^\vee = A$. Then there is a long exact sequence*

$$\ldots \to H_{k+1}(s) \to H_k(t) \xrightarrow{u} H_k(t) \to H_k(s) \to H_{k-1}(t) \to \ldots .$$

Proof. Each $\Lambda^k E$ splits into a direct sum of $\Lambda^k F$ and $\Lambda^{k-1} F$, which determines a short exact sequence of Koszul complexes

$$\begin{array}{ccccccccc}
 & & \vdots & & \vdots & & \vdots & & \\
 & & \downarrow & & \downarrow & & \downarrow & & \\
0 & \to & \Lambda^k F & \longrightarrow & \Lambda^k E & \longrightarrow & \Lambda^{k-1} F & \to & 0 \\
 & & \downarrow & & \downarrow & & \downarrow & & \\
0 & \to & \Lambda^{k-1} F & \to & \Lambda^{k-1} E & \to & \Lambda^{k-2} F & \to & 0 \\
 & & \downarrow & & \downarrow & & \downarrow & & \\
 & & \vdots & & \vdots & & \vdots & &
\end{array}$$

where the complexes are written vertically in the diagram. The required long exact sequence is the long exact homology sequence arising from this short exact sequence of complexes. One checks that, up to sign, the boundary map from $H_{k-1}(t)$ to $H_{k-1}(t)$ is multiplication by u. □

Lemma A.5.2. *Let $E = F \oplus G$, with F, G free, and let $s = t \oplus u$, $t \in F^\vee$, $u \in G^\vee$. Let $\bar{A} = A/I(t)$, $\bar{G} = G/I(t)G$, and let $\bar{u}$ be the homomorphism from $\bar{G}$ to $\bar{A}$ induced by u. If t is A-regular and $\bar{u}$ is $\bar{A}$-regular, then s is A-regular. The converse is true if A is local with maximal ideal m, and $u \in m G^\vee$.*

Proof. By induction on the rank of G, one is reduced to the case $G = A$. From the exact sequence of Lemma A.5.1, $H_k(s)$ vanishes for all $k > 0$ if and only if multiplication by u on $H_k(t)$ is an isomorphism for all $k > 0$ and injective for $k = 0$. This proves the first statement. For the converse, since $u \in m$, multiplication by u on a finitely generated A-module is surjective only if the module is zero (Nakayama's Lemma), from which the assertion follows. □

Lemma A.5.3. *Let $A \to B$ be a flat ring homomorphism. Let E be a free B-module, $s \in E^\vee$. Assume s is B-regular, and the induced homomorphism $A \to B/I(s)$ is flat. Then for any ring homomorphism $A \to A'$, the induced section of $E^\vee \otimes_A A'$ is $B \otimes_A A'$-regular.*

Proof. The complex

$$0 \to \Lambda^n(s) \to \ldots \to \Lambda^0(s) \to B/I(s) \to 0$$

is an exact sequence of flat A-modules. It follows from Lemma A.4.2 that this complex remains exact after tensoring with A' over A, which proves the claim. □

Example A.5.1. Let $E = F \oplus A$, $s = t \oplus u$. Then $\chi_A(s)$ is defined if and only if $e_A(u, H_k(t))$ is defined for all k, and then

$$\chi_A(s) = \sum_{k \geqq 0} (-1)^k e_A(u, H_k(t)) .$$

(From Lemma A.5.1 there are short exact sequences

$$0 \to H_k(t)_u \to H_k(s) \to {}_uH_{k-1}(t) \to 0 .$$

Then apply Lemma A.1.1.)

Example A.5.2. Let B be a commutative A-algebra, E a B-module, s a section of $E^\vee$, so one has the Koszul complex $\Lambda^\bullet(s)$ of B-modules. For any B-module M, let $H_k(s, M)$ be the kth homology group of the complex $\Lambda^\bullet(s) \otimes_B M$. If these homology groups have finite length as A-modules, define $\chi_A(s, M)$ by

$$\chi_A(s, M) = \sum (-1)^k \, l_A(H_k(s, M)) .$$

If E is a free B-module, $\chi_A(s, _)$ is additive for exact sequences of B-modules, and Lemmas A.5.1, A.5.2, and Example A.5.1 generalize to this context. If $E = B^d$, s is given by a sequence $\varphi_1, \ldots, \varphi_d$ of elements of B. In this case set

$$e_A(\varphi_1, \ldots, \varphi_d, M) = \chi_A(s, M) .$$

Note that to give an A-module M together with d commuting endomorphisms is the same as to give M the structure of a B-module, for $B = A[T_1, \ldots, T_d]$. For $d = 1$, φ an endomorphism of the A-module M, and $e_A(\varphi, M)$ is the same as the multiplicity discussed in § A.2. For $d = 2$, let φ and

ψ be commuting endomorphisms of the A-module M. Let ψ_φ (resp. ${}_\varphi\psi$) denote the endomorphism of M_φ (resp. ${}_\varphi M$) induced by ψ. Then, by Example A.5.1,

$$e_A(\varphi, \psi, M) = e_A(\psi_\varphi, M_\varphi) - e_A({}_\varphi\psi, {}_\varphi M),$$

the two sides being simultaneously defined. In particular, the right side is symmetric in φ and ψ, which generalizes Lemma A.2.8.

The positivity assertion of Example A.2.2 may also be generalized (see Serre (4) IV. App. 2). If M is finitely generated over A, and $e_A(\varphi_1, \dots, \varphi_d, M)$ is defined, then

$$0 \leqq e_A(\varphi_1, \dots, \varphi_d, M) \leqq l_A(\bar{M})$$

where $\bar{M}$ is the cokernel of the map from M^d to M given by $(\varphi_1, \dots, \varphi_d)$.

When $B = A$, M an A-module, $E = A^d$, $s \in E^\vee$ is given by a sequence $(a_1, \dots, a_d)$ of elements of A, and $e_A(a_1, \dots, a_d, M)$ is the *multiplicity* of M with respect to $a_1, \dots, a_d$.

Example A.5.3. If $s: E \to A$ is regular, and σ is an A-linear automorphism of E, then $s \circ \sigma: E \to A$ is also regular.

Example A.5.4. If A is a discrete valuation ring, an A-module M is flat if and only if M is torsion-free.

Example A.5.5. If s is an A-regular section of $E^\vee$, and $A \to B$ is a flat homomorphism, then the induced section of $E^\vee \otimes_A B$ is B-regular.

A.6 Regular Sequences

Definition A.6. A sequence of elements $a_1, \dots, a_d$ of a ring A is called a *regular sequence* if the ideal I generated by $a_1, \dots, a_d$ is a proper ideal of A and the image of a_i in $A/(a_1, \dots, a_{i-1})$ is a non-zero-divisor, for $i = 1, \dots, d$. Let $E = A^d$, s the section of $E^\vee$ determined by $a_1, \dots, a_d$. If $a_1, \dots, a_d$ is a regular sequence, then s is a regular section of $E^\vee$; the converse is true if A is local and all a_i belong to the maximal ideal of A (Lemma A.5.2). In this local case, it follows that the regularity of a sequence is independent of the order (cf. Example A.6.1).

There is a canonical epimorphism of graded rings

$$\alpha: A/I[X_1, \dots, X_d] \to \bigoplus_{n \geqq 0} I^n/I^{n+1}$$

which takes X_i to the image of a_i in I/I^2.

If a_1 is a non-zero-divisor in A, then $A[a_2/a_1, \dots, a_d/a_1]$ is a subring of the total ring of fractions of A, and there is a canonical epimorphism of rings

$$\beta: A[T_2, \dots, T_d]/J \to A[a_2/a_1, \dots, a_d/a_1]$$

which takes T_i to a_i/a_1; J is the ideal generated by $L_2, \dots, L_d$, with $L_i = a_1 T_i - a_i$.

Lemma A.6.1. *If $a_1, \ldots, a_d$ is a regular sequence, then α and β are isomorphisms, and the sequence $L_2, \ldots, L_d$ is a regular sequence in $A[T_2, \ldots, T_d]$.*

Proof (from Davis (1)). First one shows by induction on d, that β is an isomorphism, and $L_2, \ldots, L_d$ is a regular sequence. If $d = 2$, and $F(T_2) \in A[T_2]$ with $F(a_2/a_1) = 0$, dividing by L_2 gives an equation

$$a_1^m F(T_2) = G(T_2) \cdot (a_1 T_2 - a_2),$$

for some $m > 0$ and some $G(T_2) \in A[T_2]$. Since a_1^m, a_2 is a regular sequence, one deduces from this equation that all coefficients of $G(T_2)$ are divisible by a_1^m, so $F \in J$. Similarly if $G(T_2) \cdot L_2 = 0$, the coefficients of G must all vanish, so L_2 is a non-zero-divisor.

For $d > 2$, consider the composite

$$A[T_2, \ldots, T_d] \to A'[T_3, \ldots, T_d] \to A[a_2/a_1, \ldots, a_1]$$

where $A' = A[a_2/a_1]$. Since $A'/a_1 A' = A'/(a_1, a_2) A' \cong A/(a_1, a_2) A[T_2]$, $a_1, a_3, \ldots, a_d$ is a regular sequence in A'. By the case $d = 2$, L_2 is a non-zero-divisor generating the kernel of the first displayed homomorphism; by induction, $L_3, \ldots, L_d$ is a regular sequence generating the kernel of the second. Therefore $L_2, \ldots, L_d$ is a regular sequence generating the kernel of the composite, as stated.

To see that α is an isomorphism, it suffices to show that if F is a homogeneous polynomial in $A[X_1, \ldots, X_d]$ with $F(a_1, \ldots, a_d) = 0$, then F is in $IA[X_1, \ldots, X_d]$. But if $F(a_1, \ldots, a_d) = 0$, then $F(1, a_2/a_1, \ldots, a_d/a_1) = 0$; since β is an isomorphism, $F(1, T_2, \ldots, T_d)$ is in J. Hence all the coefficients of F are in I, as required. □

Remark A.6. Micali (1) has shown that

$$A[Y_1, \ldots, Y_d]/K \xrightarrow{\sim} \mathrm{Sym}_A(I) \xrightarrow{\sim} \oplus I^n,$$

where $Y_i \to a_i \in I$, and K is generated by $a_i Y_j - a_j Y_i$. This also implies that α is an isomorphism.

Lemma A.6.2. *Let A be a d-dimensional local ring with maximal ideal m, $k = A/m$. The following are equivalent:*

(i) $\dim_k(m/m^2) = d$;
(ii) *m has d generators;*
(iii) *m is generated by a regular sequence (of d elements);*
(iv) *$\oplus m^n/m^{n+1} \cong k[X_1, \ldots, X_d]$ as graded k-algebras.*

Proof. By Nakayama's lemma, $\dim_k(m/m^2)$ is the minimum number of generators of m. Let $a_1, \ldots, a_e$ be minimal generators of m, and map $k[X_1, \ldots, X_e]$ onto $\oplus m^n/m^{n+1}$ by sending X_i to a_i. This gives a closed imbedding of the projective cone $P(C)$ to $\mathrm{Spec}(k)$ in $\mathrm{Spec}(A)$, in $\mathbb{P}_k^{e-1}$ (see Appendix B.5). But $P(C)$ is always at least $(d-1)$-dimensional: if $\mathrm{Spec}(k) \subset V_1 \subset \ldots \subset V_d = \mathrm{Spec}(A)$ is a chain of subvarieties, then the blow-ups $\hat{V}_i$ of V_i along $\mathrm{Spec}(k)$ give a chain of subvarieties of $\hat{X}$ which meet the divisor $P(C)$. Therefore $e \geqq d$, with equality if and only if $P(C) = \mathbb{P}_k^{e-1}$. The latter holds when the map from $k[X_1, \ldots, X_e]$ to $\oplus\, m^n/m^{n+1}$ is an isomorphism in large degrees; since the polynomial ring has no zero divisors, the kernel must be zero. From this one has the equivalence of (i), (ii), and (iv).

From the fact that $\oplus\, m^n/m^{n+1}$ is a domain it follows readily that A is a domain. If $p_i = (a_1, \ldots, a_i)$, it follows by descending induction on d that A/p_i has dimension $d-i$ and satisfies the condition of (ii). In particular, p_i is a prime ideal, so $(a_1, \ldots, a_d)$ is a regular sequence, and (ii) $\Rightarrow$ (iii). And (iii) $\Rightarrow$ (iv) by Lemma A.6.1. (See Serre (4)IV.D for a purely algebraic proof of Lemma A.6.2.) □

A ring satisfying the conditions of Lemma A.6.2 is a *regular* local ring.

Example A.6.1. (i) (Matsumura (1) p. 102). Let $A = k[X, Y, Z]$, $a_1 = X$, $a_2 = Y(1-X)$, $a_3 = Z(1-X)$. Then a_1, a_2, a_3 is a regular sequence, but a_2, a_3, a_1 is not.

(ii) Let $A = \mathbb{C}[X, Y, Z]$, $a_1 = X(1-2XYZ)(1-3XYZ)$, $a_2 = Y(1-XYZ)(1-3XYZ)$, $a_3 = Z(1-XYZ)(1-2XYZ)$. Then (a_1, a_2, a_3) is a regular section of A^3, but no permutation of a_1, a_2, a_3 is a regular sequence.

Note that a sequence $a_1, \ldots, a_d$ in a ring A determine a regular section of A^d if and only if $(a_1, \ldots, a_d)$ is a regular sequence in A_p, for each prime ideal p of A which contains $a_1, \ldots, a_d$.

A.7 Depth

Definition A.7. The *depth* of a local ring A is the maximal length of a regular sequence in its maximal ideal. Every maximal regular sequence in the maximal ideal has depth (A) elements (for a particularly elementary proof of this fact, see Northcott-Rees (1).) The depth of A is no larger than the dimension of A; when $\operatorname{depth} A = \dim A$, A is called *Cohen-Macaulay*. Every regular local ring is Cohen-Macaulay (Lemma A.6.2). The following lemma gives a useful criterion for the regularity of a sequence.

Lemma A.7.1. *Let A be a local ring, with maximal ideal m, and let $s \in m E^\vee$, E free of rank d. Then*

$$\dim(A/I(s)) \geqq \dim(A) - d\,. \tag{*}$$

If A is Cohen-Macaulay (for example regular), then equality holds in (*) *if and only if s is A-regular.*

For homological proofs of the preceding facts, see Serre (4)IV.B or Matsumura (1) Ch. 6. For a simple elementary account, see Kunz (1)VI. □

A Noetherian ring A is called *Cohen-Macaulay* if all its localizations are Cohen-Macaulay. The following result has been proved by Hochster. We refer to Hochster (1) Thm. 3.1*, or to Laksov (1) for a proof; cf. also Arbarello-Cornalba-Griffiths-Harris (1), or De Concini-Eisenbud-Procesi (1); the latter discusses analogues for symmetric and skew-symmetric matrices.

Lemma A.7.2. *Let A be a polynomial ring in indeterminates x_{ij}, $1 \leqq i \leqq f$, $1 \leqq j \leqq e$, over a field K. Let $1 \leqq a_1 < \ldots < a_d \leqq e$ be a sequence of integers, and let I be the ideal of A generated by all $(a_k - k + 1)$ by $(a_k - k + 1)$ minors of*

$$\begin{pmatrix} x_{11} & \ldots & x_{1a_k} \\ \vdots & & \vdots \\ x_{f1} & \ldots & x_{fa_k} \end{pmatrix}$$

for $k = 1, \ldots, d$. Assume that $f - a_d + d \geqq 0$. Then A/I is a Cohen-Macaulay, normal domain, of dimension

$$ef - \sum_{k=1}^{d} (f - a_k + k).$$

A.8 Normal Domains

A normal domain is an integral domain which is integrally closed in its fraction field.

Lemma A.8.1. *Let A be a normal domain, $a \in A$, $a \neq 0$. Then all prime ideals associated to A/aA are minimal primes containing a.*

Proof. Localizing at an associated prime, we may assume A is local and the maximal ideal m is associated to A/aA. Then $m = (aA : b)$ for some $b \in A$. Therefore $(b/a)\, m \subset A$. If $(b/a)\, m \subset m$, then, since m is a finitely generated A-module, b/a is integral over A (cf. Zariski-Samuel (1) Ch.V. § 1); then $b/a \in A$ since A is integrally closed, so $b \in aA$ and $m = A$. Therefore $(b/a)\, m = A$, and $m = (a/b) A$ is a principal ideal, which implies the minimality of m. □

A.9 Determinantal Identities

Given formal power series $c^{(i)} = \sum c_j^{(i)} t^j$, with coefficients $c_j^{(i)}$, $1 \leqq i \leqq h$, $-\infty < j < \infty$, in some commutative ring, and a partition $\lambda = (\lambda_1, \ldots, \lambda_n)$, $\lambda_1 \geqq \lambda_2 \geqq \ldots \geqq \lambda_n \geqq 0$, set (cf. § 14 Notation)

$$\Delta_\lambda(c^{(1)}, \ldots, c^{(n)}) = \left| c^{(i)}_{\lambda_i + j - i} \right|,$$

the determinant of the n by n matrix whose ij entry is $c^{(i)}_{\lambda_i + j - i}$. If $c^{(1)} = \ldots = c^{(n)} = c$, denote this simply $\Delta_\lambda(c)$.

Lemma A.9.1. *Let $k \geqq 0$, $m \geqq 0$ be integers. Let $e = n + k$, $f = m + k$, and let $\lambda = (m, \ldots, m)$ with m repeated n times. Let $\beta_1, \ldots, \beta_f$, $\alpha_1, \ldots, \alpha_e$ be variables. Set*

$$c^{(i)} = \prod_{j=1}^{f} (1 + \beta_j t) \Big/ \prod_{j=1}^{k+i} (1 + \alpha_j t),$$

and let $c = c^{(n)}$. *Then:*

(i) $\Delta_\lambda(c^{(1)}, \ldots, c^{(n)}) = \Delta_\lambda(c)$.

(ii) *If* $k=0$, $\Delta_\lambda(c) = \prod_{i=1}^{n}\prod_{j=1}^{m}(\beta_j - \alpha_i)$.

Proof. For (i), note that $c^{(i)} + \alpha_{k+i}\, t\, c^{(i)} = c^{(i-1)}$, i.e.,

$$c_j^{(i)} + \alpha_{k+i}\, c_{j-1}^{(i)} = c_j^{(i-1)}.$$

By elementary row operations it follows that

$$\Delta_\lambda(c^{(i-1)}, \ldots, c^{(i-1)}, c^{(i)}, \ldots, c^{(n)}) = \Delta_\lambda(c^{(i)}, \ldots, c^{(i)}, c^{(i)}, \ldots, c^{(n)}),$$

from which (i) follows inductively. Both sides of (ii) are homogeneous of degree nm in the variables $\alpha_1, \ldots, \alpha_n, \beta_1, \ldots, \beta_m$. Setting $\alpha_1 = \ldots = \alpha_n = 0$, both sides give $\left(\prod_{j=1}^{m}\beta_j\right)^n$. It therefore suffices to show that the left side vanishes when $\alpha_i = \beta_j$; by symmetry we may assume $i = j = 1$. But this is clear from (i), since $c^{(1)}$ is then a polynomial of degree less than m, forcing the top row of the matrix whose determinant is $\Delta_\lambda(c^{(1)}, \ldots, c^{(n)})$ to vanish. □

Lemma A.9.2. *Let* $c = \sum_{i=0}^{\infty} c_i t^i$, $s = \sum_{i=0}^{\infty} s_i t^i$ *be related by the identity* $c(t) \cdot s(-t) = 1$, $c_0 = s_0 = 1$. *For any partition* λ, *if* $\tilde{\lambda}$ *is the conjugate to* λ (cf. § 14.4), *then*

$$\Delta_\lambda(s) = \Delta_{\tilde{\lambda}}(c).$$

Proof. Let $\mu = \tilde{\lambda}$, and write $\lambda = (\lambda_1, \ldots, \lambda_n)$, $\lambda_n > 0$, $\mu = (\mu_1, \ldots, \mu_m)$, $\mu_m > 0$. Note that $\lambda_1 = m$, $\mu_1 = n$. The fact that λ and μ are conjugate implies that the sets of integers

$$\{m - \lambda_i + i \mid 1 \leqq i \leqq n\} \quad \text{and} \quad \{m + \mu_j - j + 1 \mid 1 \leqq j \leqq m\}$$

form complementary sets in the set $\{1, 2, \ldots, n+m\}$. Indeed, the first set is increasing, the second decreasing. If the ith member of the first set equals the jth member of the second, then

$$\lambda_i + \mu_j = i + j - 1.$$

But conjugacy implies that if $\lambda_i \geqq j$ (resp. $\lambda_i < j$) then $\mu_j \geqq i$ (resp. $\mu_j < i$).

Let $\tilde{c}_i = (-1)^i c_i$. Consider the product of two $m+n$ by $m+n$ matrices:

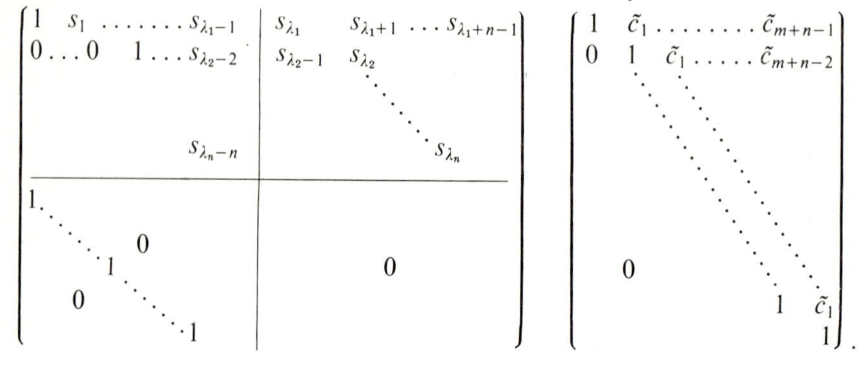

The determinant of the left (resp. right) matrix is $(-1)^{mn}\Delta_\lambda(s)$ (resp. 1). Multiplying these matrices, one obtains the matrix

$$\begin{pmatrix} 1 & 0 & & \cdots & & & \\ 0 \; \cdots & & 0 & 1 \; 0 \; \cdots & & & \\ & & & & & & \\ 0 \; \cdots & & & & 1 \; \cdots \; 0 & & \\ 1 & \tilde{c}_1 & & \cdots & & \tilde{c}_{m+n-1} & \\ 0 & 1 & \tilde{c}_1 & \cdots & & & \\ & & & & & & \\ 0 \; \cdots & & & 1 \; \cdots & & \tilde{c}_n & \end{pmatrix}$$

where, for $i=1,\ldots,n$, the ith row has a 1 in the $(m-\lambda_i+i)^{\text{th}}$ place, and zeros elsewhere. The determinant is, up to sign, the determinant of the m by m matrix obtained by removing the rows and columns containing these ones. Using the complementarity of the two sets of integers obtained at beginning of the proof, one obtains the matrix

$$\begin{pmatrix} \tilde{c}_{\mu_m} & \tilde{c}_{\mu_{m-1}+1} & \cdots & \tilde{c}_{\mu_1+m-1} \\ \tilde{c}_{\mu_m-1} & & & \\ & & & \\ \tilde{c}_{\mu_m-m+1} & & \cdots & \tilde{c}_{\mu_1} \end{pmatrix}.$$

The sign in front of this determinant is $(-1)^\varepsilon$, with

$$\varepsilon=\sum_{i=1}^{n}(m-\lambda_i)=m\,n-|\lambda|=m\,n-|\mu|,$$

where $|\lambda|=\sum_{i=1}^{n}\lambda_i$, $|\mu|=\sum_{j=1}^{m}\mu_j$. Equating the determinant of the product with the product of determinants, and replacing the above matrix by a reflection in a diagonal, one has

$$\begin{aligned}\Delta_\lambda(s)&=(-1)^{|\mu|}\Delta_\mu(\tilde{c})\\ &=(-1)^{|\mu|}\,|(-1)^{\mu_i-i}(-1)^j c_{\mu_i+j-i}|\\ &=(-1)^{|\mu|}(-1)^{\varepsilon'}\Delta_\mu(c),\end{aligned}$$

where $\varepsilon'=\sum_{i=1}^{m}(\mu_i-i)+\sum_{j=1}^{m}j=|\mu|$, which concludes the proof. □

Let $x_1,\ldots,x_n$ be commuting variables, and set

$$c=\sum_{i=0}^{n}c_i t^i=\prod_{j=1}^{n}(1+x_j t),$$

$$s=\sum_{i=0}^{\infty}s_i t^i=\prod_{j=1}^{n}(1/1-x_j t)=\left(\sum_{i=0}^{n}(-1)^i c_i t^i\right)^{-1}.$$

For a partition λ set

$$s_\lambda(x)=s_\lambda(x_1,\ldots,x_n)=\Delta_\lambda(s).$$

Lemma A.9.3 (Jacobi, Trudi). *With this notation,*

$$s_\lambda(x) = |x_j^{\lambda_i+n-i}| / |x_j^{n-i}| .$$

Proof. In the identity $\prod_{i=1}^{n}(t-x_i) = t^n - c_1 t^{n-1} + \ldots + (-1)^n c_n$, set $t = x_j$ and multiply by x_j^{m-n}, obtaining

(i) $$x_j^m - c_1 x_j^{m-1} + \ldots + (-1)^n c_n x_j^{m-n} = 0$$

for all $m \geqq n$. From the relation between s and c, for any integer q with $0 \leqq q \leqq n-1, m \geqq n$,

(ii) $$s_{m-q} - c_1 s_{m-q-1} + \ldots + (-1)^n c_n s_{m-q-n} = 0 .$$

The recursion equations (i) and (ii) are the same. Solving inductively, this means there are, for all $m \geqq 0$, and $1 \leqq k \leqq n$, universal polynomials $a(m,k)$ in the variables $c_1, \ldots, c_n$ such that

$$x_j^m = \sum_{k=1}^{n} a(m,k)\, x_j^{n-k} \qquad \text{for all } 1 \leqq j \leqq n ,$$

$$s_{m-q} = \sum_{k=1}^{n} a(m,k)\, s_{n-k-q} \qquad \text{for all } 0 \leqq q \leqq n-1 .$$

For any non-negative integers $\lambda_1, \ldots, \lambda_n$, this gives matrix identities

$$(x_j^{\lambda_i+n-i})_{ij} = (a(\lambda_i+n-i,k))_{ik} \cdot (x_j^{n-k})_{kj}$$

$$(s_{\lambda_i+j-i})_{ij} = (a(\lambda_i+n-i,k))_{ik} \cdot (s_{j-k})_{kj} .$$

(The matrix subscript ij indicates how the ij entry of the matrix is formed.) The proof concludes by taking determinants in these two equations, noting that $|s_{j-k}| = 1$ since $(s_{j-k})_{jk}$ is a triangular matrix with ones on the diagonal. Note also that

$$|x_j^{n-k}| = \prod_{j<k}(x_j - x_k) \qquad \text{(Vandermonde)}$$

is not zero. □

Lemma A.9.4. *Let* $d = \sum_{i=0}^{\infty} d_i t^i$, $d_0 = 1$, $\lambda = (\lambda_1, \ldots, \lambda_n)$ *a partition. Then for all* $m \geqq 0$,

$$d_m \cdot \Delta_\lambda(d) = \sum_\mu \Delta_\mu(d) ,$$

the sum over all $\mu = (\mu_1, \ldots, \mu_{n+1})$ *with* $\mu_1 \geqq \lambda_1 \geqq \ldots \geqq \mu_n \geqq \lambda_n \geqq \mu_{n+1} \geqq 0$, *and* $\sum_{i=1}^{n+1} \mu_i = m + \sum_{i=1}^{n} \lambda_i$.

Proof. Since $\Delta_\lambda(d)$ does not change if an arbitrary string of zeros is added to λ, we may assume $n \geqq i$ for any d_i which appears in the identity to be proved, and $\lambda_n = 0$. We may therefore assume $d_i = s_i$, s_i as in Lemma A.9.3. Set $a_i = \lambda_i + n - i$, $b_i = \mu_i + n - i$, and set

(i) $$\delta(a_1, \ldots, a_n) = |x_j^{a_i}|_{1 \leqq i, j \leqq n} .$$

By Lemma A.9.3, and adding over m, we are reduced to proving the identity

(ii) $$\delta(a_1, \ldots, a_n) \cdot \prod_{i=1}^{n} (1 - x_i)^{-1} = \sum \delta(b_1, \ldots, b_n),$$

the sum over all $(b_1, \ldots, b_n)$ with $b_1 \geqq a_1 > b_2 \geqq a_2 > \ldots > b_n \geqq a_n = 0$. For any integers $e_2, \ldots, e_n$, let $\delta_k(e_2, \ldots, e_n)$ be the determinant formed as in (i), but using integers $e_2, \ldots, e_n$ and variables $x_1, \ldots, \hat{x}_k, \ldots, x_n$. Thus, expanding along the top row,

$$\delta(a_1, \ldots, a_n) = \sum_{k=1}^{n} (-1)^{k+1} x_k^{a_1} \delta_k(a_2, \ldots, a_n).$$

Therefore

$$\delta(a_1, \ldots, a_n) \cdot \prod_{i=1}^{n} (1 - x_i)^{-1}$$

$$= \sum_{k=1}^{n} (-1)^{k+1} x_k^{a_1} (1 - x_k)^{-1} \delta_k(a_2, \ldots, a_n) \prod_{i \neq k} (1 - x_i)^{-1}.$$

Assuming the result for $n - 1$, this expression becomes

$$\sum_{k=1}^{n} (-1)^{k+1} \sum_{b_1 \geqq a_1} x_k^{b_1} \Big(\sum_{b_2 \geqq a_2 > \ldots \geqq a_n} \delta_k(b_2, \ldots, b_n) \Big) = \sum \delta(b_1, \ldots, b_n),$$

the sum over all $(b_1, \ldots, b_n)$ with $b_1 \geqq a_1$ and $b_2 \geqq a_2 > \ldots \geqq a_n$. To conclude the proof, one must verify that the sum of all terms with $b_2 \geqq a_1$ is zero. Together with any such $(b_1, b_2, \ldots, b_n)$, with $b_1 \neq b_2$, occurs also $(b_2, b_1, \ldots)$, and

$$\delta(b_1, b_2, \ldots, b_n) + \delta(b_2, b_1, \ldots, b_n) = 0$$

by the alternating property of determinants. Similarly $\delta(b_1, b_2, \ldots, b_n)$ vanishes if $b_1 = b_2$. □

The *Littlewood-Richardson rule* (cf. § 14.4)

$$\Delta_\lambda(c) \Delta_\mu(c) = \sum N_{\lambda, \mu, \varrho} \Delta_\varrho(c)$$

for multiplying general S-functions was given in Littlewood and Richardson (1), although complete proofs have only recently appeared. For a proof along the original lines we recommend Macdonald (3)I.9. Other proofs may be found in Schützenberger (1) or Akin-Buchsbaum-Weyman (1).

Example A.9.1 (cf. Lascoux (7), Macdonald (3) p. 30, 31). For partitions $\lambda = (\lambda_1, \ldots, \lambda_n)$, $\mu = (\mu_1, \ldots, \mu_n)$, write $\mu \subset \lambda$ if $\mu_i \leqq \lambda_i$ for $1 \leqq i \leqq n$. Set

$$d_{\lambda\mu} = \left| \binom{\lambda_i + n - i}{\mu_j + n - j} \right|_{1 \leqq i, j \leqq n}.$$

Let $x_1, \ldots, x_n$ be commuting variables. Then

(a) $$s_\lambda(1 + x_1, \ldots, 1 + x_n) = \sum_{\mu \subset \lambda} d_{\lambda\mu} s_\mu(x_1, \ldots, x_n).$$

(By Lemma A.9.3, $s_\lambda(1 + x_1, \ldots, 1 + x_n) = |(1 + x_j)^{\lambda_i + n - i}| / |(1 + x_j)^{n-i}|$. The denominator equals $|x_j^{n-i}|$. Expand the numerator and compute coefficients of $s_\mu(x)$.)

Let $\varepsilon = (n, n-1, \ldots, 1)$, $\delta = (n-1, n-2, \ldots, 0)$. Then

(b) $$s_\varepsilon(x_1, \ldots, x_n) = x_1 \cdot \ldots \cdot x_n \cdot \prod_{i<j} (x_i + x_j)$$

(c) $$s_\delta(x_1, \ldots, x_n) = \prod_{i<j} (x_i + x_j) .$$

Example A.9.2 (cf. Macdonald (3) p. 35, 37). Let $x_1, \ldots, x_n, y_1, \ldots, y_m$ be variables. Then

(a) $$\prod_{i=1}^{n} \prod_{j=1}^{m} (1 + x_i y_j) = \sum_\lambda s_\lambda(x)\, s_{\tilde\lambda}(y) ,$$

the sum over all partitions $\lambda : \lambda_1 \geqq \ldots \geqq \lambda_n \geqq 0$ with $\lambda_1 \leqq m$; $\tilde\lambda$ is the conjugate of λ.

(b) $$\prod_{i=1}^{n} \prod_{j=1}^{m} (x_i + y_j) = \sum_\lambda s_\lambda(x)\, s_{\tilde\lambda'}(y) ,$$

the sum over the same λ, with $\tilde\lambda' = (n - \tilde\lambda_m, \ldots, n - \tilde\lambda_1)$.

(c) $$\prod_{i=1}^{n} \prod_{j=1}^{m} (1 + x_i + y_j) = \sum_{\lambda, \mu} d_{\lambda\mu}\, s_\mu(s)\, s_{\tilde\lambda'}(y) ,$$

the sum over partitions $\mu \subset \lambda$, λ as above, $d_{\lambda\mu}$ as in the preceding example.

Example A.9.3. (a) Let $a_0, \ldots, a_d$ be non-negative integers. Then

$$|1/(a_i + j)!|_{0 \leqq i,j \leqq d} = \prod_{i<j} (a_i - a_j) \Big/ \prod_{i=0}^{d} (a_i + d)! .$$

(Multiply by $\prod_{i=0}^{d} (a_i + d)!$ and use elementary row operations.)

(b) If $c(t) = e^t = \sum_{i=0}^{\infty} (1/i!)\, t^i$, then (cf. § 14 Notation)

$$\Delta_f^{(e)}(c) = \frac{0!\, 1! \cdot \ldots \cdot (e-1)!}{f! \cdot \ldots \cdot (f+e-1)!} .$$

Example A.9.4 (cf. Harris-Tu (1)). Let $a_0, \ldots, a_d$ and n be integers, with $n \geqq a_i - d \geqq 0$ for all i. Then

$$\left| \binom{n}{a_i - j} \right|_{0 \leqq i,j \leqq d} = \prod_{i=0}^{d} \frac{(n+i)!}{(n+d-a_i)!\, a_i!} \cdot \prod_{i<j} (a_j - a_i) .$$

Example A.9.5. Let $f = 1 + a_1 t + \ldots + a_n t^n$, $g = 1 + b_1 t + \ldots + b_m t^m$. Then

$$\Delta_n^{(m)}(g/f) = R(f, g) ,$$

where $R(f, g)$ is the resultant of f and g. (This follows from Lemma A.9.1 (ii).)

Notes and References

We make no attempt to trace the development of multiplicity theory in commutative and homological algebra, other than mentioning the names of a few of the important contributors: Auslander, Buchsbaum, Nagata, Northcott, Rees, Samuel, and Serre. The results of this appendix on length, Koszul complexes, regular sequences, and flatness have been known for some time. More general results, from a more sophisticated point of view, may be found in Serre (4). Other references for this material are Atiyah-MacDonald (1), Bourbaki (1), (2), [EGA]IV, Kunz (1), Matsumara (1), Nagata (2), Northcott (2), and Zariski-Samuel (1).

For the case of a domain, the important Lemma A.2.6 was proved in Chevalley (2)I.2.3 and [EGA]IV. 21.10.17.3 by reducing to the case of a discrete valuation ring. The elegant proof given here, motivated by *K*-theory, was shown to me by B. Iversen. The extension to general rings is apparently new. The relation of this lemma to resultants (Example A.2.1) I learned from L. Gruson.

The determinantal identities in § A.9 have an even longer history, going back at least to Jacobi. For details, see Ledermann (1) and Macdonald (3).

Appendix B. Algebraic Geometry (Glossary)

The aim of this glossary is to fix terminology and notation from algebraic geometry which are used in the text, and to give proofs or references for some of the basic facts which are needed. Although a few scheme-theoretic constructions are needed, we have favored a more classical geometric language. For example, "points" are always closed points, and, for a vector bundle E, $P(E)$ is the bundle of lines in E. The basic references for facts not proved here will be: Grothendieck-Dieudonné (1), denoted [EGA], and Hartshorne (5), denoted [H], and Berthelot-Grothendieck-Illusie, et al., denoted [SGA6]; other useful references are Shafarevich (1), Altman-Kleiman (1), and Mumford (2).

B.1 Algebraic Schemes

B.1.1. An *algebraic scheme* over a field K is a scheme X, together with a morphism of finite type from X to $\operatorname{Spec}(K)$. In other words, X has a finite covering by affine open sets whose coordinate rings are finitely generated K-algebras ([EGA]I.6.5, [H]II.3). The coordinate ring of an affine open set U may be denoted $A(U)$.

In Chaps. 1–19, the word *scheme* means an algebraic scheme over some field.

A closed subscheme Y of a scheme X is defined by an *ideal sheaf* $\mathscr{I}(Y)$ in the structure sheaf $\mathscr{O}_X$ of X; for an affine open covering of X, Y corresponds to an ideal in each coordinate ring of X. A closed subscheme Y of X comes equipped with a closed imbedding $Y \to X$. In general, an *imbedding*, or a *subscheme* is assumed to be *closed*, unless prefixed by "open", or "locally closed". The notation $Y \hookrightarrow X$ is used to indicate that Y is a closed subscheme of X.

If Y is a closed subscheme of a scheme X, $X - Y$ denotes the open subscheme of X which is the complement of the support of Y.

B.1.2. A *variety* is a reduced and irreducible (integral) algebraic scheme.

A *subvariety* V of a scheme X is a reduced and irreducible closed subscheme of X; a subvariety V corresponds to a prime ideal in the coordinate ring of any affine open set meeting V. The *local ring* of X along V, denoted $\mathscr{O}_{V,X}$, is the localization of such a coordinate ring at the corresponding prime ideal ([H]I.3.13); its maximal ideal is denoted $\mathscr{M}_{V,X}$. In Grothendieck's language, $\mathcal{O}_{V,X}$ is the stalk of the structure sheaf $\mathcal{O}_X$ of X at the generic point of V.

The *function field* of a variety V is denoted $R(V)$. If V is a subvariety of X, $R(V)$ is the residue field $\mathcal{O}_{V,X}/\mathcal{M}_{V,X}$.

B.1.3. The *dimension* of a scheme X, denoted $\dim X$, is the maximum length n of a chain

$$\emptyset \neq V_0 \subsetneqq V_1 \subsetneqq \ldots \subsetneqq V_n \subset X$$

of subvarieties of X. If X is a variety, $\dim X$ is the transcendence degree of $R(X)$ over the ground field. The notation X^n may be used to signify that X is an n-dimensional variety. A scheme X is *pure-dimensional* if all irreducible components of X have the same dimension.

If V is a subvariety of a scheme X, the *codimension* of V in X, denoted $\operatorname{codim}(V, X)$, is the maximum length d of a chain of subvarieties

$$V = V_0 \subsetneqq V_1 \subsetneqq \ldots \subsetneqq V_d \subset X.$$

A *point* on a scheme X is a 0-dimensional subvariety of X. A point P is *rational* over the ground field K i $R(P) = K$. We often write $\varkappa(P)$ or $K(P)$ in place of $R(P)$. A point P is a *regular* point of X if $\mathcal{O}_{P,X}$ is a regular local ring (Appendix A.6). The open set of regular points in X may be denoted X_{reg}.

B.1.4. Affine n-space, denoted $\mathbb{A}^n$, or $\mathbb{A}^n_K$, is the affine variety whose coordinate ring is the polynomial ring $K[x_1, \ldots, x_n]$. For $n = 1, 2, 3, 4$ coordinates (t), (x, y), (x, y, z), (w, x, y, z) are often used. Projective n-space is denoted $\mathbb{P}^n$, or $\mathbb{P}^n_K$. Unless otherwise labelled, $x_0, \ldots, x_n$ will be homogeneous coordinates on $\mathbb{P}^n$, $(x_0 : \ldots : x_n)$ the point in $\mathbb{P}^n$ with homogeneous coordinates $x_0, \ldots, x_n$. We identify $\mathbb{A}^n$ as the open subscheme of $\mathbb{P}^n$ where $x_0 \neq 0$. The point $(1:0)$ in $\mathbb{P}^1$ is called *zero* point and denoted 0, while thc point $(0:1)$ is the point at *infinity*, and denoted ∞.

The subscheme of $\mathbb{A}^n$ defined by an ideal $I = (f_1, \ldots, f_m)$ in $K[x_1, \ldots, x_n]$ is denoted $V(I)$, or $V(f_1, \ldots, f_m)$. Similarly, if I is a homogeneous ideal in $K[x_0, \ldots, x_n]$, generated by forms $f_1, \ldots, f_m$, $V(I)$ or $V(f_1, \ldots, f_m)$ may denote the subscheme of $\mathbb{P}^n$ determined by I.

B.2 Morphisms

B.2.1. A *morphism* $f: X \to Y$ of algebraic schemes is assumed to be compatible with the structure morphism to $\operatorname{Spec}(K)$, K the ground field. If f maps an affine open subset U' of X into an affine open subset U of Y, then f corresponds to a homomorphism $f^*: A(U) \to A(U')$ of K-algebras.

The *identity* morphism of a scheme X is denoted id_X, or 1_X, or simply 1. The *composite* of morphisms $f: X \to Y$ and $g: Y \to Z$ is denoted $g \circ f$, or gf.

B.2.2. A morphism $f: V \to W$ of varieties is *dominant* if the image of f is dense in W; equivalently, the induced homomorphisms f^*, described above, are injective. If: $X \to Y$ is a morphism, and V is a subvariety of X, there is a unique subvariety W of Y such that f maps V dominantly to W; f induces a local homomorphism

$$f^* : \mathcal{O}_{W,Y} \to \mathcal{O}_{V,X}.$$

The induced homomorphism on residue fields is an imbedding $f^* : R(W) \to R(V)$. If $\dim V = \dim W$, then $R(V)$ is a finite field extension of $R(W)$ ([EGA]IV.5.5.6).

A morphism $f: X \to Y$ of varieties is *birational* if it is dominant and f^* maps $R(Y)$ isomorphically onto $R(X)$. More generally, a morphism $f: X \to Y$ of schemes is birational if X and Y have the same number of irreducible components, each component X_i of X is mapped dominantly to a distinct component Y_i of Y, and the induced morphisms from $\mathcal{O}_{Y_i,Y}$ to $\mathcal{O}_{X_i,X}$ are isomorphisms.

B.2.3. If $f: X \to S$, $g: Y \to S$ are morphisms, the *fibre product* of X and Y over S is denoted $X \times_S Y$. If X, Y, and S are affine, with coordinate rings A, B and Λ, then $X \times_S Y$ is the affine scheme with coordinate ring $A \otimes_\Lambda B$; in general $X \times_S Y$ is constructed by patching together such affine schemes ([H]II.3, [EGA]I.3). The fibre product comes equipped with *projections* $p: X \times_S Y \to X$ and $q: X \times_X Y \to Y$. For any scheme Z with morphisms $u: Z \to X$, $v: Z \to Y$ such that $f \circ u = g \circ v$, there is a unique morphism, denoted (u, v), from Z to $X \times_S Y$ such that $p \circ (u, v) = u$, $q \circ (u, v) = v$. This universal property:

$$\mathrm{Hom}_S(Z, X) \times \mathrm{Hom}_S(Z, Y) = \mathrm{Hom}_S(Z, X \times_S Y)$$

characterizes $X \times_S Y$ up to canonical isomorphism.

A commutative square of morphisms

$$\begin{array}{ccc} Z & \xrightarrow{q} & Y \\ {\scriptstyle p}\downarrow & & \downarrow{\scriptstyle g} \\ X & \xrightarrow[f]{} & S \end{array}$$

is called a *fibre square* if Z is the fibre product of X and Y, and p and q are the canonical projections. In a *fibre diagram*, all squares appearing in the diagram are required to be fibre squares.

When $S = \mathrm{Spec}(K)$, K the ground field, we write $X \times Y$ in place of $X \times_S Y$, and call it the *Cartesian product* of X and Y.

If $f: X \to Y$ is a morphism, and Z is a closed subscheme of Y, the *inverse image* scheme, denoted $f^{-1}(Z)$, may be identified with the fibre product $X \times_Y Z$. If $\mathcal{I}$ is the ideal sheaf of Z in Y, then $f^{-1}(Z)$ is defined by the ideal sheaf $f^{-1}(\mathcal{I}) \cdot \mathcal{O}_X$. In case f is a closed imbedding, $X \times_Y Z = f^{-1}(Z)$ is the *intersection scheme* $X \bigcap Z$, the closed subscheme of Y defined by the sum of the ideal sheaves of X and Z. If $V_1, \ldots, V_r$ are closed subschemes of Y, we write $\bigcap V_i$ or $V_1 \bigcap \ldots \bigcap V_r$ for the subscheme of Y defined by the sum of the ideal sheaves of the V_i.

A morphism $f: X \to Y$ is *separated* if the diagonal morphism from X to $X \times_Y X$ is a closed imbedding. For the valuative criterion of separatedness, see [EGA]II.7.2 or [H]II.4.3. An algebraic scheme is separated if its structural morphism to Spec (K) is separated. The reader who so wishes may assume all schemes and morphisms are separated.

B.2.4. A morphism $f: X \to Y$ is *proper* if it is separated, and universally closed, i.e., for all $Y' \to Y$, the induced morphism from $X \times_Y Y'$ to Y' takes closed sets to closed sets. For the valuative criterion of properness, see [EGA]II.7.3 or [H]II.4.7.

A scheme is *complete* if the structural morphism to Spec (K) is proper.

A morphism $f: X \to Y$ is *finite* if for each affine open subset $U \subset Y$, the inverse image $U' = f^{-1}(U)$ is affine, and the induced homomorphism of coordinate rings make $A(U')$ a finitely generated $A(U)$-module. A finite morphism is proper with finite fibres (and conversely, cf. [EGA]III.4.4.2).

If $f: X \to Y$ is a proper surjective morphism of varieties, then f factors into $g \circ f'$, where $f': Y \to Y'$ has connected fibres, and $g: Y' \to Y$ is finite. If $\dim X = \dim Y$, there is a non-empty open set of X which maps isomorphically onto an open set of Y'. (These facts, proved in [EGA]III.4.3.1, 4.4.1, are used only in an alternative proof in § 1.4, and in § 20.)

B.2.5. A morphism $f: X \to Y$ is *flat* if for $U \subset Y$, $U' \subset X$ affine open sets with $f(U') \subset U$, the induced map $f^*: A(U) \to A(U')$ makes $A(U')$ a flat $A(U)$-module. Equivalently, for all subvarieties V of X, with $W = \overline{f(V)}$, $\mathcal{O}_{V,X}$ is a flat $\mathcal{O}_{W,Y}$-module.

A morphism $f: X \to Y$ has *relative dimension n* if for all subvarieties V of Y, and all irreducible components V' of $f^{-1}(V)$, $\dim V' = \dim V + n$. If f is flat, Y is irreducible, and X has pure dimension equal to $\dim Y + n$, then f has relative dimension n, and all base extensions $X \times_Y Y' \to Y'$ have relative dimension n (cf. [H]III.9.6, [EGA]IV.14.2).

Convention. In the text, unless otherwise stated, a flat morphism is assumed to have a relative dimension.

B.2.6. A variety is *normal* if the coordinate ring of any affine open subset is integrally closed in its function field. Any variety X has a *normalization*, which is a normal variety $\tilde{X}$ together with a finite, birational morphism from $\tilde{X}$ to X. If X is affine with coordinate ring A, then $\tilde{X}$ is affine with coordinate ring the integral closure of A in $R(X)$; in general X is constructed by gluing together these affine normalizations ([EGA]II.6.3.8).

B.2.7. If $f: X \to Y$ is a morphism, the sheaf of *relative differentials* is denoted $\Omega^1_{X/Y}$. When $Y = \mathrm{Spec}(K)$, we write Ω^1_X. If $g: Y \to S$ is a morphism, there is an exact sequence of sheaves on X:

$$f^* \Omega^1_{Y/S} \to \Omega^1_{X/S} \to \Omega^1_{X/Y} \to 0$$

([EGA]IV.16.4.19, [H]II.8.11).

We say that a morphism $f: X \to Y$ is *smooth* if f is flat of some relative dimension n, and $\Omega^1_{X/Y}$ is a locally free sheaf of rank n. It follows that for any $Y' \to Y$ the base change $X \times_Y Y' \to Y'$ is also smooth of relative dimension n.

A scheme X is called *non-singular*, or *smooth*, if it is smooth over Spec (K). Such a scheme is a disjoint union of irreducible n-dimensional varieties, for some n. A *simple* point of a scheme X is a point in the open subscheme of X which is smooth over $\operatorname{Spec}(K)$.

Convention. Unless otherwise stated, a smooth morphism will be assumed to be separated. A non-singular scheme will be assumed to be irreducible.

If $f: X \to Y$ is smooth of relative dimension n, the *relative tangent bundle*, denoted $T_{X/Y}$, is the vector bundle whose sheaf of sections is the dual bundle to $\Omega^1_{X/Y}$ (§ B.3.2). When $Y = \operatorname{Spec}(K)$, we write simply T_X, the *tangent bundle* of X.

B.3 Vector Bundles

B.3.1. A vector bundle E of rank r on a scheme X is a scheme E equipped with a morphism $\pi: E \to X$, satisfying the following condition. There must be an open covering $\{U_i\}$ of X and isomorphisms φ_i of $\pi^{-1}(U_i)$ with $U_i \times \mathbb{A}^r$ over U_i, such that over $U_i \cap U_j$ the composites $\varphi_i \circ \varphi_j^{-1}$ are linear, i.e., given by a morphism

$$g_{ij}: U_i \cap U_j \to \mathrm{GL}(r, K)\,.$$

These transitions functions satisfy: $g_{ik} = g_{ij}\, g_{jk}$, $g_{ij}^{-1} = g_{ji}$, $g_{ii} = 1$. Conversely, any such transition functions determine a vector bundle. Data (U'_i, φ'_i) determine an isomorphic bundle if all composites $\varphi'_i \circ \varphi_j$ are linear on $U'_i \cap U_j$.

B.3.2. A *section* of E is a morphism $s: X \to E$ such that $\pi \circ s = \mathrm{id}_X$. If E is determined by transition functions g_{ij}, a section of E is determined by a collection of morphisms $s_i: U_i \to \mathbb{A}^r$, such that

$$s_i = g_{ij}\, s_j$$

on $U_i \cap U_j$. The *sheaf of sections* of E is a localy free sheaf $\mathscr{E}$ of $\mathscr{O}_X$-modules of rank r. Conversely, a locally free sheaf $\mathscr{E}$ (of constant rank) comes from a vector bundle E, unique up to isomorphism. This may be seen by using transition functions. For an affine open set $U \subset X$ with coordinate ring A, $\pi^{-1}(U)$ is an affine open set in E, with coordinate ring the symmetric algebra

$$\operatorname{Sym}_A \Gamma(U, \mathscr{E}^\vee)\,,$$

where $\mathscr{E}^\vee = \mathscr{H}om_{\mathscr{O}_X}(\mathscr{E}, \mathscr{O}_X)$, and $\Gamma(U, \mathscr{E}^\vee) = H^0(U, \mathscr{E}^\vee)$ is the space of sections.

If s is a section of E, the *zero scheme* of s, denoted $Z(s)$, is defined as follows. Let $s_i: U_i \to \mathbb{A}^r$ determine s on U_i, $s_i = (s_{i1}, \ldots, s_{ir})$, s_{im} in the coordinate ring of U_i; then $Z(s)$ is defined in U_i by the ideal generated by $s_{i1}, \ldots, s_{ir}$.

B.3.3. Several basic operations are defined for vector bundles, compatibly with the corresponding notions for sheaves: direct sum $E \oplus F$, tensor product $E \otimes F$, exterior product $\Lambda^i E$, symmetric product $S^i E$ or $\operatorname{Sym}^i E$, dual bundle $E^\vee$, pull-back $f^* E = X' \times_X E$ for $f: X' \to X$ a morphism. The trivial bundle of rank one on X is often denoted simply 1 or $\mathscr{O}$.

A homomorphism of vector bundles $E \to F$ corresponds to a homomorphism $\mathscr{E} \to \mathscr{F}$ of corresponding locally free sheaves. To give such a homomorphism is equivalent to giving a section of the bundle $\operatorname{Hom}(E, F) = E^\vee \otimes F$. A sequence

$$0 \to E_n \xrightarrow{d_n} E_{n-1} \to \ldots \to E_1 \xrightarrow{d_0} E_0 \to 0$$

of vector bundle homomorphisms is a *complex* if $d_{i-1} \circ d_i = 0$ for $i = 1, \ldots, n$. The complex is *exact* at $x \in X$ if the corresponding complex of sheaves is exact at x; equivalently, the induced complex of vector spaces $E_{.}(x)$ over the residue field $\varkappa(x)$ is exact.

If Γ is a finite-dimensional space of sections of a vector bundle E on X, there is a canonical homomorphism from the trivial bundle $X \times \Gamma$ to E. The sections Γ *generate* E if this homomorphism is surjective.

B.3.4 If s is a section of E, there is a *Koszul complex* $\Lambda^{\cdot}(s)$:

$$0 \to \Lambda^r E^\vee \to \Lambda^{r-1} E^\vee \to \ldots \to \Lambda^2 E^\vee \to E^\vee \to 1 \to 0$$

which is exact on $X - Z(s)$. For the corresponding sheaf $\mathscr{E}$, the image of $\mathscr{E}^\vee \to \mathscr{O}_X$ is the ideal sheaf of $Z(s)$, and one has a complex of sheaves on X

$$(*) \qquad 0 \to \Lambda^r \mathscr{E}^\vee \to \ldots \to \mathscr{E}^\vee \to \mathscr{O}_X \to \mathscr{O}_{Z(s)} \to 0\,,$$

obtained by globalizing the construction of Appendix A.5.

The section s is called a *regular section* of E if this last sequence is exact. If $x \in Z(s)$, and E is trivialized near x, so s is given by a sequence $(s_1, \ldots, s_r)$, $s_i \in \mathscr{O}_{x,X}$, then $(*)$ is exact at s if and only if $s_1, \ldots, s_r$ is a regular sequence of elements in $\mathscr{O}_{x,X}$ (Appendix A.5, A.6).

We usually identify a vector bundle with its locally free sheaf of sections, unless there is a particular reason to distinguish between them.

B.3.5. A *line bundle* is a vector bundle L of rank one. For $n \in \mathbb{Z}$, the line bundle $L^{\otimes n}$ is defined to be the n-fold tensor product of L if $n > 0$, the $(-n)$-fold tensor product of $L^\vee$ if $n < 0$, and the trivial line bundle 1 if $n = 0$.

B.3.6. In studying vector bundles, there is no loss of generality in assuming that the base space X is connected. Indeed, prescribing a vector bundle of rank r on X is equivalent to prescribing a vector bundle of rank r on each connected component of X. It is sometimes convenient to allow the rank of a vector bundle or locally free sheaf to vary on different connected components; for the same reason, this makes no essential difference.

B.4 Cartier Divisors

References for this section are [EGA]IV.20, Mumford (2), Kleiman (9).

B.4.1. Let X be an algebraic scheme. For each affine open set U of X, let $K(U)$ be the total quotient ring of the coordinate ring $A(U)$, i.e. the localization of $A(U)$ at the multiplicative system of elements which are not zero divisors. This determines a presheaf on X, whose associated sheaf is denoted

$\mathcal{K}$. Let $\mathcal{K}^*$ denote the (multiplicative) sheaf of invertible elements in $\mathcal{K}$, and $\mathcal{O}^*$ the sheaf of invertible elements in $\mathcal{O}=\mathcal{O}_X$.

A *Cartier divisor* D on X is a section of the sheaf $\mathcal{K}^*/\mathcal{O}^*$. A Cartier divisor is determined by a collection of affine open sets U_i which cover X, and elements f_i in $K(U_i)$, such that f_i/f_j is a section of $\mathcal{O}^*$ over $U_i \cap U_j$. Such f_i are called *local equations* for D. The Cartier divisors on X form a group $\operatorname{Div}(X)$, which is written additively.

B.4.2. The *support* of D, denoted $\operatorname{Supp}(D)$, or sometimes $|D|$, is the subset of X consisting of points x such that a local equation for D is not in $\mathcal{O}^*_{x,X}$. The support of D, like the support of the section of any sheaf, is a closed subset of X.

B.4.3. A Cartier divisor is *principal* if the corresponding section of $\mathcal{K}^*/\mathcal{O}^*$ is the image of a global section of $\mathcal{K}^*$.

If X is a variety, $\mathcal{K}$ is the constant sheaf $R(X)$. The principal divisor of $r \in R(X)^*$ is denoted $\operatorname{div}(r)$. Since the support of $\operatorname{div}(r)$ is a proper closed subset of X, there are only a finite number of subvarieties V of codimension one in X such that $f \notin \mathcal{O}^*_{V,X}$.

B.4.4. A Cartier divisor D on a scheme X determines a line bundle on X, denoted $\mathcal{O}_X(D)$, or $\mathcal{O}(D)$. The sheaf of sections of $\mathcal{O}(D)$ may be defined to be the $\mathcal{O}_X$-subsheaf of $\mathcal{K}$ generated on U_i as above by f_i^{-1}. Equivalently, transition functions for $\mathcal{O}(D)$, with respect to the covering U_i, are $g_{ij}=f_i/f_j$.

A *canonical divisor* K_X on a non-singular n-dimensional variety X is a divisor whose line bundle $\mathcal{O}(K_X)$ is $\Omega^n_X = \Lambda^n(T^\vee_X)$.

B.4.5. A Cartier divisor D is *effective* if local equations f_i are sections of $\mathcal{O}$ on U_i. In this case there is a *canonical section* of $\mathcal{O}(D)$, which we denote by s_D. Regarding $\mathcal{O}(D)$ as a subsheaf of $\mathcal{K}$, s_D corresponds to the section 1; with respect to the covering U_i, s_D is given by the collection of functions f_i, which clearly satisfies $f_i = g_{ij} f_j$ on $U_i \cap U_j$. The section s_D vanishes only on the support of D.

For an arbitrary Cartier divisor D on a scheme X, if U is the complement of the support of D, there is a canonical, nowhere vanishing, section of $\mathcal{O}(D)$ over U, which we also denote by s_D. (This section extends canonically to a "meromorphic" section of $\mathcal{O}(D)$ on X, with poles on the locus where D is not effective, cf. [EGA]IV.21.2.11.)

B.5 Projective Cones and Bundles

References for this section are [EGA]II.8, [H]II.7, and Lascu-Scott (1).

B.5.1. Let $S^\cdot = S^0 \oplus S^1 \oplus \ldots$ be a graded sheaf of $\mathcal{O}_X$-algebras on a scheme X, such that the canonical map from $\mathcal{O}_X$ to S^0 is an isomorphism, and $S^\cdot$ is (locally) generated as an $\mathcal{O}_X$-algebra by S^1. To $S^\cdot$ we associate two schemes over X: the *cone* of $S^\cdot$

$$C = \operatorname{Spec}(S^\cdot)\,, \quad \pi : C \to X\,;$$

and the *projective cone* of $S^\cdot$, $\operatorname{Proj}(S^\cdot)$, with projection p to X. The latter is also called the projective cone of C, and denoted $P(C)$:

$$P(C) = \operatorname{Proj}(S^\cdot)\,, \quad p : P(C) \to X\,.$$

On $P(C)$ there is a canonical line bundle, denoted $\mathcal{O}(1)$, or $\mathcal{O}_C(1)$. The morphism p is proper ([EGA]II.5.5.3, [H]II.7.10).

If X is affine, with coordinate ring A, then $S^\cdot$ is determined by a graded A-algebra, which we denote also by $S^\cdot$. If $x_0, \ldots, x_n$ are generators for S^1, then $S^\cdot = A[x_0, \ldots, x_n]/I$ for a homogeneous ideal I. In this case C is the affine subscheme of $X \times \mathbb{A}^{n+1}$ defined by the ideal I, and $P(C)$ is the subscheme of $X \times \mathbb{P}^n$ defined by I; the bundle $\mathcal{O}_C(1)$ is the bull-back of the standard line bundle on $\mathbb{P}^n$. In general $\operatorname{Proj}(S^\cdot)$ is constructed by gluing together this local construction.

If $S^\cdot \to S^{\cdot\prime}$ is a surjective, graded homomorphism of such graded sheaves of $\mathcal{O}_X$-algebras, and $C = \operatorname{Spec}(S^\cdot)$, $C' = \operatorname{Spec}(S^{\cdot\prime})$, then there are closed imbeddings $C' \hookrightarrow C$, and $P(C') \hookrightarrow P(C)$, such that $\mathcal{O}_C(1)$ restricts to $\mathcal{O}_{C'}(1)$.

The *zero section* imbedding of X in C is determined by the augmentation homomorphism from $S^\cdot$ to $\mathcal{O}_X$, which vanishes on S^i for $i > 0$, and is the canonical isomorphism of S^0 with $\mathcal{O}_X$.

If $C = \operatorname{Spec}(S^\cdot)$ is a cone on X, and $f : Z \to X$ is a morphism, the *pull-back* $f^*C = C \times_X Z$ is the cone on Z defined by the sheaf of $\mathcal{O}_Z$-algebras $f^*S^\cdot$. If $Z \subset X$ we write $C|_Z$.

Each section of the sheaf S^1 on X determines a section of the line bundle $\mathcal{O}_C(1)$ on $P(C)$. Let $\mathcal{O}(n)$ or $\mathcal{O}_C(n)$ denote the line bundle $\mathcal{O}_C(1)^{\otimes n}$.

B.5.2. Let z be a variable, $S^\cdot[z]$ the graded algebra whose n^{th} graded piece is

$$S^n \oplus S^{n-1} z \oplus \ldots \oplus S^1 z^{n-1} \oplus S^0 z^n\,.$$

The corresponding cone is denoted $C \oplus 1$. The cone $P(C \oplus 1)$ is called the *projective completion* of C. The element z in $(S^\cdot[z])^1$ determines a regular section of $\mathcal{O}_{C \oplus 1}(1)$ on $P(C \oplus 1)$ whose zero-scheme is canonically isomorphic to $P(C)$. The complement to $P(C)$ in $P(C \oplus 1)$ is canonically isomorphic to C. With this imbedding in $P(C \oplus 1)$, $P(C)$ is called the *hyperplane at infinity.*

B.5.3. If C is a cone over X, each irreducible component D of C is a cone over an irreducible subvariety V of X, which is called the *support* of D. If X is affine with coordinate ring A, then D is defined by a homogeneous prime ideal $P^\cdot$ in $S^\cdot$ (cf. Zariski-Samuel (1)VII.2); then V is the subvariety of X defined by the prime ideal P^0, and $D = \operatorname{Spec}(S^\cdot/P^\cdot)$, with $S^\cdot/P^\cdot$ an algebra over $A/P^0 = S^0/P^0$. This construction patches together to define the cone structure and support of D when X is not affine. Since $S^0 = \mathcal{O}_X$, the union of the supports of the irreducible components of C is X.

Since C is dense in $P(C \oplus 1)$ ([EGA]II.8.3.2), there is a one-to-one correspondence between the irreducible components of C and of $P(C \oplus 1)$, and the geometric multiplicities of corresponding components coincide. If D is an irreducible component of C, there is a canonical closed imbedding of $P(D \oplus 1)$ in $P(C \oplus 1)$, which identifies $P(D \oplus 1)$ with the corresponding irreducible component of $P(C \oplus 1)$; the support of D is the image of $P(D \oplus 1)$ by the projection of $P(C \oplus 1)$ to X.

B.5.4. More generally, if $S^{\cdot}$ is a graded sheaf of $\mathcal{O}_X$-algebras, generated by S^1, such that the canonical map from $\mathcal{O}_X$ to S^0 is surjective, then $S^{\cdot}$ determines a *cone* C, with a morphism from C to X. If $\bar{X}$ is the subscheme of X defined by the kernel of the homomorphism from $\mathcal{O}_X$ to S^0, then by the previous considerations, $S^{\cdot}$ defines a cone C over $\bar{X}$; we call $\bar{X}$ the *support* of C. It is convenient to call such C also a *cone over* X. Of course, in this case, the zero section imbedding of $\bar{X}$ in C may not extend to a morphism from X to C.

B.5.5. A vector bundle E on X is the cone associated to the graded sheaf $\mathrm{Sym}(\mathscr{E}^\vee)$, where $\mathscr{E}$ is the sheaf of sections of E. The *projective bundle* of $\mathscr{E}$ is

$$P(E) = \mathrm{Proj}(\mathrm{Sym}\,\mathscr{E}^\vee) .$$

There is a canonical surjection $p^* E^\vee \to \mathcal{O}_E(1)$ on $P(E)$, which gives an imbedding

$$\mathcal{O}_E(-1) \to p^* E .$$

Thus $P(E)$ is the projective bundle of lines in E, and $\mathcal{O}_E(-1)$ is the *universal*, or tautological line sub-bundle. More generally, given a morphism $f: T \to X$, to factor f into $p \circ \tilde{f}$ is equivalent to specifying a line sub-bundle (namely, $\tilde{f}^* \mathcal{O}_E(-1)$) of $f^* E$.

If E is a vector bundle on X, L a line bundle, there is a canonical isomorphism $\varphi: P(E) \to P(E \otimes L)$, commuting with projections to X, with $\varphi^* \mathcal{O}_{E \otimes L}(-1) = \mathcal{O}_E(-1) \otimes p^*(L)$.

Note. We have adopted the "old-fashioned" geometric notation for $P(E)$. With $\mathscr{E}$ as above, our $P(E)$ is the $\mathbb{P}(\mathscr{E}^\vee)$ of [EGA]II.8.

B.5.6. If E is a sub-bundle of a vector bundle F, with quotient bundle $G = F/E$, there is a canonical imbedding of $P(E)$ in $P(F)$. If $p: P(F) \to X$ is the projection, the composite of the canonical maps $\mathcal{O}_F(-1) \to p^* F$ and $p^* F \to p^* G$ corresponds to a section of $p^* G \otimes \mathcal{O}_F(1)$. This section is regular, and its zero-scheme is $P(E)$.

B.5.7. If E is a vector bundle of rank r on X, and $0 < d < r$, there is a Grassmann bundle, denoted $\mathrm{Grass}_d(E)$ or $G_d(E)$, *of d-planes in E*, with a projection $p: G_d(E) \to X$, and a *universal* rank *d sub-bundle* S of $p^* E$; S is also called the *tautological* bundle on G. The bundle $Q = p^* E/S$ is called the *universal quotient bundle*, and

$$0 \to S \to p^* E \to Q \to 0$$

the *universal exact sequence*. This is characterized by the universal property that for $f: T \to X$, factoring f through $G_d E$ is equivalent to specifying a rank d sub-bundle of $f^* E$. There is a canonical Plücker imbedding of $G_d E$ in $P(\Lambda^d E)$ corresponding to the line sub-bundle $\Lambda^d S$ of $p^* \Lambda^d E$. For the construction of $G_d E$, see [EGA]I.9.7 or Kleiman (3). Note that $G_1 E = P(E)$.

B.5.8. With $p: P(E) \to X$ a projective bundle, the imbedding $\mathscr{O}_E(-1) \subset p^* E$ corresponds to an imbedding of $\mathscr{O}_{P(E)}$ in $p^* E \otimes \mathscr{O}_E(1)$. The cokernel of this imbedding is the relative tangent bundle of $P(E)$ over X:

$$0 \to \mathscr{O}_{P(E)} \to p^* E \otimes \mathscr{O}_E(1) \to T_{P(E)/X} \to 0$$

(cf. Manin (2), Lascu-Scott (1)). More generally, if $G = G_d E$, with universal sub and quotient bundles S and Q, then

$$T_{G/X} = \operatorname{Hom}(S, Q) = S^\vee \otimes Q .$$

Indeed, the universal exact sequence on G determines a "second fundamental form" homomorphism (cf. Altman-Kleiman (1) I.3) from S to $\Omega^1_{G/X} \otimes Q$. Dualizing, this gives a homomorphism from $T_{G/X}$ to $S^\vee \otimes Q$, which is checked by local coordinates to be an isomorphism.

B.6 Normal Cones and Blowing Up

The reference for this section is [EGA] II.8.

B.6.1. Let X be a closed subscheme of a scheme Y, defined by an ideal sheaf $\mathscr{I}$. The *normal cone* $C_X Y$ *to* X *in* Y is the cone over X defined by the graded sheaf of $\mathscr{O}_X$ algebras $\oplus \mathscr{I}^n/\mathscr{I}^{n+1}$:

$$C_X Y = \operatorname{Spec}\Big(\bigoplus_{n \geqq 0} \mathscr{I}^n/\mathscr{I}^{n+1}\Big).$$

If $f: Y' \to Y$ is a morphism, $X' = f^{-1}(X)$, and g is the induced morphism from X' to X, there is a canonical closed imbedding

$$C_{X'} Y' \subset g^* C_X Y = C_X Y \times_X X' .$$

Indeed, there is a canonical surjection of $f^* \mathscr{I}$ onto the ideal sheaf $\mathscr{I}'$ of X' in Y', which gives a surjection of $\oplus\, g^* (\mathscr{I}^n/\mathscr{I}^{n+1})$ onto $\oplus\, \mathscr{I}'^n/\mathscr{I}'^{n+1}$.

B.6.2. If the imbedding of X in Y is a regular imbedding of codimension d, then $C_X Y$ is a vector bundle of rank d on X, and is denoted also $N_X Y$ (cf. § B.7.1); the sheaf of sections of $N_X Y$ is $(\mathscr{I}/\mathscr{I}^2)^\vee$.

In particular, if $i: X \to Y$ imbeds X as a Cartier divisor on Y, then

$$C_X Y = N_X Y = i^* \mathscr{O}_Y(X) .$$

B.6.3. The blow-up of Y along X, denoted $B\, l_X Y$, is the projective cone over Y of the sheaf of $\mathscr{O}_Y$-algebras $\oplus \mathscr{I}^n$:

$$B\, l_X Y = \operatorname{Proj}\Big(\bigoplus_{n \geqq 0} \mathscr{I}^n\Big).$$

Let $\tilde{Y} = B\, l_X Y$, and let π denote the projection from $\tilde{Y}$ to Y. The canonical invertible sheaf (line bundle) $\mathscr{O}(1)$ on the projective cone $\tilde{Y}$ is the ideal sheaf of $\pi^{-1}(X)$, which is therefore a Cartier divisor on $\tilde{Y}$, called the *exceptional divisor.* Let $E=\pi^{-1}(X)$. By construction E is the projective cone of $(\oplus \mathscr{I}^n)\otimes_{\mathscr{O}_Y}\mathscr{O}_X=\oplus \mathscr{I}^n/\mathscr{I}^{n+1}$, so

$$E=P(C_X Y)$$

is the projective normal cone to X in Y. From this description one sees that

$$N_E\tilde{Y}=\mathscr{O}_{\tilde{Y}}(E)|_E=\mathscr{O}_C(-1)\,,$$

where $C=C_X Y$. Let η be the projection from $E=P(C)$ to X. If the imbedding of X in Y is regular, then the canonical imbedding of normal cones $N_E\tilde{Y}\subset \eta^* N_X Y$ is the imbedding of the universal line bundle $\mathscr{O}_N(-1)$ in $\eta^* N$, $N=N_X Y$.

In general, π induces an isomorphism from $\tilde{Y}-E$ onto $Y-X$.

B.6.4. If Y is a variety, $X\subset Y$ a closed subscheme, then $B\,l_X Y$ is also a variety. This follows from the fact that if I is any ideal in a domain A, then $\oplus I^n$ is also a domain.

B.6.5. If X is nowhere dense in Y, then $\pi:\tilde{Y}\to Y$ is birational. Indeed, since E is a Cartier divisor on $\tilde{Y}$, no irreducible component of $\tilde{Y}$ can be contained in E, and since X contains no irreducible components of Y by assumption, all irreducible components of $\tilde{Y}$ and Y meet the open sets $\tilde{Y}-E$ and $Y-X$ respectively, which are isomorphic by π.

B.6.6. If Y has pure dimension k, then $C_X Y$ also has pure dimension k, for any closed subscheme $X\subset Y$. To see this, consider the imbedding of X in $Y\times \mathbb{A}^1$ which is the composite of the given imbedding of X in Y and the imbedding of Y in $Y\times\mathbb{A}^1$ at $0\in\mathbb{A}^1$. The normal cone to X in $Y\times\mathbb{A}^1$ is $C\oplus 1$, where $C=C_X Y$. Since X is nowhere dense in $Y\times\mathbb{A}^1$, $B\,l_X(Y\times\mathbb{A}^1)$ is birational to $Y\times\mathbb{A}^1$, so has pure dimension $k+1$. Since the exceptional divisor $P(C\oplus 1)$ is a Cartier divisor on $B\,l_X(Y\times\mathbb{A}^1)$, it must have pure dimension k, and C is an open subscheme of $P(C\oplus 1)$.

B.6.7. If Y is flat over a non-singular curve T, then $B\,l_X Y$ is also flat over T, for any $X\subset Y$. Indeed, all ideal sheaves in $\mathscr{O}_Y$ are torsion free over $\mathscr{O}_T$ (cf. Example A.5.4), so $\oplus\mathscr{I}^n$ is flat over $\mathscr{O}_T$; hence $B\,l_X Y$, whose local rings are localizations of $\oplus\mathscr{I}^n$, is flat over T.

B.6.8. If X is Cartier divisor on Y, then $\tilde{Y}=Y$. More generally, suppose $X\subset Y$ is defined by an ideal sheaf $\mathscr{I}$, and D is an effective Cartier divisor on Y, defined by an ideal sheaf $\mathscr{J}$. If Z is the subscheme of Y defined by the ideal sheaf $\mathscr{I}\mathscr{J}$, then there is a canonical isomorphism of $B\,l_Z Y$ with $B\,l_X Y$ over Y, such that the exceptional divisor in $B\,l_Z Y$ corresponds to the divisor $E+\pi^* D$ in $B\,l_X Y$, with E the exceptional divisor in $B\,l_X Y$.

B.6.9. If $X\subset Y$ is a closed imbedding, and $f:Y'\to Y$ is a morphism, set $X'=f^{-1}(X)$, $g:X'\to X$ the induced morphism. Then there is a closed imbedding

$$B\,l_{X'}Y'\subset B\,l_X Y\times_Y Y'$$

constructed as in B.6.1. If $\tilde{f}$ is the induced morphism from $B\,l_{X'}Y'$ to $B\,l_X Y$, then $\tilde{f}^{-1}(E)=E'$, where E and E' are the exceptional divisors.

In particular, if $X \subset Y \subset Z$ are closed imbeddings, there is a canonical imbedding of $Bl_X Y$ in $Bl_X Z$, with the exceptional divisor of $Bl_X Z$ restricting to the exceptional divisor of $Bl_X Y$.

B.6.10. In case $X \subset Y$ and $Y \subset Z$ are regular imbeddings (cf. B.7), let $\tilde{Z} = Bl_X Z$, F the exceptional divisor in $\tilde{Z}$, ϱ the projection from $\tilde{Z}$ to Z. Let $\tilde{Y} = Bl_X Y$. Then $\tilde{Y} \subset \varrho^{-1}(Y)$, $F \subset \varrho^{-1}(Y)$, and $\tilde{Y}$ is the residual scheme to F in $\varrho^{-1}(Y)$, i.e., the ideal sheaves of $\tilde{Y}$, F and $\varrho^{-1}(Y)$ in $\tilde{Z}$ are related by

$$\mathscr{I}(\tilde{Y}) \cdot \mathscr{I}(F) = \mathscr{I}(\varrho^{-1}(Y)) .$$

In addition, the canonical imbedding of $\tilde{Y}$ in $\tilde{Z}$ is a regular imbedding, with normal bundle

$$N_{\tilde{Y}} \tilde{Z} \cong \pi^* N_Y Z \otimes \mathscr{O}(-E) ,$$

where π is the projection from $\tilde{Y}$ to Y, and E is the exceptional divisor on $\tilde{Y}$. To prove this, one may assume Z is affine with coordinate ring A, Y is defined by a regular sequence $t_1, \ldots, t_d$ in A, and X is defined by the regular sequence $t_1, \ldots, t_n$, $n > d$. If $T_1, \ldots, T_n$ are homogeneous coordinates on $\mathbb{P}^{n-1}$, then (Lemma A.6.1) $\tilde{Z}$ is the subscheme of $Z \times \mathbb{P}^{n-1}$ defined by the equations $t_i T_j - t_j T_i = 0$, all $i < j$, and $\tilde{Y}$ is defined in $Y \times \mathbb{P}^{n-1}$ by the equations $\bar{t}_i T_j - \bar{t}_j T_i$, $d < i < j$, where $\bar{t}_i$ is the image of t_i in $A/(t_1, \ldots, t_d)$. On the affine open set U_k of $\tilde{Z}$ where $T_k \neq 0$, the coordinate ring of U_k is

$$A[x_1, \ldots, \hat{x}_k, \ldots, x_n]/(\{t_i - t_k x_i | i \neq k\}) ,$$

and $\varrho^{-1}(Y)$ is defined in U_k by the ideal I_k generated by $(t_1, \ldots, t_d)$. If $k \leqq d$, $I_k = (t_k)$, which is the ideal of F in $\tilde{Z}$. If $k > d$, $I_k = (t_k x_1, \ldots, t_k x_d) = t_k \cdot (x_1, \ldots, x_d)$; since t_k defines F on U_k, and $(x_1, \ldots, x_d)$ defines $\tilde{Y}$, $\tilde{Y}$ is residual to F in $\varrho^{-1}(Y)$. Since $x_1, \ldots, x_d$ is a regular sequence in the ring of U_k, $k > d$, the imbedding of $\tilde{Y}$ in $\tilde{Z}$ is regular. The asserted relation among normal bundles follows from the relation among the ideal sheaves (cf. Example 9.2.2).

B.7 Regular Imbeddings and l.c.i. Morphisms

Basic references for these concepts are: [EGA]IV.16.9,17, 19.1, [SGA6]VIII, and Grothendieck (5)II.4. Here we consider only regular imbeddings which are closed imbeddings, and l.c.i. morphisms which admit factorizations into closed imbeddings followed by smooth morphisms.

B.7.1. A closed imbedding $i: X \to Y$ of schemes is a *regular imbedding of codimension d* if every point in X has an affine neighborhood U in Y, such that if A is the coordinate ring of U, I the ideal of A defining X, then I is generated by a regular sequence of length d. If $\mathscr{I}$ is the ideal sheaf of X in Y, it follows (Lemma A.6.1) that the conormal sheaf $\mathscr{I}/\mathscr{I}^2$ is a locally free sheaf on X of rank d. The *normal bundle* to X in Y, denoted $N_X Y$, is the vector bundle on X whose sheaf of sections is dual to to $\mathscr{I}/\mathscr{I}^2$. The normal bundle $N_X Y$ is canonically isomorphic to the normal cone $C_X Y$. Indeed, by Lemma A.6.1, the canonical map from $\operatorname{Sym}(\mathscr{I}/\mathscr{I}^2)$ to $\oplus \mathscr{I}^n/\mathscr{I}^{n+1}$ is an isomorphism.

B.7.2. Let $i: X \to Y$ be a closed imbedding, and let $\mathscr{I}$ be the ideal sheaf of X in Y. If $g: Y \to X$ is a morphism, there is an exact sequence of sheaves on X:

(i) $$\mathscr{I}/\mathscr{I}^2 \to i^* \Omega^1_{Y/S} \to \Omega^1_{X/S} \to 0 .$$

If X is smooth over S, then i is a regular imbedding if and only if Y is smooth over S in some neighborhood of X. ([EGA]IV.17.12.1). In this case the three sheaves are locally free, and the above sequence is also exact on the left; this corresponds to an exact sequence of vector bundles on X:

(ii) $$0 \to T_{X/S} \to i^* T_{Y/S} \to N_X Y \to 0 .$$

B.7.3. If $g: Y \to S$ is a (separated) smooth morphism of relative dimension n, and $i: S \to Y$ is a section, i.e. $gi = \mathrm{id}_S$, it follows that i is a (closed) regular imbedding with normal bundle canonically isomorphic to $i^* T_{Y/S}$. More generally, if $f: X \to Y$ is any morphism, and $\gamma: X \to X \times_S Y$ is the graph of f, i.e. $\gamma = (1, f)$, then γ is a regular imbedding of codimension n, with normal bundle $f^* T_{Y/S}$. (This follows since the base extension $X \times_S Y \to X$ of the smooth morphism g is smooth.) In particular, the diagonal imbedding $\delta: Y \to Y \times_S Y$ is a regular imbedding with normal bundle $T_{Y/S}$.

B.7.4. If $i: X \to Y$ and $j: Y \to Z$ are regular imbeddings, then the composite $k = j \circ i$ is a regular imbedding, and there is an exact sequence of vector bundles on X:

(iii) $$0 \to N_X Y \to N_X Z \to i^* N_Y Z \to 0$$

([EGA]IV.19.1.5).

If $Y_i \subset Z$ are regular imbeddings of codimension d_i, $i = 1, \ldots, r$, and the imbedding of $X = \bigcap Y_i$ in Z is regular of codimension $\sum d_i$, then

$$N_X Z = \bigoplus_{i=1}^{r} (N_{Y_i} Z)|_X .$$

In particular, if $Y \to S$ is smooth, the normal bundle to the r-fold diagonal imbedding of Y in $Y \times_S \ldots \times_S Y$ is the direct sum of $r-1$ copies of $T_{Y/S}$.

If $i: X \to Y$ is a regular imbedding, and $f: Y' \to Y$ is flat, then the imbedding of $X' = f^{-1}(X)$ in Y' is regular, with

$$N_{X'} Y' = g^* N_X Y ,$$

where g is the induced morphism from X' to X (Example A.5.5).

B.7.5. If $i: X \to Y$ is a closed imbedding, $g: Y \to Z$ smooth, such that $g\,i$ is a closed imbedding, then i is a regular imbedding if and only if $g\,i$ is a regular imbedding ([SGA6]VIII.1.3). In this case there is an exact sequence

(iv) $$0 \to i^* T_{Y/Z} \to N_X Y \to N_X Z \to 0$$

of vector bundles on X. (Apply B.7.3, B.7.4 to the imbeddings $X \hookrightarrow X \times_Z Y \hookrightarrow Y$.)

B.7.6. A morphism $f: X \to Y$ will be called a *local complete intersection* (l.c.i.) morphism of codimension d if f admits a factorization into a (closed) regular imbedding of some codimension e, followed by a smooth morphism of relative dimension $d+e$. It follows from the preceding paragraph that if $f = g\, i$ is any factorization with $i: X \to P$ a closed imbedding and $g: P \to Y$ smooth of relative dimension n, then i is a regular imbedding of codimension $n-d$.

If $f = g\, i$ is such a factorization of the l.c.i. morphism f, the *virtual tangent bundle*, denoted T_f, is defined to be the difference of the bundles $i^* T_{P/Y}$ and $N_X P$ in the Grothendieck group of vector bundles on X:

$$T_f = [i^* T_{P/Y}] - [N_X P] \in K^\circ X .$$

It follows from the exact sequences (iii) and (iv) that T_f is independent of the choice of factorization (see the proof of Proposition 6.6 (a), or [SGA 6] VIII.2.5)

Note: In contrast to standard usage, we require a l.c.i. morphism to admit a global factorization into a closed imbedding followed by a smooth morphism. If X admits a closed imbedding j into a scheme M smooth over Spec(K), then any morphism $f: X \to Y$ admits such a factorization: $X \xrightarrow{i} Y \times M \xrightarrow{p} Y$, where $i = (f, j)$, p the projection.

B.8 Bundles on Imbeddable Schemes

B.8.1. If X is a scheme which admits a closed imbedding in a non-singular scheme, and $\mathscr{F}$ is a coherent sheaf on X, then there is a locally free sheaf E on X and a surjection $E \to \mathscr{F}$. In case X is quasi-projective, some tensor product of $\mathscr{F}$ by a line bundle $\mathscr{O}(n)$ is generated by its sections, and one may take E to be a sum of copies of $\mathscr{O}(-n)$, cf. Serre (1) § 55. In general, one may assume X is non-singular, in which case one may find such E which is a direct sum of bundles $\mathscr{O}(D_i)$, D_i divisors on X, a result of S. Kleiman. The essential point is that for an affine open covering U_i of X, the complements $X - U_i$ are supports of divisors. For details see Borelli (1), or [SGA 6] II.2.2.6.

B.8.2. If $X \hookrightarrow Y$ is a closed imbedding, and Y admits a closed imbedding in a non-singular scheme, then there is a vector bundle E on Y so that the projection from $B l_X Y$ to Y factors into a closed imbedding in $P(E)$ followed by a projection from $P(E)$ to Y. Indeed, if $\mathscr{E}$ is locally free and surjects onto the ideal sheaf $\mathscr{I}$ of X in Y, then Sym$\mathscr{E}$ surjects onto $\oplus \mathscr{I}^n$, which imbeds Proj$(\oplus \mathscr{I}^n)$ in Proj(Sym$\mathscr{E}$). One may take E to be the bundle whose sheaf of sections is $\mathscr{E}^\vee$. In particular, $B l_X Y$ also admits a closed imbedding into a non-singular scheme. For if $Y \subset M$, M non-singular, $B l_X Y \subset B l_X M$, and by the preceding assertion, $B l_X M$ imbeds in a projective bundle over M, which is non-singular.

The homomorphism $\mathscr{E} \to \mathscr{I} \subset \mathscr{O}_Y$ corresponds to a section s of E whose zero scheme $Z(s)$ is X.

B.8.3. The fact that the canonical homomorphism from $K^\circ X$ to $K_\circ X$ is an isomorphism for X non-singular, follows from several facts:

(i) For any coherent sheaf $\mathscr{F}$ on a non-singular variety X there is a locally free sheaf E_0 and a surjection $E_0 \to \mathscr{F}$ (B.8.1).

(ii) If X is non-singular and n-dimensional, and

$$0 \to \mathscr{G} \to E_{n-1} \to E_1 \to E_0 \to \mathscr{F} \to 0$$

is an exact sequence of coherent sheaves on X, with $E_0, \ldots, E_{n-1}$ locally free, then $\mathscr{G}$ is locally free. (This is a local assertion, and follows from the fact that finitely generated modules over a regular local ring have finite free resolutions, cf. Serre (4)IV. th. 8.)

(iii) If $0 \to E_n'' \to \ldots \to E_0'' \to \mathscr{F}'' \to 0$ is a resolution of $\mathscr{F}''$ by locally free sheaves E_i'', and $\mathscr{F} \to \mathscr{F}''$ is a surjection, there is a compatible resolution $E.$ of $\mathscr{F}$, i.e., with a commutative diagram

$$\begin{array}{ccccccccccccc} 0 & \to & E_n & \to & E_{n-1} & \to & \ldots & \to & E_1 & \to & E_0 & \to & \mathscr{F} & \to 0 \\ & & \downarrow & & \downarrow & & & & \downarrow & & \downarrow & & \downarrow & \\ 0 & \to & E_n'' & \to & E_{n-1}'' & \to & \ldots & \to & E_1'' & \to & E_0'' & \to & \mathscr{F}'' & \to 0 \end{array}$$

with the vertical maps surjective. (To achieve this, choose E_0 to map onto the kernel of the canonical map from $E_0'' \oplus \mathscr{F}$ to $\mathscr{F}''$. Having constructed the diagram through E_r, let $\mathscr{Z}_r$ be the kernel of the map from E_r to E_{r-1}, and $\mathscr{Z}_r''$ the kernel of $E_r'' \to E_{r-1}''$; then choose E_{r+1} to map onto the kernel of the canonical map from $E_{r+1}'' \oplus \mathscr{Z}_r$ to $\mathscr{Z}_r''$.)

(iv) Let $\mathscr{F} \to \mathscr{F}''$, $E. \to E.''$ be as in (iii). Then the kernels E_i' of the vertical maps give a resolution of the kernel $\mathscr{F}'$ of $\mathscr{F} \to \mathscr{F}''$, and

$$\sum (-1)^i [E_i] = \sum (-1)^i [E_i'] + \sum (-1)^i [E_i'']$$

in $K^\circ X$.

(v) A resolution $E.$ of $\mathscr{F}$ is said to *dominate* a resolution $E.''$ of $\mathscr{F}$ if they can be related by a commutative diagram as in (iii), with $\mathscr{F}'' = \mathscr{F}$, $\mathscr{F} \to \mathscr{F}''$ the identity. Given any two resolutions of $\mathscr{F}$, there is a third which dominates them both. (The argument is similar to that in (iii), cf. Borel-Serre (1) Lemma 11.)

From (i) and (iii) it follows that any coherent sheaf $\mathscr{F}$ has a finite resolution $E.$ by locally free sheaves. From (iv) and (v), the class $\sum (-1)^i [E_i]$ in $K^\circ X$ is independent of the resolution. From (iv), the resulting map $[\mathscr{F}] \to \sum (-1)^i [E_i]$ is well-defined on the Grothendieck group $K_\circ X$. This determines a homomorphism $K_\circ X \to K^\circ X$, inverse to the canonical homomorphism $K^\circ X \to K_\circ X$. For details see Borel-Serre (1) and Borelli (1).

B.9 General Position

B.9.1. The following is a variation of a lemma of Serre.

Lemma. *Let E be a vector bundle of rank r on a scheme X over an algebraically closed field, π the projection from E to X. Let Γ be a finite* *dimensional vector space of sections of E which generates E, and let $C_1, \ldots, C_p$ be closed subsets of E, and $Z_1, \ldots, Z_q$ closed subsets of X. Then there is a non-empty Zariski open subset Γ° of Γ such that for all $s \in \Gamma^\circ$, and all i, j, either $s^{-1}(C_i)$ is disjoint from Z_j, or*

$$\dim (s^{-1}(C_i) \cap Z_j) \leqq \dim (C_i \cap \pi^{-1}(Z_j)) - r .$$

Proof. Since Γ generates E, the canonical morphism

$$\varphi : X \times \Gamma \to E$$

is surjective and smooth of relative dimension $v - r$, where $v = \dim \Gamma$. For each irreducible component D_{ijk} of $C_i \cap \pi^{-1}(Z_j)$, $\varphi^{-1}(D_{ijk})$ is an irreducible variety with

$$\dim \varphi^{-1}(D_{ijk}) \leqq \dim (C_i \cap \pi^{-1}(Z_j)) + v - r .$$

Let p be the projection from $X \times \Gamma$ to Γ. By the theorem on the dimension of fibres of a morphism ([EGA]IV.13, [H]II Ex. 3.22), there is a non-empty open set $\Gamma_{ijk} \subset \Gamma$ such that, for all $s \in \Gamma_{ijk}$, either $p^{-1}(s)$ is disjoint from $\varphi^{-1}(D_{ijk})$, or

$$\dim (p^{-1}(s) \cap \varphi^{-1}(D_{ijk})) = \dim \varphi^{-1}(D_{ijk}) - v .$$

The intersection of all the Γ_{ijk} is the desired set Γ°. (See Example 12.1.11 for a refinement.) □

B.9.2. The following is proved in Kleiman (6), cf. [H]III,10.8.

Lemma. *Suppose a connected algebraic group G acts transitively on a variety X, over an algebraically closed field K. Let $f : Y \to X$, $g : Z \to X$ be morphisms of varieties Y, Z to X. For each point σ in G, let Y^σ denote Y with the morphism $\sigma \circ f$ from Y to X.*

(a) *There is a non-empty open set $G^\circ \subset G$ such that for all σ in G°, $Y^\sigma \times_X Z$ is either empty or of pure dimension*

$$\dim (Y) + \dim (Z) - \dim (X) .$$

(b) *If Y and Z are non-singular, and* $\operatorname{char}(K) = 0$, *then there is a non-empty open set $G^\circ \subset G$ such that for all σ in G°, $Y^\sigma \times_X Z$ is non-singular.*

For analogues of (b) in characteristic p, see Kleiman (6) and Vainsencher (3).

Bibliography

The notation [EGA] refers to Grothendieck, Dieudonné (1), while [SGA6] is Berthelot, Grothendieck, Illusie, et al. (1).

Akin, K., Buchsbaum, D. A., Weyman, J.
(1) Schur functors and Schur complexes, Advances in Math. *44* (1982), 207–278

Altman, A. B., Kleiman, S. L.
(1) *Introduction to Grothendieck duality theory,* Springer Lecture Notes *146,* 1970
(2) Foundations of the theory of Fano schemes, Composito Math. *34* (1977), 3–47

Arakelov, S. J.
(1) Theory of intersections on the arithmetic surface, Proc. Intern. Cong. Math., Vancouver, 1974, vol. 1, 405–408

Arason, J. K., Pfister, A.
(1) Quadratische Formen über affinen Algebren und ein algebraischer Beweis des Satzes von Borsuk-Ulam, J. Reine Angew. Math. *331* (1982), 181–184

Arbarello, E., Cornalba, M., Griffiths, P., Harris, J.
(1) *Geometry of algebraic curves,* Springer-Verlag, to appear

Artin, M., Nagata, M.
(1) Residual intersection in Cohen Macauley rings, J. Math. Kyoto Univ. *12* (1972), 307–323

Atiyah, M. F., Bott, R., Shapiro, A.
(1) Clifford modules, Topology *3* (Suppl. 1) (1964), 3–38

Atiyah, M. F., Hirzebruch, F.
(1) Cohomologie-Operationen und charakteristische Klassen, Math. Z. *77* (1961), 149–187
(2) Analytic cycles on complex manifolds, Topology *1* (1962), 25–45
(3) The Riemann-Roch theorem for analytic imbeddings, Topology *1* (1962), 151–166

Atiyah, M. F., Singer, I.
(1) The index of elliptic operators on compact manifolds, Bull. Amer. Math. Soc. *69* (1963), 422–433

Baker, H. F.
(1) *Principles of Geometry,* Vol. V, Analytical principles of the theory of curves, Cambridge Univ. Press, 1933
(2) *Principles of Geometry,* Vol. VI, Introduction to the theory of algebraic surfaces, Cambridge Univ. Press, 1933

Baldassarri, M.
(1) *Algebraic varieties,* Ergebnisse der Math., Springer-Verlag, 1956

Barton, C., Clemens, C. H.
(1) A result on the integral Chow ring of a generic principally polarized complex abelian variety of dimension four, Compositio Math. *34* (1977), 49–67

Bass, H., Connell, E. H., Wright, D.
(1) The Jacobian conjecture: reduction of degree and formal expansion of the inverse, Bull. Amer. Math. Soc. *7* (1982), 287–330

Baum, P., Fulton, W., MacPherson, R.
(1) Riemann-Roch for singular varieties, Publ. Math. I.H.E.S. *45* (1975), 101–145
(2) Riemann-Roch and topological *K*-theory for singular varieties, Acta Math. *143* (1979), 155–192

Baum, P., Fulton, W., Quart, G.
(1) Lefschetz-Riemann-Roch for singular varieties, Acta Math. *143* (1979), 193–211

Beauville, A.
(1) Variétés de Prym et jacobiennes intermédiaires, Ann. scient. Éc. Norm. Sup. *10* (1977), 309–391
(2) *Surfaces algébriques complexes,* Astérisque *54,* Soc. Math. France, 1978
(3) Diviseurs speciaux et intersection de cycles dans la jacobienne d'une courbe algébrique, in *Enumerative and classical geometry,* Nice 1981, Progress in Math. 24, Birkhäuser (1982), 133–142
(4) Les singularités du diviseur Θ de la jacobienne intermediaire de l'hypersurface cubique dans $\mathbb{P}^4$, Springer Lecture Notes *947* (1982), 190–208

Behrend, F.
(1) Über Systeme reeller algebraischer Gleichungen, Compositio Math. *7* (1939), 1–19

Behrens, E.-A.
(1) Zur Schnittmultiplizität uneigentlicher Komponenten in der algebraischen Geometrie, Math. Z. *55* (1952), 199–215

Bernstein, I. N., Gel'fand, I. M., Gel'fand, S. I.
(1) Schubert cells and cohomology of the spaces G/P, Russ. Math. Surveys *28* (1973), 1–26

Berthelot, P., Grothendieck, A., Illusie, L. et al.
(1) *Theorie des Intersections et Théorème de Riemann-Roch,* SGA6, 1966/67, Springer Lecture Notes *225,* 1971

Berzolari, L.
(1) Sulle intersezioni di tre superficie algebriche, Ann. di Mat. (2) *24* (1896), 165–191
(2) Théorie générale des courbes planes algébriques, Encyclopédie des sciences mathématiques, III.3, Gauthier-Villars, Paris (1915), 257–304
(3) Algebraische Transformationen und Korrespondenzen, Encyklopädie der Mathematischen Wissenschaften III C 11, Teubner, Leipzig (1933), 1781–2218

Bézout, É.
(1) *Théorie générale des équations algébriques,* Ph.-D. Pierres, Paris 1779

Bloch, S.
(1) K_2 and algebraic cycles, Annals of Math. *99* (1974), 349–379
(2) K_2 of Artinian Q-algebras, with applications to algebraic cycles, Comm. Algebra *3* (1975) 405–428
(3) Torsion cycles and a theorem of Roĭtman, Compositio Math. *39* (1979), 107–127
(4) *Lectures on algebraic cycles,* Duke Univ. Math. Series, 1980

Bloch, S., Gieseker, D.
(1) The positivity of the Chern classes of an ample vector bundle, Inventiones Math. *12* (1971), 112–117

Bloch, S., Kas, A., Lieberman, D.
(1) Zero cycles on surfaces with $p_g = 0$, Compositio Math. *33* (1976), 135–145

Bloch, S., Murre, J. P.
(1) On the Chow group of certain types of Fano threefolds, Compositio Math. *39* (1979), 47–105

Bloch, S., Ogus, A.
(1) Gersten's conjecture and the homology of schemes, Ann. scient. Éc. Norm. Sup. *7* (1974), 181–202

Boda, E., Vogel, W.
(1) On system of parameters, local intersection multiplicity and Bézout's theorem, Proc. Amer. Math. Soc. *78* (1980), 1–7

de Boer, J. H.
(1) The intersection multiplicity of a connected component, Math. Ann. *172* (1967), 238–246

Borel, A.
(1) *Linear Algebraic Groups,* Math. Lecture Note Series, W. A. Benjamin, 1969

Borel, A., Haefliger, A.
(1) La classe d'homologie fundamentale d'une espace analytique, Bull. Soc. Math. France *89* (1961), 461–513

Borel, A., Moore, J.
(1) Homology theory for locally compact spaces, Michigan Math. J. *7* (1960), 137–159
Borel, A., Serre, J.-P.
(1) Le théorème de Riemann-Roch (d'après Grothendieck), Bull. Soc. Math. France *86* (1958), 97–136
Borelli, M.
(1) Some results on ampleness and divisorial schemes, Pacfic J. Math. *23* (1967), 217–227
Bott, R., Tu, L.
(1) *Differential forms in algebraic topology*, Springer-Verlag, 1982
Bourbaki, N.
(1) Algèbre commutative, Éléments de Mathématique, Chap. 1–7, Hermann, Paris, 1961–1965
(2) *Algèbre homologique*, Éléments de Mathématique, Algèbre, Chap. 10, Masson, Paris, 1980
Briney, R. E.
(1) Intersection theory on quotients of algebraic varieties, Amer. J. Math. *84* (1962), 217–238
Brylinski, J.-L., Kashiwara, M.
(1) Kazhdan-Lusztig conjecture and holonomic systems, Inventiones Math. *65* (1981), 387–410
Carlson, J., Green, M., Griffiths, P., Harris, J.
(1) Infinitesimal variation of Hodge structures, Compositio Math., to appear
Carrell, J. B.
(1) Chern classes of the Grassmannian and Schubert calculus, Topology *17* (1978), 177–182
Carrel, J. B., Lieberman, D. I.
(1) Vector fields, Chern classes, and cohomology, Proc. Sympos. Pure Math. 30, Part 1, Amer. Math. Soc. (1977), 251–254
Cayley, A.
(1) On skew surfaces, otherwise scrolls, Phil. Trans. Royal Soc. London *153* (1863), 453–483; Collected Math. Papers V, 168–200
(2) On the curves which satisfy given conditions, Phil. Trans Royal Soc. London *158* (1868), 75–172; Collected Math. Papers VI, 191–262
(3) A memoir on the theory of reciprocal surfaces, Phil. Trans. Royal Soc. London *119* (1869), 201–229; Collected Math Papers VI, 329–358
Ceresa, G.
(1) C is not algebraically equivalent to C^- in its Jacobian, Annals of Math. *117* (1983), 285–291
Ceresa, G., Collino, A.
(1) Some remarks on algebraic equivalence of cycles, Pacific J. Math. *105* (1983), 285–290
Chasles, M.
(1) Principle de correspondance entre deux variables, qui peut être d'un grand usage en Géométrie, C. R. Acad. Sci. *41* (1855), 1097–1107
(2) Determination du nombre des sections coniques qui doivent toucher cinq courbes d'ordre quelconque, ou satisfaire à diverses autres conditions, C. R. Acad. Sci. *58* (1864), 222–226
(3) Construction des coniques qui satisfont à cinq conditions. Nombre des solutions dans chaque question, C. R. Acad. Sci. *58* (1864), 297–308, continued in vol. *58* (1864), 425–431, vol. *59* (1865), 7–15, 93–97
(4) Considérations sur la méthode générale exposée dans la séance de 15 février. – Différences entre cette méthode et la méthode analytique. – Procédés généraux de démonstration, C. R. Acad. Sci. *58* (1864), 1167–1175
(5) Relations entre les deux caractéristiques d'un système de courbes d'ordre quelconque, C. R. Acad. Sci. *62* (1866), 325–334
(6) Détermination immédiate par le principle de correspondance, du nombre des points d'intersection de deux courbes, d'ordre quelconque, qui se trouvent à distance finie, C. R. Acad. Sci. *75* (1872), 736–744, *76* (1873), 126–132

Chevalley, C.
(1) Intersections of algebraic and algebroid varieties, Trans. Amer. Math Soc. *57* (1945), 1–85
(2) Les classes d'équivalence rationelle, I, II, Séminaire C. Chevalley, 2e année, *Anneaux de Chow et Applications*, Secr. Math. Paris, 1958
Chow, W.-L.
(1) On the equivalence classes of cycles in an algebraic variety, Annals of Math. *64* (1956), 450–479
(2) Algebraic varieties with rational dissections, Proc. Nat. Acad. Sci. *42* (1956), 116–119
Claborn, L., Fossum, R.
(1) Generalizations of the notion of class group, Illinois J. Math. *12* (1968), 218–253
Clemens, H.
(1) Homological equivalence, modulo algebraic equivalence, is not finitely generated, preprint, 1982
Clemens, H., Griffiths, G.
(1) The intermediate Jacobian of a cubic three-fold, Annals of Math. *95* (1972), 281–356
Coleff, N., Herrera, M., Lieberman, D.
(1) Algebraic cycles as residues of meromorphic forms, Math. Ann. *254* (1980), 73–87
Collino, A.
(1) The rational equivalence ring of symmetric products of curves, Illinois J. Math. *19* (1975), 567–583
(2) Quillen's *K*-theory and algebraic cycles on almost non-singular varieties, Illinois J. Math. *25* (1981), 654–666
Colliot-Thélène, J.-L.
(1) Formes quadratiques multiplicatives et variétés algébriques: deux compléments, Bull. Soc. Math. France *108* (1980), 213–227
Colliot-Thélène, J.-L., Coray, D.
(1) L'équivalence rationelle sur les points fermés des surfaces rationnelles fibrées en coniques, Compositio Math. *39* (1979), 301–332
Colliot-Thélène, J.-L., Ischebeck, F.
(1) L'équivalence rationnelle sur les cycles de dimension zéro des variétés algébriques réelles, C. R. Acad. Sci. *292* (1981), 723–725
De Concini, C., Eisenbud, D., Procesi, C.
(1) *Hodge Algebras*, Astérisque *91*, 1982
Conforto, F.
(1) Lo stato attuale della teoria dei sistemi di equivalenza e della teoria delle correspondenze algebriche tra varietà, Atti del Convegno Matematico, Roma 1942, 49–83
Coolidge, J. L.
(1) *A treatise on algebraic plane curves*, 1931, Dover Publ. New York (1959)
(2) *A history of geometrical methods*, 1940, Dover Publ. New York (1963)
Damon, J.
(1) The Gysin homomorphism for flag bundles, Amer. J. Math. *95* (1973), 643–659; *96* (1974), 248–260
Davis, E. D.
(1) Ideals of the principal class, *R*-sequences and a certain monoidal transformation, Pacific J. Math. *20* (1967), 197–205
Deligne, P.
(1) Formes modulaires et représentations l-adiques, Sém. Bourbaki 355, 1968/69, Springer Lecture Notes *179* (1971), 139–172
(2) *Cohomologie Etale*, SGA 4½, Springer Lecture Notes *569*, 1977
Deligne, P., Katz, N.
(1) *Groupes de Monodromie en Géométrie Algébrique*, SGA 7 II, 1967–1969, Springer Lecture Notes *340*, 1973
Deligne, P., Milne, J. S., Ogus, A., Shih, K.-y.
(1) *Hodge cycles, motives, and Shimura varieties*, Springer Lecturc Notes *900*, 1982
Demazure, M.
(1) Désingularisation des variétés de Schubert généralisées, Ann scient. Éc. Norm. Sup. *7* (1974), 52–88

Dieudonné, J.

(1) *Cours de géométrie algébrique* 1, Aperçu historique sur le développement de la géométrie algébrique, Presses Universitaires de France, 1974

(2) Schur functions and group representations, in *Tableaux de Young et foncteurs de Schur en algèbre et géométrie,* Conférence internationale Torún, Pologne 1980, Astérisque *87–88* (1981), 7–19

Dold, A.

(1) *Lectures on Algebraic Topology,* Springer-Verlag, 1972

Dolgachev, I.

(1) The Euler characteristic of a family of algebraic varieties, Math. USSR Sb. *18* (1972), 303–319

Dolgachev, I., Parshin, A. N.

(1) The different and discriminant of regular mappings, Math. Notes *4* (1968), 802–804

Donagi, R., Smith, R.

(1) The degree of the Prym map onto the moduli space of five dimensional abelian varieties, in *Journées de géométrie algébrique,* Angers/France 1979, Sijthoff and Noordhoff (1980), 143–155

Douady, A.

(1) Cycles analytiques, d'aprés Atiayh et Hirzebruch, Sém. Bourbaki 223, 1961/1962

Douady, A., Verdier, J.-L.

(1) *Séminaire de Géométrie Analytique de l'Ecole Normal Supérieure* 1974/1975, Astérisque *36–37,* 1976

Draper, R.

(1) Intersection theory in analytic geometry, Math. Ann. *180* (1969), 175–204

Dubson, A. S.

(1) Théorie d'intersection pour les cycles algébriques ou $\mathbb{C}$-analytiques réels, C. R. Acad. Sci. *286* (1978), 755–758

Dutta, S., Hochster, M., McLaughlin, J. E.

(1) Modules of finite length and finite projective dimension with negative intersection multiplicities, preprint, 1982

Ehresmann, C.

(1) Sur la topologie de certains espaces homogènes, Annals of Math. *35* (1934), 396–443

Eisenbud, D., Evans, E. G.

(1) Every algebraic set in n-space is the intersection of n hypersurfaces, Inventiones Math. *19* (1973), 107–112

Eisenbud, D., Harris, J.

(1) Divisors on general curves and cuspidal rational curves, Inventiones Math., to appear

Enriques, F.

(1) Sulle intersezioni di due varietà algebriche, Rend. R. Accad. Sci. Ist. Bologna *19* (1915), 90–92

Enriques, F., Chrisini, O.

(1) *Lezioni sulla teoria geometrica delle equazioni e delle funzioni algebriche,* vol. III, N. Zanichelli, Bologna, 1924

Fiorentini, M., Lascu, A. T.

(1) Una formula di geometria numerative, Ann. Univ. Ferrara *27* (1981), 201–227

Fogarty, J., Rim, D. S.

(1) Serre sequences and Chern classes, J. Algebra *10* (1968), 436–447

Freudenthal, H.

(1) La géométrie énumérative, Colloques Intern. du CNRS XII, Topologie Algébrique, Paris 1947, 17–33

Fulton, W.

(1) *Algebraic Curves,* Math. Lecture Note Series, W. A. Benjamin, 1969

(2) Rational equivalence on singular varieties, Publ. Math. I.H.E.S. *45* (1975), 147–167

(3) Ample vector bundles, Chern classes, and numerical criteria, Inventiones Math. *32* (1976), 171–178

(4) A Hirzebruch-Riemann-Roch formula for analytic spaces and non-projective algebraic varieties, Compositio Math. *34* (1977), 279–283

(5) A fixed point formula for varieties over finite fields, Math. Scand. *42* (1978), 189–196
(6) A note on residual intersections and the double point formula, Acta. Math. *140* (1978), 93–101
(7) A note on the arithmetic genus, Amer. J. Math. *101* (1979), 1355–1363

Fulton, W., Gillet, H.
(1) Riemann-Roch for general algebraic schemes, Bull. Soc. Math. France, to appear

Fulton, W., Hansen, J.
(1) A connectedness theorem for projective varieties, with applications to intersections and singularities of mappings, Annals of Math. *110* (1979), 159–166

Fulton, W., Johnson, K.
(1) Canonical classes on singular varieties, Manu. Math. *32* (1980), 381–389

Fulton, W., Kleiman, S., MacPherson, R.
(1) About the enumeration of contacts, to appear in Proceedings of the conference on open questions in algebraic geometry, Ravello, 1982

Fulton, W., Laksov, D.
(1) Residual intersections and the double point formula, in *Real and Complex Singularities,* Oslo 1976, P. Holm (ed.) Sijthoff and Noordhoff (1977), 171–178

Fulton, W., Lazarsfeld, R.
(1) Connectivity and its applications in algebraic geometry, Springer Lecture Notes *862* (1981), 26–92
(2) On the connectedness of degeneracy loci and special divisors, Acta. Math. *146* (1981), 271–283
(3) Positive polynomials for ample vector bundles, Annals of Math, *118* (1983), 35–60
(4) Positivity and excess intersections, in *Enumerative and classical geometry,* Nice 1981, Progress in Math. 24, Birkhäuser (1982), 97–105

Fulton, W., MacPherson, R.
(1) Intersecting cycles on an algebraic variety, in *Real and Complex Singularities,* Oslo 1976, P. Holm (ed.), Sijthoff and Noordhoff (1977), 179–197
(2) Defining algebraic intersections, in *Algebraic Geometry* Proceedings, Tromsø 1977, Springer Lecture Notes *687* (1978), 1–30
(3) *Categorical framework for the study of singular spaces,* Mem. Amer. Math. Soc. *243,* 1981

Gaeta, F.
(1) Sul calculo effectivo della forma associata $F(W^{g\,l}_{\alpha+\beta-n})$ all'intersezione de due cicle effettivi pure U^g_α, V^l_β di S_n, in funzione delle $F(U^g_\alpha)$, $F(V^l_\beta)$ relative ai cicli secanti, Atti Accad. Naz. Linc. (8) *24* (1958), 269–276

Gaffney, T., Lazarsfeld, R.
(1) On the ramification of branched coverings of $\mathbb{P}^n$, Inventiones Math. *59* (1980), 53–58

Galbură, G.
(1) Sur les variétés canoniques et cycles caractéristique d'une varietá algébrique (Roumanian), Acad. Repub. Pop. Romîne Bul. Sti. Sect. Mat. Fiz. *6* (1954), 61–64
(2) Sui covarianti di immersione, Roma Univ. Ist. Mat. Rend. (5) *25* (1966), 239–247

Galbură, G., Lascu, A.
(1) Éclatement des classes de Chern d'une variété algébrique, Rev. Rom. Math. Pures et Appl. *12* (1967), 1255–1258

Gamkrelidze, R. V.
(1) Computation of the Chern cycles of algebraic manifolds (Russian), Dokl. Akad. Nauk, SSSR *90* (1953), 719–722
(2) Chern cycles of complex algebraic manifolds (Russian) Isv. Akad. Nauk. SSSR *20* (1956), 685–706

Gerstenhaber, M.
(1) On the deformations of rings and algebras: II, Annals of Math. *84* (1966), 1–19

Giambelli, G. Z.
(1) Rizoluzione del problema degli spazi secanti, Mem. R. Accad. Torino *52* (1902), 171–211

(2) Ordine di una varietà più ampia di quella rappresentata coll'annullare tutti i minori di dato ordine estratti da una data matrice generica di forme, Mem. R. Ist. Lombardo (3) *11* (1904), 101–135
(3) Sulle varietá rappresentata coll'annullare determinanti minori contenuti in un determinante simmetrico od emisimmetrico generico di forme, Atti, R. Accad. Torino *41* (1906), 102–125

Gillet, H.
(1) Comparison of K-theory spectral sequences, with applications, Springer Lecture Notes *854* (1981), 141–167
(2) Riemann-Roch theorems for higher algebraic K-theory, Advances in Math. *40* (1981), 203–289
(3) Universal cycle classes, Compositio Math. *49* (1983), 3–49

Gonzalez-Sprinberg, G.
(1) L'obstruction locale d'Euler et le théorème de MacPherson, Astérisque *82–83* (1981), 7–32

Goresky, R. M.
(1) Whitney stratified chains and cochains, Trans. Amer. Math. Soc. *261* (1981), 175–196

Goresky, R. M., MacPherson, R.
(1) Intersection homology theory, Topology *19* (1980), 135–162
(2) Intersection homology II, Inventiones Math. *72* (1983), 77–129

Grayson, D.
(1) Products in K-theory and intersecting algebraic cycles, Inventiones Math. *47* (1978), 71–83
(2) Projections, cycles, and algebraic K-theory, Math. Ann. *234* (1978), 69–72
(3) Coincidence formulas in enumerative geometry, Comm. Algebra *7* (1979), 1685–1711

Griffiths, P. A.
(1) Hermitian differential geometry, Chern classes, and positive vector bundles, in *Global Analysis*, D. Spencer and S. Iyanaga (eds.), Princeton Math. Series No. *29*, Tokyo (1969), 185–251
(2) On the periods of certain rational integrals: I, II, Annals of Math. *90* (1969), 460–541
(3) Some transcendental methods in the study of algebraic cycles, Springer Lecture Notes *185* (1971), 1–46
(4) Some results on algebraic cycles on algebraic manifolds, in *Algebraic Geometry* (Bombay Colloquim), Oxford (1969), 93–191
(5) Variations on a theorem of Abel, Inventiones Math. *35* (1976), 321–390

Griffiths, P., Harris, J.
(1) *Principles of Algebraic Geometry*, Wiley-Interscience, 1978
(2) On the variety of special linear systems on a general algebraic curve, Duke Math. J. *47* (1980), 233–272

Gröbner, W.
(1) *Idealtheoretischer Aufbau der algebraischen Geometrie*, Teubner, Leipzig, 1941

Grothendieck, A.
(1) Sur quelques propriétés fondamentales en théorie des intersections, Séminaire C. Chevalley, 2ᵉ année, *Anneaux de Chow et applications.* Secr. Math. Paris, 1958
(2) La théorie des classes de Chern, Bull. Soc. Math. France *86* (1958), 137–154
(3) Sur une note de Mattuck-Tate, J. Reine Angew. Math. *200* (1958), 208–215
(4) The cohomology theory of abstract algebraic varieties, Proc. Intern. Cong. Math., Edinburgh 1958, 103–118
(5) *Revêtements étales et groupe fondamental,* SGA 1, 1960–61, Springer Lecture Notes *224*, 1971
(6) Standard conjectures on algebraic cycles, in *Algebraic Geometry,* (Bombay Colloquium), Oxford (1969), 193–199

Grothendieck, A., Dieudonné, J.
(1) *Eléments de Géométrie Algébrique* (EGA) I, Grundlehren der Math. Wissenschaften 166, Springer-Verlag (1971); II, Publ. Math. I.H.E.S. *8* (1961); III, Ibid. *11* (1961), *17* (1963); IV, Ibid. *20* (1964), *24* (1965), *28* (1966), *32* (1967)

Halphen, G.-H.
(1) Sur une question d'élimination ou sur l'intersection de deux courbes en un point singulier, Bull. Soc. Math. France *3* (1875), 76–92

Hansen, J.
(1) Singularities under projections – another proof of the double point formula, Aarhus Preprint (1980/81), No. 24

Harris, J.
(1) On the Kodaira dimension of the moduli space of curves, II: the even genus case, Inventiones Math., to appear

Harris, J., Mumford, D.
(1) On the Kodaira dimension of the moduli space of curves, Inventiones Math. *67* (1982), 23–86

Harris, J., Tu, L.
(1) On symmetric and skew-symmetric determinantal varieties, Topology, to appear.
(2) Chern numbers of kernel and cokernel bundles, Inventiones Math., to appear

Hartshorne, R.
(1) Ample vector bundles, Publ. Math. I.H.E.S. *29* (1966), 63–94
(2) *Residues and Duality*, Springer Lecture Notes *20*, 1966
(3) *Ample Subvarieties of Algebraic Varieties*, Springer Lecture Notes *156*, 1970
(4) Equivalence relations on algebraic cycles and subvarieties of small codimension, Proc. Sympos. Pure Math. *29*, Amer. Math. Soc. (1975), 129–164
(5) *Algebraic Geometry*, Graduate Texts in Math, Springer-Verlag, 1977

Harthsorne, R., Rees, E., Thomas, E.
(1) Non-smoothing of algebraic cycles on Grassmann varieties, Bull. Amer. Math. Soc. *80* (1974), 847–851

Harvey, R.
(1) Holomorphic chains and their boundaries, Proc. Sympos. Pure Math. 30, Part 1, Amer. Math. Soc. (1977), 309–382

Herbert, J.
(1) *Multiple points of immersed manifolds*, Mem. Amer. Math. Soc. *250*, 1981

Hiller, H.
(1) Schubert calculus of a Coxeter group, l'Enseign. Math. *27* (1981), 57–84
(2) Combinatorics and intersections of Schubert varieties, Comm. Math. Helv. *57* (1982), 41–59
(3) *Geometry of Coxeter groups*, Research Notes in Math. 54, Pitman Publ., 1982

Hironaka, H.
(1) Smoothing of algebraic cycles of small dimension, Amer. J. Math. *90* (1968), 1–54
(2) Triangulations of algebraic sets, Proc. Sympos. Pure Math. 29, Amer. Math. Soc. (1975), 165–185

Hirzebruch, F.
(1) *Topological methods in algebraic geometry*, 1956, Grundlehren der math. Wissenschaften, Vol. 131, Third enlarged edition, Springer-Verlag, 1966

Hochster, M.
(1) Grassmannians and their Schubert subvarieties are arithmetically Cohen-Macaulay, J. Algebra *25* (1973), 40–57

Hodge, W. V. D.
(1) The characteristic classes on algebraic varieties, Proc. London Math. Soc. *3* (1951), 138–151

Hodge, W. V. D., Pedoe, D.
(1) *Methods of Algebraic Geometry*, Vol. 2, Cambridge Univ. Press, 1952

Holme, A.
(1) Embedding-obstructions for smooth projective varieties I, in Studies in Algebraic Topology, Advances in Math. Suppl. Studies *5* (1979), 39–67
(2) Embedding obstructions for singular algebraic varieties in $\mathbb{P}^n$, Acta. Math. *135* (1975), 155–185

Horrocks, G.

(1) On the relation of S-functions to Schubert varieties, Proc. London Math. Soc. *7* (1957), 265–280

Horrocks, G., Mumford, D.

(1) A rank 2 vector bundle on $\mathbb{P}^4$ with 15,000 symmetries, Topology *12* (1973), 63–81

Hoyt, W.

(1) On the moving lemma for rational equivalence, J. Indian Math. Soc. *35* (1971),47–66

Huneke, C.

(1) On the associated graded ring of an ideal, Illinois J. Math. *26* (1982), 121–137

(2) A remark concerning multiplicities, Proc. Amer. Math. Soc. *85* (1982), 331–332

Hurwitz, A.

(1) Über algebraische Korrespondenzen und das verallgemeinerte Korrespondenzprinzip, Math. Ann. *28* (1887), 561–585

Ilori, S. A.

(1) A generalization of the Gysin homomorphism for flag bundles, Amer. J. Math *100* (1978), 621–630

Ilori, S. A., Ingleton, A. W., Lascu, A. T.

(1) On a formula of D. B. Scott, J. London Math. Soc. *8* (1974), 539–544

Ingleton, A. S., Scott, D. B.

(1) The tangent direction bundle of an algebraic variety, Ann. di Mat. *56* (1961) 359–374

Iversen, B.

(1) Numerical invariants and multiple planes, Amer. J. Math. *92* (1970), 968–996

(2) Critical points of an algebraic function, Inventiones Math. *12* (1971), 210–224

(3) Local Chern classes, Ann. scient. Éc. Norm. Sup. *9* (1976), 155–169

(4) *Cohomology of Sheaves,* to appear.

Jacobi, C. G. J.

(1) De relationibus, quae locum habere debent inter puncta intersectionis duarum curvarum vel trium superficierum algebraicarum dati ordinis, simul cum enodatione parodoxi algebraici, J. Reine Angew. Math. *15* (1836), 285–308

James, C. G. F.

(1) On the intersection of constructs in space of three or four dimensions, with special reference to the matrix representation of curves and surfaces, Proc. Cambridge Phil. Soc. *21* (1923), 435–462

Johnson, K. W.

(1) Immersion and embedding of projective varieties, Acta Math. *140* (1978), 49–74

Jouanolou, J. P.

(1) Cohomologie de quelques schémas classiques et théorie cohomologique des classes de Chern, in SGA 5, 1965–66, Springer Lecture Notes *589* (1977), 282–350

(2) Riemann-Roch sans dénominateurs, Inventiones Math. *11* (1970), 15–26

(3) Singularités rationelles du résultant, in *Algebraic Geometry,* Proceedings, Copenhagen 1978, Springer Lecture Notes *732* (1979), 183–213

Józefiak, T., Lascoux, A., Pragacz, P.

(1) Classes of determinantal varieties associated with symmetric and skew-symmetric matrices, Math. USSR Izv. *18* (1982), 575–586

Kawai, S.

(1) A note on Steenrod reduced powers of algebraic cycles, in *Complex analysis and algebraic geometry,* Iwanami Shoten, Tokyo (1977), 375–381

Kempf, G.

(1) Schubert methods with an application to algebraic curves, Publ. Math. Centrum, Amsterdam, 1971

(2) On the geometry of a theorem of Riemann, Annals of Math. *98* (1973), 178–185

Kempf, G., Laksov, D.

(1) The determinantal formula of Schubert calculus, Acta. Math. *132* (1974), 153–162

King, J. R.

(1) The currents defined by analytic varieties, Acta Math. *127* (1971), 185–220

(2) Global residues and intersections on a complex manifold, Trans. Amer. Math. Soc. *192* (1974), 163–199

(3) Refined residues, Chern forms, and intersections, in *Value distribution theory,* Part A, Dekker, New York (1974), 169–190

Kirby, D.

(1) Isolated intersections of a set of n primals in n-space, J. London Math. Soc. *33* (1958), 185–196

Kleiman, S. L.

(1) Toward a numerical theory of ampleness, Annals of Math. *84* (1966), 293–344

(2) Algebraic cycles and the Weil conjectures, in *Dix Exposés sur la cohomologie des schémas,* North Holland (1968), 359–386

(3) Geomtery on Grassmannians and applications to splitting bundles and smoothing cycles, Publ. Math. I.H.E.S. *36* (1969), 281–298

(4) Finiteness theorems for algebraic cycles, Proc. Intern. Cong. Math., Nice 1970, vol. 1, 445–449

(5) Motives, in *Algebraic Geometry,* Oslo 1970, F. Oort (ed.), Wolters-Noordhoff Publ., Groningen (1972), 53–82

(6) The transversality of a general translate, Compositio Math. *38* (1974), 287–297

(7) Rigorous foundations of Schubert's enumerative calculus, Proc. Sympos. Pure Math. 28, Amer. Math. Soc. (1976), 445–482

(8) The enumerative theory of singularities, in *Real and Complex Singularities,* Oslo 1976, P. Holm (ed.), Sijthoff and Noordhoff (1977), 297–396

(9) Misconceptions about K_X, l'Enseign. Math. *25* (1979), 203–206

(10) Chasles's enumerative theory of conics: a historical introduction, in *Studies in algebraic geometry,* Math. Assoc. Amer. Stud. Math. *20* (1980), 117–138

(11) Concerning the dual variety, in *18th Scandinavian Congress of Mathematicians,* 1980, E. Balslev (ed.). Progress in Math. *11,* Birkhäuser (1981), 386–396

(12) Multiple-point formulas I: iteration, Acta. Math. *147* (1981), 13–49

(13) Multiple-point formulas II: the Hilbert scheme, to appear

Kleiman, S. L., Laksov, D.

(1) Schubert calculus, Amer. Math. Monthly *79* (1972), 1061–1082

(2) On the existence of special divisors, Amer. J. Math. *93* (1972), 431–436

(3) Another proof of the existence of special divisors, Acta. Math. *132* (1974), 163–176

Kleiman, S. L., Landolfi, J.

(1) Geometry and deformation of special Schubert varieties, in *Algebraic Geometry,* Oslo 1970, F. Oort (ed.), Wolters-Noordhoff Publ., Groningen (1972), 97–124

Knebusch, M.

(1) On algebraic curves over real closed fields, Math. Z. *150* (1976), 49–70; *151* (1976), 189–205; *155* (1977), 299

(2) An algebraic proof of the Borsuk-Ulam theorem for polynomial mappings, Proc. Amer. Math. Soc. *84* (1982), 29–32

Kodaira, K.

(1) On a differential-geometric in the theory of algebraic stacks, Proc. Nat. Acad. Sci. USA *39* (1953), 1268–1273

Kumar, N. M., Murthy, M. P.

(1) Algebraic cycles and vector bundles over affine three-folds, Annals of Math. *116* (1982), 579–591

Kunz, E.

(1) *Einführung in die kommutative Algebra und algebraische Geometrie,* Friedr. Vieweg & Sohn, Braunschweig, 1980

Lakshmibai, V., Musili, C., Seshadri, C. S.

(1) Geometry of G/P, Bull. Amer. Math. Soc. *1* (1979), 432–435

Laksov, D.

(1) The arithmetic Cohen-Macaulay character of Schubert schemes, Acta. Math. *129* (1972), 1–9

(2) Algebraic cycles on Grassmann varieties, Advances in Math. *9* (1972), 267–295

(3) Residual intersections and Todd's formula for the double locus of a morphism, Acta. Math. *140* (1978), 75–92

(4) Weierstrass points on curves, Astérisque *87–88* (1981), 221–247
(5) Wronskians and Plücker formulas for linear systems on curves, Institut Mittag-Leffler, Report No. 11, 1981

Lang, S.
(1) The theory of real places, Annals of Math. *57* (1953), 378–391

Lascoux, A.
(1) Calcul de certains polynômes de Thom, C. R. Acad. Sci. *278* (1974), 889–891
(2) Polynômes symétriques et coefficients d'intersection de cycles de Schubert, C. R. Acad. Sci. *279* (1974), 201–204
(3) Puissance extérieures, déterminants et cycles de Schubert, Bull. Soc. Math. France *102* (1974), 161–179
(4) Fonctions de Schur et grassmanniennes, C. R. Acad. Sci. *281* (1975), 813–815, 851–854
(5) Tableaux de Young gauches, in *Séminaire Pisot,* Théorie des Nombres, Année 1974–1975, exposé n° 4, M. Belgodère ed., Paris, 1975
(6) Sistemi lineari di divisori sulle curve e sulle superficie, Ann. di Mat. *114* (1977), 141–153
(7) Classes de Chern d'une produit tensoriel, C. R. Acad. Sci. *286* (1978), 385–387

Lascu, A. T.
(1) The self-intersection formula and the 'formule-clef', Proc. Cambridge Phil. Soc. *78* (1975), 117–123

Lascu, A. T., Scott, D. B.
(1) An algebraic correspondence with applications to blowing up Chern classes, Ann. di Mat. *102* (1975), 1–36
(2) A simple proof of the formula for the blowing up of Chern classes, Amer. J. Math. *100* (1978), 293–301

Laumon, G.
(1) Homologie étale, Astérisque *36–37* (1976), 163–188

Lazarsfeld, R.
(1) Excess intersection of divisors, Compositio Math. *43* (1981), 281–296

Lê, D. T., Teissier, B.
(1) Variétés polaires locales et classes de Chern des variétés singulières, Annals of Math. *114* (1981), 457–491

Le Barz, P.
(1) Géométrie énumérative pour les multisécantes, Springer Lecture Notes *683* (1977), 116–167
(2) Validité de certaines formules de géométrie énumérative, C. R. Acad. Sci. *289* (1979), 755–758
(3) Quadrisécantes d'une surface de $\mathbb{P}^5$, C. R. Acad. Sci. *291* (1980), 639–642
(4) Formules pour les multisécantes de surfaces, C. R. Acad. Sci. *292* (1981), 797–800
(5) Une application de la formule de Fulton-MacPherson, preprint, 1982

Lech, C.
(1) On the associativity formula for multiplicities, Arkiv. Math. *3* (1956), 301–314

Ledermann, W.
(1) *Introduction to group characters,* Cambridge Univ. Press, 1977

Lefschetz, S.
(1) Intersections and transformations of complexes and manifolds, Trans. Amer. Math. Soc. *28* (1926), 1–49
(2) *Topology,* Amer. Math. Soc. Colloq. Publ. 12, 1930
(3) *Algebraic Topology,* Amer. Math. Soc. Colloq. Publ. 27, 1942
(4) *Algebraic Geometry,* Princeton Univ. Press, 1953

Lemoyne, T.
(1) *Recherches de géométrie contemporaine,* Tome 1, 2, A. Blanchard, Paris, 1968

Lesieur, L.
(1) Les problèmes d'intersection sur une variété de Grassmann, C. R. Acad. Sci. *225* (1947), 916–917

Levi, B.
(1) Dell'intersezione di due varietà, Atti R. Acad. Sci. Torino *34* (1899), 3–17
Lichtenbaum, S.
(1) Curves over discrete valuation rings, Amer. J. Math. *90* (1968), 380–405
Lieberman, D.
(1) Numerical and homological equivalence of algebraic cycles on Hodge manifolds, Amer. J. Math. *90* (1968), 366–374
(2) Higher Picard varieties, Amer. J. Math. *90* (1968), 1165–1199
Littlewood, D. E., Richardson, A. R.
(1) Group characters and algebra, Phil. Trans. Royal Soc. London A *233* (1934), 99–141
Lomadze, V. G.
(1) On the intersection index of divisors, Math. USSR Izv. *17* (1981), 343–352
Lübke, M.
(1) Beweis einer Vermutung von Hartshorne für den Fall homogener Mannigfaltigkeiten, J. Reine Angew. Math. *316* (1980), 215–220
Macdonald, I. G.
(1) Some enumerative formulae for algebraic curves, Proc. Camb. Phil. Soc. *54* (1958), 399–416
(2) Symmetric products of an algebraic curve, Topology 1 (1962), 319–343
(3) *Symmetric functions and Hall polynomials*, Oxford Univ. Press, 1979
MacPherson, R.
(1) Chern classes of singular varieties, Annals of Math. *100* (1974), 423–432
Macaulay, F. S.
(1) *Algebraic theory of modular systems*, Cambridge Tracts Math., Cambridge Univ. Press, 1916
Malliavin, M.-P.
(1) Une remarque sur les anneaux locaux réguliers, Séminaire P. Dubreil 1970/71, Fasc. 2, Exp. 13
Manin, J. I.
(1) Correspondences, motifs, and monoidal transformations, Math. USSR Sb. *6* (1968), 439–470
(2) Lectures on the K-functor in algebraic geometry, Russ. Math. Survey 24, No. *5* (1969), 1–89
Marlin, R.
(1) Comparison de l'anneau de Chow et de l'anneau de Grothendieck, Astérisque *36–37* (1976), 229–240
Martinelli, E.
(1) Sulle intersezioni delle curve analitiche complesse, Rend. Mat. Univ. Roma (5) *14* (1955), 422–430
(2) Sulle varietà delle faccette p-dimensionali di S_r, Atti. Accad. Italia, Mem. (6) *12* (1941), 917–943
Matsumura, H.
(1) *Commutative Algebra*, Second edition, Math. Lecture Note Series, Benjamin/Cummings, Reading, 1980
Matsusaka, T.
(1) The criteria for algebraic equivalence and the torsion group, Amer. J. Math. *79* (1957), 53–66
(2) *Theory of Q-varieties*, Math. Soc. of Japan, 1964
Mattuck, A. P.
(1) Symmetric products and Jacobians, Amer. J. Math. *83* (1961), 189–206
(2) On symmetric products of curves, Proc. Am. Math. Soc. *13* (1962), 82–87
(3) Secant bundles on symmetric products, Amer. J. Math. *87* (1965), 779–797
Mattuck, J., Tate, J.
(1) On the inequality of Castelnuovo-Severi, Abb. Math. Sem. Univ. Hamb. *22* (1958), 295–299
Micali, A.
(1) Sur les algèbres universelles, Ann. Inst. Fourier, Grenoble, *14, 2* (1964), 33–88

Milne, J. S.
(1) Zero cycles on algebraic varieties in non-zero characteristic: Rojtman's theorem, Compositio Math. *47* (1982), 271–287
Milnor, J.
(1) Some consequences of a theorem of Bott, Annals of Math. *68* (1958), 444–449
(2) On the Betti numbers of real varieties, Proc. Amer. Math. Soc. *15* (1964), 275–280
(3) *Singular points of complex hypersurfaces*, Annals of Math. Studies 61, Princeton Univ. Press, 1968
Milnor, J., Stasheff, J.
(1) *Characteristic classes*, Annals of Math. Studies 76, Princeton Univ. Press, 1974
Monk, D.
(1) The geometry of flag manifolds, Proc. London Math. Soc. *9* (1959), 253–286
Mori, S.
(1) Projective manifolds with ample tangent bundles, Annals of Math. *10* (1979), 593–606
Mumford, D.
(1) Topology of normal singularities and a criterion for simplicity, Publ. Math. I.H.E.S. *9* (1961), 5–22
(2) *Lectures on curves on an algebraic surface*, Annals of Math. Studies *59*, Princeton Univ. Press, 1966
(3) Rational equivalences of 0-cycles on surfaces, J. Math. Kyoto Univ. *9* (1969), 195–204
(4) *Abelian varieties*, Oxford Univ. Press, 1970
(5) *Algebraic Geometry I, Complex Projective Varieties*, Grundlehren der math. Wissenschaften 221, Springer-Verlag, 1976
(6) Stability of projective varieties, l'Enseign. Math. *24* (1977), 39–110
(7) Towards an enumerative geometry of the moduli space of curves, to appear in a volume dedicated to I. Shafarevich
Murre, J. P.
(1) Intersection multiplicities of maximal connected bunches, Amer. J. Math. *80* (1958), 311–330
(2) Algebraic equivalence modulo rational equivalence on a cubic threefold, Compositio Math. *25* (1972), 161–206
(3) Reduction of the proof of the non-rationality of a non-singular cubic threefold to a result of Mumford, Compositio Math. *27* (1973), 63–82
Murthy, M. P., Swan, R. G.
(1) Vector bundles over affine surfaces, Inventiones Math. *36* (1976), 125–165
Nagata, M.
(1) The theory of multiplicity in general local rings, in *Proceedings of the International Symposium*, Tokyo-Nikko 1955, Tokyo (1956), 191–226
(2) *Local Rings*, Interscience Tracts in Pure and Appl. Math., Wiley, New York, 1962
Nakano, S.
(1) Tangential vector bundle and Todd canonical systems of an algebraic variety, Mem. Coll. Sci. Univ. Kyoto (A) *29* (1955), 145–149
Namba, M.
(1) *Families of Meromorphic Functions on Compact Riemann Surfaces*, Springer Lecture Notes *767*, 1979
Nastold, H.-J.
(1) Zur Serreschen Multiplizitätstheorie in der arithmetischen Geometrie, Math. Ann. *143* (1961), 333–343
Néron, A.
(1) Hauteurs et théorie des intersections, in *Questions on Algebraic Varieties*, Cremonese, Roma (1970), 101–120
Noether, M.
(1) Zur Theorie des eindeutigen Entsprechens algebraischer Gebilde, Zweiter Aufsatz, Math. Ann. *8* (1875), 495–533
(2) Rationale Ausführung der Operationen in der Theorie der algebraischen Functionen, Math. Ann. *23* (1884), 311–358

Northcott, D. G.
(1) Hilbert's function in a local ring, Quart. J. Math. Oxford *4* (1953), 67–80
(2) *Lessons on rings, modules and multiplicities*, Cambridge Univ. Press, 1968
Northcott, D. G., Rees, D.
(1) Extensions and simplifications of the theory of regular local rings, Proc. Cambridge Phil. Soc. *57* (1961), 483–488
O'Brian, N. R., Toledo, D., Tong, Y. L. L.
(1) The trace map and characteristic classes for coherent sheaves, Amer. J. Math. *103* (1981), 225–252
(2) Hirzebruch-Riemann-Roch for coherent sheaves, Amer. J. Math. *103* (1981), 253–271
(3) Grothendieck-Riemann-Roch for complex manifolds, Bull. Amer. Math. Soc. *5* (1981), 182–184
Orlik, P.
(1) The multiplicity of a holomorphic map at an isolated critical point, in *Real and complex singularities*, Oslo 1976, P. Holm (ed.), Sitjhoff and Noordhoff (1977), 405–474
Patil, D. P., Vogel, W.
(1) Remarks on the algebraic approach to intersection theory, preprint, 1982
Perron, O.
(1) Studien über den Vielfachheitsbegriff und den Bézoutschen Satz, Math. Z. *49* (1944), 654–680
Persson, U.
(1) *On degenerations of algebraic surfaces*, Mem. Amer. Math. Soc. *189*, 1977
Peskine, C., Szpiro, L.
(1) Liaison des variétés algébriques, Inventiones Math. *26* (1973), 271–302
(2) Syzygies et multiplicités, C. R. Acad. Sci. *278* (1974), 1421–1424
Peters, C. A. M., Simonis, J.
(1) A secant formula, Quart. J. Math. Oxford (2) *27* (1976), 181–189
Pfister, A.
(1) Systems of quadratic forms, Bull. Soc. Math. France, Mémore *59* (1979), 115–123
Picard, É.
(1) Sur le nombre des racines communes à plusieurs équations simultanées, J. Math. Pures App. *8* (1892), 5–24
Piene, R.
(1) Numerical characters of a curve in projective n-space, in *Real and complex singularities*, Oslo 1976, P. Holm (ed.), Sijthoff and Noordhoff (1977), 475–495
(2) Some formulas for a surface in $\mathbb{P}^3$, in *Algebraic Geometry* Proceedings, Tromsø, 1977, Springer Lecture Notes *687* (1978), 196–235
(3) Polar classes of singular varieties, Ann. scient. Éc. Norm. Sup. *11* (1978), 247–276
(4) Ideals associated to a desingularization, in *Algebraic Geometry* Proceedings, Copenhagen 1978, Springer Lecture Notes *732* (1979), 503–517
(5) A proof of Noether's formula for the arithmetic genus of an algebraic surface, Compositio Math. *38* (1979), 113–119
Piene, R., Ronga, F.
(1) A geometric approach to the arithmetic genus of a projective manifold of dimension three, Topology *20* (1981), 179–190
Pieri, M.
(1) Formule di coincidenza per le serie algebriche ∞^n di coppie di punti dello spazio a n dimensioni, Rend. Circ. Mat. Palermo *5* (1891), 252–268
Pohl, W. F.
(1) Extrinsic complex projective geometry, in *Proceedings of the Conference on Complex Analysis*, Minneapolis 1964, Springer-Verlag (1965), 18–29
Poncelet, J. V.
(1) *Traité des propriétés projectives des figures*, 1822, Gauthier-Villars, Paris, 1865
Porteous, I. R.
(1) Blowing up Chern classes, Proc. Cambridge Phil. Soc. *56* (1960), 118–124

(2) Simple singularities of maps, Springer Lecture Notes *192* (1971), 286–307
(3) Todd's canonical classes, Springer Lecture Notes *192* (1971), 308–312

Quillin, D.
(1) Elementary proofs of some results of cobordism theory using Steenrod operations, Advances in Math. *7* (1971), 29–56
(2) Higher algebraic K-theory: I, Springer Lecture Notes *341* (1973), 85–147

Raisonnier, J.
(1) Formes de Chern et résidus raffinés de J. R. King, Bull. Sci. Math. *102* (1978), 145–154

Ramanujam, C. P.
(1) On a geometric interpretation of multiplicity, Inventiones Math. *22* (1973), 63–67

Ran, Z.
(1) On subvarieties of abelian varieties, Inventiones Math. *62* (1981), 459–479
(2) Curvilinear enumerative geometry, preprint, 1983

Rao, A. Prabhakar
(1) Liaison among curves in $\mathbb{P}^3$, Inventiones Math. *50* (1979), 205–217

Raynaud, M., Gruson, L.
(1) Critères de platitude et de projectivité, Inventiones Math. *13* (1971), 1–89

Rees, D.
(1) A note on valuations associated with a local domain, Proc. Cambridge Phil. Soc. *51* (1955), 252–253
(2) Valuations associated with ideals (II), J. London Math. Soc. *31* (1956), 221–235

Reeve, J. F.
(1) A summary of results in the topological classification of plane algebroid singularities, Rend. Sem. Mat. Torino *14* (1954–55), 159–187
(2) A note on fractional intersection multiplicities, Rend. Circ. Mat. Palermo *7* (1958), 167–184

Roberts, J.
(1) Chow's moving lemma, in *Algebraic Geometry*, Oslo 1970, F. Oort (ed.), Wolters-Noordhoff Publ., Groningen (1972), 89–96

Roĭtman, A. A.
(1) Rational equivalence of zero-cycles, Math. USSR Sb. *18* (1972), 571–588
(2) The torsion of the group of 0-cycles modulo rational equivalence, Annals of Math. *111* (1980), 553–569

Ronga, F.
(1) Les calcul des classes duales aux singularités de Boardman d'ordre deux, Comm. Math. Helv. *47* (1972), 15–35
(2) La classe duale aux points double d'une application, Compositio Math. *27* (1973), 223–232
(3) On multiple points of smooth immersions, Comm. Math. Helv. *55* (1980), 521–527

Rosati, C.
(1) Sulle corrispondenze fra curve algebriche, Proc. Intern. Cong. Math., Bologna 1928, vol. 4, 79–91

Sabbah, C.
(1) Quelques remarques sur la théorie des classes de Chern des espaces analytiques singuliers, preprint, 1983

Salmon, G.
(1) On the degree of a surface reciprocal to a given one, Cambridge and Dublin Math. J. *2* (1847), 65–73
(2) *Lessons introductory to the modern higher algebra*, 1859, Reprint of the 5th ed., Chelsea, New York, 1964
(3) *A treatise on the analytic geometry of three dimensions*, Vol. II, 1862, Reprint of the 5th edition, Chelsea, New York, 1965

Salmon, G., Fiedler, W.
(1) *Analytische Geometrie des Raumes*, II, 3. Auflage, Teubner, Leipzig, 1880

Samuel, P.
(1) La notion de multiplicité en algèbre et en géométrie algébrique, J. Math. Pures Appl. *30* (1951), 159–274
(2) *Méthodes d'algèbre abstraite en géométrie algébrique,* Ergebnisse der Math., Springer-Verlag, 1955
(3) Rational equivalence of arbitrary cycles, Amer. J. Math. *78* (1956), 383–400
(4) Relations d'équivalence en géométrie algébrique, Proc. Intern. Cong. Math., Edinburgh 1958, 470–487
(5) *Seminaire sur l'équivalence rationelle,* Paris-Orsay, 1971, Faculté des Sciences, Orsay

Schubert, H. C. H.
(1) *Kalkül der abzählenden Geometrie,* 1879, reprinted with an introduction by S. L. Kleiman, Springer-Verlag, 1979
(2) Anzahlbestimmungen für lineare Räume beliebiger Dimension, Acta Math. *8* (1866), 97–118
(3) Beziehungen zwischen den linearen Räumen auferlegbaren charakterischen Bedingungen, Math. Ann. *38* (1891), 598–602
(4) Allgemeine Anzahlfunctionen für Kegelschnitte, Flächen und Räume zweiten Grades in *n* Dimensionen, Math. Ann. *45* (1894), 153–206

Schützenberger, M.-P.
(1) La correspondance de Robinson, Springer Lecture Notes *579* (1977), 59–113

Schwarzenberger, R. L. E.
(1) Jacobians and symmetric products, Illinois J. Math. *7* (1963), 257–268
(2) The secant bundle of a projective variety, Proc. London Math. Soc. *14* (1964), 369–384

Scott, D. B.
(1) Natural lifts and the covariant systems of Todd, J. London Math. Soc. *1* (1969), 709–718
(2) An idea of Beniamino Segre, Rend. Sem. Mat. Fis. Milano *41* (1971), 9–17

Segre, B.
(1) Nuovi contributi alla geometria sulle varietà algebriche, Mem. R. Acad. d'Italia *5* (1934), 479–576
(2) Intorno ad teorema di Hodge sulla teoria della base per le curve di una superficie algebrica, Ann. di Mat. *16* (1937), 157–163
(3) On limits of algebraic varieties, Proc. London Math. Soc. *47* (1940), 351–403
(4) Nuovi methodi e resultati nella geometria sulle varietà algebriche, Ann. di Mat. *35* (1953), 1–128
(5) Geometry upon an algebraic variety, Proc. Intern. Cong. Math., Amsterdam 1954, vol. 3, 497–513
(6) Dilatazioni e varietà canoniche sulle varietà algebriche, Ann. di Mat. *37* (1954), 139–155
(7) Invarianti topologico-differenziali, varietà di Veronese e moduli di forme algebriche, Ann. di Mat. *41* (1956), 113–138
(8) *Prodomi di geometria algebrica,* Roma, Edizioni Cremonese, 1972

Segre, C.
(1) Le multiplicità nelle intersezioni delle curve piane algebriche con alcuni applicazioni ai principii della theoria di tali curve, 1898, Opere, Vol. 1, 380–429

Semple, J. G.
(1) On complete quadrics, I, II, J. London Math. Soc. *23* (1948), 258–267; *27* (1952), 280–287

Semple, J. G., Roth, L.
(1) *Introduction to algebraic geometry,* Clarendon Press, Oxford, 1949

Serre, J-P.
(1) Faisceaux algébriques cohérents, Annals of Math. *61* (1955), 197–278
(2) Un théorème de dualité, Comm. Math. Helv. *29* (1955), 9–26
(3) Géométrie algébrique et géométrie analytique, Ann. Inst. Fourier, Grenoble, *6* (1956), 1–42
(4) *Algèbre locale · Multiplicités,* 1957/58, Second edition, Springer Lecture Notes, 1965

(5) *Groupes algébriques et corps de classes*, Hermann, Paris, 1959
(6) *Corps locaux*, Hermann, Paris, 1962

Severi, F.
(1) Le coincidenze di una serie algebrica $\infty^{(k+1)(r-k)}$ di coppie di spazi a k dimensioni, immersi nello spazio ad r dimensioni, Rend. Accad. Linc. *9* (1900), 321–326
(2) Sulle intersezioni delle varietà algebriche e sopra i loro caratteri e singolarità proiettive, Mem. Accad. Sci. Torino (2) *52* (1902), 61–118
(3) Fondamenti per le geometria sulle varietà algebriche, Rend. Circ. Mat. Palermo *28* (1909), 33–87
(4) Sul principio della conservazione del numero, Rend. Circ. Mat. Palermo *33* (1912), 313–327
(5) *Vorlesungen über Algebraische Geometrie*, Teubner, Leipzig, 1921
(6) La serie canonica e la teoria delle serie principali di gruppi di punti sopra una superficie algebrica, Comm. Math. Helv. *4* (1932), 268–326
(7) Le rôle de la géométrie algébrique dans les mathématiques, Proc. Intern. Cong. Math., Zürich 1932, vol. 1, 209–220
(8) La teoria delle corrispondenze a valenza sopra una superficie algebrica, Rend. Accad. Linc. *17* (1933), 681–685, 759–764, 869–881
(9) Über die Grundlagen der algebraischen Geometrie, Abh. Math. Sem. Hamburg Univ. *9* (1933), 335–364
(10) La teoria generale delle corrispondenze fra due varietà algebriche e i sistemi d'equivalenza, Abh. Math. Sem. Hamburg Univ. *13* (1939), 101–112
(11) I fondamenti della geometria numerativa, Ann. di Mat. (4) *19* (1940), 153–242
(12) Sul limite dell'intersezione di due curve variabili sopra una superficie, le quali tendano ad avere una parte comune, Rend. R. Accad. d'Italia (7) *3* (1942), 410–414
(13) *Serie, sistemi d'equivalenza, e corrispondenze algebriche sulle varietà algebriche*, Roma, 1942
(14) Il punto di vista gruppale nei vari tipi di equivalenza sulle varietà algebriche, Comm. Math. Helv. *21* (1948), 189–224
(15) Il concetto generale di multiplicità delle soluzioni pei sistemi di equazioni algebriche e la teoria dell'eliminazione, Ann. di Mat. (4) *26* (1947), 221–270
(16) La géométrie algébrique italienne, Centre Belge Rech. Math., Colloque de géométrie algébrique, Liège (1950), 9–55
(17) Sulle molteplicità d'intersezione delle varietà algebriche ed analitiche e sopra una teoria geometrica dell'eliminazione, Math. Z. *52* (1950), 827–851
(18) Problèmes résolus et problèmes nouveaux dans la théorie des systémes d'équivalence, Proc. Intern. Cong. Math., Amsterdam 1954, vol. 3, 529–541
(19) Le equivalenze razionali, Rend. Accad. Linc. (8) *18* (1955), 443–451
(20) *Il theorema di Riemann-Roch per curve, superficie et varietà*, Ergebnisse der Math. 17, Springer-Verlag, 1958

Shafarevich, I. R.
(1) *Basic Algebraic Geometry*, Springer-Verlag, New York, 1977

Shimura, G.
(1) Reduction of algebraic varieties with respect to a discrete valuation of the base field, Amer. J. Math. *77* (1955), 134–176
(2) *Introduction to the arithmetic theory of automorphic functions*, Princeton Univ. Press, 1971

Snapper, E.
(1) Polynomials associated with divisors, J. Math. and Mech. *9* (1960), 123–129

Sommese, A.
(1) Submanifolds of Abelian varieties, Math. Ann. *233* (1978), 229–256

Srinivas, V.
(1) Vector bundles on the cone over a curve, Compositio Math. *47* (1982), 249–269

Stanley, R. P.
(1) Some combinatorial aspects of the Schubert calculus, in *Combinatoire et représentation du groupe symétrique*, Strasbourg 1976, Springer Lecture Notes *579* (1977), 217–251

Stienstra, J.
(1) The formal completion of the second Chow group, a K-theoretic approach, Astérisque *64* (1979), 149–168
Study, E.
(1) Über das sogenannte Prinzip von der Erhaltung der Anzahl, Archiv der Math. und Phys. (3) *8* (1905), 271–278
Szpiro, L.
(1) Sur la théorie des complexes parfaits, in *Commutative Algebra*, Durham 1981, London Math. Soc. Lecture Note Series *72* (1982), 83–90
Teissier, B.
(1) Sur une inégalité à la Minkowski pour les multiplicités, Annals of Math. *106* (1977), 40–44
(2) On a Minkowski-type inequality for multiplicities – II, in *C. P. Ramanujam – A Tribute*, Studies in Math. 8, Tata Institute (1978), 347–361
Tennison, B. R.
(1) Intersection multiplicities and tangent cones, Math. Proc. Cambridge Phil. Soc. *85* (1979), 33–42
Terjanian, G.
(1) Dimension arithmétique d'un corps, J. Algebra *22* (1972), 517–545
Tjurin, A. N.
(1) The geometry of singularities of a generic quadratic form, Math. USSR Izv. *17* (1981), 413–422
Todd, J. A.
(1) On Giambelli's formulae for incidences of linear spaces, J. London Math. Soc. *6* (1931), 209–216
(2) Some group-theoretic considerations in algebraic geometry, Annals of Math. *35* (1934), 702–704
(3) Intersection of loci on an algebraic V_4, Proc. Cambridge Phil. Soc. *33* (1937), 425–437
(4) The arithmetical invariants of algebraic loci, Proc. London Math. Soc. *43* (1937), 190–225
(5) The geometric invariants of algebraic loci, Proc. London Math. Soc. *45* (1939), 410–434
(6) Invariant and covariant systems on an algebraic variety, Proc. London Math. Soc. *46* (1980), 199–230
(7) Birational transformations with a fundamental surface, Proc. London Math. Soc. *47* (1941), 81–100
(8) Canonical systems on algebraic varieties, Bol. Soc. Mat. Mexicana *2* (1957), 26–44
Toledo, D., Tong, Y. L. L.
(1) A parametrix for $\bar{\partial}$ and Riemann-Roch in Cěch Theory, Topology *15* (1976), 273–301
(2) Duality and intersection theory in complex manifolds, I; II, Math. Ann. *237* (1978), 41–77; Annals of Math. *108* (1978), 519–538
Tyrell, J. A.
(1) Complete quadrics and collineations in S_n, Mathematika *3* (1956), 69–79
Vainsencher, I.
(1) On a formula of Ingleton and Scott, Atti. Accad. Linc. (8) *60* (1976), 629–631
(2) Curves tangent to a given curve, Notas Comun. Mat. Recife 82, 1977
(3) Conics in characteristic 2, Compositio Math. *36* (1978), 101–112
(4) Counting divisors with prescribed singularities, Trans. Amer. Math. Soc. *267* (1981), 399–422
Van de Ven, A. J. H. M.
(1) Characteristic classes and monoidal transformations, Indigationes Math. *18* (1956), 571–578
Verdier, J.-L.
(1) Dualité dans la cohomologie des espaces localement compacts, Sém. Bourbaki 300 (1965)
(2) Le théorème de Le Potier, Asterisque *17* (1974), 68–78

(3) Le théorème de Riemann-Roch pour les variétés algébriques éventuellement singulières, Sém. Bourbaki 464, Springer Lecture Notes *514* (1956), 159–175, or Astérisque *36–37* (1976), 5–20
(4) Class d'homologie associée a un cycle, Astérisque *36–37* (1976), 101–151
(5) Le théorème de Riemann-Roch pour les intersections complètes, Astérisque *36–37* (1976), 189–228
(6) Spécialisation des classes de Chern, Astérisque *82–83* (1981), 149–159

Vogel, W.
(1) On Bézout's theorem, in *Seminar D. Eisenbud/B. Singh/W. Vogel*, Teubner-Texte zur Math., Band 29, Leipzig (1980), 113–144

van der Waerden, B. L.
(1) Der Multiplizitätsbegriff der algebraischen Geometrie, Math. Ann. *97* (1927), 756–774
(2) Eine Verallgemeinerung des Bézoutschen Theorems, Math. Ann. *99* (1928), 497–541
(3) Topologische Begründung des Kalküls der abzählenden Geometrie, Math. Ann. *102* (1930), 337–362
(4) *Modern Algebra*, I, II, Ungar, New York, 1950
(5) On the definition of rational equivalence of cycles on a variety, Proc. Intern. Cong. Math., Amsterdam 1954, vol. 3, 545–549
(6) The theory of equivalence systems of cycles on a variety, Symp. Math. INDAM V (1971), 255–262
(7) On varieties in multiple-projective spaces, Indigationes Math. *51* (1978), 303–312

Walker, R. J.
Algebraic Curves, Princeton Univ. Press, 1950

Washnitzer, G.
(1) The characteristic classes of an algebraic fibre bundle, Proc. Nat. Acad. Sci. USA *42* (1956), 433–436
(2) Geometric syzygies, Amer. J. Math. *81* (1959), 171–248

Weil, A.
(1) Généralisation des fonctions abéliennes, J. Math. Pures Appl. *17* (1938), 47–87
(2) *Foundations of Algebraic Geometry*, 1946, Revised and enlarged edition, Amer. Math. Soc. Colloq. Publ. 29, 1962
(3) Sur la théorie des formes différentielles attachées à une variété analytique complexe, Comm. Math. Helv. *20* (1947), 110–116
(4) *Sur les courbes algébriques et les variétés qui s'en déduisent*, Hermann, Paris, 1948
(5) Sur les critères d'équivalence en géométrie algébrique, Math. Ann. *128* (1954), 95–127
(6) *Variétés kählériennes*, Hermann, Paris, 1958

Yuzhakov, A. N., Tsikh, A. K.
(1) The multiplicity of zero of a system of holomorphic functions, Siberian Math. J. *19* (1978), 489–492

Zariski, O.
(1) *Algebraic surfaces*, 1935, Second supplemented edition, Ergebnisse der Math. 61, Springer-Verlag, 1971
(2) Generalized weight properties of the resultant of $n+1$ polynomials in n indeterminates, Trans. Amer. Math. Soc. *41* (1937), 249–265
(3) Complete linear systems on normal varieties and a lemma of Enriques-Severi, Annals of Math. *55* (1952), 552–592
(4) The theorem of Riemann-Roch for high multiples of an effective divisor on an algebraic surface, Annals of Math. *76* (1962), 560–615

Zariski, O., Samuel, P.
(1) *Commutative Algebra*, van Nostrand, New York, 1958

Zeuthen, H. G.
(1) Nouvelle démonstration de théorèmes sur des séries de points correspondants sur deux courbes, Math. Ann. *3* (1871), 150–156
(2) Exemple d'une correspondance sans werthigkeit, Proc. Intern. Cong. Math., Rome 1908, vol. 2, 227–230
(3) *Lehrbuch der abzählenden Methoden der Geometrie*, Teubner, Leipzig, 1914

Zeuthen, H. G., Pieri, M.
(1) Géométrie énumérative, *Encyclopédie des sciences mathématiques* III.2, Gauthier-Villars, Paris (1915), 260–331
Zobel, A.
(1) On the contacts between the varieties of two systems, Rend. di Mat. (3–4) *17* (1958), 415–422
(2) Intersection theory on an open variety, Ann. di Mat. (4) *46* (1958), 1–17
(3) On the non-specialization of intersections on a singular variety, Mathematika *8* (1961), 39–44
(4) A condition calculus on an open variety, Rend. di Mat. *19* (1960), 72–94
Zucker, S.
(1) Generalized intermediate Jacobians and the theorem on normal functions, Inventiones Math. *33* (1976), 185–222

Notation

Index

Forthcoming titles:

W. Barth, C. A. M. Peters, A. van de Ven
Compact Complex Surfaces
ISBN 3-540-12172-2

M. Beeson
Foundations of Constructive Mathematics
Metamathematical Studies
ISBN 3-540-12173-0

K. Diederich, J. E. Fornaess, R. P. Pflug
Convexity in Complex Analysis
ISBN 3-540-12174-9

E. Freitag, R. Kiehl
Etale Kohomologietheorie und Weil-Vermutung
ISBN 3-540-12175-7

M. Gromov
Partial Differential Relations
ISBN 3-540-12177-3

J. C. Jantzen
Einhüllende Algebren halbeinfacher Lie-Algebren
ISBN 3-540-12178-1

G. A. Margulis
Discrete Subgroups of Lie Groups
ISBN 3-540-12179-X